Acclaim for

James Gleick's

GENIUS

"Written with grace and passion...All those wonderful Feynman stories are there, interwoven with touching moments in his private life....Scientists, and especially science students, ought to read James Gleick's engaging portrait." —*Boston Globe*

"A thorough and masterful portrait of one of the great minds of the century... Gleick succeeds in giving us a rare insight into the scientific community, its values, and its mentality....[He] brings to *Genius* high intelligence, a strong sense of narrative and excellent prose." —*The New York Review of Books*

"A thorough overview of modern physics...inspired." —*Newsweek*

"A noteworthy book...Gleick demonstrates a great ability to portray scientific people and places and to dramatize the emergence of new ideas.... One can thoroughly enjoy this well-researched biography for its picture of Feynman and his world." —*The New York Times Book Review*

"Not until now have we been given a full account of Feynman's extraordinary career and no less extraordinary personality....[It is] splendidly written, scrupulously documented.... Gleick...seems to have read every paper and personal letter and to have talked to everyone who ever knew Feynman.... A readable, accurate account of Feynman's great contributions to quantum mechanics." —Martin Gardner, *Raleigh News and Observer*

ALSO BY JAMES GLEICK
Chaos: Making a New Science

James Gleick

GENIUS

James Gleick was born in New York City in 1954. He graduated from Harvard College and worked for ten years as an editor and reporter at *The New York Times*. His first book, *Chaos: Making a New Science*, was a 1987 National Book Award nominee for non-fiction and has been translated into eighteen languages. He lives in New York with his wife and their son.

GENIUS

÷

The Life and Science of Richard Feynman

James Gleick

VINTAGE BOOKS
A Division of Random House, Inc.
New York

FIRST VINTAGE BOOKS EDITION, OCTOBER 1993

Copyright © 1992 by James Gleick

All rights reserved under International and Pan-American Copyright Conventions. Published in the United States by Vintage Books, a division of Random House, Inc., New York, and simultaneously in Canada by Random House of Canada Limited, Toronto. Originally published in hardcover by Pantheon Books, a division of Random House, Inc., New York, in 1992.

Grateful acknowledgment is made to the following for permission to reprint previously published material:

British Broadcasting Corporation: Excerpt from "Fun to Imagine," an interview between Richard Feynman and Christopher Sykes as broadcast by the British Broadcasting Corporation in 1983. Used by permission.

David Higham Associates: Four lines from "Tattered Serenade," from *Collected Poems* by Edith Sitwell, published by Macmillan London. Reprinted by permission of David Higham Associates.

Acknowledgments for the illustrations are on page 532.

Library of Congress Cataloging-in-Publication Data

Gleick, James.
Genius: the life and science of Richard Feynman / James Gleick.—1st Vintage Books ed.
p.   cm.
Originally published: New York: Pantheon Books, © 1992.
Includes bibliographical references and index.
ISBN 0-679-74704-4
1. Feynman, Richard Phillips.   2. Physics—History—20th century.   3. Physicists—United States—Biography.   I. Title.
QC16.F49G55   1993
530'.092—dc20
[B]   93-7838
CIP

Author photograph © Nancy Buirsky

Manufactured in the United States of America
10 9 8 7 6 5 4 3 2 1

For my mother and father,
Beth and Donen

I was born not knowing
and have only had a little time to change that here and there.
—Richard Feynman

CONTENTS

GENIUS

PROLOGUE

÷

Nothing is certain. This hopeful message went to an Albuquerque sanatorium from the secret world at Los Alamos. *We lead a charmed life.*

Afterward demons afflicted the bomb makers. J. Robert Oppenheimer made speeches about his shadowed soul, and other physicists began to feel his uneasiness at having handed humanity the power of self-destruction. Richard Feynman, younger and not so responsible, suffered a more private grief. He felt he possessed knowledge that set him alone and apart. It gnawed at him that ordinary people were living their ordinary lives oblivious to the nuclear doom that science had prepared for them. Why build roads and bridges meant to last a century? If only they knew what he knew, they surely would not bother. The war was over, a new era of science was beginning, and he was not at ease. For a while he could hardly work—by day a boyish and excitable professor at Cornell University, by night wild in love, veering from freshman mixers (where women sidled away from this rubber-legged dancer claiming to be a scientist who had made the atomic bomb) to bars and brothels. Meanwhile new colleagues, young physicists and mathematicians of his own age, were seeing him for the first time and forming their quick impressions. "Half genius and half buffoon," Freeman Dyson, himself a rising prodigy, wrote his parents back in England.

Feynman struck him as uproariously American—unbuttoned and burning with physical energy. It took him a while to realize how obsessively his new friend was tunneling into the very bedrock of modern science.

In the spring of 1948, still in the shadow of the bomb they had made, twenty-seven physicists assembled at a resort hotel in the Pocono Mountains of northern Pennsylvania to confront a crisis in their understanding of the atom. With Oppenheimer's help (he was now more than ever their spiritual leader) they had scraped together the thousand-odd dollars needed to cover their rooms and train fare, along with a small outlay for liquor. In the annals of science it was the last time but one that such men would meet in such circumstances, without ceremony or publicity. They were indulging a fantasy, that their work could remain a small, personal, academic enterprise, invisible to most of the public, as it had been a decade before, when a modest building in Copenhagen served as the hub of their science. They were not yet conscious of how effectively they had persuaded the public and the military to make physics a mission of high technology and expense. This meeting was closed to all but the few invited participants, the elite of physics. No transcript was kept. Next year most of these men would meet once more, hauling their two blackboards and eighty-two cocktail and brandy glasses in Oppenheimer's station wagon, but by then the modern era of physics had begun in earnest, science conducted on a scale the world had not seen, and never again would its chiefs come together privately, just to work.

The bomb had shown the aptness of physics. The scientists had found enough sinew behind their penciled abstractions to change history. Yet in the cooler days after the war's end, they realized how fragile their theory was. They thought that quantum mechanics gave a crude, perhaps temporary, but at least workable way to make calculations about light and matter. When pressed, however, the theory gave wrong results. And not merely wrong—they were senseless. Who could love a theory that worked so neatly at first approximation and then, when a scientist tried to make the results more exact, broke down so grotesquely? The Europeans who had invented quantum physics had tried everything they could imagine to shore up the theory, without success.

How were these men to know anything? The mass of the electron? Up for grabs: a quick glance gave a reasonable number, a hard look gave infinity—nonsense. The very idea of mass was unsettled: mass was not exactly stuff, but not exactly energy, either. Feynman toyed with an extreme view. On the last page of his tiny olive-green dime-store address book, mostly for phone numbers of women (annotated *dancer beauty* or *call when her nose is not red*), he scrawled a near haiku.

> Principles
> You can't say A is made of B
> or vice versa.
> All mass is interaction.

Even when quantum physics worked, in the sense of predicting nature's behavior, it left scientists with an uncomfortable blank space where their picture of reality was supposed to be. Some of them, though never Feynman, put their faith in Werner Heisenberg's wistful dictum, "The equation knows best." They had little choice. These scientists did not even know how to visualize the atom they had just split so successfully. They had created and then discarded one sort of picture, a picture of tiny particles orbiting a central nucleus as planets orbit the sun. Now they had nothing to replace it. They could write numbers and symbols on their pads, but their mental picture of the substance beneath the symbols had been reduced to a fuzzy unknown.

As the Pocono meeting began, Oppenheimer had reached the peak of his public glory, having risen as hero of the atomic bomb project and not yet having fallen as the antihero of the 1950s security trials. He was the meeting's nominal chairman, but more accomplished physicists were scattered about the room: Niels Bohr, the father of the quantum theory, on hand from his institute in Denmark; Enrico Fermi, creator of the nuclear chain reaction, from his laboratory in Chicago; Paul A. M. Dirac, the British theorist whose famous equation for the electron had helped set the stage for the present crisis. It went without saying that they were Nobel laureates; apart from Oppenheimer almost everyone in the room either had won or would win this honor. A few Europeans were absent, as was Albert Einstein, settling into his statesmanlike retirement, but with these exceptions the Pocono conclave represented the whole priesthood of modern physics.

Night fell and Feynman spoke. Chairs shifted. The priesthood had trouble following this brash young man. They had spent most of the day listening to an extraordinary virtuoso presentation by Feynman's exact contemporary, Julian Schwinger of Harvard University. This had been difficult to follow (when published, Schwinger's work would violate the *Physical Review*'s guidelines limiting the sprawl of equations across the width of the page) but convincing nonetheless. Feynman was offering fewer and less meticulous equations. These men knew him from Los Alamos, for better and for worse. Oppenheimer himself had privately noted that Feynman was the most brilliant young physicist at the atomic bomb project. Why he had acquired such a reputation none of them could say precisely. A few knew of his contribution to the key equation for the efficiency of a nuclear explosion (still classified forty years later, although the spy Klaus Fuchs had transmitted it promptly to his incredulous masters in the Soviet Union) or his theory of predetonation, measuring the probability that a lump of uranium might explode too soon. If they could not describe his actual scientific work, nevertheless they had absorbed an intense image of an original mind. They remembered him organizing the world's first large-scale computing system, a hybrid of new electro-mechanical business calculators and teams of women with color-coded cards; or delivering a hypnotic lecture on, of all things, elementary arithmetic; or frenetically twisting a control knob in a game whose object was to crash together a pair of electric trains; or sitting defiantly upright, for once motionless, in an army weapons carrier lighted by the purple-white glare of the century's paradigmatic explosion.

Facing his elders in the Pocono Manor sitting room, Feynman realized that he was drifting deeper and deeper into confusion. Uncharacteristically, he was nervous. He had not been able to sleep. He, too, had heard Schwinger's elegant lecture and feared that his own presentation seemed unfinished by comparison. He was trying to put across a new program for making the more exact calculations that physics now required—more than a program, a vision, a dancing, shaking picture of particles, symbols, arrows, and fields. The ideas were unfamiliar, and his slightly reckless style irritated some of the Europeans. His vowels were a raucous urban growl. His consonants slurred in a way that struck them as lower-class. He shifted his weight back and forth and twirled a piece of chalk rapidly between his fingers, around and around and end over end. He was a few weeks shy of his thirtieth birthday, too old now to pass for a boy wonder. He was trying to skip some details that would seem controversial—but too late. Edward Teller, the contentious Hungarian physicist, on his way to heading the postwar project to build the Super, the hydrogen bomb, interrupted with a question about basic quantum physics: "What about the exclusion principle?"

Feynman had hoped to avoid this. The exclusion principle meant that only one electron could inhabit a particular quantum state; Teller thought he had caught him pulling two rabbits from a single hat. Indeed, in Feynman's scheme particles did seem to violate this cherished principle by coming into existence for a ghostly instant. "It doesn't make any difference——" he started to reply.

"How do you know?

"I know, I worked from a——"

"How could it be!" Teller said.

Feynman was drawing unfamiliar diagrams on the blackboard. He showed a particle of antimatter going backward in time. This mystified Dirac, the man who had first predicted the existence of antimatter. Dirac now asked a question about causality: "Is it unitary?" Unitary! What on earth did he mean?

"I'll explain it to you," Feynman said, "and then you can see how it works, then you can tell me if it's unitary." He went on, and from time to time he thought he could still hear Dirac muttering, "Is it unitary?"

Feynman—mystifyingly brilliant at calculating, strangely ignorant of the literature, passionate about physics, reckless about proof—had for once overestimated his ability to charm and persuade these great physicists. Yet in truth he had now found what had eluded all of his elders, a way to carry physics forward into a new era. He had created a private new science that brought past and future together in a starkly majestic tapestry. His new friend Dyson at Cornell had glimpsed it—"this wonderful vision of the world as a woven texture of world lines in space and time, with everything moving freely," as Dyson described it. "It was a unifying principle that would either explain everything or explain nothing." Twentieth-century physics had reached an edge. Older men were looking for a way beyond an obstacle to their calculations. Feynman's listeners were eager for the new ideas of young physicists, but they were wedded to a certain view of the atomic world—or rather, a series of different views, each freighted with private confusion. Some were thinking mostly about waves—mathematical waves carrying the past into the present. Often, of course, the waves behaved as particles, like the particles whose trajectories Feynman sketched and erased on the blackboard. Some merely took refuge in the mathematics, chains of difficult calculations using symbols as stepping stones on a march through fog. Their systems of equations represented a submicroscopic world defying the logic of everyday objects like baseballs and water waves, ordinary objects with, "thank God," as W. H. Auden put it (in a poem Feynman detested):

> sufficient mass
> To be altogether there,
> Not an indeterminate gruel
> Which is partly somewhere else.

The objects of quantum mechanics were always partly somewhere else. The chicken-wire diagrams that Feynman had etched on the blackboard seemed, by contrast, quite definite. Those trajectories looked classical in their precision. Niels Bohr stood up. He knew this young physicist from Los Alamos—Feynman had argued freely and vehemently with Bohr. Bohr had sought Feynman's private counsel there, valuing his frankness, but now he was disturbed by the evident implications of those crisp lines. Feynman's particles seemed to be following paths neatly fixed in space and time. This they could not do. The uncertainty principle said so.

"Already we know that the classical idea of the trajectory in a path is not a legitimate idea in quantum mechanics," he said, or so Feynman thought—Bohr's soft voice and notoriously vague Danish tones kept his listeners straining to understand. He stepped forward and for many minutes, with Feynman standing unhappily to the side, delivered a humiliating lecture on the uncertainty principle. Afterward Feynman kept his despair to himself. At Pocono a generation of physics was melting into the next, and the passing of generations was neither as clean nor as inevitable as it later seemed.

Architect of quantum theories, brash young group leader on the atomic bomb project, inventor of the ubiquitous Feynman diagram, ebullient bongo player and storyteller, Richard Phillips Feynman was the most brilliant, iconoclastic, and influential physicist of modern times. He took the half-made conceptions of waves and particles in the 1940s and shaped them into tools that ordinary physicists could use and understand. He had a lightning ability to see into the heart of the problems nature posed. Within the community of physicists, an organized, tradition-bound culture that needs heroes as much as it sometimes mistrusts them, his name took on a special luster. It was permitted in connection with Feynman to use the word *genius*. He took center stage and remained there for forty years, dominating the science of the postwar era—forty years that turned the study of matter and energy down an unexpectedly dark and spectral road.

The work that made its faltering appearance at Pocono tied together in an experimentally perfect package all the varied phenomena at work in light, radio, magnetism, and electricity. It won Feynman a Nobel Prize. At least three of his later achievements might also have done so: a theory of superfluidity, the strange, frictionless behavior of liquid helium; a theory of weak interactions, the force at work in radioactive decay; and a theory of partons, hypothetical hard particles inside the atom's nucleus, that helped produce the modern understanding of quarks. His vision of particle interaction kept returning to the forefront of physics as younger scientists explored esoteric new domains. He continued to find new puzzles. He could not, or would not, distinguish between the prestigious problems of elementary particle physics and the apparently humbler everyday questions that seemed to belong to an earlier era. No other physicist since Einstein so ecumenically accepted the challenge of all nature's riddles. Feynman studied friction on highly polished surfaces, hoping—and mostly failing—to understand how friction worked. He tried to make a theory of how wind makes ocean waves grow; as he said later, "We put our foot in a swamp and we pulled it up muddy." He explored the connection between the forces of atoms and the elastic properties of the crystals they form. He assembled experimental data and theoretical ideas on the folding of strips of paper into peculiar shapes called flexagons. He made influential progress—but not enough to satisfy himself—on the quantum theory of gravitation that had eluded Einstein. He struggled for years, in vain, to penetrate the problem of turbulence in gases and liquids.

Feynman developed a stature among physicists that transcended any raw sum of his actual contributions to the field. Even in his twenties, when his published work amounted to no more than a doctoral thesis (profoundly original but little understood) and a few secret papers in the Los Alamos archives, his legend was growing. He was a master calculator: in a group of scientists he could create a dramatic impression by slashing his way through a difficult problem. Thus scientists—believing themselves to be unforgiving meritocrats—found quick opportunities to compare themselves unfavorably to Feynman. His mystique might have belonged to a gladiator or a champion arm-wrestler. His personality, unencumbered by dignity or decorum, seemed to announce: Here is an unconventional mind. The English writer C. P. Snow, observing the community of physicists, thought Feynman lacked the *"gravitas"* of his seniors. "A little bizarre . . . He would grin at himself if guilty of stately behaviour. He is a showman and enjoys it . . . rather as though Groucho Marx was suddenly standing in for a great scientist." It made Snow think of Einstein, now so shaded and dignified that few remembered the "merry boy" he had been in his creative time. Perhaps Feynman, too, would grow into a stately personage. Perhaps not. Snow predicted, "It will be interesting for young men to meet Feynman in his later years."

One team of physicists, assembled for the Manhattan Project, met him for the first time in Chicago, where he solved a problem that had baffled them for a month. It was "a shallow way to judge a superb mind," one of them admitted later, but they had to be impressed, by the unprofessorial manner as much as the feat itself: "Feynman was patently not struck in the prewar mold of most young academics. He had the flowing, expressive postures of a dancer, the quick speech we thought of as Broadway, the pat phrases of the hustler and the conversational energy of a finger snapper." Physicists quickly got to know his bounding theatrical style, his way of bobbing sidelong from one foot to the other when he lectured. They knew that he could never sit still for long and that when he did sit he would slouch comically before leaping up with a sharp question. To Europeans like Bohr his voice was as American as any they had heard, a sort of musical sandpaper; to the Americans it was raw, unregenerate New York. No matter. "We got the indelible impression of a star," another young physicist noted. "He may have emitted light as well as words. . . . Isn't *areté* the Greek word for that shining quality? He had it."

Originality was his obsession. He had to create from first principles—a dangerous virtue that sometimes led to waste and failure. He had the cast of mind that often produces cranks and misfits: a willingness, even eagerness, to consider silly ideas and plunge down wrong alleys. This strength could have been a crippling weakness had it not been redeemed, time and again, by a powerful intelligence. "Dick could get away with a lot because he was so goddamn smart," a theorist said. "He really *could* climb Mont Blanc barefoot." Isaac Newton spoke of having stood on the shoulders of giants. Feynman tried to stand on his own, through various acts of contortion, or so it seemed to the mathematician Mark Kac, who was watching Feynman at Cornell:

There are two kinds of geniuses, the "ordinary" and the "magicians." An ordinary genius is a fellow that you and I would be just as good as, if we were only many times better. There is no mystery as to how his mind works. Once we understand what they have done, we feel certain that we, too, could have done it. It is different with the magicians. They are, to use mathematical jargon, in the orthogonal complement of where we are and the working of their minds is for all intents and purposes incomprehensible. Even after we understand what they have done, the process by which they have done it is completely dark. They seldom, if ever, have students because they cannot be emulated and it must be terribly frustrating for a brilliant young mind to cope with the mysterious ways in which the magician's mind works. Richard Feynman is a magician of the highest caliber.

Feynman resented the polished myths of most scientific history, submerging the false steps and halting uncertainties under a surface of orderly intellectual progress, but he created a myth of his own. When he had ascended to the top of the physicists' mental pantheon of heroes, stories of his genius and his adventures became a sort of art form within the community. Feynman stories were clever and comic. They gradually created a legend from which their subject (and chief purveyor) seldom emerged. Many of them were transcribed and published in the eighties in two books with idiosyncratic titles, *Surely You're Joking, Mr. Feynman!* and *What Do You Care What Other People Think?* To the surprise of their publisher these became popular best-sellers. After his death in 1988 his sometime friend, collaborator, office neighbor, foil, competitor, and antagonist, the acerbic Murray Gell-Mann, angered his family at a memorial service by asserting, "He surrounded himself with a cloud of myth, and he spent a great deal of time and energy generating anecdotes about himself." These were stories, Gell-Mann added, "in which he had to come out, if possible, looking smarter than anyone else." In these stories Feynman was a gadfly, a rake, a clown, and a naïf. At the atomic bomb project he was the thorn in the side of the military censors. On the commission investigating the 1986 space-shuttle explosion he was the outsider who pushed aside red tape to uncover the true cause. He was the enemy of pomp, convention, quackery, and hypocrisy. He was the boy who saw the emperor with no clothes. So he was in life. Yet Gell-Mann spoke the truth, too. Amid the legend were misconceptions about Feynman's accomplishments, his working style, and his deepest beliefs. His own view of himself worked less to illuminate than to hide the nature of his genius.

The reputation, apart from the person, became an edifice standing monumentally amid the rest of the scenery of modern science. Feynman diagrams, Feynman integrals, and Feynman rules joined Feynman stories in the language that physicists share. They would say of a promising young colleague, "He's no Feynman, but . . ." When he entered a room where physicists had gathered—the student cafeteria at the California Institute of Technology, or the auditorium at any scientific meeting—with him would come a shift in the noise level, a disturbance of the field, that seemed to radiate from where he was carrying his tray or taking his front-row seat. Even his senior colleagues tried to look without looking. Younger physicists were drawn to Feynman's rough glamour. They practiced imitating his handwriting and his manner of throwing equations onto the blackboard. One group held a half-serious debate on the question, Is Feynman human? They envied the inspiration that came (so it seemed to them) in flashes. They admired him for other qualities as well: a faith in nature's simple truths, a skepticism about official wisdom, and an impatience with mediocrity.

He was widely considered a great educator. In fact few physicists of even the middle ranks left behind so small a cadre of students, or so assiduously shirked ordinary teaching duties. Although science remained one of the few domains of true apprenticeship, with students learning their craft at the master's side, few learned this way from Feynman. He did not have the patience to guide a student through a research problem, and he raised high barriers against students who sought him as a thesis adviser. Nevertheless when Feynman did teach he left a deep imprint on the subject. Although he never actually wrote a book, books bearing his name began to appear in the sixties—*Theory of Fundamental Processes* and *Quantum Electrodynamics*, lightly edited versions of lectures transcribed by students and colleagues. They became influential. For years he offered a mysterious noncredit course called Physics X, for undergraduates only, in a small basement room. Some physicists years later remembered this unpredictable free-form seminar as the most intense intellectual experience of their education. Above all in 1961 he took on the task of reorganizing and teaching the introductory physics course at Caltech. For two years the freshmen and sophomores, along with a team of graduate-student teaching assistants, struggled to follow a tour de force, the universe according to Feynman. The result was published and became famous as "the red books"—*The Feynman Lectures on Physics*. They reconceived the subject from the bottom up. Colleges that adopted the red books dropped them a few years later: the texts proved too difficult for their intended readers. Instead, professors and working physicists found Feynman's three volumes reshaping their own conception of their subject. They were more than just authoritative. A physicist, citing one of many celebrated passages, would dryly pay homage to "Book II, Chapter 41, Verse 6."

Authoritative, too, were Feynman's views of quantum mechanics, of the scientific method, of the relations between science and religion, of the role of beauty and uncertainty in the creation of knowledge. His comments on such subjects were mostly expressed offhand in technical contexts, but also in two slim models of science writing, again distilled from lectures: *The Character of Physical Law* and *QED: The Strange Theory of Light and Matter*. Feynman was widely quoted by scientists and science writers (although he seldom submitted to interviews). He despised philosophy as soft and unverifiable. Philosophers "are always on the outside making stupid remarks," he said, and the word he pronounced *philozawfigal* was a mocking epithet, but his influence was philosophical anyway, particularly for younger physicists. They remembered, for example, his Gertrude Stein-like utterance on the continuing nervousness about quantum mechanics— or, more precisely, the "world view that quantum mechanics represents":

It has not yet become obvious to me that there's no real problem. I cannot define the real problem, therefore I suspect there's no real problem, but I'm not sure there's no real problem.

or, similarly, what may have been the literature's most quoted mixed metaphor:

Do not keep saying to yourself, if you can possibly avoid it, "But how can it be like that?" because you will get "down the drain," into a blind alley from which nobody has yet escaped. Nobody knows how it can be like that.

In private, with pencil on scratch paper, he labored over aphorisms that he later delivered in spontaneous-seeming lectures:

Nature uses only the longest threads to weave her patterns, so each small piece of her fabric reveals the organization of the entire tapestry.

Why is the world the way it is? Why is science the way it is? How do we discover new rules for the flowering complexity around us? Are we reaching toward nature's simple heart, or are we merely peeling away layers of an infinitely deep onion? Although he sometimes retreated to a stance of pure practicality, Feynman gave answers to these questions, philosophical and unscientific though he knew they were. Few noticed, but his answer to the starkest of science's metaphysical questions—Is there a meaning, a simplicity, a comprehensibility at the core of things?—underwent a profound change in his lifetime.

Feynman's reinvention of quantum mechanics did not so much explain how the world was, or why it was that way, as tell how to confront the world. It was not knowledge of or knowledge about. It was knowledge how to. How to compute the emission of light from an excited atom. How to judge experimental data, how to make predictions, how to construct new tool kits for the new families of particles that were about to proliferate through physics with embarrassing fecundity.

There were other kinds of scientific knowledge, but pragmatic knowledge was Feynman's specialty. For him knowledge did not describe; it acted and accomplished. Unlike many of his colleagues, educated scientists in a cultivated European tradition, Feynman did not look at paintings, did not listen to music, did not read books, even scientific books. He refused to let other scientists explain anything to him in detail, often to their immense frustration. He learned anyway. He pursued knowledge without prejudice. During a sabbatical he learned enough biology to make a small but genuine contribution to geneticists' understanding of mutations in DNA. He once offered (and then awarded) a one-thousand-dollar prize for the first working electric motor less than one sixty-fourth of an inch long, and his musing on the possibilities of tiny machinery made him, a generation later, the intellectual father of a legion of self-described nanotechnologists. In his youth he experimented for months on end with trying to observe his unraveling stream of consciousness at the point of falling asleep. In his middle age he experimented with inducing out-of-body hallucinations in a sensory-deprivation tank, with and without marijuana. His lifetime saw a stratification of the branch of knowledge called physics. Those specializing in the understanding of elementary particles came to control much of the field's financing and much of its public rhetoric. With the claim that particle physics was the most fundamental science, they scorned even subdisciplines like solid-state physics—"squalid-state" was Gell-Mann's contemptuous phrase. Feynman embraced neither the inflating language of Grand Unified Theories nor the disdain for other sciences.

Democratically, as if he favored no skill above any other, he taught himself how to play drums, to give massages, to tell stories, to pick up women in bars, considering all these to be crafts with learnable rules. With the gleeful prodding of his Los Alamos mentor Hans Bethe ("Don't you know how to take squares of numbers near 50?") he taught himself the tricks of mental arithmetic, having long since mastered the more arcane arts of mental differentiation and integration. He taught himself how to make electroplated metal stick to plastic objects like radio knobs, how to keep track of time in his head, and how to make columns of ants march to his bidding. He had no difficulty learning to make an impromptu xylophone by filling water glasses; nor had he any shyness about playing them, all evening, at a dinner party for an astonished Niels Bohr. At the same time, when he was engrossed in the physicists' ultimate how-to endeavor, the making of an atomic bomb, he digressed to learn how to defeat the iron clamp of an old-fashioned soda machine, how to pick Yale locks, and then how to open safes—a mental, not physical, skill, though his colleagues mistakenly supposed he could feel the vibrations of falling tumblers in his fingertips (as well they might, after watching him practice his twirling motion day after day on their office strongboxes). Meanwhile, dreamily wondering how to harness atomic power for rockets, he worked out a nuclear reactor thrust motor, not quite practical but still plausible enough to be seized by the government, patented, and immediately buried under an official secrecy order. With no less diligence, much later, having settled into a domestic existence complete with garden and porch, he taught himself how to train dogs to do counterintuitive tricks—for example, to pick up a nearby sock not by the direct route but by the long way round, circling through the garden, in the porch door and back out again. (He did the training in stages, breaking the problem down until after a while it was perfectly obvious to the dog that one did not go directly to the sock.) Then he taught himself how to find people bloodhound-style, sensing the track of their body warmth and scent. He taught himself how to mimic foreign languages, mostly a matter of confidence, he found, combined with a relaxed willingness to let lips and tongue make silly sounds. (Why then, his friends wondered, could he never learn to soften his Far Rockaway accent?) He made islands of practical knowledge in the oceans of personal ignorance that remained: knowing nothing about drawing, he taught himself to make perfect freehand circles on the blackboard; knowing nothing about music, he bet his girlfriend that he could teach himself to play one piece, "The Flight of the Bumblebee," and for once failed dismally; much later he learned to draw after all, after a fashion, specializing in sweetly romanticized female nudes and letting his friends know that a concomitant learned skill thrilled him even more—how to persuade a young woman to disrobe. In his entire life he could never quite teach himself to feel a difference between right and left, but his mother finally pointed out a mole on the back of his left hand, and even as an adult he checked the mole when he wanted to be sure. He taught himself how to hold a crowd with his not-jazz, not-ethnic improvisational drumming; and how to sustain a two-handed polyrhythm of not just the usual three against two and four against three but—astonishing to classically trained musicians—seven against six and thirteen against twelve. He taught himself how to write Chinese, a skill acquired specifically to annoy his sister and limited therefore to the characters for "elder brother also speaks." In the era when high-energy particle accelerators came to dominate theoretical physics, he taught himself how to read the most modern of hieroglyphics, the lacy starburst photographs of particle collisions in cloud chambers and bubble chambers—how to read them not for new particles but for the subtler traces of experimental bias and self-deception. He taught himself how to discourage autograph seekers and refuse lecture invitations; how to hide from colleagues with administrative requests; how to force everything from his field of vision except for his research problem of the moment; how to hold off the special terrors of aging that shadow scientists; then how to live with cancer, and how to surrender to it.

After he died several colleagues tried to write his epitaph. One was Schwinger, in a certain time not just his colleague but his preeminent rival, who chose these words: "An honest man, the outstanding intuitionist of our age, and a prime example of what may lie in store for anyone who dares to follow the beat of a different drum." The science he helped create was like nothing that had come before. It rose as his culture's most powerful achievement, even as it sometimes sent physicists down the narrowing branches of an increasingly obscure tunnel. When Feynman was gone, he had left behind—perhaps his chief legacy—a lesson in what it meant to know something in this most uncertain of centuries.

FAR ROCKAWAY

÷

Eventually the art went out of radio tinkering. Children forgot the pleasures of opening the cabinets and eviscerating their parents' old Kadettes and Clubs. Solid electronic blocks replaced the radio set's messy innards—so where once you could learn by tugging at soldered wires and staring into the orange glow of the vacuum tubes, eventually nothing remained but featureless ready-made chips, the old circuits compressed a thousandfold or more. The transistor, a microscopic quirk in a sliver of silicon, supplanted the reliably breakable tube, and so the world lost a well-used path into science.

In the 1920s, a generation before the coming of solid-state electronics, one could look at the circuits and see how the electron stream flowed. Radios had valves, as though electricity were a fluid to be diverted by plumbing. With the click of the knob came a significant hiss and hum, just at the edge of audibility. Later it was said that physicists could be divided into two groups, those who had played with chemistry sets and those who had played with radios. Chemistry sets had their appeal, but a boy like Richard Feynman, loving diagrams and maps, could see that the radio was its own map, a diagram of itself. Its parts expressed their function, once he learned to break the code of wires, resistors, crystals, and capacitors.

He assembled a crystal set, attached oversized earphones from a rummage sale, and listened under the bedcovers until he fell asleep. Sometimes his parents would tiptoe in and take the earphones off their sleeping boy. When atmospheric conditions were right, his radio could pull in signals from far away—Schenectady in upstate New York or even station WACO from Waco, Texas. The mechanism responded to the touch. To change channels he moved a little wire across the crystal. Still, the radio was not like a watch, with gears and wheels. It was already one step removed from the mechanical world. Its essential magic was invisible after all. The crystal, motionless, captured waves of electromagnetic radiation from the ether.

Yet there was no ether—no substance bearing these waves. If scientists wished to imagine radio waves propagating with the unmistakable undulating rhythm of waves in a pond, they nonetheless had to face the fact that these waves were not *in* anything. Not in the era of relativity: Einstein was showing that if an ether existed it would have to be motionless with respect to any and all observers—though they themselves moved in different directions. This was impossible. "It seems that the aether has betaken itself to the land of the shades in a final effort to elude the inquisitive search of the physicist!" the mathematician Hermann Weyl wrote in 1918, the year Feynman was born. Through what medium, then, were radio waves sweeping in their brief journey from the aerials of downtown New York to Feynman's second-story bedroom in a small frame house on the city's outskirts? Whatever it was, the radio wave was only one of the many sorts of oscillations disturbing every region of space. Waves of light, physically identical to radio waves but many times shorter, crisscrossing hectically; infrared waves, perceptible as heat on the skin; the ominously named X rays; the ultra-high-frequency gamma rays, with wavelengths smaller than atoms—all these were just different guises of one phenomenon, electromagnetic radiation. Already space was an electromagnetic babel, and human-built transmitters were making it busier still. Fragmented voices, accidental clicks, slide-whistle drones: strange noises passed through one another, more waves in a well-corrugated waviness. These waves coexisted not in the ether but in a rather more abstract medium, the precise nature of which was posing difficulties for physicists. They could not imagine what it was—a problem that was only mildly allayed by the fact that they had a name for it, the electromagnetic field, or just the field. The field was merely a continuous surface or volume across which some quantity varied. It had no substance, yet it shook; it vibrated. Physicists were discovering that the vibrations sometimes behaved like particles, but this just complicated the issue. If they were particles, they were nonetheless particles with an undeniably wavelike quality that enabled boys like Feynman to tune in to certain desirable wavelengths, the ones carrying "The Shadow" and "Uncle Don" and advertisements for Eno Effervescent Salts. The scientific difficulties were obscure, known only to a handful of scientists more likely to speak German than English. The essence of the mystery, however, was clear to amateurs who read about Einstein in the newspapers and pondered the simple magic of a radio set.

No wonder so many future physicists started as radio tinkerers, and no wonder, before *physicist* became a commonplace word, so many of them grew up thinking they might become electrical engineers, professionals known to earn a good wage. Richard, called Ritty by his friends, seemed to be heading single-mindedly in that direction. He accumulated tube sets and an old storage battery from around the neighborhood. He assembled transformers, switches, and coils. A coil salvaged from a Ford automobile made showy sparks that burned brown-black holes in newspaper. When he found a leftover rheostat, he pushed 110-volt electricity through it until it overloaded and burned. He held the stinking, smoking thing outside his second-floor window, as the ashes drifted down to the grassy rear yard. This was standard emergency procedure. When a pungent odor drifted in downstairs during his mother's bridge game, it meant that Ritty was dangling his metal wastebasket out the window, waiting for the flames to die out after an abortive experiment with shoe polish—he meant to melt it and use the liquid as black paint for his "lab," a wooden crate roughly the size of a refrigerator, standing in his bedroom upstairs in the rear of the house. Screwed into the crate were various electrical switches and lights that Ritty had wired, in series and in parallel. His sister, Joan, nine years younger, served eagerly as a four-cents-a-week lab assistant. Her duties included putting a finger into a spark gap and enduring a mild shock for the entertainment of Ritty's friends.

It had already occurred to psychologists that children are innate scientists, probing, puttering, experimenting with the possible and impossible in a confused local universe. Children and scientists share an outlook on life. *If I do this, what will happen?* is both the motto of the child at play and the defining refrain of the physical scientist. Every child is observer, analyst, and taxonomist, building a mental life through a sequence of intellectual revolutions, constructing theories and promptly shedding them when they no longer fit. The unfamiliar and the strange—these are the domain of all children and scientists.

None of which could fully account for the presence of laboratory, rheostat, and lab assistant—tokens of a certain vivid cultural stereotype. Richard Feynman was relentless in filling his bedroom with the trappings and systems of organized science.

Neither Country nor City

Charmed lives were led by the children of Far Rockaway, a village that amounted to a few hundred acres of frame houses and brick apartment blocks on a spit of beach floating off Long Island's south shore. The neighborhood had been agglomerated into the political entity of New York City as one of the more than sixty towns and neighborhoods that merged as the borough of Queens in 1898. The city was investing generously in these neighborhoods, spending tens of millions of dollars on the laying of water mains, sewers, and roadways and the construction of grand public buildings. Still, in the first part of the twentieth century, before the IND subway line reached out across the marshes of Jamaica Bay, the city seemed a faraway place. Commuters took the Long Island Rail Road. Beyond Far Rockaway's eastern border lay the small towns of Nassau County, Long Island. To the northwest, across marshy tongues of ocean called Mott Basin and Hassock Channel, lay a flat expanse that later became Idlewild Airport and then Kennedy International Airport. On foot or on their bicycles, Far Rockaway's children had free run of a self-contained world: ivy-covered houses, fields, and vacant lots. No one has yet isolated the circumstances that help a child grow whole and independent, but they were present. At some point in a town's evolution, houses and fences grow dense enough to form a connected barrier. When that critical point is reached, movement is mostly restricted to public streets. In Far Rockaway boys and girls still percolated through the neighborhood and established their own paths through backyards and empty lots behind the houses and streets. They were autonomous and enterprising in play, roaming far from their parents' immediate oversight, riding their bicycles without accounting for their whereabouts. They could wander through fields on the way to the shore, and then they could rent boats and row them up and down the protected inlets. Richard walked to the library and, sitting on the stone steps, watched people go by in all directions. Distant as New York seemed, he felt bound enough to the great city to look down on the outsiders living a few blocks away, in Cedarhurst,

Long Island. But he also knew that his neighborhood was a place apart.

"When I was a child I thought we lived at the end of the world," wrote another New Yorker, the critic Alfred Kazin; he grew up in Brownsville, a Brooklyn neighborhood a little poorer and almost as remote, another district of Jewish immigrants and children of immigrants occupying that unusual boundary between the urban and the rural. "There were always raw patches of unused city land all around us filled with 'monument works' where they cut and stored tombstones, as there were still on our street farmhouses and the remains of old cobbled driveways," he wrote—"most of it dead land, neither country nor city. . . . That was the way to the ocean we always took summer evenings—through silent streets of old broken houses whose smoky red Victorian fronts looked as if the paint had clotted like blood and had then been mixed with soot—past infinite weedy lots. . . ."

For Ritty Feynman the beach was best of all—the long southern strand stretching almost unbroken to the far east end of Long Island, framed by its boardwalk and summer hotels, cottages and thousands of private lockers. Far Rockaway was a summer resort with beach clubs for people from the city: the Ostend Baths, Roche's (for a long time Richard thought this was named after the insect), the Arnold. There were wooden pavilions and changing rooms for rent by the season, with shiny locks and keys. For the local children, though, the beach served its purpose the year round. They splashed in the light surf, attenuated by a long breakwater pale beneath the waves. At the height of the summer's crowds the pink and green of bathing suits dotted the sand like gumdrops. It was his favorite place. He usually rode his bicycle the four thousand feet from his house (a distance that expanded in his later memory to two miles). He went with friends or alone. The sky was larger there than anywhere else in the city's confines; the ocean tempted his imagination as it does any child's. All those waves, all that space, the boats crawling like apparitions along the horizon toward New York Harbor, Europe and Africa lying far beyond, at the end of a long uninterrupted vector curving downward below the sky. It sometimes seemed that the things near the sea were the only things that were any good.

The dome of the sky stretched upward. The arcs of the sun and moon crossed directly ahead, rising and falling with the season. He could splash his heels in the surf and recognize a line that formed the tripartite boundary between earth, sea, and air. At night he would take his flashlight. For teenagers the beach was a site for social mixing between boys and girls; he did his best, though he sometimes felt gawky. He often swam. When he was forty-three, setting out nearly everything he knew about physics in the historic two-year undergraduate course that became *The Feynman Lectures on Physics*, he stood before a hall of freshmen and tried to place them mentally at the beach. "If we stand on the shore and look at the sea," he said, "we see the water, the waves breaking, the foam, the sloshing motion of the water, the sound, the air, the winds and the clouds, the sun and the blue sky, and light; there is sand and there are rocks of various hardness and permanence, color and texture. There are animals and seaweed, hunger and disease, and the observer on the beach; there may even be happiness and thought." Nature was elemental there, though for Feynman *elemental* did not mean simple or austere. The questions he considered within the physicist's purview—the fundamental questions—arose on the beach. "Is the sand other than the rocks? That is, is the sand perhaps nothing but a great number of very tiny stones? Is the moon a great rock? If we understood rocks, would we also understand the sand and the moon? Is the wind a sloshing of the air analogous to the sloshing motion of the water in the sea?"

The great European migration to America was ending. For the Jews of Russia, Eastern Europe, and Germany, for the Irish and the Italians, the first-hand and first-generation memories would now recede. The outer neighborhoods of New York flourished in the generations before World War II and then began to wane. In Far Rockaway not much changed visibly in the sixty-nine years of Feynman's lifetime. When Feynman returned on a visit with his children a few years before his death, everything seemed shrunken and forlorn, the fields and vacant lots were gone, but it was the same beach with its boardwalk, the same high school, the same house he had wired for radio broadcasts—the house now divided, to accommodate a tenant, and not nearly so spacious as in memory. He did not ring the bell. The village's main street, Central Avenue, seemed shabby and narrow. The population had become largely Orthodox Jewish, and Feynman was vaguely disturbed to see so many yarmulkes, or, as he actually said, "those little hats that they wear"—meaning: *I don't care what things are called.* And casually repudiating the culture that hung as thick in the air of his childhood as the smoke of the city or the salt of the ocean.

The Judaism of Far Rockaway took in a liberal range of styles of belief, almost broad enough to encompass atheists like Richard's father, Melville. It was a mostly Reform Judaism, letting go the absolutist and fundamental traditions for the sake of a gentle, ethical humanism, well suited for fresh Americans pinning their hopes on children who might make their way into the mainstream of the New World. Some households barely honored the Sabbath. In some, like Feynman's, Yiddish would have been a foreign language. The Feynmans belonged to the neighborhood temple. Richard went to Sunday school for a while and belonged to a Shaaray Tefila youth group that organized after-school activities. Religion remained part of the village's ethical core. Families like the Feynmans, in neighborhoods all around greater New York City, produced in the first half of the twentieth century an outpouring of men and women who became successful in many fields, but especially science. These hundred-odd square miles of the planet's surface were disproportionately fertile in the spawning of Nobel laureates. Many families, as Jews, were embedded in a culture that prized learning and discourse; immigrants and the children of immigrants worked to fulfill themselves through their own children, who had to be sharply conscious of their parents' hopes and sacrifices. They shared a sense that science, as a profession, rewarded merit. In fact, the best colleges and universities continued to raise barriers against Jewish applicants, and their science faculties remained determinedly Protestant, until after World War II. Science nevertheless offered the appearance of a level landscape, where the rules seemed mathematical and clear, free from the hidden variables of taste and class.

As a town Far Rockaway had a center that even Cedarhurst lacked. When Richard's mother, Lucille, walking down to Central Avenue, headed for stores like Nebenzahl's and Stark's, she appreciated the centralization. She knew her children's teachers personally, helped get the school lunchroom painted, and joined her neighbors in collecting the set of red glassware given out as a promotion by a local movie theater. This village looked inward as carefully as the shtetl that remained in some memories. There was a consistency of belief and behavior. To be honest, to be principled, to study, to save money against hard times—the rules were not so much taught as assumed. Everyone worked hard. There was no sense of poverty— certainly not in Feynman's family, though later he realized that two families had shared one house because neither could get by alone. Nor in his friend Leonard Mautner's, even after the father had died and an older brother was holding the family together by selling eggs and butter from house to house. "That was the way the world was," Feynman said long afterward. "But now I realize that everybody was struggling like mad. Everybody was struggling and it didn't seem like a struggle." For children, life in such neighborhoods brought a rare childhood combination of freedom and moral rigor. It seemed to Feynman that morality was made easy. He was allowed to surrender to a natural inclination to be honest. It was the downhill course.

A Birth and a Death

Melville Feynman (he pronounced his surname like the more standard variants: Fineman or Feinman) came from Minsk, Byelorussia. He immigrated with his parents, Louis and Anne, in 1895, at the age of five, and grew up in Patchogue, Long Island. He had a fascination with science but, like other immigrating Jews of his era, no possible means to fulfill it. He studied a fringe version of medicine called homeopathy; then he embarked on a series of businesses, selling uniforms for police officers and mail carriers, selling an automobile polish called Whiz (for a while the Feynmans had a garage full of it), trying to open a chain of cleaners, and finally returning to the uniform business with a company called Wender & Goldstein. He struggled for much of his business life.

His wife had grown up in better circumstances. Lucille was the daughter of a successful milliner who had emigrated as a child from Poland to an English orphanage, where he acquired the name Henry Phillips. From there Lucille's father came to the United States, where he got his first job selling needles and thread from a pack on his back. He met Johanna Helinsky, a daughter of German-Polish immigrants, when she repaired his watch in a store on the Lower East Side of New York. Henry and Johanna not only married but also went into business together. They had an idea that rationalized the trimming of the elaborate hats that women wore before World War I, and their millinery business thrived. They moved to a town house well uptown on the East Side, on 92d Street near Park Avenue, and there Lucille, the youngest of their five children, was born in 1895.

Like many well-off, assimilating Jews, Lucille Phillips attended the Ethical Culture School (an institution whose broad humanist ethos soon left its mark on J. Robert Oppenheimer, nine years her junior). She prepared to teach kindergarten. Instead, soon after graduating, still a teenager, she met Melville. The introduction to her future husband came through her best friend. Melville was the friend's date; Lucille was invited to accompany a friend of Melville's. They went for a drive, with Lucille joining Melville's friend sitting in the back seat. On the return trip, it was Lucille and Melville who sat together.

A few days later he said, "Don't get married to anybody else." This was not quite a proposal, and her father would not allow her to marry Melville until three years later, when she turned twenty-one. They moved into an inexpensive apartment in upper Manhattan in 1917, and Richard was born in a Manhattan hospital the next year.

A later family legend held that Melville announced in advance that, if the baby was a boy, he would be a scientist. Lucille supposedly replied, Don't count your chickens before they hatch. But Richard's father undertook to help his prophecy along. Before the baby was out of his high chair, he brought home some blue and white floor tiles and laid them out in patterns, blue-white-blue-white or blue-white-white-blue-white-white, trying to coax the baby to recognize visual rhythms, the shadow of mathematics. Richard had walked at an early age, but he was two before he talked. His mother worried for months. Then, as late talkers so often do, Richard became suddenly and unstoppably voluble. Melville bought the *Encyclopaedia Britannica*, and Richard devoured it. Melville took his son on trips to the American Museum of Natural History, with its animal tableaux in glass cases and its famous, towering, bone-and-wire dinosaurs. He described dinosaurs in a way that taught a lesson about expressing dimensions in human units: "twenty-five feet high and the head is six feet across" meant, he explained, that "if he stood in our front yard he would be high enough to put his head through the window but not quite because the head is a little bit too wide and it would break the window"—a vivid enough illustration for any small boy.

Melville's gift to the family was knowledge and seriousness. Humor and a love of storytelling came from Lucille. At any rate, that was how family lore tended to apportion their influence. Melville liked to laugh at the stories his wife and children told, at dinner and afterward, when the family regularly read aloud. He had a surprising giggle, and his son acquired an eerily exact facsimile. Comedy, for Lucille, was a high calling and a way of defying misfortune: the hard reality of her grandparents' lives in a Polish ghetto, and tragedy in her own family. Her mother suffered from epilepsy and her eldest sister from schizophrenia. Except for another sister, Pearl, her brothers and sisters died young.

Early death also came to her new household. In the winter Richard was five, she gave birth to a second son, named Henry Phillips Feynman, after her father, who had died a year before. Four weeks later the baby came

down with a fever. A fingernail had been bleeding and never quite healed.
Within days the baby was dead, probably from spinal meningitis. The grief,
the quick turning of happiness into despair—and surely for Richard the
fear as well—darkened their home for a long time. He had waited for a
brother. Now he had a lesson in human precariousness, in the cruelty of
nature's untamed accidents. Later he almost never spoke of the harsh death
that dominated this year. He had no brother or sister again until finally,
when he was nine, Joan was born. Henry's presence remained a shadow
in the household. Richard knew—even Joan knew—that their mother
always kept a birth certificate and a hat that had once belonged to a boy
whose remains now lay in the vault of the family mausoleum five miles
away, behind a stone plate inscribed, "HENRY PHILLIPS FEYNMAN JANUARY
24, 1924–FEBRUARY 25, 1924."

The Feynmans moved several times, leaving Manhattan for the small
towns straddling the city border: first to Far Rockaway; then from Far Rockaway to Baldwin, Long Island; then to Cedarhurst, when Richard was about
ten, and then back to Far Rockaway. Lucille's father owned a house there,
and they moved in—a two-story house of stucco the color of sand, on a
small lot at 14 New Broadway. There were front and rear yards and a double
driveway. They shared the house with Lucille's sister Pearl and her family—
her husband, Ralph Lewine, a boy, Robert, just older than Richard, and
a girl, Frances, just younger. A rail of white wood ringed the porch. The
ground floor held two living rooms, one for show and one for general use,
with gas logs in a fireplace for cold days. The bedrooms were small, but
there were eight of them. Richard's, on the second floor, overlooked the
back yard, with its forsythia and peach tree. Some evenings the adults
would come home to find his cousin, Frances, shivering at the upstairs
landing, unable to sleep because Richard, as chief baby-sitter, had told
ghost stories drawing their mood from the old Gothic panels that lined the
stairs.

The household had two other members during those pre-Depression
years, a German immigrant couple, Ludwig and Marie, easing their passage
into the United States by working as household servants for room and
board. Marie cooked; Ludwig said wryly that he was gardener, chauffeur,
and butler, serving meals in a formal white coat. They also arranged some
serious and inventive play. With Ludwig's help the north window of the
garage became the North Fenster Bank. Everyone took turns playing teller
and customer. As Ludwig and Marie learned English they taught the chil-

dren other routines: the protocols of gardening and formal table manners. If Feynman acquired such skills, he carefully shed them later.

To Joan, the youngest of all the children, it seemed like a well-run household where things happened when they were supposed to happen. Late one night, however, when she was three or four, her brother shook her awake in violation of the routine. He said he had permission to show her something rare and wonderful. They walked, holding hands, onto Far Rockaway's small golf course, away from the illuminated streets. "Look up," Richard said. There, far above them, the streaky wine-green curtains of the aurora borealis rippled against the sky. One of nature's surprises. Somewhere in the upper atmosphere solar particles, focused by the earth's magnetosphere, ripped open trails of luminous high-voltage ionization. It was a sight that the street lights of a growing city would soon cast out forever.

It's Worth It

The mathematics and the tinkering developed separately. At home the scientific inventory expanded to include chemicals from chemistry sets, lenses from a telescope, and photographic developing equipment. Ritty wired his laboratory into the electrical circuits of the entire house, so that he could plug his earphones in anywhere and make impromptu broadcasts through a portable loudspeaker. His father declared—something he had heard—that electrochemistry was an important new field, and Ritty tried in vain to figure out what electrochemistry was: he made piles of dry chemicals and set live wires in them. A jury-rigged motor rocked his baby sister's crib. When his parents came home late one night, they opened the door to a sudden clang-clang-clang and Ritty's shout: "It works!" They now had a burglar alarm. If his mother's bridge partners asked how she could tolerate the noise, or the chemical smoke, or the not-so-invisible ink on the good linen hand towels, she said calmly that it was worth it. There were no second thoughts in the middle-class Jewish families of New York about the value of ambition on the children's behalf.

The Feynmans raised their children according to a silent creed shared with many of their neighbors. Only rarely did they express its tenets, but they lived by them. They were sending their children into a world of hardships and dangers. A parent does all he or she can to bring a child up "so that he can better face the world and meet the intense competition of others for existence," as Melville once put it. The child will have to find a niche in which he can live a useful and fruitful life. The parents' motives are selfish—for nothing can magnify parents in the eyes of their neighbors as much as the child's success. "When a child does something *good* and unusual," Melville wrote, "it is the parents chest that swells up and who looks around and says to his neighbors (without actually speaking, of course) 'See what I have wrought? Isn't he wonderful? What have you got that can equal what I can show?' And the neighbors help the ego of the parent along by acclaiming the wonders of the child and by admiring the parent for *his* success . . ." A life in the business world, "the commercial world," is arid and exhausting; turn rather to the professions, the world of learning and culture. Ultimately, for the sacrifices of his parents a child owes no debt— or rather the debt is paid to his own children in turn.

The adult Richard Feynman became an adept teller of stories about himself, and through these stories came a picture of his father as a man transmitting a set of lessons about science. The lessons were both naïve and wise. Melville Feynman placed a high value on curiosity and a low value on outward appearances. He wanted Richard to mistrust jargon and uniforms; as a salesman, he said, he saw the uniforms empty. The pope himself was just a man in a uniform. When Melville took his son on walks, he would turn over stones and tell him about the ants and the worms or the stars and the waves. He favored process over facts. His desire to explain such things often outstripped his knowledge of them; much later Feynman recognized that his father must have invented sometimes. The gift of these lessons, as Feynman expressed it in his two favorite stories about his father, was a way of thinking about scientific knowledge.

One was the story about birds. Fathers and sons often walked together on summer weekends in the Catskill Mountains of New York, and one day a boy said to Richard, "See that bird? What kind of bird is that?"

I said, "I haven't the slightest idea what kind of bird it is."

He says, "It's a brown-throated thrush. Your father doesn't teach you anything!"

But it was the opposite. He had already taught me: "See that bird?" he says. "It's a Spencer's warbler." (I knew he didn't know the real name.) "Well, in Italian, it's a *Chutto Lapittida*. In Portuguese, it's a *Bom da Peida*. In Chinese, it's a *Chung-long-tah*, and in Japanese, it's a *Katano Tekeda*. You can know the name of that bird in all the languages of the world, but when you're finished, you'll know absolutely nothing whatever about the bird. You'll only know about humans in different places and what they call the bird. So let's look at the bird and see what it's *doing*—that's what counts."

The second story also carried a moral about the difference between the name and the thing named. Richard asks his father why, when he pulls his red wagon forward, a ball rolls to the back.

"That," he says, "nobody knows. The general principle is that things that are moving try to keep on moving, and things that are standing still tend to stand still, unless you push on them hard." And he says, "This tendency is called inertia, but nobody knows why it's true." Now that's a deep understanding.

Deeper than Melville could have known: few scientists or educators recognized that even a complete Newtonian understanding of force and inertia leaves the *why* unanswered. The universe does not have to be that way. It is hard enough to explain inertia to a child; to recognize that the ball actually moves forward slightly with respect to the ground while moving backward sharply with respect to the wagon; to see the role of friction in transferring the force; to see that *every body perseveres in its state of being at rest or of moving uniformly straight forward, except insofar as it is compelled to change its state by forces impressed upon it*. It is hard enough to convey all that without adding an almost scholastically subtle lesson about the nature of explanation. Newton's laws do explain why balls roll to the back of wagons, why baseballs travel in wind-bent parabolas, and even why crystals pick up radio waves, up to a point. Later Feynman became acutely conscious of the limits of such explanations. He agonized over the difficulty of truly explaining how a magnet picks up an iron bar or how the earth imparts the force called gravity to a projectile. The Feynman who developed an agnosticism about such concepts as inertia had a stranger physics in mind as well, the physics being born in Europe while father and son talked about wagons. Quantum mechanics imposed a new sort of doubt on science, and Feynman expressed that doubt often, in many different ways. *Do not ask how it can be like that. That, nobody knows.*

Even when he was young, absorbing such wisdom, Feynman sometimes glimpsed the limits of his father's understanding of science. As he was going to bed one night, he asked his father what algebra was.

"It's a way of doing problems that you can't do in arithmetic," his father said.

"Like what?"

"Like a house and a garage rents for $15,000. How much does the garage rent for?"

Richard could see the trouble with that. And when he started high school, he came home upset by the apparent triviality of Algebra 1. He went into his sister's room and asked, "Joanie, if 2^x is equal to 4 and x is an unknown number, can you tell me what x is?" Of course she could, and Richard wanted to know why he should have to learn anything so obvious in high school. The same year, he could see just as easily what x must be if 2^x was 32. The school quickly switched him into Algebra 2, taught by Miss Moore, a plump woman with an exquisite sense of discipline. Her class ran as a roundelay of problem solving, the students making a continual stream to and from the blackboard. Feynman was slightly ill at ease among the older students, but he already let friends know that he thought he was smarter. Still, his score on the school IQ test was a merely respectable 125.

At School

The New York City public schools of that era gained a reputation later for high quality, partly because of the nostalgic reminiscences of famous alumni. Feynman himself thought that his grammar school, Public School 39, had been stultifyingly barren: "an intellectual desert." At first he learned more at home, often from the encyclopedia. Having trained himself in rudimentary algebra, he once concocted a set of four equations with four unknowns and showed it off to his arithmetic teacher, along with his methodical solution. She was impressed but mystified; she had to take it to the principal to find out whether it was correct. The school had one course in general science, for boys only, taught by a blustering, heavyset man called Major Connolly—evidently his World War I rank. All Feynman remembered from the course was the length of a meter in inches, 39.37, and a futile argument with the teacher over whether rays of light from a single source come out radially, as seemed logical to Richard, or in parallel, as in the conventional textbook diagrams of lens behavior. Even in grade school he had no doubt that he was right about such things. It was just obvious, physically—not the sort of argument that could be settled by an appeal to authority. At home, meanwhile, he boiled water by running 110-volt house current through it and watched the lines of blue and yellow sparks that flow when the current breaks. His father sometimes described the beauty of the flow of energy through the everyday world, from sunlight to plants to muscles to the mechanical work stored in the spring of a windup toy. Assigned at school to write verse, Richard applied this idea to a fancifully bucolic scene with a farmer plowing his field to make food, grass, and hay:

> . . . Energy plays an important part
> And it's used in all this work;
> Energy, yes, energy with power so great,
> A kind that cannot shirk.
>
> If the farmer had not this energy,
> He would be at a loss,
> But it's sad to think, this energy
> Belongs to a little brown horse.

Then he wrote another poem, brooding self-consciously about his own obsession with science and with the idea of science. Amid some borrowed apocalyptic imagery he expressed a feeling that science meant skepticism about God—at least about the standardized God to whom he had been exposed at school. Over the Feynmans' rational and humanistic household God had never held much sway. "Science is making us wonder," he began— then on second thought he scratched out the word *wonder*.

> Science is making us ~~wonder~~ wander,
> Wander, far and wide;
> And know, by this time,
> Our face we ought to hide.
>
> Some day, the mountain shall wither,
> While the valleys get flooded with fire;
> Or men shall be driven like horses,
> And stamper, like beasts, in the mire.
>
> And we say, "The earth was thrown from the sun,"
> Or, "Evolution made us come to be
> And we come from lowest of beasts,
> Or one step back, the ape and monkey."

Our minds are thinking of science,
And science is in our ears;
Our eyes are seeing science,
And science is in our fears.

Yes, we're wandering from the Lord our God,
Away from the Holy One;
But now we cannot help it,
For it is already done.

But poetry was (Richard thought) "sissy-like." This was no small problem. He suffered grievously from the standard curse of boy intellectuals,
the fear of being thought, or of being, a sissy. He thought he was weak and
physically awkward. In baseball he was inept. The sight of a ball rolling
toward him across a street filled him with dread. Piano lessons dismayed
him, too, not just because he played so poorly, but because he kept playing
an exercise called "Dance of the Daisies." For a while this verged on
obsession. Anxiety would strike when his mother sent him to the store for
"peppermint patties."

As a natural corollary he was shy about girls. He worried about getting
in fights with stronger boys. He tried to ingratiate himself with them by
solving their school problems or showing how much he knew. He endured
the canonical humiliations: for example, watching helplessly while some
neighborhood children turned his first chemistry set into a brown, useless,
sodden mass on the sidewalk in front of his house. He tried to be a good
boy and then worried, as good boys do, about being too good—"goody-
good." He could hardly retreat from intellect to athleticism, but he could
hold off the taint of sissiness by staying with the more practical side of the
mental world, or so he thought. The practical man—that was how he saw
himself. At Far Rockaway High School he came upon a series of mathematics primers with that magical phrase in the title—*Arithmetic for the
Practical Man*; *Algebra for the Practical Man*—and he devoured them. He
did not want to let himself be too "delicate," and poetry, literature, drawing,
and music were too delicate. Carpentry and machining were activities for
real men.

For students whose competitive instincts could not be satisfied on the
baseball field, New York's high schools had the Interscholastic Algebra
League: in other words, math team. In physics club Feynman and his
friends studied the wave motions of light and the odd vortex phenomenon
of smoke rings, and they re-created the already classic experiment of the

California physicist Robert Millikan, using suspended oil drops to measure
the charge of a single electron. But nothing gave Ritty the thrill of math
team. Squads of five students from each school met in a classroom, the
two teams sitting in a line, and a teacher would present a series of problems.
These were designed with special cleverness. By agreement they could
require no calculus—nothing more than standard algebra—yet the routines
of algebra as taught in class would never suffice within the specified time.
There was always some trick, or shortcut, without which the problem would
just take too long. Or else there was no built-in shortcut; a student had to
invent one that the designer had not foreseen.

According to the fashion of educators, students were often taught that
using the proper methods mattered more than getting the correct answer.
Here only the answer mattered. Students could fill the scratch pads with
gibberish as long as they reached a solution and drew a circle around it.
The mind had to learn indirection and flexibility. Head-on attacks were
second best. Feynman lived for these competitions. Other boys were president and vice president, but Ritty was team captain, and the team always
won. The team's number-two student, sitting directly behind Feynman,
would calculate furiously with his pencil, often beating the clock, and
meanwhile he had a sensation that Feynman, in his peripheral vision, was
not writing—never wrote, until the answer came to him. You are rowing
a boat upstream. The river flows at three miles per hour; your speed against
the current is four and one-quarter. You lose your hat on the water. Forty-
five minutes later you realize it is missing and execute the instantaneous,
acceleration-free about-face that such puzzles depend on. How long does
it take to row back to your floating hat?

A simpler problem than most. Given a few minutes, the algebra is
routine. But a student whose head starts filling with 3s and 4¼s, adding
them or subtracting them, has already lost. This is a problem about reference
frames. The river's motion is irrelevant—as irrelevant as the earth's motion
through the solar system or the solar system's motion through the galaxy.
In fact all the velocities are just so much foliage. Ignore them, place your
point of reference at the floating hat—think of yourself floating like the
hat, the water motionless about you, the banks an irrelevant blur—now
watch the boat, and you see at once, as Feynman did, that it will return
in the same forty-five minutes it spent rowing away. For all the best competitors, the goal was a mental flash, achieved somewhere below consciousness. In these ideal instants one did not strain toward the answer so
much as relax toward it. Often enough Feynman would get this unstudied

insight while the problem was still being read out, and his opponents, before they could begin to compute, would see him ostentatiously write a single number and draw a circle around it. Then he would let out a loud sigh. In his senior year, when all the city's public and private schools competed in the annual championship at New York University, Feynman placed first.

For most people it was clear enough what mathematics was—a cool body of facts and rote algorithms, under the established headings of arithmetic, algebra, geometry, trigonometry, and calculus. A few, though, always managed to find an entry into a freer and more colorful world, later called "recreational" mathematics. It was a world where rowboats had to ferry foxes and rabbits across imaginary streams in nonlethal combinations; where certain tribespeople always lied and others always told the truth; where gold coins had to be sorted from false-gold in just three weighings on a balance scale; where painters had to squeeze twelve-foot ladders around inconveniently sized corners. Some problems never went away. When an eight-quart jug of wine needed to be divided evenly, the only measures available were five quarts and three. When a monkey climbed a rope, the end was always tied to a balancing weight on the other side of a pulley (a physics problem in disguise). Numbers were prime or square or perfect. Probability theory suffused games and paradoxes, where coins were flipped and cards dealt until the head spun. Infinities multiplied: the infinity of counting numbers turned out to be demonstrably smaller than the infinity of points on a line. A boy plumbed geometry exactly as Euclid had, with compass and straightedge, making triangles and pentagons, inscribing polyhedra in circles, folding paper into the five Platonic solids. In Feynman's case, the boy dreamed of glory. He and his friend Leonard Mautner thought they had found a solution to the problem of trisecting an angle with the Euclidean tools—a classic impossibility. Actually they had misunderstood the problem: they could trisect one side of an equilateral triangle, producing three equal segments, and they mistakenly assumed that the lines joining those segments to the far corner mark off equal angles. Riding around the neighborhood on their bicycles, Ritty and Len excitedly imagined the newspaper headlines: "Two Children in High School First Learning Geometry Solve the Age-Old Problem of the Trisection of the Angle."

This cornucopian world was a place for play, not work. Yet unlike its stolid high-school counterpart it actually connected here and there to real, adult mathematics. Illusory though the feeling was at first, Feynman had the sense of conducting research, solving unsolved problems, actively ex-

ploring a live frontier instead of passively receiving the wisdom of a dead
era. In school every problem had an answer. In recreational mathematics
one could quickly understand and investigate problems that were open.
Mathematical game playing also brought a release from authority. Recognizing some illogic in the customary notation for trigonometric functions,
Feynman invented a new notation of his own: $\overline{Sx}$ for sin $\overline{Cx}$ for
cos (x), $\overline{Tx}$ for tan (x). He was free, but he was also extremely methodical.
He memorized tables of logarithms and practiced mentally deriving values
in between. He began to fill notebooks with formulas, continued fractions
whose sums produced the constants π and e.

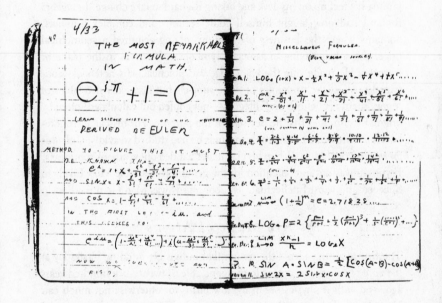

A page from one of Feynman's teenage notebooks.

A month before he turned fifteen he covered a page with an elated inch-high scrawl:

THE MOST REMARKABLE
FORMULA
IN MATH.

$e^{i\pi} + 1 = 0$

(FROM SCIENCE HISTORY OF THE UNIVERSE)

By the end of this year he had mastered trigonometry and calculus, both differential and integral. His teachers could see where he was heading. After three days of Mr. Augsbury's geometry class, Mr. Augsbury abdicated, putting his feet up on his desk and asking Richard to take charge. In algebra Richard had now taught himself conic sections and complex numbers, domains where the business of equation solving acquired a geometrical tinge, the solver having to associate symbols with curves in the plane or in space. He made sure the knowledge was practical. His notebooks contained not just the principles of these subjects but also extensive tables of trigonometric functions and integrals—not copied but calculated, often by original techniques that he devised for the purpose. For his calculus notebook he borrowed a title from the primers he had studied so avidly, *Calculus for the Practical Man*. When his classmates handed out yearbook sobriquets, Feynman was not in contention for the genuinely desirable Most Likely to Succeed and Most Intellectual. The consensus was Mad Genius.

All Things Are Made of Atoms

The first quantum idea—the notion that indivisible building blocks lay at the core of things—occurred to someone at least twenty-five hundred years ago, and with it physics began its slow birth, for otherwise not much can be understood about earth or water, fire or air. The idea must have seemed dubious at first. Nothing in the blunt appearance of dirt, marble, leaves, water, flesh, or bone suggests that it is so. But a few Greek philosophers in the fifth century B.C. found themselves hard pressed to produce any other satisfactory possibilities. Things change—crumble, fade, wither, or grow—yet they remain the same. The notion of immutability seemed to require some fundamental immutable parts. Their motion and recombination might give the appearance of change. On reflection, it seemed worthwhile to regard the basic constituents of matter as unchanging and indivisible: *atomos*—uncuttable. Whether they were also uniform was disputed. Plato thought of atoms as rigid blocks of pure geometry: cubes, octahedrons, tetrahedrons, and icosahedrons for the four pure elements, earth, air, fire, and water. Others imagined little hooks holding the atoms together (of what, though, could these hooks be made?).

Experiment was not the Greek way, but some observations supported the notion of atoms. Water evaporated; vapor condensed. Animals sent forth invisible messengers, their scents on the wind. A jar packed with ashes could still accept water; the volumes did not sum properly, suggesting interstices within matter. The mechanics were troubling and remained so. How did these grains move? How did they bind? "Cloudy, cloudy is the stuff of stones," wrote the poet Richard Wilbur, and even in the atomic era it was hard to see how the physicist's swarming clouds of particles could give rise to the hard-edged world of everyday sight and touch.

Someone who trusts science to explain the everyday must continually make connections between textbook knowledge and real knowledge, the knowledge we receive and the knowledge we truly own. We are told when we are young that the earth is round, that it circles the sun, that it spins on a tilted axis. We may accept the knowledge on faith, the frail teaching of a modern secular religion. Or we may solder these strands to a frame of understanding from which it may not so easily be disengaged. We watch the sun's arc fall in the sky as winter approaches. We guess the time from the shadow of a lamppost. We walk across a merry-go-round and strain against the sideways Coriolis force, and we try to connect the sensation to our received knowledge of the habits of earthly cyclones: northern hemisphere, low pressure, counterclockwise. We time the vanishing point of a tall-masted ship below the horizon. The sun, the winds, the waves all join in preventing our return to a flat-earth world, where we could watch the tides follow the moon without understanding.

All things are made of atoms—how much harder it is to reconcile this received fact with the daily experience of solid tables and chairs. Glancing at the smooth depressions worn in the stone steps of an office building, we seldom recognize the cumulative loss of invisibly small particles struck off by ten million footfalls. Nor do we connect the geometrical facets of a jewel to a mental picture of atoms stacked like cannonballs, favoring a particular crystalline orientation and so forcing regular angles visible to the naked eye. If we do think about the atoms in us and around us, the persistence of solid stone remains a mystery. Richard Feynman asked a high-school teacher (and never heard a satisfactory reply), "How do sharp things stay sharp all this time if the atoms are always jiggling?"

The adult Feynman asked: If all scientific knowledge were lost in a cataclysm, what single statement would preserve the most information for the next generations of creatures? How could we best pass on our understanding of the world? He proposed, *"All things are made of atoms—little particles that move around in perpetual motion, attracting each other when they are a little distance apart, but repelling upon being squeezed into one another,"* and he added, "In that one sentence, you will see, there is an enormous amount of information about the world, if just a little imagination and thinking are applied." Although millennia had passed since natural philosophers broached the atomic idea, Feynman's lifetime saw the first generations of scientists who truly and universally believed in it, not just as a mental convenience but as a hard physical reality. As late as 1922 Bohr, delivering his Nobel Prize address, felt compelled to remind his listeners that scientists "believe the existence of atoms to be proved beyond a doubt." Richard nevertheless read and reread in the Feynmans' *Encyclopaedia Britannica* that "pure chemistry, even to-day, has no very conclusive arguments for the settlement of this controversy." Stronger evidence was at hand from the newer science, physics: the phenomenon called radioactivity seemed to involve the actual disintegration of matter, so discretely as to produce audible pings or visible blips. Not until the eighties could people say that they had finally seen atoms. Even then the seeing was indirect, but it stirred the imagination to see shadowy globules arrayed in electron-microscope photographs or to see glowing points of orange light in the laser crossfire of "atom traps."

Not solids but gases began to persuade seventeenth- and eighteenth-century scientists of matter's fundamental granularity. In the heady aftermath of Newton's revolution scientists made measurements, found constant quantities, and forged mathematical relationships that a philosophy without numbers had left hidden. Investigators made and unmade water, ammonia, carbonic acid, potash, and dozens of other compounds. When they carefully weighed the ingredients and end products, they discovered regularities. Volumes of hydrogen and oxygen vanished in a neat two-to-one ratio in the making of water. Robert Boyle found in England that, although one could vary both the pressure and the volume of air trapped at a given temperature in a piston, one could not vary their product. Pressure multiplied by volume was a constant. These measures were joined by an invisible rod—why? Heating a gas increased its volume or its pressure. Why?

Heat had seemed to flow from one place to another as an invisible fluid— "phlogiston" or "caloric." But a succession of natural philosophers hit on a less intuitive idea—that heat was motion. It was a brave thought, because no one could see the things in motion. A scientist had to imagine uncountable corpuscles banging invisibly this way and that in the soft pressure of wind against his face. The arithmetic bore out the guess. In Switzerland Daniel Bernoulli derived Boyle's law by supposing that pressure was precisely the force of repeated impacts of spherical corpuscles, and in the same way, assuming that heat was an intensification of the motion hither and thither, he derived a link between temperature and density. The corpuscularians advanced again when Antoine-Laurent Lavoisier, again with painstaking care, demonstrated that one could keep reliable account books of the molecules entering and exiting any chemical reaction, even when gases joined with solids, as in rusting iron.

"Matter is unchangeable, and consists of points that are perfectly simple, indivisible, of no extent"—that the atom could itself contain a crowded and measurable universe remained for a later century to guess—"& separated from one another." Ruggiero Boscovich, an eighteenth-century mathematician and director of optics for the French navy, developed a view of atoms with a strikingly prescient bearing, a view that Feynman's single-sentence credo echoed two centuries later. Boscovich's atoms stood not so much for substance as for forces. There was so much to explain: how matter compresses elastically or inelastically, like rubber or wax; how objects bounce or recoil; how solids hold together while liquids congeal or release vapors; "effervescences & fermentations of many different kinds, in which the particles go & return with as many different velocities, & now approach towards & now recede from one another."

The quest to understand the corpuscles translated itself into a need to understand the invisible attractions and repulsions that gave matter its visible qualities. *Attracting each other when they are a little distance apart, but repelling upon being squeezed into one another,* Feynman would say simply. That mental picture was already available to a bright high-school student in 1933. Two centuries had brought more and more precise inquiry into the chemical behavior of substances. The elements had proliferated. Even a high-school laboratory could run an electric current through a beaker of water to separate it into its explosive constituents, hydrogen and oxygen. Chemistry as packaged in educational chemistry sets seemed to have reduced itself to a mechanical collection of rules and recipes. But the fundamental questions remained for those curious enough to ask, How do solids stay solid, with atoms always "jiggling"? What forces control the fluid motions of air and water, and what agitation of atoms engenders fire?

A Century of Progress

By then the search for forces had produced a decade of reinterpretation of the nature of the atom. The science known as chemical physics was giving way rapidly to the sciences that would soon be known as nuclear and high-energy physics. Those studying the chemical properties of different substances were trying to assimilate the first startling findings of quantum mechanics. The American Physical Society met that summer in Chicago. The chemist Linus Pauling spoke on the implications of quantum mechanics for complex organic molecules, primitive components of life. John C. Slater, a physicist from the Massachusetts Institute of Technology, struggled to make a connection between the quantum mechanical view of electrons and the energies that chemists could measure. And then the meeting spilled onto the fairgrounds of the spectacular 1933 Chicago World's Fair, "A Century of Progress." Niels Bohr himself spoke on the unsettling problem of measuring anything in the new physics. Before a crowd of visitors both sitting and standing, his ethereal Danish tones often smothered by crying babies and a balking microphone, he offered a principle that he called "complementarity," a recognition of an inescapable duality at the heart of things. He claimed revolutionary import for this notion. Not just atomic particles, but all reality, he said, fell under its sway. "We have been forced to recognize that we must modify not only all our concepts of classical physics but even the ideas we use in everyday life," he said. He had lately been meeting with Professor Einstein (their discussions were actually more discordant than Bohr now let on), and they had found no way out. "We have to renounce a description of phenomena based on the concept of cause and effect."

Elsewhere amid the throngs at the fairground that summer, enduring the stifling heat, were Melville, Lucille, Richard, and Joan Feynman. For the occasion Joan had been taught to eat bacon with a knife and fork; then the Feynmans strapped suitcases to the back of a car and headed off cross-country, a seemingly endless drive on the local roads of the era before interstate highways. On the way they stayed at farmhouses. The fair spread across four hundred acres on the shore of Lake Michigan, and the emblems of science were everywhere. Progress indeed: the fair celebrated a public sense of science that was reaching a crest. *Knowledge Is Power*—that earnest motto adorned a book of Richard's called *The Boy Scientist*. Science was invention and betterment; it changed the way people lived. The eponymous business enterprises of Edison, Bell, and Ford were knotting the countryside with networks of wire and pavement—an altogether positive good, it seemed. How wonderful were these manifestations of the photon and the electron, lighting lights and bearing voices across hundreds of miles!

Even in the trough of the Depression the wonder of science fueled an optimistic faith in the future. Just over the horizon were fast airships, half-mile-high skyscrapers, and technological cures for diseases of the human body and the body politic. Who knew where the bright young students of today would be able to carry the world? One New York writer painted a picture of his city fifty years in the future: New York in 1982 would hold a magnificent fifty million people, he predicted, the East River and much of the Hudson River having been "filled in." "Traffic arrangements will no doubt have provided for several tiers of elevated roadways and noiseless railways—built on extended balconies flanking the enormous sky-scrapers . . ." Nourishment will come from concentrated pellets. Ladies' dress will be streamlined to something like the 1930s bathing suit. The hero of this fantasy was the "high-school genius (who generally knows more than anybody else)." There was no limit to the hopes vested in the young.

Scientists, too, struggled to assimilate the new images pouring into the culture from the laboratory. Electricity powered the human brain itself, a University of Chicago researcher announced that summer; the brain's central switchboard used vast numbers of connecting lines to join brain cells, each one of which could be considered both a tiny chemical factory and electric battery. Chicago's business community made the most of these symbols, too. In an opening-day stunt, technicians at four astronomical observatories used faint rays of starlight from Arcturus, forty light-years distant, focused by telescopes and electrically amplified, to turn on the lights of the exposition. "Here are gathered the evidences of man's achievements in the realm of physical science, proofs of his power to prevail over all the perils that beset him," declared Rufus C. Dawes, president of the fair corporation, as loud projectiles released hundreds of American flags in the sky over the fairgrounds. Life-size dinosaurs awed visitors. A robot gave lectures. Visitors less interested in science could pay to see an unemployed actress named Sally Rand dance with ostrich-feather fans. The Feynmans, though, took the Sky Ride, suspended on cables between two six-hundred-foot towers, and visited the Hall of Science, where a 151-word wall motto summed up the history of science from Pythagoras to Euclid to Newton to Einstein.

The Feynmans had never heard of Bohr or any of the other physicists gathering in Chicago, but, like most other American newspaper readers, they knew Einstein's name well. That summer he was traveling in Europe, uprooted, having left Germany for good, preparing to arrive in New York Harbor in October. For fourteen years America had been in the throes of a publicity craze over this "mathematician." The *New York Times*, the Feynmans' regular paper, had led a wave of exaltation with only one precedent, the near deification of Edison a generation earlier. No theoretical scientist, European or American, before or since, ignited such a fever of adulation. A part of the legend, the truest part, was the revolutionary import of relativity for the way citizens of the twentieth century should conceive their universe. Another part was Einstein's supposed claim that only twelve people worldwide could understand his work. "Lights All Askew in the Heavens," the *Times* reported in a 1919 classic of headline writing. "Einstein Theory Triumphs. Stars Not Where They Seem or Were Calculated to Be, but Nobody Need Worry. A Book for 12 Wise Men. No More in the World Could Comprehend It, Said Einstein." A series of editorials followed. One was titled "Assaulting the Absolute." Another declared jovially, "Apprehensions for the safety of confidence even in the multiplication table will arise."

The presumed obscurity of relativity contributed heavily to its popularity. Yet had Einstein's message really been incomprehensible it could hardly have spread so well. More than one hundred books arrived to explain the mystery. The newspapers mixed tones of reverence and self-deprecating amusement about the mystery of relativity's paradoxes; in actuality, they and their readers correctly understood the elements of this new physics. Space is curved—curved where gravity warps its invisible fabric. The ether is banished, along with the assumption of an absolute frame of reference for space and time. Light has a fixed velocity, measured at 186,000 miles per second, and its path bends in the sway of gravity. Not long after the general theory of relativity was transmitted by underwater cable to eager New York newspapers, schoolchildren who could barely compute the hypotenuse of a right triangle could nevertheless recite a formula of Einstein's, E equals MC squared, and some could even report its implication: that matter and energy are theoretically interchangeable; that within the atom lay unreleased a new source of power. They sensed, too, that the universe had shrunk. It was no longer merely *everything*—an unimaginable totality. Now it might be bounded, thanks to four-dimensional curvature, and somehow it began to seem artificial. As the English physicist J. J. Thomson said unhappily, "We have Einstein's space, de Sitter's space, expanding universes, contracting universes, vibrating universes, mysterious universes.

In fact the pure mathematician may create universes just by writing down an equation . . . he can have a universe of his own."

There will never be another Einstein—just as there will never be another Edison, another Heifetz, another Babe Ruth, figures towering so far above their contemporaries that they stood out as legends, heroes, half-gods in the culture's imagination. There will be, and almost certainly have already been, scientists, inventors, violinists, and baseball players with the same raw genius. But the world has grown too large for such singular heroes. When there are a dozen Babe Ruths, there are none. In the early twentieth century, millions of Americans could name exactly one contemporary scientist. In the late twentieth century, anyone who can name a scientist at all can name a half-dozen or more. Einstein's publicists, too, belonged to a more naïve era; icons are harder to build in a time of demythologizing, deconstruction, and pathography. Those celebrating Einstein had the will and the ability to remake the popular conception of scientific genius. It seemed that Edison's formula favoring perspiration over inspiration did not apply to this inspired, abstracted thinker. Einstein's genius seemed nearly divine in its creative power: he imagined a certain universe and this universe was born. Genius seemed to imply a detachment from the mundane, and it seemed to entail wisdom. Like sports heroes in the era before television, he was seen exclusively from a distance. Not much of the real person interfered with the myth. By now, too, he had changed from the earnest, ascetic-looking young clerk whose genius had reached its productive peak in the first and second decades of the century. The public had hardly seen that man at all. Now Einstein's image drew on a colorful and absentminded appearance—wild hair, ill-fitting clothes, the legendary socklessness. The mythologizing of Einstein occasionally extended to others. When Paul A. M. Dirac, the British quantum theorist, visited the University of Wisconsin in 1929, the *Wisconsin State Journal* published a mocking piece about "a fellow they have up at the U. this spring . . . who is pushing Sir Isaac Newton, Einstein and all the others off the front page." An American scientist, the reporter said, would be busy and active, "but Dirac is different. He seems to have all the time there is in the world and his heaviest work is looking out the window." Dirac's end of the dialogue was suitably mono-syllabic. (The *Journal*'s readers must have assumed he was an ancient eminence; actually he was just twenty-seven years old.)

"Now doctor will you give me in a few words the low-down on all your investigations?"

"No."

"Good. Will it be all right if I put it this way—'Professor Dirac solves all the problems of mathematical physics, but is unable to find a better way of figuring out Babe Ruth's batting average'?"

"Yes."

. . .

"Do you go to the movies?"

"Yes."

"When?"

"In 1920—perhaps also 1930."

The genius was otherworldly and remote. More than the practical Americans whose science meant gizmos and machines, Europeans such as Einstein and Dirac also incarnated the culture's standard oddball view of the scientist. "Is he the tall, backward boy . . . ?" Barbara Stanwyck's character asked in *The Lady Eve* about Henry Fonda's, an ophiologist roughly Feynman's age.

—He isn't backward, he's a scientist.
—Oh, is that what it is. I knew he was *peculiar*.

"Peculiar" meant harmless. It meant that brilliant men paid for their gifts with compensating, humanizing flaws. There was an element of self-defense in the popular view. And there was a little truth. Many scientists did walk through the ordinary world seeming out of place, their minds elsewhere. They sometimes failed to master the arts of dressing carefully or making social conversation.

Had the *Journal*'s reporter solicited Dirac's opinion of the state of American science, he might have provoked a longer comment. "There are no physicists in America," Dirac had said bitingly, in more private company. It was too harsh an assessment, but the margin of his error was only a few years, and when Dirac spoke of physics he meant something new. Physics was not about vacuum cleaners or rayon or any of the technological wonders spreading in that decade; it was not about lighting lights or broadcasting radio waves; it was not even about measuring the charge of the electron or the frequency spectra of glowing gases in laboratory experiments. It was about a vision of reality so fractured, accidental, and tenuous that it frightened those few older American physicists who saw it coming.

"I feel that there is a real world corresponding to our sense perceptions,"

Yale University's chief physicist, John Zeleny, defensively told a Minneapolis audience. "I believe that Minneapolis is a real city and not simply a city of my dreams." What Einstein had (or had not) said about relativity was truer of quantum mechanics: a bare handful of people had the mathematics needed to understand it.

Richard and Julian

Summer brought a salty heat to Far Rockaway, the wind rising across the beaches. The asphalt shimmered with refractive air. In winter, snow fell early from low, gray clouds; then dazzlingly white hours would pass, the sky too bright to see clearly. Free and impudent times—Richard lost himself in his notebooks, or roamed to the drugstore, where he would play a mean-spirited optical-hydrodynamical trick on the waitress by inverting a glass of water over a one-penny tip on the smooth tabletop.

On the beach some days he watched a particular girl. She had warm, deep blue eyes and long hair that she wore deftly knotted up in a braid. After swimming she would comb it out, and boys Richard knew from school would flock around her. Her name was Arline (for a long time Richard thought it was spelled the usual way, "Arlene") Greenbaum, and she lived in Cedarhurst, Long Island, just across the city line. He dreamed about her. He thought she was wonderful and beautiful, but getting to know girls seemed hopeless enough, and Arline, he discovered, already had a boyfriend. Even so, he followed her into an after-school social league sponsored by the synagogue. Arline joined an art class, so Richard joined the art class, overlooking a lack of aptitude. Shortly he found himself lying on the floor and breathing through a straw, while another student made a plaster cast of his face.

If Arline noticed Richard, she did not let on. But one evening she arrived at a boy-girl party in the middle of a kissing session. An older boy was teaching couples the correct lip angles and nose positions, and in this instructional context a certain amount of practice was under way. Richard himself was practicing, with a girl he hardly knew. When Arline came in, there was a little commotion. Almost everyone got up to greet her—everyone, it seemed to her, but one horribly rude boy, off in the corner, who ostentatiously kept on kissing.

Occasionally Richard went on dates with other girls. He could never rid himself of a sense that he was a stranger engaging in a ritual the rules to which he did not know. His mother taught him some basic manners. Even so, the waiting in a girl's parlor with her parents, the procedures for cutting in at dances, the stock phrases ("Thank you for a lovely evening") all left him feeling inept, as if he could not quite decipher a code everyone else had mastered.

He stayed not quite conscious of the hopes his parents had for him. He was not quite aware of the void left by the death of his infant brother—his mother still thought of the baby often—or of his mother's social descent to the lower middle class, in increasingly tight circumstances. With the coming of the Depression the Feynmans had to give up the house and yard on New Broadway and move to a small apartment, where they used a dining room and a breakfast room as bedrooms. Melville was often on the road now, selling. When he was home, he would read the *National Geographic* magazines that he collected secondhand. On Sundays he would go outdoors and paint woodland scenery or flowers. Or he and Richard would take Joan into the city to the Metropolitan Museum of Art. They went to the Egyptian section, first studying glyphs in the encyclopedia so that they could stand and decode bits of the chiseled artifacts, a sight that made people stare.

Richard still had some tinkering and probing to do. The Depression broadened the market for inexpensive radio repair, and Richard found himself in demand. In just over a decade of full-scale commercial production, the radio had penetrated nearly half of American households. By 1932 the average price of a new set had fallen to $48, barely a third of the price just three years before. "Midget" sets had arrived, just five tubes compactly arranged within an astonishing six-pound box, containing its own built-in aerial and a shrunken loudspeaker the size of a paper dollar. Some receivers offered knobs that would let the user adjust the high and low tones separately; some advertised high style, like the "satin-finished ebony black Durez with polished chromium grille and trimmings."

Broken radios confronted Richard with a whole range of pathologies in the circuits he had learned so well. He rewired a plug or climbed a neighbor's roof to install an antenna. He looked for clues, wax on a condenser or telltale charcoal on a burned-out resistor. Later he made a story out of it—"He Fixes Radios by Thinking!" The hero was an exaggeratedly young boy, with a comically large screwdriver sticking out of his back pocket, who solved an ever-more-challenging sequence of puzzles. The last and best broken radio—the one that established his reputation—made a blood-curdling howl when first turned on. Richard paced back and forth, thinking, while the curmudgeonly owner badgered him: "What are you doing? Can you fix it?" Richard thought about it. What could be making a noise that changed with time? It must have something to do with the heating of the tubes—first some extraneous signal was swollen into a shriek; then it settled back to normal. Richard stopped pacing, went back to the set, pulled out one tube, pulled out a second tube, and exchanged them. He turned on the set, and the noise had vanished. *The boy who fixes radios by thinking*—that was how he saw himself, reflected in the eyes of his customers in Far Rockaway. Reason worked. Equations could be trusted; they were more than schoolbook exercises. The heady rush of solving a puzzle, of feeling the mental pieces shift and fade and rearrange themselves until suddenly they slid into their grooves—the sense of power and sheer rightness—these pleasures sustained an addiction. Luxuriating in the buoyant joy of it, Feynman could sink into a trance of concentration that even his family found unnerving.

Knowledge was rarer then. A secondhand magazine was an occasion. For a Far Rockaway teenager merely to find a mathematics textbook took will and enterprise. Each radio program, each telephone call, each lecture in a local synagogue, each movie at the new Gem theater on Mott Avenue carried the weight of something special. Each book Richard possessed burned itself into his memory. When a primer on mathematical methods baffled him, he worked through it formula by formula, filling a notebook with self-imposed exercises. He and his friends traded mathematical tidbits like baseball cards. If a boy named Morrie Jacobs told him that the cosine of 20 degrees multiplied by the cosine of 40 degrees multiplied by the cosine of 80 degrees equaled exactly one-eighth, he would remember that curiosity for the rest of his life, and he would remember that he was standing in Morrie's father's leather shop when he learned it.

Even with the radio era in full swing, one's senses encountered nothing like the bombardment of images and sounds that television would bring—accelerated, flash-cut, disposable knowledge. For now, knowledge was scarce and therefore dear. It was the same for scientists. The currency of scientific information had not yet been devalued by excess. For a young student, that meant that the most timely questions were surprisingly close to hand. Feynman recognized early the special, distinctive feeling of being close to the edge of knowledge, where people do not know the answers. Even in grade school, when he would haunt the laboratory late in the afternoon, playing with magnets and helping a teacher clean up, he recognized the pleasure of asking questions that the teacher could not handle.

Now, graduating from high school, he could not tell how near or how far he was from science's active frontier, where scientists pulled fresh problems like potatoes from the earth, and in fact he was not far. The upheaval caused by quantum mechanics had laid the fundamental issues bare. Physics was still a young science, more obscure than any human knowledge to date, yet still something of a family business. Its written record remained small, even as whole new scientific frameworks—nuclear physics, quantum field theory—were being born. The literature sustained just a handful of journals, still mostly in Europe. Richard knew nothing of these.

Across town, another precocious teenager, named Julian Schwinger, had quietly inserted himself into the world of the new physics. He was already as much a creature of the city as Feynman was of the city's outskirts: the younger son of a well-to-do garment maker, growing up in Jewish Harlem and then on Riverside Drive, where dark, stately apartment buildings and stone town houses followed the curve of the Hudson River. The drive was built for motor traffic, but truck horses still pulled loads of boxes to the merchants of Broadway, a few blocks east. Schwinger knew how to find books; he often prowled the used-book stores of lower Fourth and Fifth Avenues for advanced texts on mathematics and physics. He attended Townsend Harris High School, a nationally famous institution associated with the City College of New York, and even before he entered City College, in 1934, when he was sixteen, he found out what physics was—the modern physics. With his long, serious face and slightly stooped shoulders he would sit in the college's library and read papers by Dirac in the *Proceedings of the Royal Society of London* or the *Physikalische Zeitschrift der Sowjetunion*. He also read the *Physical Review*, now forty years past its founding; it had advanced from monthly to biweekly publication in hopes of competing more nimbly with the European journals. Schwinger struck his teachers as intensely shy. He carried himself with a premature elegant dignity.

That year he carefully typed out on six legal-size sheets his first real physics paper, "On the Interaction of Several Electrons," and the same elegance was evident. It assumed for a starting point the central new tenet of field theory: "that two particles do not interact directly but, rather the interaction is explained as being caused by one of the particles influencing the field in its vicinity, which influence spreads until it reaches the second particle." Electrons do not simply bounce off one another, that is. They plow through that magnificent ether substitute, the field; the waves they make then swish up against other electrons. Schwinger did not pretend to break ground in this paper. He showed his erudition by adopting "the quantum

electrodynamics of Dirac, Fock, and Podolsky," the "Heisenberg representation" of potentials in empty space, the "Lorentz-Heaviside units" for
expressing such potentials in relatively compact equations. This was heavy
machinery in soft terrain. The field of Maxwell, which brought electricity
and magnetism together so effectively, now had to be quantized, built up
from finite-size packets that could be reduced no further. Its waves were
simultaneously smooth and choppy. Schwinger, in his first effort at professional physics, looked beyond even this difficult electromagnetic field to a
more abstract field still, a field twice removed from tangible substance,
buoying not particles but mathematical operators. He pursued this conception through a sequence of twenty-eight equations. Once, at equation 20,
he was forced to pause. A fragment of the equation had grown unman-
ageable—infinite, in fact. To the extent that this fragment corresponded to
something physical, it was the tendency of an electron to act on itself. Having shaken its field, the electron is shaken back, with (so the mathematics
insisted) infinite energy. Dirac and the others had grudgingly settled on a response to this difficulty, and Schwinger handled it in the prescribed manner:
he simply discarded the offending term and moved on to equation 21.

Julian Schwinger and Richard Feynman, exact contemporaries, obsessed
as sixteen-year-olds with the abstract mental world of a scientist, had already
set out on different paths. Schwinger studying the newest of the new physics,
Feynman filling schoolboy notebooks with standard mathematical formulas,
Schwinger entering the arena of his elders, Feynman still trying to impress
his peers with practical jokes, Schwinger striving inward toward the city's
intellectual center, Feynman haunting the beaches and sidewalks of its
periphery—they would hardly have known what to say to each other. They
would not meet for another decade; not until Los Alamos. Long afterward,
when they were old men, after they had shared a Nobel Prize for work
done as rivals, they amazed a dinner party by competing to see who could
most quickly recite from memory the alphabetical headings on the spines
of their half-century-old edition of the *Encyclopaedia Britannica*.

As his childhood ended, Richard worked at odd jobs, for a neighborhood
printer or for his aunt, who managed one of the smaller Far Rockaway
resort hotels. He applied to colleges. His grades were perfect or near perfect
in mathematics and science but less than perfect in other subjects, and
colleges in the thirties enforced quotas in the admission of Jews. Richard
spent fifteen dollars on a special entrance examination for Columbia University, and after he was turned down he long resented the loss of the fifteen
dollars. MIT accepted him.

MIT

÷

A seventeen-year-old freshman, Theodore Welton, helped some of the older students operate the wind-tunnel display at the Massachusetts Institute of Technology's Spring Open House in 1936. Like so many of his classmates he had arrived at the Tech knowing all about airplanes, electricity, and chemicals and revering Albert Einstein. He was from a small town, Saratoga Springs, New York. With most of his first year behind him, he had lost none of his confidence. When his duties ended, he walked around and looked at the other exhibits. A miniature science fair of current projects made the open house a showcase for parents and visitors from Boston. He wandered over to the mathematics exhibit, and there, amid a crowd, his ears sticking out noticeably from a very fresh face, was what looked like another first-year boy, inappropriately taking charge of a complex, suitcase-size mechanical-mathematical device called a harmonic analyzer. This boy was pouring out explanations in a charged-up voice and fielding questions like a congressman at a press conference. The machine could take any arbitrary wave and break it down into a sum of simple sine and cosine waves. Welton, his own ears burning, listened while Dick Feynman rapidly explained the workings of the Fourier transform, the advanced mathematical technique for analyzing complicated wave forms, a piece of privileged knowledge that Welton until that moment had felt sure no other freshman possessed.

Welton (who liked to be called by his initials, T. A.) already knew he was a physics major. Feynman had vacillated twice. He began in mathematics. He passed an examination that let him jump ahead to the second-year calculus course, covering differential equations and integration in three-dimensional space. This still came easily, and Feynman thought he should have taken the second-year examination as well. But he also began to wonder whether this was the career he wanted. American professional mathematics of the thirties was enforcing its rigor and abstraction as never before, disdaining what outsiders would call "applications." To Feynman—having finally reached a place where he was surrounded by fellow tinkerers and radio buffs—mathematics began to seem too abstract and too far removed.

In the stories modern physicists have made of their own lives, a fateful moment is often the one in which they realize that their interest no longer lies in mathematics. Mathematics is always where they begin, for no other school course shows off their gifts so clearly. Yet a crisis comes: they experience an epiphany, or endure a slowly building disgruntlement, and plunge or drift into this other, hybrid field. Werner Heisenberg, seventeen years older than Feynman, experienced his moment of crisis at the University of Munich, in the office of the local statesman of mathematics, Ferdinand von Lindemann. For some reason Heisenberg could never forget Lindemann's horrid yapping black dog. It reminded him of the poodle in *Faust* and made it impossible for him to think clearly when the professor, learning that Heisenberg was reading Weyl's new book about relativity theory, told him, "In that case you are completely lost to mathematics." Feynman himself, halfway through his freshman year, reading Eddington's book about relativity theory, confronted his own department chairman with the classic question about mathematics: What is it good for? He got the classic answer: If you have to ask, you are in the wrong field. Mathematics seemed suited only for teaching mathematics. His department chairman suggested calculating actuarial probabilities for insurance companies. This was not a joke. The vocational landscape had just been surveyed by one Edward J. v. K. Menge, Ph.D., Sc.D., who published his findings in a monograph titled *Jobs for the College Graduate in Science*. "The American mind is taken up largely with applications rather than with fundamental principles," Menge noticed. "It is what is known as 'practical.' " This left little room for would-be mathematicians: "The mathematician has little opportunity of employment except in the universities in some professorial capacity. He may become a practitioner of his profession, it is true, if he acts as an actuary for some large insurance company. . . ." Feynman changed to electrical engineering. Then he changed again, to physics.

Not that physics promised much more as a vocation. The membership of the American Physical Society still fell shy of two thousand, though it had doubled in a decade. Teaching at a college or working for the government in, most likely, the Bureau of Standards or the Weather Bureau, a physicist might expect to earn a good wage of from three thousand to six thousand dollars a year. But the Depression had forced the government and the leading corporate laboratories to lay off nearly half of their staff scientists. A Harvard physics professor, Edwin C. Kemble, reported that finding jobs for graduating physicists had become a "nightmare." Not many arguments could be made for physics as a vocation.

Menge, putting his pragmatism aside for a moment, offered perhaps the only one: Does the student, he asked, "feel the craving of adding to the sum total of human knowledge? Or does he want to see his work go on and on and his influence spread like the ripples on a placid lake into which a stone has been cast? In other words, is he so fascinated with simply *knowing* the subject that he cannot rest until he learns all he can about it?"

Of the leading men in American physics MIT had three of the best, John C. Slater, Philip M. Morse, and Julius A. Stratton. They came from a more standard mold—gentlemanly, homebred, Christian—than some of the physicists who would soon eclipse them, foreigners like Hans Bethe and Eugene Wigner, who had just arrived at Cornell University and Princeton University, respectively, and Jews like I. I. Rabi and J. Robert Oppenheimer, who had been hired at Columbia University and the University of California at Berkeley, despite anti-Semitic misgivings at both places. Stratton later became president of MIT, and Morse became the first director of the Brookhaven National Laboratory for Nuclear Research. The department head was Slater. He had been one of the young Americans studying overseas, though he was not as deeply immersed in the flood tides of European physics as, for example, Rabi, who made the full circuit: Zurich, Munich, Copenhagen, Hamburg, Leipzig, and Zurich again. Slater had studied briefly at Cambridge University in 1923, and somehow he missed the chance to meet Dirac, though they attended at least one course together.

Slater and Dirac crossed paths intellectually again and again during the decade that followed. Slater kept making minor discoveries that Dirac had made a few months earlier. He found this disturbing. It seemed to Slater furthermore that Dirac enshrouded his discoveries in an unnecessary and somewhat baffling web of mathematical formalism. Slater tended to mistrust them. In fact he mistrusted the whole imponderable miasma of philosophy now flowing from the European schools of quantum mechanics: assertions about the duality or complementarity or "Jekyll-Hyde" nature of things; doubts about time and chance; the speculation about the interfering role of the human observer. "I do not like mystiques; I like to be definite," Slater said. Most of the European physicists were reveling in such issues. Some felt an obligation to face the consequences of their equations. They recoiled from the possibility of simply putting their formidable new technology to work without developing a physical picture to go along with it. As they manipulated their matrices or shuffled their differential equations, questions kept creeping in. *Where is that particle when no one is looking?* At the ancient stone-built universities philosophy remained the coin of the realm. A theory about the spontaneous, whimsical birth of photons in the energy decay of excited atoms—an effect without a cause—gave scientists a sledgehammer to wield in late-evening debates about Kantian causality. Not so in America. "A theoretical physicist in these days asks just one thing of his theories," Slater said defiantly soon after Feynman arrived at MIT. The theories must make reasonably good predictions about experiments. That is all.

> He does not ordinarily argue about philosophical implications. . . . Questions about a theory which do not affect its ability to predict experimental results correctly seem to me quibbles about words, . . . and I am quite content to leave such questions to those who derive some satisfaction from them.

When Slater spoke for common sense, for practicality, for a theory that would be experiment's handmaid, he spoke for most of his American colleagues. The spirit of Edison, not Einstein, still governed their image of the scientist. Perspiration, not inspiration. Mathematics was unfathomable and unreliable. Another physicist, Edward Condon, said everyone knew what mathematical physicists did: "they study carefully the results obtained by experimentalists and rewrite that work in papers which are so mathematical that they find them hard to read themselves." Physics could really only justify itself, he said, when its theories offered people a means of predicting the outcome of experiments—and at that, only if the predicting took less time than actually carrying out the experiments.

Unlike their European counterparts, American theorists did not have their own academic departments. They shared quarters with the experimenters, heard their problems, and tried to answer their questions pragmatically. Still, the days of Edisonian science were over and Slater knew it. With a mandate from MIT's president, Karl Compton, he was assembling a physics department meant to bring the school into the forefront of American science and meanwhile to help American science toward a less humble world standing. He and his colleagues knew how unprepared the United States had been to train physicists in his own generation. Leaders of the nation's rapidly growing technical industries knew it, too. When Slater arrived, the MIT department sustained barely a dozen graduate students. Six years later, the number had increased to sixty. Despite the Depression the institute had completed a new physics and chemistry laboratory with money from the industrialist George Eastman. Major research programs had begun in the laboratory fields devoted to using electromagnetic radiation as a probe into the structure of matter: especially spectroscopy, analyzing the signature frequencies of light shining from different substances, but also X-ray crystallography. (Each time physicists found a new kind of "ray" or particle, they put it to work illuminating the interstices of molecules.) New vacuum equipment and finely etched mirrors gave a high precision to the spectroscopic work. And a monstrous new electromagnet created fields more powerful than any on the planet.

Julius Stratton and Philip Morse taught the essential advanced theory course for seniors and graduate students, Introduction to Theoretical Physics, using Slater's own text of the same name. Slater and his colleagues had created the course just a few years before. It was the capstone of their new thinking about the teaching of physics at MIT. They meant to bring back together, as a unified subject, the discipline that had been subdivided for undergraduates into mechanics, electromagnetism, thermodynamics, hydrodynamics, and optics. Undergraduates had been acquiring their theory piecemeal, in ad hoc codas to laboratory courses mainly devoted to experiment. Slater now brought the pieces back together and led students toward a new topic, the "modern atomic theory." No course yet existed in quantum mechanics, but Slater's students headed inward toward the atom with a grounding not just in classical mechanics, treating the motion of solid objects, but also in wave mechanics—vibrating strings, sound waves bouncing around in hollow boxes. The instructors told the students at the outset that the essence of theoretical physics lay not in learning to work out the mathematics, but in learning how to apply the mathematics to the real phenomena that could take so many chameleon forms: moving bodies, fluids, magnetic fields and forces, currents of electricity and water, and waves of water and light. Feynman, as a freshman, roomed with two seniors who took the course. As the year went on he attuned himself to their chatter and surprised them sometimes by joining in on the problem solving. "Why don't you try Bernoulli's equation?" he would say—mispronouncing *Bernoulli* because, like so much of his knowledge, this came from reading the encyclopedia or the odd textbooks he had found in Far Rockaway. By sophomore year he decided he was ready to take the course himself.

The first day everyone had to fill out enrollment cards: green for seniors and brown for graduate students. Feynman was proudly aware of the sophomore-pink card in his own pocket. Furthermore he was wearing an ROTC uniform; officer's training was compulsory for first- and second-year students. But just as he was feeling most conspicuous, another uniformed, pink-card–carrying sophomore sat down beside him. It was T. A. Welton. Welton had instantly recognized the mathematics whiz from the previous spring's open house.

Feynman looked at the books Welton was stacking on his desk. He saw Tullio Levi-Civita's *Absolute Differential Calculus*, a book he had tried to get from the library. Welton, meanwhile, looked at Feynman's desk and realized why he had not been able to find A. P. Wills's *Vector and Tensor Analysis*. Nervous boasting ensued. The Saratoga Springs sophomore claimed to know all about general relativity. The Far Rockaway sophomore announced that he had already learned quantum mechanics from a book by someone called Dirac. They traded several hours' worth of sketchy knowledge about Einstein's work on gravitation. Both boys realized that, as Welton put it, "cooperation in the struggle against a crew of aggressive-looking seniors and graduate students might be mutually beneficial."

Nor were they alone in recognizing that Introduction to Theoretical Physics now harbored a pair of exceptional young students. Stratton, handling the teaching chores for the first semester, would sometimes lose the thread of a string of equations at the blackboard, the color of his face shifting perceptibly toward red. He would then pass the chalk, saying, "Mr. Feynman, how did you handle this problem," and Feynman would stride to the blackboard.

The Best Path

A law of nature expressed in a strange form came up again and again that term: the principle of least action. It arose in a simple sort of problem. A lifeguard, some feet up the beach, sees a drowning swimmer diagonally ahead, some distance offshore and some distance to one side. The lifeguard can run at a certain speed and swim at a certain lesser speed. How does one find the fastest path to the swimmer?

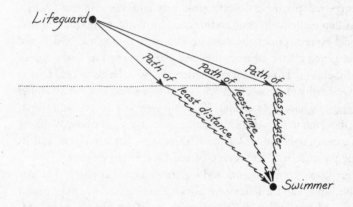

The path of least time. The lifeguard travels faster on land than in water; the best path is a compromise. Light—which also travels faster through air than through water—seems somehow to choose precisely this path on its way from an underwater fish to the eye of an observer.

A straight line, the shortest path, is not the fastest. The lifeguard will spend too much time in the water. If instead he angles far up the beach and dives in directly opposite the swimmer—the path of least water—he still wastes time. The best compromise is the path of least time, angling up the beach and then turning for a sharper angle through the water. Any calculus student can find the best path. A lifeguard has to trust his instincts. The mathematician Pierre de Fermat guessed in 1661 that the bending of a ray of light as it passes from air into water or glass—the refraction that makes possible lenses and mirages—occurs because light behaves like a lifeguard with perfect instincts. It follows the path of least time. (Fermat, reasoning backward, surmised that light must travel more slowly in denser media. Later Newton and his followers thought they had proved the opposite: that light, like sound, travels faster through water than through air. Fermat, with his faith in a principle of simplicity, was right.)

Theology, philosophy, and physics had not yet become so distinct from one another, and scientists found it natural to ask what sort of universe God would make. Even in the quantum era the question had not fully disappeared from the scientific consciousness. Einstein did not hesitate to invoke His name. Yet when Einstein doubted that God played dice with the world, or when he uttered phrases like the one later inscribed in the stone of Fine Hall at Princeton, "The Lord God is subtle, but malicious he is not," the great man was playing a delicate game with language. He had found a formulation easily understood and imitated by physicists, religious or not. He could express convictions about how the universe ought to be designed without giving offense either to the most literal believers in God or to his most disbelieving professional colleagues, who were happy to read *God* as a poetic shorthand for *whatever laws or principles rule this flux of matter and energy we happen to inhabit*. Einstein's piety was sincere but neutral, acceptable even to the vehemently antireligious Dirac, of whom Wolfgang Pauli once complained, "Our friend Dirac, too, has a religion, and its guiding principle is 'There is no God and Dirac is His prophet.' "

Scientists of the seventeenth and eighteenth centuries also had to play a double game, and the stakes were higher. Denying God was still a capital offense, and not just in theory: offenders could be hanged or burned. Scientists made an assault against faith merely by insisting that knowledge— some knowledge—must wait on observation and experiment. It was not so obvious that one category of philosopher should investigate the motion of falling bodies and another the provenance of miracles. On the contrary, Newton and his contemporaries happily constructed scientific proofs of God's existence or employed God as a premise in a chain of reasoning. Elementary particles must be indivisible, Newton wrote in his *Opticks*, "so very hard as never to wear or break in pieces; no ordinary power being able to divide what God himself made one in the first creation." Elementary particles cannot be indivisible, René Descartes wrote in his *Principles of Philosophy*:

There cannot be any atoms or parts of matter which are indivisible of their own nature (as certain philosophers have imagined). . . . For though God had rendered the particle so small that it was beyond the power of any creature to divide it, He could not deprive Himself of the power of division, because it was absolutely impossible that He should lessen His own omnipotence. . . .

Could God make atoms so flawed that they could break? Could God make atoms so perfect that they would defy His power to break them? It was only one of the difficulties thrown up by God's omnipotence, even before relativity placed a precise upper limit on velocity and before quantum mechanics placed a precise upper limit on certainty. The natural philosophers wished to affirm the presence and power of God in every corner of the universe. Yet even more fervently they wished to expose the mechanisms by which planets swerved, bodies fell, and projectiles recoiled in the absence of any divine intervention. No wonder Descartes appended a blanket disclaimer: "At the same time, recalling my insignificance, I affirm nothing, but submit all these opinions to the authority of the Catholic Church, and to the judgment of the more sage; and I wish no one to believe anything I have written, unless he is personally persuaded by the evidence of reason."

The more competently science performed, the less it needed God. There was no special providence in the fall of a sparrow; just Newton's second law, $f = ma$. Forces, masses, and acceleration were the same everywhere. The Newtonian apple fell from its tree as mechanistically and predictably as the moon fell around the Newtonian earth. Why does the moon follow its curved path? Because its path is the sum of all the tiny paths it takes in successive instants of time; and because at each instant its forward motion is deflected, like the apple, toward the earth. God need not choose the path. Or, having chosen once, in creating a universe with such laws, He need not choose again. A God that does not intervene is a God receding into a distant, harmless background.

Yet even as the eighteenth-century philosopher scientists learned to compute the paths of planets and projectiles by Newton's methods, a French geometer and *philosophe*, Pierre-Louis Moreau de Maupertuis, discovered a strangely magical new way of seeing such paths. In Maupertuis's scheme a planet's path has a logic that cannot be seen from the vantage point of someone merely adding and subtracting the forces at work instant by instant. He and his successors, and especially Joseph Louis Lagrange, showed that the paths of moving objects are always, in a special sense, the most economical. They are the paths that minimize a quantity called *action*—a quantity based on the object's velocity, its mass, and the space it traverses. No matter what forces are at work, a planet somehow chooses the cheapest, the simplest, the best of all possible paths. It is as if God—a parsimonious God—were after all leaving his stamp.

None of which mattered to Feynman when he encountered Lagrange's method in the form of a computational shortcut in Introduction to Theoretical Physics. All he knew was that he did not like it. To his friend Welton and to the rest of the class the Lagrange formulation seemed elegant and useful. It let them disregard many of the forces acting in a problem and cut straight through to an answer. It served especially well in freeing them from the right-angle coordinate geometry of the classical reference frame required by Newton's equations. Any reference frame would do for the Lagrangian technique. Feynman refused to employ it. He said he would not feel he understood the real physics of a system until he had painstakingly isolated and calculated all the forces. The problems got harder and harder as the class advanced through classical mechanics. Balls rolled down inclines, spun in paraboloids—Feynman would resort to ingenious computational tricks like the ones he learned in his mathematics-team days, instead of the seemingly blind, surefire Lagrangian method.

Feynman had first come on the principle of least action in Far Rockaway, after a bored hour of high-school physics, when his teacher, Abram Bader, took him aside. Bader drew a curve on the blackboard, the roughly parabolic shape a ball would take if someone threw it up to a friend at a second-floor window. If the time for the journey can vary, there are infinitely many such paths, from a high, slow lob to a nearly straight, fast trajectory. But if you know how long the journey took, the ball can have taken only one path. Bader told Feynman to make two familiar calculations of the ball's energy: its kinetic energy, the energy of its motion, and its potential energy, the energy it possesses by virtue of its presence high in a gravitational field. Like all high-school physics students Feynman was used to adding those energies together. An airplane, accelerating as it dives, or a roller coaster, sliding down the gravity well, trades its potential energy for kinetic energy: as it loses height it gains speed. On the way back up, friction aside, the airplane or roller coaster makes the same conversion in reverse: kinetic energy becomes potential energy again. Either way, the total of kinetic and potential energy never changes. The total energy is conserved.

Bader asked Feynman to consider a less intuitive quantity than the sum of these energies: their difference. Subtracting the potential energy from the kinetic energy was as easy as adding them. It was just a matter of changing signs. But understanding the physical meaning was harder. Far from being conserved, this quantity—the *action*, Bader said—changed constantly. Bader had Feynman calculate it for the ball's entire flight to the window. And he pointed out what seemed to Feynman a miracle. At any particular moment the action might rise or fall, but when the ball arrived at its destination, the path it had followed would always be the path for which the total action was least. For any other path Feynman might try drawing on the blackboard—a straight line from the ground to the window, a higher-arcing trajectory, or a trajectory that deviated however slightly from the fated path—he would find a greater average difference between kinetic and potential energy.

It is almost impossible for a physicist to talk about the principle of least action without inadvertently imputing some kind of volition to the projectile. The ball seems to *choose* its path. It seems to *know* all the possibilities in advance. The natural philosophers started encountering similar minimum principles throughout science. Lagrange himself offered a program for computing planetary orbits. The behavior of billiard balls crashing against each other seemed to minimize action. So did weights swung on a lever. So, in a different way, did light rays bent by water or glass. Fermat, in plucking his principle of least time from a pristine mathematical landscape, had found the same law of nature.

Where Newton's methods left scientists with a feeling of comprehension, minimum principles left a sense of mystery. "This is not quite the way one thinks in dynamics," the physicist David Park has noted. One likes to think that a ball or a planet or a ray of light makes its way instant by instant, not that it follows a preordained path. From the Lagrangian point of view the forces that pull and shape a ball's arc into a gentle parabola serve a higher law. Maupertuis wrote, "It is not in the little details . . . that we must look for the supreme Being, but in phenomena whose universality suffers no exception and whose simplicity lays them quite open to our sight." The universe wills simplicity. Newton's laws provide the mechanics; the principle of least action ensures grace.

The hard question remained. (In fact, it would remain, disquieting the few physicists who continued to ponder it, until Feynman, having long since overcome his aversion to the principle of least action, found the answer in quantum mechanics.) Park phrased the question simply: How does the ball know which path to choose?

Socializing the Engineer

"Let none say that the engineer is an unsociable creature who delights only in formulae and slide rules." So pleaded the MIT yearbook. Some administrators and students did worry about the socialization of this famously awkward creature. One medicine prescribed by the masters of student life was Tea, compulsory for all freshmen. ("But after they have conquered their initial fears and learned to balance a cup on a saucer while conversing with the wife of a professor, compulsion is no longer necessary.") Students also refined their conversational skills at Bull Session Dinners and their other social skills at an endless succession of dances: Dormitory Dinner Dances, the Christmas Dance and the Spring Dance, a Monte Carlo Dance featuring a roulette wheel and a Barn Dance offering sleigh rides, dances to attract students from nearby women's colleges like Radcliffe and Simmons, dances accompanied by the orchestras of Nye Mayhew and Glenn Miller, the traditional yearly Field Day Dance after the equally traditional Glove Fight, and, in the fraternity houses that provided the most desirable student quarters, formal dances that persuaded even Dick Feynman to put on a tuxedo almost every week.

The fraternities at MIT, as elsewhere, strictly segregated students by religion. Jews had a choice of just two, and Feynman joined the one called Phi Beta Delta, on Bay State Road in Boston, in a neighborhood of town houses just across the Charles River from campus. One did not simply "join" a fraternity, however. One enjoyed a wooing process that began the summer before college at local smokers and continued, in Feynman's case, with insistent offers of transportation and lodging that bordered on kidnapping. Having chosen a fraternity, one instantly underwent a status reversal, from an object of desire to an object of contempt. New pledges endured systematic humiliation. Their fraternity brothers drove Feynman and the other boys to an isolated spot in the Massachusetts countryside, abandoned them beside a frozen lake, and left them to find their way home. They submitted to wrestling matches in mud and allowed themselves to be tied down overnight on the wooden floor of a deserted house—though Feynman, still secretly afraid that he would be found out as a sissy, made a surprising show of resisting his sophomore captors by grabbing at their legs and trying to knock them over. These rites were tests of character, after all, mixed with schoolboy sadism that colleges only gradually learned to restrain. The hazing left many boys with emotional bonds both to their tormentors and to their fellow victims.

Walking into the parlor floor of the Bay State Road chapter house of Phi Beta Delta, a student could linger in the front room with its big bay windows overlooking the street or head directly for the dining room, where Feynman ate most of his meals for four years. The members wore jackets and ties to dinner. They gathered in the anteroom fifteen minutes before and waited for the bell that announced the meal. White-painted pilasters rose toward the high ceilings. A stairway bent gracefully up four flights. Fraternity members often leaned over the carved railing to shout down to those below, gathered around the wooden radio console in one corner or waiting to use the pay telephone on an alcove wall. The telephone provided an upperclassman with one of his many opportunities to harass freshmen: they were obliged to carry nickels for making change. They also carried individual black notebooks for keeping a record of their failures, among other things, to carry nickels. Feynman developed a trick of catching a freshman nickelless, making a mark in his black book, and then punishing the same freshman all over again a few minutes later. The second and third floors were given over entirely to study rooms, where students worked in twos and threes. Only the top floor was for sleeping, in double-decker bunks crowded together.

Compulsory Tea notwithstanding, some members argued vehemently that other members lacked essential graces, among them the ability to dance and the ability to invite women to accompany them to a dance. For a while this complaint dominated the daily counsel of the thirty-odd members of Phi Beta Delta. A generation later the ease of postwar life made a place for words like "wonk" and "nerd" in the collegiate vocabulary. In more class-bound and less puritanical cultures the concept flowered even earlier. Britain had its boffins, working researchers subject to the derision of intellectual gentlemen. At MIT in the thirties the nerd did not exist; a penholder worn in the shirt pocket represented no particular gaucherie; a boy could not become a figure of fun merely by studying. This was fortunate for Feynman and others like him, socially inept, athletically feeble, miserable in any but a science course, risking laughter every time he pronounced an unfamiliar name, so worried about the other sex that he trembled when he had to take the mail out past girls sitting on the stoop. America's future scientists and engineers, many of them rising from the working class, valued studiousness without question. How could it be otherwise, in the knots that gathered almost around the clock in fraternity study rooms, filling dappled cardboard notebooks with course notes to be handed down to generations? Even so, Phi Beta Delta perceived a problem. There did seem to be a connection between hard studying and failure to dance. The fraternity made a cooperative project of enlivening the potential dull boys. Attendance at dances became mandatory for everyone in Phi Beta Delta. For those who could not find dates, the older boys arranged dates. In return, stronger students tutored the weak. Dick felt he got a good bargain. Eventually he astonished even the most sociable of his friends by spending long hours at the Raymore-Playmore Ballroom, a huge dance hall near Boston's Symphony Hall with a mirrored ball rotating from the ceiling.

The best help for his social confidence, however, came from Arline Greenbaum. She was still one of the most beautiful girls he knew, with dimples in her round, ruddy face, and she was becoming a distinct presence in his life, though mostly from a distance. On Saturdays she would visit his family in Far Rockaway and give Joan piano lessons. She was the kind of young woman that people called "talented"—musical and artistic in a well-rounded way. She danced and sang in the Lawrence High School revue, "America on Her Way." The Feynmans let her paint a parrot on the inside door of the coat closet downstairs. Joan started to think of her as an especially benign older sister. Often after their piano lesson they went for walks or rode their bicycles to the beach.

Arline also made an impression on the fraternity boys when she started visiting on occasional weekends and spared Dick the necessity of finding a date from among the students at the nearby women's colleges or (to the dismay of his friends) from among the waitresses at the coffee shop he frequented. Maybe there was hope for Dick after all. Still, they wondered whether she would succeed in domesticating him before he found his way to the end of her patience. Over the winter break he had some of his friends home to Far Rockaway. They went to a New Year's Eve party in the Bronx, taking the long subway-train ride across Brooklyn and north through Manhattan and returning, early in the morning, by the same route. By then Dick had decided that alcohol made him stupid. He avoided it with unusual earnestness. His friends knew that he had drunk no wine or liquor at the party, but all the way home he put on a loud, staggering drunk act, reeling off the subway car doors, swinging from the overhead straps, leaning over the seated passengers, and comically slurring nonsense at them. Arline watched unhappily. She had made up her mind about him, however. Sometime in his junior year he suggested that they become engaged. She agreed. Long afterward he discovered that she considered that to have been not his first but his second proposal of marriage—he had once said (offhandedly, he thought) that he would like her to be his wife.

Her well-bred talents for playing the piano, singing, drawing, and conversing about literature and the arts met in Feynman a bristling negatively charged void. He resented art. Music of all kinds made him edgy and uncomfortable. He felt he had a feeling for rhythm, and he had fallen into a habit of irritating his roommates and study partners with an absentminded drumming of his fingers, a tapping staccato against walls and wastebaskets. But melody and harmony meant nothing to him; they were sand in the mouth. Although psychologists liked to speculate about the evident mental links between the gift for mathematics and the gift for music, Feynman found music almost painful. He was becoming not passively but aggressively uncultured. When people talked about painting or music, he heard nomenclature and pomposity. He rejected the bird's nest of traditions, stories, and knowledge that cushioned most people, the cultural resting place woven from bits of religion, American history, English literature, Greek myth, Dutch painting, German music. He was starting fresh. Even the gentle, hearth-centered Reform Judaism of his parents left him cold. They had sent him to Sunday school, but he had quit, shocked at the discovery that those stories—Queen Esther, Mordechai, the Temple, the Maccabees, the oil that burned eight nights, the Spanish inquisition, the Jew who sailed with Christopher Columbus, the whole pastel mosaic of holiday legends and morality tales offered to Jewish schoolchildren on Sundays—mixed fiction with fact. Of the books assigned by his high-school teachers he read almost none. His friends mocked him when, forced to read a book, any book, in preparing for the New York State Regents Examination, he chose *Treasure Island*. (But he outscored all of them, even in English, when he wrote an essay on "the importance of science in aviation" and padded his sentences with what he knew to be redundant but authoritative phrases like "eddies, vortices, and whirlpools formed in the atmosphere behind the aircraft . . .")

He was what the Russians derided as *nekulturniy*, what Europeans refused to permit in an educated scientist. Europe prepared its scholars to register knowledge more broadly. At one of the fateful moments toward which Feynman's life was now beginning to speed, he would stand near the Austrian theorist Victor Weisskopf, both men watching as a light flared across the southern New Mexico sky. In that one instant Feynman would see a great ball of flaming orange, churning amid black smoke, while Weisskopf would hear, or think he heard, a Tchaikovsky waltz playing over the radio. That was strangely banal accompaniment for a yellow-orange sphere surrounded by a blue halo—a color that Weisskopf thought he had seen before, on an altarpiece at Colmar painted by the medieval master

Matthias Grünewald to depict (the irony was disturbing) the ascension of Christ. No such associations for Feynman. MIT, America's foremost technical school, was the best and the worst place for him. The institute justified its required English course by reminding students that they might someday have to write a patent application. Some of Feynman's fraternity friends actually liked French literature, he knew, or actually liked the lowest-common-denominator English course, with its smattering of great books, but to Feynman it was an intrusion and a pain in the neck.

In one course he resorted to cheating. He refused to do the daily reading and got through a routine quiz, day after day, by looking at his neighbor's answers. English class to Feynman meant arbitrary rules about spelling and grammar, the memorization of human idiosyncrasies. It seemed like supremely useless knowledge, a parody of what knowledge ought to be. Why didn't the English professors just get together and straighten out the language? Feynman got his worst grade in freshman English, barely passing, worse than his grades in German, a language he did not succeed in learning. After freshman year matters eased. He tried to read Goethe's *Faust* and felt he could make no sense of it. Still, with some help from his fraternity friends he managed to write an essay on the limitations of reason: problems in art or ethics, he argued, could not be settled with certainty through chains of logical reasoning. Even in his class themes he was beginning to assert a moral viewpoint. He read John Stuart Mill's *On Liberty* ("Whatever crushes individuality is despotism") and wrote about the despotism of social niceties, the white lies and fake politesse that he so wanted to escape. He read Thomas Huxley's "On a Piece of Chalk," and wrote, instead of the analysis he was assigned, an imitation, "On a Piece of Dust," musing on the ways dust makes raindrops form, buries cities, and paints sunsets. Although MIT continued to require humanities courses, it took a relaxed view of what might constitute humanities. Feynman's sophomore humanities course, for example, was Descriptive Astronomy. "Descriptive" meant "no equations." Meanwhile in physics itself Feynman took two courses in mechanics (particles, rigid bodies, liquids, stresses, heat, the laws of thermodynamics), two in electricity (electrostatics, magnetism, . . .), one in experimental physics (students were expected to design original experiments and show that they understood many different sorts of instruments), a lecture course and a laboratory course in optics (geometrical, physical, and physiological), a lecture course and a laboratory course in electronics (devices, thermionics, photoemission), a course in X rays and crystals, a course and a laboratory in atomic structure (spectra, radioactivity, and a physicist's view of the periodic table), a special seminar on the new nuclear theory, Slater's advanced theory course, a special seminar on quantum theory, and a course on heat and thermodynamics that worked toward statistical mechanics both classical and quantum; and then, his docket full, he listened in on five more advanced courses, including relativity and advanced mechanics. When he wanted to round out his course selection with something different, he took metallography.

Then there was philosophy. In high school he had entertained the conceit that different kinds of knowledge come in a hierarchy: biology and chemistry, then physics and mathematics, and then philosophy at the top. His ladder ran from the particular and ad hoc to the abstract and theoretical—from ants and leaves to chemicals, atoms, and equations and then onward to God, truth, and beauty. Philosophers have entertained the same notion. Feynman did not flirt with philosophy long, however. His sense of what constituted a proof had already developed into something more hard-edged than the quaint arguments he found in Descartes, for example, whom Arline was reading. The Cartesian proof of God's perfection struck him as less than rigorous. When he parsed *I think, therefore I am*, it came out suspiciously close to *I am and I also think*. When Descartes argued that the existence of imperfection implied perfection, and that the existence of a God concept in his own fuzzy and imperfect mind implied the existence of a Being sufficiently perfect and infinite as to create such a conception, Feynman thought he saw the obvious fallacy. He knew all about imperfection in science—"degrees of approximation." He had drawn hyperbolic curves that approached an ideal straight line without ever reaching it. People like Descartes were stupid, Richard told Arline, relishing his own boldness in defying the authority of the great names. Arline replied that she supposed there were two sides to everything. Richard gleefully contradicted even that. He took a strip of paper, gave it a half twist, and pasted the ends together: he had produced a surface with one side.

No one showed Feynman, in return, the genius of Descartes's strategy in proving the obvious—obvious because he and his contemporaries were supposed to take their own and God's existence as given. The Cartesian master plan was to reject the obvious, reject the certain, and start fresh from a state of total doubt. Even *I* might be an illusion or a dream, Descartes declared. It was the first great suspension of belief. It opened a door to the skepticism that Feynman now savored as part of the modern scientific method. Richard stopped reading, though, long before giving himself the pleasure of rejecting Descartes's final, equally unsyllogistic argument for the existence of God: that a perfect being would certainly have, among other excellent features, the attribute of existence.

Philosophy at MIT only irritated Feynman more. It struck him as an industry built by incompetent logicians. Roger Bacon, famous for introducing *scientia experimentalis* into philosophical thought, seemed to have done more talking than experimenting. His idea of experiment seemed closer to mere *experience* than to the measured tests a twentieth-century student performed in his laboratory classes. A modern experimenter took hold of some physical apparatus and performed certain actions on it, again and again, and generally wrote down numbers. William Gilbert, a less well-known sixteenth-century investigator of magnetism, suited Feynman better, with his credo, "In the discovery of secret things and in the investigation of hidden causes, stronger reasons are obtained from sure experiments and demonstrated arguments than from probable conjectures and the opinions of philosophical speculators of the common sort." That was a theory of knowledge Feynman could live by. It also stuck in his mind that Gilbert thought Bacon wrote science "like a prime minister." MIT's physics instructors did nothing to encourage students to pay attention to the philosophy instructors. The tone was set by the pragmatic Slater, for whom philosophy was smoke and perfume, free-floating and untestable prejudice. Philosophy set knowledge adrift; physics anchored knowledge to reality.

"Not from positions of philosophers but from the fabric of nature"— William Harvey three centuries earlier had declared a division between science and philosophy. Cutting up corpses gave knowledge a firmer grounding than cutting up sentences, he announced, and the gulf between two styles of knowledge came to be accepted by both camps. What would happen when scientists plunged their knives into the less sinewy reality inside the atom remained to be seen. In the meantime, although Feynman railed against philosophy, an instructor's cryptic comment about "stream of consciousness" started him thinking about what he could learn of his own mind through introspection. His inward looking was more experimental than Descartes's. He would go up to his room on the fourth floor of Phi Beta Delta, pull down the shades, get into bed, and try to watch himself fall asleep, as if he were posting an observer on his shoulder. His father years before had raised the problem of what happens when one falls asleep. He liked to prod Ritty to step outside himself and look afresh at his usual way of thinking: he asked how the problem would look to a Martian who arrived in Far Rockaway and starting asking questions. What if Martians never slept? What would they want to know? How does it feel to fall asleep? Do you simply turn off, as if someone had thrown a switch? Or do your ideas come slower and slower until they stop? Up in his room, taking midday naps for the sake of philosophy, Feynman found that he could follow his consciousness deeper and deeper toward the dissolution that came with sleep. His thoughts, he saw, did not so much slow down as fray apart, snapping from place to place without the logical connectives of waking brain work. He would suddenly realize he had been imagining his bed rising amid a contraption of pulleys and wires, ropes winding upward and catching against one another, Feynman thinking, the tension of the ropes will hold . . . and then he would be awake again. He wrote his observations in a class paper, concluding with a comment in the form of doggerel about the hall-of-mirrors impossibility of true introspection: "I wonder why I wonder why. I wonder why I wonder. I wonder why I wonder *why* I wonder why I wonder!"

After his instructor read his paper aloud in class, poem and all, Feynman began trying to watch his dreams. Even there he obeyed a tinkerer's impulse to take phenomena apart and look at the works inside. He was able to dream the same dream again and again, with variations. He was riding in a subway train. He noticed that kinesthetic feelings came through clearly. He could feel the lurching from side to side, see colors, hear the whoosh of air through the tunnel. As he walked through the car he passed three girls in bathing suits behind a pane of glass like a store window. The train kept lurching, and suddenly he thought it would be interesting to see how sexually excited he could become. He turned to walk back toward the window—but now the girls had become three old men playing violins. He could influence the course of a dream, but not perfectly, he realized. In another dream Arline came by subway train to visit him in Boston. They met and Dick felt a wave of happiness. There was green grass, the sun was shining, they walked along, and Arline said, "Could we be dreaming?"

"No, sir," Dick replied, "no, this is not a dream." He persuaded himself of Arline's presence so forcibly that when he awoke, hearing the noise of the boys around him, he did not know where he was. A dismayed, disoriented moment passed before he realized that he had been dreaming after all, that he was in his fraternity bedroom and that Arline was back home in New York.

The new Freudian view of dreams as a door to a person's inner life had no place in his program. If his subconscious wished to play out desires too frightening or confusing for his ego to contemplate directly, that hardly mattered to Feynman. Nor did he care to think of his dream subjects as symbols, encoded for the sake of a self-protective obscurity. It was his ego, his "rational mind," that concerned him. He was investigating his mind as an intriguingly complex machine, one whose tendencies and capabilities mattered to him more than almost anything else. He did develop a rudimentary theory of dreams for his philosophy essay, though it was more a theory of vision: that the brain has an "interpretation department" to turn jumbled sensory impressions into familiar objects and concepts; that the people or trees we think we see are actually created by the interpretation department from the splotches of color that enter the eye; and that dreams are the product of the interpretation department running wild, free of the sights and sounds of the waking hours.

His philosophical efforts at introspection did nothing to soften his dislike of the philosophy taught at MIT as The Making of the Modern Mind. Not enough sure experiments and demonstrated arguments; too many probable conjectures and philosophical speculations. He sat through lectures twirling a small steel drill bit against the sole of his shoe. *So much stuff in there, so much nonsense,* he thought. *Better I should use my modern mind.*

The Newest Physics

The theory of the fast and the theory of the small were narrowing the focus of the few dozen men with the suasion to say what *physics* was. Most of human experience passed in the vast reality that was neither fast nor small, where relativity and quantum mechanics seemed unnecessary and unnatural, where rivers ran, clouds flowed, baseballs soared and spun by classical means—but to young scientists seeking the most fundamental knowledge about the fabric of their universe, classical physics had no more to say. They could not ignore the deliberately disorienting rhetoric of the quantum mechanicians, nor the unifying poetry of Einstein's teacher Hermann Minkowski: "Space of itself and time of itself will sink into mere shadows, and only a kind of union between them shall survive."

Later, quantum mechanics suffused into the lay culture as a mystical fog. It was uncertainty, it was acausality, it was the Tao updated, it was the century's richest fount of paradoxes, it was the permeable membrane between the observer and the observed, it was the funny business sending shudders up science's all-too-deterministic scaffolding. For now, however, it was merely a necessary and useful contrivance for accurately describing the behavior of nature at the tiny scales now accessible to experimenters.

Nature had seemed so continuous. Technology, however, made discreteness and discontinuity a part of everyday experience: gears and rachets creating movement in tiny jumps; telegraphs that digitized information in dashes and dots. What about the light emitted by matter? At everyday temperatures the light is infrared, its wavelengths too long to be visible to the eye. At higher temperatures, matter radiates at shorter wavelengths: thus an iron bar heated in a forge glows red, yellow, and white. By the turn of the century, scientists were struggling to explain this relationship between temperature and wavelength. If heat was to be understood as the motion of molecules, perhaps this precisely tuned radiant energy suggested an internal oscillation, a vibration with the resonant tonality of a violin string. The German physicist Max Planck pursued this idea to its logical conclusion and announced in 1900 that it required an awkward adjustment to the conventional way of thinking about energy. His equations produced the desired results only if one supposed that radiation was emitted in lumps, discrete packets called quanta. He calculated a new constant of nature, the indivisible unit underlying these lumps. It was a unit, not of energy, but of the product of energy and time—the quantity called action.

Five years later Einstein used Planck's constant to explain another puzzle, the photoelectric effect, in which light absorbed by a metal knocks electrons free and creates an electric current. He, too, followed the relationship between wavelength and current to an inevitable mathematical conclusion: that light itself behaves not as a continuous wave but as a broken succession of lumps when it interacts with electrons.

These were dubious claims. Most physicists found Einstein's theory of special relativity, published the same year, more palatable. But in 1913 Niels Bohr, a young Dane working in Ernest Rutherford's laboratory in Manchester, England, proposed a new model of the atom built on these quantum underpinnings. Rutherford had recently imagined the atom as a solar system in miniature, with electrons orbiting the nucleus. Without a quantum theory, physicists would have to accept the notion of electrons gradually spiraling inward as they radiated some of their energy away. The result would be continuous radiation and the eventual collapse of the atom in on itself. Bohr instead described an atom whose electrons could inhabit only certain orbits, prescribed by Planck's indivisible constant. When an electron absorbed a light quantum, it meant that in that instant it jumped to a higher orbit: the soon-to-be-proverbial quantum jump. When the electron jumped to a lower orbit, it emitted a light quantum at a certain frequency. Everything else was simply forbidden. What happened to the electron "between" orbits? One learned not to ask.

These new kinds of lumpiness in the way science conceived of energy were the essence of quantum mechanics. It remained to create a theory, a mathematical framework that would accommodate the working out of these ideas. Classical intuitions had to be abandoned. New meanings had to be assigned to the notions of probability and cause. Much later, when most of the early quantum physicists were already dead, Dirac, himself chalky-haired and gaunt, with just a trace of white mustache, made the birth of quantum mechanics into a small fable. By then many scientists and writers had done so, but rarely with such unabashed stick-figure simplicity. There were heroes and almost heroes, those who reached the brink of discovery and those whose courage and faith in the equation led them to plunge onward.

Dirac's simple morality play began with LORENTZ. This Dutch physicist realized that light shines from the oscillating charges within the atom, and he found a way of rearranging the algebra of space and time that produced a strange contraction of matter near the speed of light. As Dirac said, "Lorentz succeeded in getting correctly all the basic equations needed to establish the relativity of space and time, but he just was not able to make the final step." Fear held him back.

Next came a bolder man, EINSTEIN. He was not so inhibited. He was able to move ahead and declare space and time to be joined.

HEISENBERG started quantum mechanics with "a brilliant idea": "one should try to construct a theory in terms of quantities which are provided by experiment, rather than building it up, as people had done previously, from an atomic model which involved many quantities which could not be observed." This amounted to a new philosophy, Dirac said.

(Conspicuously a noncharacter in Dirac's fable was Bohr, whose 1913 model of the hydrogen atom now represented the old philosophy. Electrons whirling about a nucleus? Heisenberg wrote privately that this made no sense: "My whole effort is to destroy without a trace the idea of orbits." One could observe light of different frequencies shining from within the atom. One could not observe electrons circling in miniature planetary orbits, nor any other atomic structure.)

It was 1925. Heisenberg set out to pursue his conception wherever it might lead, and it led to an idea so foreign and surprising that "he was really scared." It seemed that Heisenberg's quantities, numbers arranged

in matrices, violated the usual commutative law of multiplication that says
a times *b* equals *b* times *a*. Heisenberg's quantities did not commute. There
were consequences. Equations in this form could not specify both momentum and position with definite precision. A measure of uncertainty
had to be built in.

A manuscript of Heisenberg's paper made its way to DIRAC himself. He
studied it. "You see," he said, "I had an advantage over Heisenberg because
I did not have his fears."

Meanwhile, SCHRÖDINGER was taking a different route. He had been
struck by an idea of DE BROGLIE two years before: that electrons, those
pointlike carriers of electric charge, are neither particles nor waves but a
mysterious combination. Schrödinger set out to make a wave equation, "a
very neat and beautiful equation," that would allow one to calculate electrons tugged by fields, as they are in atoms.

Then he tested his equation by calculating the spectrum of light emitted
by a hydrogen atom. The result: failure. Theory and experiment did not
agree. Eventually, however, he found that if he compromised and ignored
the effects of relativity his theory agreed more closely with observations.
He published this less ambitious version of his equation.

Thus fear triumphed again. "Schrödinger had been too timid," Dirac
said. Two other men, KLEIN and GORDON, rediscovered the more complete
version of the theory and published it. Because they were "sufficiently bold"
not to worry too much about experiment, the first relativistic wave equation
now bears their names.

Yet the Klein-Gordon equation still produced mismatches with experiments when calculations were carried out carefully. It also had what seemed
to Dirac a painful logical flaw. It implied that the probability of certain
events must be negative, less than zero. Negative probabilities, Dirac said,
"are of course quite absurd."

It remained only for Dirac to invent—or was it "design" or "discover"?—
a new equation for the electron. This was exceedingly beautiful in its formal
simplicity and the sense of inevitability it conveyed, after the fact, to sensitive
physicists. The equation was a triumph. It correctly predicted (and
so, to a physicist, "explained") the newly discovered quantity called spin,
as well as the hydrogen spectrum. For the rest of his life Dirac's equation
remained his signal achievement. It was 1927. "That is the way in which
quantum mechanics was started," Dirac said.

These were the years of *Knabenphysik*, boy physics. When they began,
Heisenberg was twenty-three and Dirac twenty-two. (Schrödinger was an

elderly thirty-seven, but, as one chronicler noted, his discoveries came "during a late erotic outburst in his life.") A new *Knabenphysik* began at MIT in the spring of 1936. Dick Feynman and T. A. Welton were hungry to make their way into quantum theory, but no course existed in this nascent science, so much more obscure even than relativity. With guidance from just a few texts they embarked on a program of self-study. Their collaboration began in one of the upstairs study rooms of the Bay State Road fraternity house and continued past the end of the spring term. Feynman returned home to Far Rockaway, Welton to Saratoga Springs. They filled a notebook, mailing it back and forth, and in a period of months they recapitulated nearly the full sweep of the 1925–27 revolution.

"Dear R. P. . . ." Welton wrote on July 23. "I notice you write your equation:

$$[(P_\mu - K_\mu)g^{\mu\nu}(P_\mu - K_\nu) + m^2c^2]\psi = 0."$$

This was the relativistic Klein-Gordon equation. Feynman had rediscovered it, by correctly taking into account the tendency of matter to grow more massive at velocities approaching the speed of light—not just quantum mechanics, but relativistic quantum mechanics. Welton was excited. "Why don't you apply your equation to a problem like the hydrogen atom, and see what results it gives?" Just as Schrödinger had done ten years before, they worked out the calculation and saw that it was wrong, at least when it came to making precise predictions.

"Here's something, the problem of an electron in the *gravitational* field of a heavy particle. Of course the electron would contribute something to the field . . ."

"I wonder if the energy would be quantized? The more I think about the problem, the more interesting it sounds. I'm going to *try* it . . .

". . . I'll probably get an equation that I can't solve anyway," Welton added ruefully. (When Feynman got his turn at the notebook he scrawled in the margin, "Right!") "That's the trouble with quantum mechanics. It's easy enough to set up equations for various problems, but it takes a mind twice as good as the differential analyzer to solve them."

General relativity, barely a decade old, had merged gravity and space into a single object. Gravity was a curvature of space-time. Welton wanted more. Why not tie electromagnetism to space-time geometry as well? "Now you see what I mean when I say, I want to make electrical phenomena a result of the metric of a space in the same way that gravitational phenomena

are. I wonder if your equation couldn't be extended to Eddington's af-
fine geometry . . ." (In response Feynman scribbled: "I tried it. No luck
yet.")

Feynman also tried to invent an operator calculus, writing rules of differentiation and integration for quantities that did not commute. The rules
would have to depend on the order of the quantities, themselves matrix
representations of forces in space and time. "Now I think I'm wrong on
account of those darn partial integrations," Feynman wrote. "I oscillate
between right and wrong."

"Now I *know* I'm right . . . In my theory there are a lot more 'fundamental' invariants than in the other theory."

And on they went. "Hot dog! after 3 wks of work . . . I have at last
found a simple proof," Feynman wrote. "It's not important to write it,
however. The only reason I wanted to do it was because I couldn't do it
and felt that there were some more relations between the A^n & their de-
rivatives that I had not discovered . . . Maybe I'll get electricity into the
metric yet! Good night, I have to go to bed."

The equations came fast, penciled across the notebook pages. Sometimes
Feynman called them "laws." As he worked to improve his techniques for
calculating, he also kept asking himself what was fundamental and what
was secondary, which were the essential laws and which were derivative.
In the upside-down world of early quantum mechanics, it was far from
obvious. Heisenberg and Schrödinger had taken starkly different routes to
the same physics. Each in his way had embraced abstraction and renounced
visualization. Even Schrödinger's waves defied every conventional picture.
They were not waves of substance or energy but of a kind of probability,
rolling through a mathematical space. This space itself often resembled the
space of classical physics, with coordinates specifying an electron's position,
but physicists found it more convenient to use momentum space (denoted
by P_α), a coordinate system based on momentum rather than position—
or based on the direction of a wavefront rather than any particular point
on it. In quantum mechanics the uncertainty principle meant that position
and momentum could no longer be specified simultaneously. Feynman in
the August after his sophomore year began working with coordinate space
(Q_α)—less convenient for the wave point of view, but more directly visu-
alizable.

"P_α is *no more fundamental* than Q_α nor vice versa—why then has P_α
played such an important rôle in theory and why don't I try Q_α instead of

P_α in certain generalizations of equations . . ." Indeed, he proved that the customary approach could be derived directly from the theory as cast in terms of momentum space.

In the background both boys were worrying about their health. Welton had an embarrassing and unexplained tendency to fall asleep in his chair, and during the summer break he was taking naps, mineral baths, and sunlamp treatments—doses of high ultraviolet radiation from a large mercury arc light. Feynman suffered something like nervous exhaustion as he finished his sophomore year. At first he was told he would have to stay in bed all summer. "I'd go nuts if it were me I," T. A. wrote in their notebook. "Anyhow, I hope you get to school all right in the fall. Remember, we're going to be taught quantum mechanics by no less an authority than Prof. Morse himself. I'm really looking forward to that." ("Me too," Feynman wrote.)

They were desperately eager to be at the front edge of physics. They both started reading journals like the *Physical Review*. (Feynman made a mental note that a surprising number of articles seemed to be coming from Princeton.) Their hope was to catch up on the latest discoveries and to jump ahead. Welton would set to work on a development in wave tensor calculus; Feynman would tackle an esoteric application of tensors to electrical engineering, and only after wasting several months did they begin to realize that the journals made poor Baedekers. Much of the work was out of date by the time the journal article appeared. Much of it was mere translation of a routine result into an alternative jargon. News did sometimes break in the *Physical Review*, if belatedly, but the sophomores were ill equipped to pick it out of the mostly inconsequential background.

Morse had taught the second half of the theoretical physics course that brought Feynman and Welton together, and he had noticed these sophomores, with their penetrating questions about quantum mechanics. In the fall of 1937 they, along with an older student, met with Morse once a week and began to fit their own blind discoveries into the context of physics as physicists understood it. They finally read Dirac's 1935 bible, *The Principles of Quantum Mechanics*. Morse put them to work calculating the properties of different atoms, using a method of his own devising. It computed energies by varying the parameters in equations known as hydrogenic radial functions—Feynman insisted on calling them *hygienic* functions—and it required more plain, plodding arithmetic than either boy had ever encountered. Fortunately they had calculators, a new kind that replaced the old hand cranks with electric motors. Not only could the calculators add, multiply, and subtract; they could divide, though it took time. They would enter numbers by turning metal dials. They would turn on the motor and watch the dials spin toward zero. A bell would ring. The chug-chug-ding-ding rang in their ears for hours.

In their spare time Feynman and Welton used the same machines to earn money through a Depression agency, the National Youth Administration, calculating the atomic lattices of crystals for a professor who wanted to publish reference tables. They worked out faster methods of running the calculator. And when they thought that they had their system perfected, they made another calculation: how long it would take to complete the job. The answer: seven years. They persuaded the professor to set the project aside.

Shop Men

MIT was still an engineering school, and an engineering school in the heyday of mechanical ingenuity. There seemed no limit to the power of lathes and cams, motors and magnets, though just a half-generation later the onset of electronic miniaturization would show that there had been limits after all. The school's laboratories, technical classes, and machine shops gave undergraduates a playground like none other in the world. When Feynman took a laboratory course, the instructor was Harold Edgerton, an inventor and tinkerer who soon became famous for his high-speed photographs, made with a stroboscope, a burst of light slicing time more finely than any mechanical shutter could. Edgerton extended human sight into the realm of the very fast just as microscopes and telescopes were bringing into view the small and the large. In his MIT workshop he made pictures of bullets splitting apples and cards; of flying hummingbirds and splashing milk drops; of golf balls at the moment of impact, deformed to an ovoid shape that the eye had never witnessed. The stroboscope showed how much had been unseen. "All I've done is take God Almighty's lighting and put it in a container," he said. Edgerton and his colleagues gave body to the ideal of the scientist as a permanent child, finding ever more ingenious ways of taking the world apart to see what was inside.

That was an American technical education. In Germany a young would-be theorist could spend his days hiking around alpine lakes in small groups, playing chamber music and arguing philosophy with an earnest *Magic*

Mountain volubility. Heisenberg, whose name would come to stand for the twentieth century's most famous kind of uncertainty, grew enraptured as a young student with his own "utter certainty" that nature expressed a deep Platonic order. The strains of Bach's D Minor Chaconne, the moonlit landscapes visible through the mists, the atom's hidden structure in space and time—all seemed as one. Heisenberg had joined the youth movement that formed in Munich after the trauma of World War I, and the conversation roamed freely: Did the fate of Germany matter "more than that of all mankind"? Can human perception ever penetrate the atom deeply enough to see why a carbon atom bonds with two but never three oxygen atoms? Does youth have "the right to fashion life according to its own values"? For such students philosophy came first in physics. The search for meaning, the search for purpose, led naturally down into the world of atoms.

Students entering the laboratories and machine shops at MIT left the search for meaning outside. Boys tested their manhood there, learning to handle the lathes and talk with the muscular authority that seemed to emanate from the "shop men." Feynman wanted to be a shop man but felt he was a faker among these experts, so easy with their tools and their working-class talk, their ties tucked in their belts to avoid catching in the chuck. When Feynman tried to machine metal it never came out quite right. His disks were not quite flat. His holes were too big. His wheels wobbled. Yet he understood these gadgets and he savored small triumphs. Once a machinist who had often teased him was struggling to center a heavy disk of brass in his lathe. He had it spinning against a position gauge, with a needle that jerked with each revolution of the off-kilter disk. The machinist could not see how to center the disk and stop the tick-tick-tick of the needle. He was trying to mark the point where the disk stuck out farthest by lowering a piece of chalk as slowly as he could toward the spinning edge. The lopsidedness was too subtle; it was impossible to hold the chalk steady enough to hit just the right spot. Feynman had an idea. He took the chalk and held it lightly above the disk, gently shaking his hand up and down in time with the rhythm of the shaking needle. The bulge of the disk was invisible, but the rhythm wasn't. He had to ask the machinist which way the needle went when the bulge was up, but he got the timing just right. He watched the needle, said to himself, *rhythm,* and made his mark. With a tap of the machinist's mallet on Feynman's mark, the disk was centered.

The machinery of experimental physics was just beginning to move beyond the capabilities of a few men in a shop. In Rome, as the 1930s began, Enrico Fermi made his own tiny radiation counters from lipstick-size aluminum tubes at his institute above the Via Panisperna. He methodically brought one element after another into contact with free neutrons streaming from samples of radioactive radon. By his hands were created a succession of new radioactive isotopes, substances never seen in nature, some with half-lives so short that Fermi had to race his samples down the corridor to test them before they decayed to immeasurability. He found a nameless new element heavier than any found in nature. By hand he placed lead barriers across the neutron stream, and then, in a moment of mysterious inspiration, he tried a barrier of paraffin. Something in paraffin—hydrogen?—seemed to slow the neutrons. Unexpectedly, the slow neutrons had a far more powerful effect on some of the bombarded elements. Because the neutrons were electrically neutral, they floated transparently through the knots of electric charge around the target atoms. At speeds barely faster than a batted baseball they had more time to work nuclear havoc. As Fermi tried to understand this, it seemed to him that the essence of the process was a kind of diffusion, analogous to the slow invasion of the still air of a room by the scent of perfume. He imagined the path they must be taking through the paraffin, colliding one, two, three, a hundred times with atoms of hydrogen, losing energy with each collision, bouncing this way and that according to laws of probability.

The neutron, the chargeless particle in the atom's core, had not even been discovered until 1932. Until then physicists supposed that the nucleus was a mixture of electrically negative and positive particles, electrons and protons. The evidence taken from ordinary chemical and electrical experiments shed little light on the nucleus. Physicists knew only that this core contained nearly all the atom's mass and whatever positive charge was needed to balance the outer electrons. It was the electrons—floating or whirling in their shells, orbits, or clouds—that seemed to matter in chemistry. Only by bombarding substances with particles and measuring the particles' deflection could scientists begin to penetrate the nucleus. They also began to split it. By the spring of 1938 not just dozens but hundreds of physics professors and students were at least glancingly aware of the ideas leading toward the creation of heavy new elements and the potential release of nuclear energies. MIT decided to offer a graduate seminar on the theory of nuclear structure, to be taught by Morse and a colleague.

Feynman and Welton, juniors, showed up in a room of excited-looking graduate students. When Morse saw them he demanded to know whether they were planning to register. Feynman was afraid they would be turned down, but when he said yes, Morse said he was relieved. Feynman and Welton brought the total enrollment to three. The other graduate students were willing only to audit the class. Like quantum mechanics, this was difficult new territory. No textbook existed. There was just one essential text for anyone studying nuclear physics in 1938: a series of three long articles in *Reviews of Modern Physics* by Hans Bethe, a young German physicist newly relocated to Cornell. In these papers Bethe effectively rebuilt this new discipline. He began with the basics of charge, weight, energy, size, and spin of the simplest nuclear particles. He moved on to the simplest compound nucleus, the deuteron, a single proton bound to a single neutron. He systematically worked his way toward the forces that were beginning to reveal themselves in the heaviest atoms known.

As he studied these most modern branches of physics, Feynman also looked for chances to explore more classical problems, problems he could visualize. He investigated the scattering of sunlight by clouds—*scattering* being a word that was taking a more and more central place in the vocabulary of physicists. Like so many scientific borrowings from plain English, the word came deceptively close to its ordinary meaning. Particles in the atmosphere scatter rays of light almost in the way a gardener scatters seeds or the ocean scatters driftwood. Before the quantum era a physicist could use the word without having to commit himself mentally either to a wave or a particle view of the phenomenon. Light simply dispersed as it passed through some medium and so lost some or all of its directional character. The scattering of waves implied a general diffusion, a randomizing of the original directionality. The sky is blue because the molecules of the atmosphere scatter the blue wavelengths more than the others; the blue seems to come from everywhere in the sky. The scattering of particles encouraged a more precise visualization: actual billiard-ball collisions and recoils. A single particle could scatter another. Indeed, the scattering of a very few particles would soon become the salient experiment of modern physics.

That clouds scattered sunlight was obvious. Close up, each wavering water droplet must shimmer with light both reflected and refracted, and the passage of the light from one drop to the next must be another kind of diffusion. A well-organized education in science fosters the illusion that when problems are easy to state and set up mathematically they are then easy to solve. For Feynman the cloud-scattering problem helped disperse the illusion. It seemed as primitive as any of hundreds of problems set out in his textbooks. It had the childlike quality that marks so many fundamental questions. It came just one step past the question of why we see clouds at all: water molecules scatter light perfectly well when they are floating as vapor, yet the light grows much whiter and more intense when the vapor condenses, because the molecules come so close together that their tiny electric fields can resonate in phase with one another to multiply the effect. Feynman tried to understand also what happened to the direction of the scattered light, and he discovered something that he could not believe at first. When the light emerges from the cloud again, caroming off billions of droplets, seemingly smeared to a ubiquitous gray, it actually retains some memory of its original direction. One foggy day he looked at a building far away across the river in Boston and saw its outline, faint but still sharp, diminished in contrast but not in focus. He thought: the mathematics worked after all.

Feynman of Course Is Jewish

Feynman's probing reached the edge of known science. His scattering calculations had immediate application to a problem that was troubling one of his professors, Manuel S. Vallarta, concerning cosmic rays. These had become a major issue. Not just specialists but also the public worried about these unknown rays of unknown origin, streaming through space at high energies and entering the atmosphere, where they left trails of electric charge. This ionization first gave their presence away. It occurred to scientists just before the turn of the century that the atmosphere, left alone, ought not to conduct electricity. Now scientists were sending forth ray-detecting equipment on ships, aircraft, and balloons all around the globe, but especially in the neighborhood of Pasadena, California, where Robert Millikan and Carl Anderson had made the California Institute of Technology the nation's focal point of cosmic ray research. Later it began to become clear that the term was a catchall for a variety of particles with different sources. In the thirties the detective work meant trying to understand which of the universe's constituents might emit them and which might influence their timing and direction as seen from earth. At MIT Vallarta was puzzling over how cosmic rays might be scattered by the magnetic fields of the galaxy's stars, just as cloud droplets scatter sunlight. Whether cosmic rays came from inside or outside the galaxy, should the scattering effect bias their apparent direction toward or away from the main body of the Milky Way? Feynman's work produced a negative answer: neither. The net effect of the scattering was zero. If cosmic rays seemed to come from all directions, it was not because the stars' interference disguised their original orientation. They wrote this up together for publication as a letter to the *Physical Review*—Feynman's first published work. Unrevolutionary though the item was, its reasoning turned on a provocative and clever idea: that the probability of a particle's emerging from a clump of scattering matter in a certain direction must be equivalent to the probability of an antiparticle's taking the reverse path. From the antiparticle's point of view, time was running backward.

Vallarta let his student in on a secret of mentor-protégé publishing: the senior scientist's name comes first. Feynman had his revenge a few years later, when Heisenberg concluded an entire book on cosmic rays with the phrase, "such an effect is not to be expected according to Vallarta and Feynman." When they next met, Feynman asked gleefully whether Vallarta had seen Heisenberg's book. Vallarta knew why Feynman was grinning. "Yes," he replied. "You're the last word in cosmic rays."

Feynman had developed an appetite for new problems—any problems. He would stop people he knew in the corridor of the physics building and ask what they were working on. They quickly discovered that the question was not the usual small talk. Feynman pushed for details. He caught one classmate, Monarch Cutler, in despair. Cutler had taken on a senior thesis problem based on an important discovery in 1938 by two professors in the optics laboratory. They found that they could transform the refracting and reflecting qualities of lenses by evaporating salts onto them, forming very thin coatings, just a few atoms thick. Such coatings became essential to reducing unwanted glare in the lenses of cameras and telescopes. Cutler was supposed to find a way of calculating what happened when different thin films were applied, one atop another. His professors wondered, for example, whether there was a way to make exceedingly pure color filters, passing only light of a certain wavelength. Cutler was stymied. Classical optics should have sufficed—no peculiarly quantum effects came into play—but no one had ever analyzed the behavior of light passing through a parade of mostly transparent films thinner than a single wavelength. Cutler told Feynman he could find no literature on the subject. He did not know where to start. A few days later Feynman returned with the solution: a formula summing an infinite series of reflections back and forth from the inner surfaces of the coatings. He showed how the combinations of refraction and reflection would affect the phase of the light, changing its color. Using Feynman's theory and many hours on the Marchant calculator, Cutler also found a way to make the color filters his professors wanted.

Developing a theory for reflection by multiple-layer thin films was not so different for Feynman from math team in the now-distant past of Far Rockaway. He could see, or feel, the intertwined infinities of the problem, the beam of light resonating back and forth between the pair of surfaces, and then the next pair, and so on, and he had a giant mental kit bag of formulas to try out. Even when he was fourteen he had manipulated series of continued fractions the way a pianist practices scales. Now he had an intuition for the translating of formulas into physics and back, a feeling for the rhythms or the spaces or the forces that a given set of symbols implied. In his senior year the mathematics department asked him to join a team of three entrants to the nation's most difficult and prestigious mathematics contest, the Putnam competition, then in its second year. (The top five finishers are named as Putnam Fellows and one receives a scholarship at Harvard.) The problems were intricate exercises in calculus and algebraic manipulation; no one was expected to complete them all satisfactorily in the allotted time. In some years the median has been zero—more than half the entrants fail to solve a single problem. One of Feynman's fraternity brothers was surprised to see him return home while the examination was still going on. Feynman learned later that the scorers had been astounded by the gap between his result and the next four. Harvard sounded him out about the scholarship, but he told them he had already decided to go elsewhere: to Princeton.

His first thought had been to remain at MIT. He believed that no other American institution rivaled it and he said so to his department chairman. Slater had heard this before from loyal students whose provincial world contained nothing but Boston and the Tech, or the Bronx and the Tech, or Flatbush and the Tech. He told Feynman flatly that he would not be allowed back as a graduate student—for his own good.

Slater and Morse communicated directly with their colleagues at Princeton in January 1939, signaling that Feynman was something special. One said his record was "practically perfect," the other that he had been "the best undergraduate student we have had in the Physics Department for five years at least." At Princeton, when Feynman's name came up in the deliberations of the graduate admissions committee, the phrase "diamond in the rough" kept materializing out of the wash of conversation. The committee had seen its share of one-sided applicants but had never before admitted a student with such low scores in history and English on the Graduate Record Examination. Feynman's history score was in the bottom fifth, his literature score in the bottom sixth; and 93 percent of those who took the test had given better answers about fine arts. His physics and mathematics scores were the best the committe had seen. In fact the physics score was perfect.

Princeton had another problem with Feynman, as the head of its department, H. D. Smyth, made clear to Morse. "One question always arises, particularly with men interested in theoretical physics," Smyth wrote.

Is Feynman Jewish? We have no definite rule against Jews but have to keep their proportion in our department reasonably small because of the difficulty of placing them.

By March no word had come and Slater was concerned enough to write Smyth again, collegially: "Dear Harry . . . definitely the best undergraduate we have had for a number of years . . . first-rate both in matters of scholarship and personality . . ." The recommendation was formal and conventional, but in a handwritten postscript that would not appear on the carbon copies Slater got to the point: "Feynman of course is Jewish . . ." He wanted to assure Smyth there were mitigating circumstances:

. . . but as compared for instance with Kanner and Eisenbud he is more attractive personally by several orders of magnitude. We're not trying to get rid of him—we want to keep him, and privately hope you won't give him anything. But he apparently has decided to go to Princeton. I guarantee you'll like him if he does.

Morse, too, reported that Feynman's "physiognomy and manner, however, show no trace of this characteristic and I do not believe the matter will be any great handicap."

On the eve of the Second World War institutional anti-Semitism remained a barrier in American science, and a higher barrier for graduate schools than colleges. At universities a graduate student, unlike an undergraduate, was as much hired as admitted to a department; he would be paid for teaching and research and would be on a track for promotion. Furthermore, graduate departments considered themselves responsible to the industries they fed, and the industrial companies that conducted most research in the applied sciences were largely closed to Jews. "We know perfectly well that names ending in 'berg' or 'stein' have to be skipped," the chairman of Harvard's chemistry department, whose name was Albert Sprague Coolidge, said in 1946. Admissions quotas had been imposed broadly in the twenties and thirties, with immigrant children seeking admission to college in greater numbers. The case against Jews rarely had to be articulated. It was understood that their striving, their pushiness, smelled of the tenement. It was unseemly. "They took obvious pride in their academic success. . . . We despised the industry of those little Jews," a Harvard Protestant wrote in 1920. Thomas Wolfe, himself despising the ambition of "the Jew boy," nevertheless understood the attraction of the scientific career: "Because, brother, he is burning in the night. He sees the class, the lecture room, the shining apparatus of gigantic laboratories, the open field of scholarship and pure research, certain knowledge and the world distinction of an Einstein name." It was also understood that a professor needed a certain demeanor to work well with students; that Jews were often soft-spoken and diffident or, contradictorily, so brilliant as to be impatient and insensitive. In the close, homogenous university communities, code words were *attractive* or *nice*. Even the longtime chairman of J. Robert Oppenheimer's department at the University of California at Berkeley, Raymond T. Birge, was quoted as saying of Oppenheimer, "New York Jews flocked out here to him, and some were not as nice as he was."

Feynman, as a New York Jew distinctly uninterested in either the faith or the sociology of Judaism, did not give voice to any awareness of anti-Semitism. Princeton did accept him, and from then on he never had occasion to worry about the contingencies of academic hiring. Still, when he was at MIT, the Bell Telephone Laboratories turned him down for summer jobs year after year, despite recommendations by William Shockley, Bell's future Nobel laureate. Bell was an institution that hired virtually no Jewish scientists before the war. Birge himself eventually had an opportunity to hire Feynman for Berkeley: a frustrated Oppenheimer was recommending him urgently, but Birge put off a decision for two years, until it was too late. In the first case anti-Semitism may have played the deciding role; in the second case perhaps a smaller role. If Feynman ever suspected that his religion might have shifted the path of his career, he declined to say so.

Forces in Molecules

Thirteen physics majors completed senior theses in 1939. The world of accumulated knowledge was still small enough that MIT could expect a thesis to represent original and possibly publishable work. The thesis should begin the scientist's normal career and meanwhile supply missing blocks in the wall of organized knowledge, by analyzing such minutiae as the spectra of singly ionized gadolinium or hydrated manganese chloride crystals. (Identifying the telltale combinations of wavelengths emitted by such substances still required patience and good experimental technique, and science seemed to be engendering new substances as fast as spectroscopists could analyze them.) Seniors could devise new laboratory instruments or investigate crystals that produced electrical currents when squeezed. Feynman's thesis began as a circumscribed problem like these. It ended as a fundamental discovery about the forces acting within the molecules of any substance. If it bore little connection to his greater work that followed— and Feynman himself dismissed it as an obvious result that he should have written in "half a line"—it nevertheless found its way into the permanent tool kit of the physics of solids.

Although he did not know it, his quantum-mechanics professor, Morse, had recommended in his junior year that the department graduate him a year early. The suggestion was turned down, and Slater himself became Feynman's thesis adviser. Slater proposed a problem that at first seemed not much deeper than most senior theses. The question could almost have come from a physics and chemistry handbook: Why does quartz expand so little when heated? Compared to metals, for example, why is its coefficient of expansion so small? Any substance expands because heat agitates its molecules—heat *is* the agitation of its molecules—but in a solid the details of the expansion depend on the actual molecular layout. A crystal, with its molecules in a regular geometrical array, can expand more along one axis than another. Typically scientists would represent a crystalline structure with a Tinkertoy model, balls stuck on rods, but real matter is not so rigid. Atoms may be more or less locked in an array, or they may swing or float more or less freely from one place to another. Electrons in a metal will swarm freely about. The color, the texture, the rigidity, the frangibility, the conductivity, the softness, the taste of a substance all depend on the local habits of atoms. Those habits in turn depend on the forces at work within a substance—forces both classical and quantum mechanical—and when

Feynman began his thesis work those forces were not well understood, even in quartz, the most common mineral on earth.

An old-fashioned steam engine was regulated by a mechanical governor: a pair of iron balls swinging outward from a spinning shaft. The faster it spun, the farther outward they would swing. But the farther they would swing, the harder they would make it to spin the shaft. Feynman started by imagining some analogous effect in the atoms of quartz, silicon dioxide, a pair of oxygen atoms clinging to each atom of silicon. Instead of spinning, the silicon atoms were vibrating; as the quartz grew warmer, he thought that the oxygen atoms might provide a mechanical force that would pull inward against the increasing agitation of the molecules, thus compensating somehow for the ordinary expansion. But how could the forces within each molecule—forces that varied in different directions—be calculated? No straightforward method seemed to exist.

He had never thought about molecular structure in such detail before. He taught himself everything he could about crystals, their standard arrangements, the geometries and the symmetries, the angles between atoms. It all came down to one unknown, he realized: the nature of the forces pressing the molecules into particular alignments. In its search for fundamental laws ever farther down the hierarchy of sizes, physics had now reached a level where molecular forces should be coming into focus. Scientists could measure how much pressure it took to squeeze quartz a given distance in a given direction. With the still-new technique of X-ray diffraction, they could look at the shadow patterns of a regular crystal and deduce its structure. As some theorists continued to look even deeper toward the atom's core, others now tried applying the quantum techniques to questions of structure and chemistry. "A science of materials as distinct from matter became possible," a scholar of structure, Cyril Stanley Smith, who worked with Feynman a few years later as the chief metallurgist on the secret project at Los Alamos, said of this time. From atomic forces to the stuff that feeds our senses—that was the connection waiting to be made. From abstract energy levels to three-dimensional forms. As Smith added epigrammatically, "Matter is a holograph of itself in its own internal radiation."

Forces or energy—that was the choice for those seeking to apply the quantum understanding of the atom to the workings of real materials. At stake was not mere terminology but a root decision about how to conceive of a problem and how to proceed in calculating.

The conception of nature in terms of forces went back to Newton. It was a direct way of dealing with the world, envisioning firsthand interactions between objects. One exerts a force on another. A distinction between force and energy did not emerge clearly until the nineteenth century, and then, gradually, energy began to take over as the fulcrum of scientists' thinking. Force is, in modern terms, a vector quantity, with both a magnitude and a direction. Energy is directionless, scalar—meaning that it has a magnitude only. With the rise of thermodynamics energy came to the fore. It began to seem more fundamental. Chemical reactions could be neatly computed as operations designed to minimize energy. Even a ball rolling down a hill—moving from a state of higher to lower potential energy— was seeking to minimize its energy. The Lagrangian approach that Feynman resisted in his sophomore-year physics class also used a minimum of energy to circumvent the laborious calculation of direct interactions. And the law of conservation of energy provided a tidy bookkeeping approach to a variety of calculations. No comparable law existed for forces.

Yet Feynman continued to seek ways of using the language of forces, and his senior thesis evolved beyond the problem Slater had posed. As Feynman conceived the structure of molecules, forces were the natural ingredients. He saw springlike bonds with varying stiffness, atoms attracting and repelling one another. The usual energy-accounting methods seemed secondhand and euphemistic. He titled his thesis—grandly—"Forces and Stresses in Molecules" and began by arguing that it would be more illuminating to attack molecular structure directly by means of forces, intractable though that approach had been considered in the past.

Quantum mechanics had begun with energy for two reasons, he contended. One was that the original quantum theorists had habitually tested their formulas against a single type of application, the calculation of the observed spectra of light emitted by atoms, where forces played no obvious part. The other was that the wave equation of Schrödinger simply did not lend itself to the calculation of vector quantities; its natural context was the directionless measurement of energy.

In Feynman's senior year, just over a decade after the three-year revolution of Heisenberg, Schrödinger, and Dirac, the applied branches of physics and chemistry had been drawn into an explosion of activity. To outsiders quantum mechanics might have seemed a nuisance, with its philosophical entanglements and computational nightmares. In the hands of those analyzing the structures of metals or chemical reactions, however, the new physics was slicing through puzzles that classical physics found impenetrable. Quantum mechanics was triumphing not because a few leading theorists found it mathematically convincing, but because hundreds of materials scientists found that it worked. It gave them insights into problems that had languished, and it gave them a renewed livelihood. One had only to understand the manipulation of a few equations and one could finally compute the size of an atom or the precise gray sheen of a pewter surface.

Chief in the new handbook was Schrödinger's wave equation. Quantum mechanics taught that a particle was not a particle but a smudge, a traveling cloud of probabilities, like a wave in that the essence was spread out. The wave equation made it possible to compute with smudges and accommodate the probability that a feature of interest might appear anywhere within a certain range. This was essential. No classical calculation could show how electrons would arrange themselves in a particular atom: classically the negatively charged electrons should seek their state of lowest energy and spiral in toward the positively charged nuclei. Substance itself would vanish. Matter would crumple in on itself. Only in terms of quantum mechanics was that impossible, because it would give the electron a definite pointlike position. Quantum-mechanical uncertainty was the air that saved the bubble from collapse. Schrödinger's equation showed where the electron clouds would find their minimum energy, and on those clouds depended all that was solid in the world.

Often enough, it became possible to gain an accurate picture of where the electrons' charge would be distributed in the three-dimensional space of a solid crystal lattice of molecules. That charge distribution in turn held the massive nuclei of the atoms in place—again, in places that kept the overall energy at a minimum. If a researcher wanted to calculate the forces working on a given nucleus, there was a way to do it—a laborious way. He had to calculate the energy, and then calculate it again, this time with the nucleus slightly shifted out of position. Eventually he could draw a curve representing the change in energy. The slope of that curve represented the sharpness of the change—the force. Each varied configuration had to be computed afresh. To Feynman this seemed wasteful and ugly.

It took him a few pages to demonstrate a better method. He showed that one could calculate the force directly for a given configuration, without having to look at nearby configurations at all. His computational technique led directly to the slope of the energy curve—the force—instead of producing the full curve and deriving the slope secondarily. The result caused a small sensation among MIT's physics faculty, many of whom had spent enough time working on applied molecular problems to appreciate Feynman's remark, "It is to be emphasized that this permits a considerable saving of labor of calculations."

Slater made him rewrite the first version. He complained that Feynman wrote the way he talked, hardly an acceptable style for a scientific paper. Then he advised him to submit a shortened version for publication. The *Physical Review* accepted it, with the title shortened as well, to "Forces in Molecules."

Not all computational devices have analogues in the word pictures that scientists use to describe reality, but Feynman's discovery did. It corresponded to a theorem that was easy to state and almost as easy to visualize: The force on an atom's nucleus is no more or less than the electrical force from the surrounding field of charged electrons—the electrostatic force. Once the distribution of charge has been calculated quantum mechanically, then from that point forward quantum mechanics disappears from the picture. The problem becomes classical; the nuclei can be treated as static points of mass and charge. Feynman's approach applies to all chemical bonds. If two nuclei act as though strongly attracted to each other, as the hydrogen nuclei do when they bond to form a water molecule, it is because the nuclei are each drawn toward the electrical charge concentrated quantum mechanically between them.

That was all. His thesis had strayed from the main line of his thinking about quantum mechanics, and he rarely thought about it again. When he did, he felt embarrassed to have spent so much time on a calculation that now seemed trivial and self-evident. As far as he knew, it was useless. He had never seen a reference to it by another scientist. So he was surprised to hear in 1948 that a controversy had erupted among physical chemists about the discovery, now known as Feynman's theorem or the Feynman-Hellmann theorem. Some chemists felt it was too simple to be true.

Is He Good Enough?

A few months before graduation, most of the thirty-two brothers of Phi Beta Delta posed for their portrait photograph. Feynman, seated at the left end of the front row, still looked smaller and younger than his classmates. He clenched his jaw, obeyed the photographer's instruction to rest his hands on his knees, and leaned gravely in toward the center. He went home at the end of the term and returned for the ceremony in June 1939. He had just learned to drive an automobile, and he drove his parents and Arline to Cambridge. On the way he became sick to his stomach—from the tension of driving, he thought. He was hospitalized for a few days, but he recovered in time to graduate. Decades later he remembered the drive. He remembered his friends teasing him when he donned his academic robe—Princeton did not know what a rough guy it was getting. He remembered Arline.

"That's all I remember of it," he told a historian. "I remember my sweet girl."

Slater left MIT not many years after Feynman. By then the urgency of war research had brought I. I. Rabi from Columbia to become the vigorous scientific personality driving a new laboratory, the Radiation Laboratory, set up to develop the use of shorter and shorter radio wavelengths for the detection of aircraft and ships through night and clouds: radar. It seemed to some that Slater, unaccustomed to the shadow of a greater colleague, found Rabi's presence unbearable. Morse, too, left MIT to take a role in the growing administrative structure of physics. Like so many scientists of the middle rank, both men saw their reputations fade in their lifetimes. Both published small autobiographies. Morse, in his, wrote about the challenges in guiding students toward a career as esoteric as physics. He recalled a visit from the father of a graduating senior named Richard. The father struck Morse as uneducated, nervous merely to be visiting a university. He did not speak well. Morse recalled his having said ("omitting his hesitations and apologies"):

My son Richard is finishing his schooling here next spring. Now he tells me he wants to go on to do more studying, to get still another degree. I guess I can afford to pay his way for another three or four years. But what I want to know is, is it worth it for him? He tells me you've been working with him. Is he good enough to deserve the extra schooling?

Morse tried not to laugh. Jobs in physics were hard to get in 1939, but he told the father that Richard would surely do all right.

PRINCETON

÷

The apostle of Niels Bohr at Princeton was a compact, gray-eyed, twenty-eight-year-old assistant professor named John Archibald Wheeler who had arrived the year before Feynman, in 1938. Wheeler had Bohr's rounded brow and soft features, as well as his way of speaking about physics in oracular undertones. In the years that followed, no physicist surpassed Wheeler in his appreciation for the mysterious or in his command of the Delphic catchphrase:

A black hole has no hair was his. In fact he coined the term "black hole."

There is no law except the law that there is no law.

I always keep two legs going, with one trying to reach ahead.

In any field find the strangest thing and then explore it.

Individual events. Events beyond law. Events so numerous and so uncoordinated that, flaunting their freedom from formula, they yet fabricate firm form.

He dressed like a businessman, his tie tightly knotted and his white cuffs starched, and he fastidiously pulled out a pocket watch when he began a session with a student (conveying a message: the professor will spare just so much time . . .). It seemed to one of his Princeton colleagues, Robert

R. Wilson, that behind the gentlemanly façade lay a perfect gentleman—and behind that façade another perfect gentleman, and on and on. "However," Wilson said, "somewhere among those polite façades there was a tiger loose; a reckless buccaneer . . . who had the courage to look at any crazy problem." As a lecturer he performed with a magnificent self-assurance, impressing his audience with elegant prose and provocative diagrams. When he was a boy, he spent many hours poring over the drawings in a book called *Ingenious Mechanisms and Mechanical Devices*. He made adding machines and automatic pistols with gears and levers whittled from wood, and his blackboard illustrations of the most foggy quantum paradoxes retained that ingenious flavor, as though the world were a wonderful silvery machine. Wheeler grew up in Ohio, the son of librarians and the nephew of three mining engineers. He went to college in Baltimore, got his graduate degree at Johns Hopkins University, and then won a National Research Council Fellowship that brought him to Copenhagen in 1934 via freighter (fifty-five dollars one way) to study with Bohr.

He and Bohr worked together again, as colleagues this time, in the first months of 1939. Princeton had hired Wheeler and promoted the distinguished Hungarian physicist Eugene Wigner in a deliberate effort to turn toward nuclear physics. MIT had remained deliberately conservative about rushing to board the wagon train; Slater and Compton preferred to emphasize well-roundedness and links to more applied fields. Not so Princeton. Wheeler still remembered the magic of his first vision of radioactivity: how he had sat in a lightless room, staring toward the black of a zinc sulfide screen, counting the intermittent flashes of individual alpha particles sent forth by a radon source. Bohr, meanwhile, had left the growing tumult of Europe to visit Einstein's institute in Princeton. When Wheeler met his ship at the pier in New York, Bohr was carrying news about what would now rapidly become the most propitious object in physics: the uranium atom.

Compared to the hydrogen atom, stark kernel with which Bohr had begun his quantum revolution, the uranium atom was a monster, the heaviest atom in nature, bulked out with 92 protons and 140-odd neutrons, so scarce in the cosmos that hydrogen atoms outnumber it by seventeen trillion to one, and unstable, given to decaying at quantum mechanically unpredictable moments down a chain of lighter elements or—this was the extraordinary news that kept Bohr at his portable blackboard all through the North Atlantic voyage—splitting, when slugged by a neutron, into odd pairs of smaller atoms, barium and krypton or tellurium and zirconium, plus a bonus of new neutrons and free energy. How was anyone to visualize this bloated nucleus? As a collection of marbles sliding greasily against one another? As a bunch of grapes squeezed together by nuclear rubber bands? Or as a "liquid drop"—the phrase that spread like a virus through the world of physics in 1939—a shimmering, jostling, oscillating globule that pinches into an hourglass and then fissures at its new waist. It was this last image, the liquid drop, that enabled Wheeler and Bohr to produce one of those unreasonably powerful oversimplifications of science, an effective theory of the phenomenon that had been named, only in the past year, fission. (The word was not theirs, and they spent a late night trying to find a better one. They thought about *splitting* or *mitosis* and then gave up.)

By any reasonable guess, a liquid drop should have served as a poor approximation for the lumpy, raisin-studded complex at the heart of a heavy atom, with each of two hundred—odd particles bound to each of the others by a strong close-range nuclear force, a force quite different from the electrical forces Feynman had analyzed on the scale of whole molecules. For smaller atoms the liquid-drop metaphor failed, but for large agglomerations like uranium it worked. The shape of the nucleus, like the shape of a liquid drop, depends on a delicate balance between the two opposing forces. Just as surface tension encourages a compact geometry in a drop, so do the forces of nuclear attraction in an atom. The electrical repulsion of the positively charged protons counters the attraction. Bohr and Wheeler recognized the unexpected importance of the slow neutrons that Fermi had found so useful at his laboratory in Rome. They made two remarkable predictions: that only the rarer uranium isotope, uranium 235, would fission explosively; and that neutron bombardment would also spark fission in a new substance, with atomic number 94 and mass 239, not found in nature and not yet created in the laboratory. To this pair of theoretical assertions would shortly be devoted the greatest technological enterprise the world had ever seen.

The laboratories of nuclear physics were spreading rapidly. Considerable American inventive spirit had gone into the development of an arsenal of machinery designed to accelerate beams of particles, smash them into metal foils or gaseous atoms, and track the collision products through chambers of ionizing gas. Princeton had one of the nation's first large "cyclotrons"— the name rang proudly of the future—completed in 1936 for the cost of a few automobiles. The university also kept smaller accelerators working daily, manufacturing rare elements and new isotopes and generating volumes of data. Almost any experimental result seemed worthwhile when hardly anything was known. With all the newly cobbled-together equipment came difficulties of measurement and interpretation, often messy and ad hoc. A student of Wheeler's, Heinz Barschall, came to him in the early fall of 1939 with a typical problem. Like so many new experimenters Barschall was using an accelerator beam to scatter particles through an ionizing chamber, where their energies could be measured. He needed to gauge the different energies that would appear at different angles of recoil. Barschall had realized that his results were distorted by the circumstances of the chamber itself. Some particles would start outside the chamber; others would start inside and run into the chamber's cylindrical wall, and in neither case would the particle have its full energy. The problem was to compensate, find a way to translate the measured energies into the true energies. It was a problem of awkward probabilities in a complicated geometry. Barschall had no idea where to start. Wheeler said that he was too busy to think about it himself but that he had a very bright new graduate student . . .

Barschall dutifully sought out Dick Feynman at the residential Graduate College. Feynman listened but said nothing. Barschall assumed that would be the end of it. Feynman was adjusting to this new world, much smaller, for a physicist, than the scientific center he had left. He shopped for supplies in the stores lining Nassau Street on the west edge of the campus, and an older graduate student, Leonard Eisenbud, saw him in the street. "You look like you're going to be a good theoretical physicist," Eisenbud said. He gestured toward Feynman's new wastebasket and blackboard eraser. "You've bought the right tools." The next time Feynman saw Barschall, he surprised him with a sheaf of handwritten pages; he had been riding on a train and had time to write out a full solution. Barschall was overwhelmed, and Feynman had added another young physicist to the growing group of his peers with a weighty private appreciation for his ability.

Wheeler himself was already beginning to appreciate Feynman, who had been assigned to him—neither of them quite knew why—as a teaching assistant. Feynman had expected to be working with Wigner. He was surprised at their first meeting to see that his professor was barely older than he was. Then he was surprised again by Wheeler's pointed display of a pocket watch. He took in the message. At their second meeting he pulled out a dollar pocket watch of his own and set it down facing Wheeler's. There was a pause; then both men laughed.

A *Quaint Ceremonious Village*

Princeton's gentility was famous: the eating clubs, the arboreal lanes, the ersatz-Georgian carved stone and stained glass, the academic gowns at dinner and punctilious courtesies at tea. No other college so keenly delineated the social status of its undergraduates as Princeton did with its club system. Although the twentieth century had begun to intrude—the graduate departments were growing in stature, and Nassau Street had been paved—Princeton before the war remained, as F. Scott Fitzgerald described it adoringly a generation earlier, "lazy and good-looking and aristocratic," an outpost for New York, Philadelphia, and Southern society. Its faculty, though increasingly professional, was still sprinkled with Fitzgerald's "mildly poetic gentlemen." Even the kindly genius who became the town's most famous resident on arriving in 1933 could not resist a gibe: "A quaint ceremonious village," Einstein wrote, "of puny demigods on stilts."

Graduate students, on track to a professional world, were partly detached from the university's more frivolous side. The physics department in particular was moving decisively with the times. It had seemed to Feynman from a distance that Princeton's physicists were disproportionately represented in the current journals. Even so he had to adjust to a place which, even more than Harvard and Yale, styled itself after the great English universities, with courtyards and residential "colleges." At the Graduate College a "porter" monitored the downstairs entranceway. The formality genuinely frightened Feynman, until slowly he realized that the obligatory black gowns hid bare arms or sweaty tennis clothes. The afternoon he arrived at Princeton in the fall of 1939, Sunday tea with Dean Eisenhart turned his edginess about social convention into anxiety. He dressed in his good suit. He walked through the door and saw—worse than he had imagined—young women. He could not tell whether he was supposed to sit. A voice behind him said, "Would you like cream or lemon in your tea, sir?" He turned and saw the dean's wife, a famous lioness of Princeton society. It was said that when the mathematician Carl Ludwig Siegel returned to Germany in 1935 after a year in Princeton he told friends that Hitler had been bad but Mrs. Eisenhart was worse.

Feynman blurted, "Both, please."

"Heh-heh-heh-heh-heh," he heard her say. "Surely you're *joking*, Mr. Feynman!" More code—the phrase evidently signaled a gaffe. Whenever he thought about it afterward, the words rang in his ears: surely you're joking. Fitting in was not easy. It bothered him that the raincoat his parents sent was too short. He tried sculling, the Ivy League sport that seemed least foreign to his Far Rockaway experience—he remembered the many happy hours spent rowing in the inlets of the south shore—and promptly fell from the impossibly slender boat into the water. He worried about money. When he entertained guests in his room they would share rice pudding and grapes, or peanut butter and jelly on crackers with pineapple juice. As a first-year teaching assistant he earned fifteen dollars a week. Cashing several savings certificates to pay a bill for $265, he spent twenty minutes calculating what combination would forfeit the least interest. The difference between the worst case and the best case, he found, came to eight cents. Outwardly, though, he cultivated his brashness. Not long after he arrived, he had his neighbors at the Graduate College convinced that he and Einstein (whom he had not met) were on regular speaking terms. They listened with awe to these supposed conversations with the great man on the pay phone in the hallway: "Yeah, I tried that . . . yeah, I did . . . oh, okay, I'll try that." Most of the time he was actually speaking with Wheeler.

As Wheeler's teaching assistant—first for a course in mechanics, then in nuclear physics—Feynman quickly found himself taking over in the professor's absence (and it began to sink in that facing a roomful of students was part of the profession he had chosen). He also met with Wheeler weekly on research problems of their own. At first Wheeler assigned the problems. Then a collaboration took shape.

The purview of physics had exploded in the first four decades of the century. Relativity, the quantum, cosmic rays, radioactivity, the nucleus— these new realms held the attention of leading physicists to the virtual exclusion of such classical topics as mechanics, thermodynamics, hydro- dynamics, statistical mechanics. To a smart graduate student fresh on the theoretical scene these traditional fields seemed like textbook science, al- ready part of history and—in their applied forms—engineering. Physics was "inward bound," as its chronicler Abraham Pais put it; into the core of the atom the theorists went. All the superlatives were here. The exper- imental apparatus was the most expensive (machines could now cost thou- sands or even tens of thousands of dollars). The necessary energies were the highest. The materials and "particles" (this word was acquiring a spe- cialized meaning) were the most esoteric. The ideas were the strangest. Relativity notoriously changed astronomers' sense of the cosmos but found its most routine application in the physics of the atom, where near-light speeds made relativistic mathematics essential. As experimenters learned to ply greater levels of energy, the basic constituents gave way to new units even more basic. Through quantum mechanics, physics had established a primacy over chemistry—itself formerly the most fundamental of sciences, if the most fundamental was the one responsible for nature's basic constituents.

As the thirties ended and the forties began, particle physics had not established its later dominance of the public relations of science. In choosing a theme for the annual Washington Conference on theoretical physics in 1940, organizers considered "The Elementary Particles" and the quaintly geophysical "Interior of the Earth"—and chose the interior of the earth. Still, neither Feynman nor Wheeler had any doubt about where a pure theorist's focus must turn. The fundamental issue in the fundamental science was the weakness in the heart of quantum mechanics. At MIT Feynman had read Dirac's 1935 text as a cliffhanger with the most thrilling possible conclusion: "It seems that some essentially new physical ideas are here needed." Dirac and the other pioneers had taken their quantum electrodynamics—the theory of the interplay of electricity, magnetism, light, and matter—as far as they could. Yet it remained incomplete, as Dirac well knew.

The difficulty concerned the electron, the fundamental speck of negative charge. As a modern concept, the electron was still young, although many high-school students now performed (as Feynman had in Far Rockaway) a tabletop experiment showing that electric charge came in discrete units. What exactly was the electron? Wilhelm Röntgen, the discoverer of X rays, forbade the use of this upstart term in his laboratories as late as 1920. The developers of quantum mechanics, attempting to describe the electron's charge or mass or momentum or energy or spin in almost every new equation, nevertheless maintained a silent agnosticism about certain issues of its existence. Particularly troubling: Was it a finite pellet or an infinitesimal point? In his model of the atom, already obsolete, Niels Bohr had imagined electrons as miniature planetoids orbiting the nucleus; now the atom's electron seemed more to reverberate in an oscillatory harmony. In some formulations it assumed a wavelike cloak, the wave representing a distribution of probabilities that it would appear in particular places at particular times. But *what* would appear? An entity, a unit—a particle?

Even before quantum mechanics, a worm had gnawed at the heart of the classical understanding. The equations linking the electron's energy (or mass) and charge implicated another quantity, its radius. As its size diminished, the electron's energy grew, just as the pressure transmitted by a carpenter's hammer becomes thousands of pounds per square inch when concentrated at the point of a nail. Furthermore, if the electron was to be imagined as a little ball of finite size, then what force or glue kept it from bursting from its own charge? Physicists found themselves manipulating a quantity called the "classical electron radius." *Classical* in this context came to mean something like *make-believe*. The problem was that the alternative—a vanishingly small, pointlike electron—left the equations of electrodynamics plagued with divisions by zero: infinities. Infinitely small nails, infinitely energetic hammers.

In a sense the equations were measuring the effect of the electron's charge on itself, its "self-energy." That effect would increase with proximity, and how much nearer could the electron be to itself? If the distance were zero, the effect would be infinite—impossible. The wave equation of quantum mechanics only made the infinities more complicated. Instead of the grade-school horror of a division by zero, physicists now contemplated equations that grew out of bounds because they summed infinitely many wavelengths, infinitely many oscillations in the field—although even now Feynman did not quite understand this formulation of the infinities problem. Temporarily, for simple problems, physicists could get reasonable answers by the embarrassing expedient of discarding the parts of the equations that diverged. As Dirac recognized, however, in concluding his *Principles of Quantum Mechanics*, the electron's infinities meant that the theory was mortally flawed. *It seems that some essentially new physical ideas are here needed.*

Feynman quietly nursed an attachment to a solution so radical and straightforward that it could only have appealed to someone ignorant of the literature. He proposed—to himself—that electrons not be allowed to act on themselves at all. The idea seemed circular and silly. As he recognized, however, eliminating self-action meant eliminating the field itself. It was the field, the totality of the charges of all electrons, that served as the agent of self-action. An electron contributed its charge to the field and was influenced by the field in turn. Suppose there was no field. Then perhaps the circularity could be broken. Each electron would act directly on another. Only the direct interaction between charges would be permitted. One would have to build a time delay into the equations, for whatever form this interaction took, it could hardly surpass the speed of light. The interaction *was* light, in the form of radio waves, visible light, X rays, or any of the other manifestations of electromagnetic radiation. "Shake this one, that one shakes later," Feynman said later. "The sun atom shakes; my eye electron shakes eight minutes later because of a direct interaction across."

No field; no self-action. Implicit in Feynman's attitude was a sense that the laws of nature were not to be discovered so much as constructed. Although language blurred the distinction, Feynman was asking not whether an electron acted on itself but whether the theorist could plausibly discard the concept; not whether the field existed in nature but whether it had to exist in the physicist's mind. When Einstein banished the ether, he was reporting the absence of something real—at least something that might have been—like a surgeon who opened a chest and reported that the bloody, pulsing heart was not to be found. The field was different. It had begun as an artifice, not an entity. Michael Faraday and James Clerk Maxwell, the nineteenth-century Britons who contrived the notion and made it into an implement no more dispensable than a surgeon's scalpel, started out apologetically. They did not mean to be taken literally when they wrote of "lines of force"—Faraday could actually see these when he sprinkled iron filings near a magnet—or "idle wheels," the pseudomechanical, invisible vortices that Maxwell imagined filling space. They assured their readers that these were analogies, though analogies with the newly formidable weight of mathematical rectitude.

The field had not been invented without reason. It had unified light and electromagnetism, establishing forever that the one was no more or less than a ripple in the other. As an abstract successor to the now-defunct ether the field was ideal for accommodating waves, and energy did seem to ripple wavelike from its sources. Anyone who played with electrical circuits and magnets as intently as Faraday and Maxwell could feel the way the "vibrations" or "undulations" could twist and spin like tubes or wheels. Crucially, the field also obviated the unpleasantly magical idea of action at a distance, objects influencing one another from afar. In the field, forces propagated sensibly and continuously from one place to the next. There was no jumping about, no sorcerous obeying of faraway orders. As Percy Bridgman, an American experimental physicist and philosopher, said, "It is felt to be more acceptable to rational thought to conceive of the gravitational action of the sun on the earth, for example, as propagated through the intermediate space by the handing on of some sort of influence from one point to its proximate neighbor, than to think of the action overleaping the intervening distance and finding its target by some sort of teleological clairvoyance." By then scientists had efficiently forgotten that the field, too, was a piece of magic—a wave-bearing nullity, or empty space that was not quite empty (and more than space). Or in the elegant phrase of a later theorist, Steven Weinberg: "the tension in the membrane, but without the membrane." The field grew so dominant in physicists' thinking that even matter itself sometimes withdrew to the status of mere appendage: a "knot" of the field, or a "blemish," or as Einstein himself said, merely a place where the field was especially intense.

Embrace the field or abhor it—either way, by the nineteen-thirties the choice seemed more one of method than reality. The events of 1926 and 1927 had made that clear. No one could be so naïve now as to ask whether Heisenberg's matrices or Schrödinger's wave functions *existed*. They were alternative ways of viewing the same processes. Thus Feynman, looking for a new eyepiece himself, began drifting back to a classical notion of un-fieldlike particle interaction. The wavelike transmission of energy and the hocus-pocus of action at a distance were issues that he would have to address. In the meantime, Wheeler, too, had reasons to be drawn toward this implausibly pure conception. Electrons might interact directly, without the mediation of the field.

Folds and Rhythms

Feynman tended to associate more with the mathematicians than the physicists at the Graduate College. Students from the two groups joined each afternoon for tea in a common lounge—more English tradition transplanted—and Feynman would listen to an increasingly alien jargon. Pure mathematics had swerved away from the fields of direct use to contemporary physicists and toward such seeming esoterica as topology, the study of shapes in two, three, or many dimensions without regard to rigid lengths or angles. An effective divorce had occurred between mathematics and physics. By the time practitioners reached the graduate level, they shared no courses and had nothing practical to say to one another. Feynman listened to the mathematicians standing in groups or sitting on the couch at tea, talking about their proofs. Rightly or wrongly he felt he had an intuition for what theorems could be derived from what lemmas, even without quite understanding the subject. He enjoyed the strange rhetoric. He enjoyed trying to guess the counterintuitive answers to their nearly unvisualizable questions, and he enjoyed applying the physicist's favorite needle, the claim that mathematicians spent their time proving the obvious. Although he teased them, he thought they were an exciting group—happy and interested in a kind of science that was getting beyond him. One friend was Arthur Stone, a patient young man attending Princeton on a fellowship from England. Another was John Tukey, who later became one of the world's leading statisticians. These men spent their leisure time in curious ways.

Stone had brought with him English-standard loose-leaf notebooks. The American-standard paper he bought at Woolworth's overhung the notebooks by an inch, so he presently found himself with a supply of inch-wide paper ribbons, suitable for folding and twisting in different configurations. He tried diagonal folds at the 60-degree angle that produced rows of equilateral triangles. Then, following these folds, he wrapped a strip into a perfect hexagon.

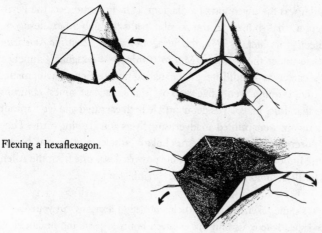

Flexing a hexaflexagon.

When he closed the loop by taping the ends together, he found that he had created an odd toy: by pinching opposite corners of the hexagon, he could perform a queer origami-like fold, producing a new hexagon with a different set of triangles exposed. Repeating the operation exposed a third face. One more "flex" brought back the original configuration. In effect, he had a flattened tube that he was steadily turning inside out.

He considered this overnight. In the morning he took a longer strip and confirmed a new hypothesis: that a more elaborate hexagon could be made to cycle through not three but six different faces. The cycling was not so straightforward this time. Three of the faces tended to come up again and again, while the other three seemed harder to find. This was a nontrivial challenge to his topological imagination. Centuries of origami had not produced such an elegantly convoluted object. Within days copies of these "flexagons"—or, as this subspecies came to be more precisely known, "hexa-hexaflexagons" (six sides, six internal faces)—were circulating across the dining hall at lunch and dinner. The steering committee of the flexagon investigation soon comprised Stone, Tukey, a mathematician named Bryant Tuckerman, and their physicist friend Feynman. Honing their dexterity with paper and tape, they made hexaflexagons with twelve faces buried amid the folds, then twenty-four, then forty-eight. The number of varieties within each species rose rapidly according to a law that was far from evident. The theory of flexigation flowered, acquiring the flavor, if not quite the substance, of a hybrid of topology and network theory. Feynman's best contribution was the invention of a diagram, called in retrospect the Feyn-man diagram, that showed all the possible paths through a hexaflexagon.

Seventeen years later, in 1956, the flexagons reached *Scientific American* in an article under the byline of Martin Gardner. "Flexagons" launched Gardner's career as a minister to the nation's recreational-mathematics underground, through twenty-five years of "Mathematical Games" columns and more than forty books. His debut article both captured and fed a minor craze. Flexagons were printed as advertising flyers and greeting cards. They inspired dozens of scholarly or semischolarly articles and several books. Among the hundreds of letters the article provoked was one from the Allen B. Du Mont Laboratories in New Jersey that began:

Sirs: I was quite taken with the article entitled "Flexagons" in your De-cember issue. It took us only six or seven hours to paste the hexahexa-flexagon together in the proper configuration. Since then it has been a source of continuing wonder.

But we have a problem. This morning one of our fellows was sitting flexing the hexahexaflexagon idly when the tip of his necktie became caught in one of the folds. With each successive flex, more of his tie vanished into the flexagon. With the sixth flexing he disappeared entirely.

We have been flexing the thing madly, and can find no trace of him, but we have located a sixteenth configuration of the hexahexaflexagon. . . .

The spirits of play and intellectual inquiry ran together. Feynman spent slow afternoons sitting in the bay window of his room, using slips of paper to ferry ants back and forth to a box of sugar he had suspended with string, to see what he could learn about how ants communicate and how much geometry they can internalize. One neighbor barged in on Feynman sitting by the window, open, on a wintry day, madly stirring a pot of Jell-O with a spoon and shouting "Don't bother me!" He was trying to see how the Jell-O would coagulate while in motion. Another neighbor provoked an argument about the motile techniques of human spermatozoa; Feynman disappeared and soon returned with a sample. With John Tukey, Feynman carried out a long, introspective investigation into the human ability to keep track of time by counting. He ran up and down stairs to quicken his heartbeat and practiced counting socks and seconds simultaneously. They discovered that Feynman could read to himself silently and still keep track of time but that if he spoke he would lose his place. Tukey, on the other hand, could keep track of the time while reciting poetry aloud but not while reading. They decided that their brains were applying different functions to the task of counting: Feynman was using an aural rhythm, hearing the numbers, while Tukey visualized a sort of tape with numbers passing behind his eyes. Tukey said years later: "We were interested and happy to be empirical, to try things out, to organize and reduce to simple things what had been observed."

Once in a while a small piece of knowledge from the world outside science would float Feynman's way and stick like a bur from a chestnut. One of the graduate students had developed a passion for the poetry of Edith Sitwell, then considered modern and eccentric because of her flamboyant diction and cacophonous, jazzy rhythms. He read some poems aloud, and suddenly Feynman seemed to catch on; he took the book and started reciting gleefully. "Rhythm is one of the principal translators between dream and reality," the poet said of her own work. "Rhythm might be described as, to the world of sound, what light is to the world of sight." To Feynman rhythm was a drug and a lubricant. His thoughts sometimes seemed to slip and flow with a variegated drumbeat that his friends noticed spilling out into his fingertips, restlessly tapping on desks and notebooks. "While a universe grows in my head,—" Sitwell wrote,

> I have dreams, though I have not a bed—
> The thought of a world and a day
> When all may be possible, still come my way.

Forward or Backward?

For a while the tea-time conversation among the physicists both at Princeton and at the Institute for Advanced Study was dominated by the image of a rotating lawn sprinkler, an S-shaped apparatus spun by the recoil of the water it sprays forth. Nuclear physicists, quantum theorists, and even pure mathematicians were consumed by the problem: What would happen if this familiar device were placed under water and made to suck water in instead of spewing it out? Would it spin in the reverse direction, because the direction of the flow was now reversed, pulling rather than pushing? Or would it spin in the same direction, because the same twisting force was exerted by the water, whichever way it flowed, as it was bent around the curve of the S? ("It's clear to me at first sight," a friend of Feynman's said to him some years later. Feynman shot back: "It's clear to *everybody* at first sight. The trouble was, some guy would think it was perfectly clear one way, and another guy would think it was perfectly clear the other way.") In an increasingly sophisticated time the simple problems still had the capacity to surprise. One did not have to probe far into physicists' understanding of Newton's laws before reaching a shallow bottom. Every action produces an equal and opposite reaction—that was the principle at work in the lawn sprinkler, as in a rocket. The inverse problem forced people to test their understanding of where, exactly, the reaction wielded its effects. At the point of the nozzle? Somewhere in the curve of the S, where the twisted metal forces the water to change course? Wheeler was asked for his own verdict one day. He said that Feynman had absolutely convinced him the day before that it went around backward; that Feynman had absolutely convinced him today that it went around forward; and that he did not yet know which way Feynman would convince him the next day.

If the mind was the most convenient of laboratories, it was not proving the most trustworthy. Because the *Gedankenexperiment* was failing, Feynman decided to bring the lawn-sprinkler problem back into the world of matter—stiff metal and wet water. He bent a piece of tubing into an S. He ran a piece of soft rubber hose into it. Now he needed a convenient source of compressed air.

The Palmer Physical Laboratory at Princeton housed a magnificent array of facilities, though not quite up to the standards of MIT. There were four large laboratories and several smaller ones, with a total floor space of more

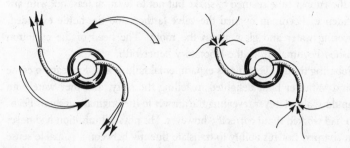

The mystery of the lawn sprinkler. When it sprays water, it spins counterclockwise. But what happens when it is made to suck water in?

than two acres. Machine shops supplied electrical charging devices, storage batteries, switchboards, chemical equipment, and diffraction gratings. The third floor was devoted to a high-voltage laboratory capable of direct currents at 400,000 volts. A low-temperature laboratory had machinery for lique-fying hydrogen. Palmer's pride, however, was its new cyclotron, built in 1936. Feynman had made a point of wandering over the day after he arrived at Princeton and had tea with the Dean. By comparison, MIT's even newer cyclotron was an elegant futuristic masterpiece of shiny metal and geo-metrically arrayed dials; when MIT had finally decided to invest in high-energy physics, it had not stinted. Princeton's gave Feynman a shock. He made his way down into the basement of Palmer, opened the door, and saw wires hanging like cobwebs from the ceiling. Safety valves for the cooling system were exposed, and water dripped from them. Tools were scattered on tables. It could not have looked less like Princeton. He thought of his wooden-crate laboratory at home in Far Rockaway.

Amid the chaos, it seemed reasonable enough for Feynman to borrow the use of an outlet for compressed air. He attached the rubber tube and pushed the end through a large cork. He lowered his miniature lawn sprinkler through the neck of a giant glass water bottle and sealed the bottle with the cork. Rather than try to suck water from the tube, he was going to pump air into the top of the bottle. That would increase the pressure of the water, which would then flow backward into the S-shaped pipe, up the rubber hose, and out the bottle.

He turned on the air valve. The apparatus gave a slight tremble, and water started to dribble from the cork. More air—the flow of water increased and the rubber tube seemed to shake but not to twist, at least not with any confidence. Feynman opened the valve farther, and the bottle exploded, showering water and glass across the room. The head of the cyclotron banished Feynman from the laboratory henceforth.

Sobering though Feynman's experimental failure was, for years to come he and Wheeler both delighted in telling the story, and they were both scrupulous about never revealing the answer to the original question. Feynman had worked it out correctly, however. His physical intuition had never been sharper, nor his ability to translate fluently between a palpable sense of the physics and the formal mathematical equations. His experiment had actually worked, until it exploded. Which way does the lawn sprinkler turn? It does not turn at all. As the nozzles suck water in, they do not pull themselves along, like a rope climber pulling himself up hand over hand. They have no purchase on the water ahead. And the idea of force exerted as a torque within the curve of the S is beside the point. In the normal version, water sprays forth in organized jets. The action and reaction are straightforward and measurable. The momentum of the water spraying in one direction equals the momentum that spins the nozzle in the opposite direction. But in the inverse case, when water is sucked in, there are no jets. The water is not organized. It enters the nozzle from all directions and therefore applies no force at all.

A development in twentieth-century entertainment technology—the motion picture—incidentally provided an advance in the technology of thought experiments. It was now natural for a scientist, in his mind's laboratory, to *play the film backward*. In the case of the lawn sprinkler, reversibility proved to be an illusion. If the flow of the water were visible, a motion picture of an ordinary lawn sprinkler played backward would look distinctly different from the sucking lawn sprinkler played forward. Filmmakers themselves had been seduced by the new, often comical insights that could be gained by taking a strip of celluloid and running it backward through the projector. Divers sprang feet first from lakes as a spray of water collapsed into the space left behind. Fires drew smoke from the air and created a trail of new-made paper. Fragmented eggshells assembled themselves around shuddering chicks.

For Feynman and Wheeler reversibility was becoming a central issue at the level of atomic processes, where spins and forces interacted more abstractly than in a lawn sprinkler. It was well known that the equations describing the motions and collisions of objects ran equally well forward and backward. They were symmetrical with respect to time, at least where just a few objects were concerned. How embarrassing, therefore, that time seemed so one-way in the real world, where a small amount of energy could scramble an egg or shatter a dish and where unscrambling and unshattering were beyond the power of science. "Time's arrow" was already the catchphrase for this directionality, so evident to common experience, yet so invisible in the equations of physicists. There, in the equations, the road from past to future looked identical to the road from future to past. "There is no signboard to indicate that it is a one-way street," complained Arthur Eddington. The paradox had been there all along, since Newton at least, but relativity had highlighted it. The mathematician Hermann Minkowski, by visualizing time as a fourth dimension, had begun to reduce past-future to the status of any pair of directions: left-right, up-down, back-front. The physicist drawing his diagrams obtains a God's-eye view. In the space-time picture a line representing the path of a particle through time simply exists, past and future visible together. The four-dimensional space-time manifold displays all eternity at once.

The laws of nature are not rules controlling the metamorphosis of what is into what will be. They are descriptions of patterns that exist, all at once, in the whole tapestry. The picture is hard to reconcile with our everyday sense that time is special. Even the physicist has his memories of the past and his aspirations for the future, and no space-time diagram quite obliterates the difference between them.

Philosophers, in whose province such speculations had usually belonged, were left with a muddy and senescent set of concepts. The distress of the philosophers of time spilled into their adverbs: *sempiternally, hypostatically, tenselessly, retrodictably*. Centuries of speculation and debate had left them unprepared for the physicists' sudden demolition of the notion of simultaneity (in the relativistic universe it meant nothing to say that two events took place at the same time). With simultaneity gone, sequentiality was foundering, causality was under pressure, and scientists generally felt themselves free to consider temporal possibilities that would have seemed farfetched a generation before.

In the fall of 1940 Feynman returned to the fundamental problem with which he had flirted since his undergraduate days. Could the ugly infinities of quantum theory be eliminated by forbidding the possibility that an electron acts on itself—by eliminating, in effect, the field? Unfortunately he had meanwhile learned what was wrong with his idea. The problem was a phenomenon that could only be explained, it seemed, in terms of the action of an electron on itself. When real electrons are pushed, they push back: an accelerating electron drains energy by radiating it away. In effect the electron feels a resistance, called radiation resistance, and extra force has to be applied to overcome it. A broadcasting antenna, radiating energy in the form of radio waves, encounters radiation resistance—extra current has to be sent through the antenna to make up for it. Radiation resistance is at work when a hot, glowing object cools off. Because of radiation resistance, an electron in an atom, alone in empty space, loses energy and dies out; the lost energy has been radiated away in the form of light. To explain why this damping takes place, physicists assumed they had no choice but to imagine a force exerted by the electron on itself. By what else, in empty space?

One day, however, Feynman walked into Wheeler's office with a new idea. He was "pie-eyed," he confessed, from struggling with an obscure problem Wheeler had given him. Instead he had turned back to self-action. What if (he thought) an electron isolated in empty space does not emit radiation at all, any more than a tree makes a sound in an empty forest. Suppose radiation were to be permitted only when there is both a source and a receiver. Feynman imagined a universe with just two electrons. The first shakes. It exerts a force on the second. The second shakes and generates a force that acts back on the first. He computed the force by a familiar field equation of Maxwell's, but in this two-particle universe there was to be no field, if the field meant a medium in which waves were freely spreading outward on their own.

He asked Wheeler, Could such a force, exerted by one particle on another and then back on the first, account for the phenomenon of radiation resistance?

Wheeler loved the idea—it was the sort of approach he might have taken, stripping a problem down to nothing but a pair of point charges and trying to build up a new theory from first principles. But he saw immediately that the numbers would come out wrong. The force coming back to the first charge would depend on how strong the second charge was, how massive it was, and how near it was. But none of those quantities influence radiation resistance. This objection seemed obvious to Feynman afterward, but at the time he was astonished by his professor's fast insight. And there was another problem: Feynman had not properly accounted for the delay in the transmission of the force to and fro. Whatever force was exerted back on the first particle would come at the wrong time, too late to match the known effect of radiation resistance. In fact Feynman suddenly realized that he had been describing a different phenomenon altogether, a painfully simple one: ordinary reflected light. He felt foolish.

Time delay had not been a feature of the original electromagnetic theory. In Maxwell's time, on the eve of relativity, it still seemed natural to assume, as Newton had, that forces acted instantaneously. An imaginative leap was needed to see that the earth swerves in its orbit not because the sun is there but because it was there eight minutes before, the time needed for gravity's influence to cross nearly a hundred million miles of space—to see that if the sun were plucked away, the earth would continue to orbit for eight minutes. To accommodate the insights of relativity, the field equations had to be amended. The waves were now *retarded* waves, held back by the finite speed of light.

Here the problem of time's symmetry entered the picture. The electromagnetic equations worked magnificently when retarded waves were correctly incorporated. They worked equally well when the sign of the time quantities was reversed, from plus to minus. Translated back from mathematics into physics, that meant *advanced* waves—waves that were received *before* they were emitted. Understandably, physicists preferred to stay with the retarded-wave solutions. An advanced wave, running backward in time, seemed peculiar. Viewed in close-up it would look like any other wave, but it would converge on its source, like a concentric ripple heading toward the center of a pond, where a rock was about to fly out—the film played backward again. Thus, despite their mathematical soundness, the advanced-wave solutions to field equations stayed in the background, an unresolved but not especially urgent puzzle.

Wheeler immediately proposed to Feynman that they consider what would happen if advanced waves were added to his two-electron model. What if the apparent time-symmetry of the equations were taken seriously? One would have to imagine a shaken electron sending its radiation outward symmetrically in time. Like a lighthouse sending its beam both north and south, an electron might shine both forward and backward to the future and the past. It seemed to Wheeler that a combination of advanced and retarded waves might cancel each other in a way that would overcome the lack of any time delay in the phenomenon of radiation resistance. (The canceling of waves was well understood. Depending on whether they were in or out of phase, waves of the same frequency would interfere either constructively or destructively. If their crests and troughs lined up exactly, the size of the waves would double. If crests lined up with troughs, then

the waves would precisely neutralize each other.) He and Feynman, calculating excitedly over the next hour, found that the other difficulties also
seemed to vanish. The energy arriving back at the original source no longer
depended on the mass, the charge, or the distance of the second particle.
Or so it seemed, in the first approximation produced by their rough computation on Wheeler's blackboard.

Feynman set to work on this possibility. He was not troubled by the
seemingly nonsensical meaning of it. His original notion contained nothing
out of the ordinary: Shake a charge here—then another charge shakes a
little later. The new notion turned paradoxical as soon as it was expressed
in words: Shake a charge here—then another charge shakes a little *earlier*.
It explicitly required an action backward in time. Where was the cause and
where was the effect? If Feynman ever felt that this was a deep thicket to
enter merely for the sake of eliminating the electron's self-action, he suppressed the thought. After all, self-action created an undeniable contradiction within quantum mechanics, and the entire profession was finding
it insoluble. At any rate, in the era of Einstein and Bohr, what was one
more paradox? Feynman already believed that it was the mark of a good
physicist never to say, "Oh, whaddyamean, how could that be?"

The work required intense calculation, working out the correct forms of
the equations, always checking to make sure that the apparent paradox
never turned into an actual mathematical contradiction. Gradually the basic
model became, not a system of two particles, but a system where the electron
interacted with a multitude of other "absorber" particles all around it. It
would be a universe where all radiation eventually reached the surrounding
absorber. As it happened, that softened the most bizarre time-reversed
tendencies of the model. For those who were squeamish about the prospect
of effects anticipating their causes, Feynman offered a barely more palatable
view: that energy is momentarily "borrowed" from empty space, and paid
back later in exact measure. The lender of this energy, the absorber, was
assumed to be a chaotic multitude of particles, moving in all directions so
that almost all its effects on a given particle would cancel one another. The
only time an electron would feel the presence of this absorbing layer would
be when it accelerated. Then the effect of the source on the absorber would
return to the source at exactly the right time, with exactly the right force,
to account for radiation resistance. Thus, given that one cosmological
assumption—that the universe has enough matter in every direction to
soak up outgoing radiation—Feynman found that a system of equations in

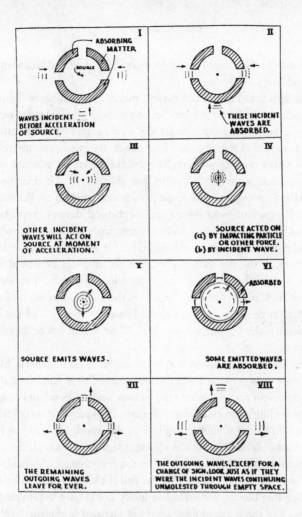

I — ABSORBING MATTER — SOURCE — WAVES INCIDENT BEFORE ACCELERATION OF SOURCE.	**II** — THESE INCIDENT WAVES ARE ABSORBED.
III — OTHER INCIDENT WAVES WILL ACT ON SOURCE AT MOMENT OF ACCELERATION.	**IV** — SOURCE ACTED ON (a) BY IMPACTING PARTICLE OR OTHER FORCE. (b) BY INCIDENT WAVE.
V — SOURCE EMITS WAVES.	**VI** — ABSORBED — SOME EMITTED WAVES ARE ABSORBED.
VII — THE REMAINING OUTGOING WAVES LEAVE FOR EVER.	**VIII** — THE OUTGOING WAVES, EXCEPT FOR A CHANGE OF SIGN, LOOK JUST AS IF THEY WERE THE INCIDENT WAVES CONTINUING UNMOLESTED THROUGH EMPTY SPACE.

Waves forward and backward in time. Wheeler and Feynman tried to work out a consistent scheme for the interactions of particles, and they embroiled themselves in paradoxes of past and future. A particle shakes; its influence spreads outward like waves from a stone thrown into a pond. To make their theory symmetrical, they also had to use inward-traveling waves—implying action backward in time.

They found that they could avoid unpleasant paradoxes because these normal and time-reserved waves ("retarded" and "advanced") canceled each other out—but only if the universe was arranged so as to guarantee that all radiation would be absorbed somewhere, sometime. A beam of light traveling forever into infinite, empty space, never striking an absorber, would foil their theory's bookkeeping. Thus cosmologists and philosophers of time continued to consider their scheme long after it had been supplanted in the mainstream of quantum theory.

which advanced and retarded waves were combined half and half seemed to withstand every objection.

He described it to his graduate student friends and challenged them to find a paradox he could not explain his way through. For example, could one design a mechanism with a target that would shut a gate when struck by a pellet, such that the advanced field closed the gate before the pellet arrived, in which case the pellet could not strike the target, in which case the advanced field would not close the gate after all . . . He imagined a Rube Goldberg contraption that might have come straight from Wheeler's old book of ingenious mechanisms and mechanical devices. Feynman's calculations suggested that the model was surprisingly immune to paradox. As long as the theory relied on probabilities, it seemed to escape fatal contradictions. It did not matter where the absorber was or how it was shaped, as long as there were absorbing particles off at some distance in every direction. Only if there were "holes" in the surrounding layer, places where radiation could go forever without being absorbed, could the advanced effects make trouble, arriving back at the source before they had been triggered.

Wheeler had his own motive for pursuing this quixotic theory. Most physicists were now persuaded that the atom embodied at least three irreconcilably different particles, electrons, protons, and neutrons, and cosmic rays were providing intimations of several more. This proliferation offended Wheeler's faith in the ultimate simplicity of the world. He continued to cherish a notion so odd that he was reluctant to discuss it aloud, the idea that a different kind of theory would reveal everything to be made of electrons after all. It was crazy, he knew. But if electrons were to be the ultimate building blocks, their radiative forces would have to provide the key, in ways that the standard theory was not prepared to explain. Within weeks he began pressing Feynman to write a preliminary paper. If they were going to make grand theories, Wheeler would make sure they publicized the work properly. Early in 1941 he told Feynman to prepare a presentation for the departmental seminar, usually a forum for distinguished visiting physicists, in February. It would be Feynman's first professional talk. He was nervous about it.

As the day approached, Wigner, who ran the colloquiums, stopped Feynman in the hall. Wigner said he had heard enough from Wheeler about the absorber theory to think it was important. Because of its implications for cosmology he had invited the great astrophysicist Henry Norris

Russell. John von Neumann, the mathematician, was also going to come. The formidable Wolfgang Pauli happened to be visiting from Zurich; he would be there. And though Albert Einstein rarely bestirred himself to the colloquiums, he had expressed interest in attending this one.

Wheeler tried to calm Feynman by promising to field questions from the audience. Wigner tried to brief him. If Professor Russell appears to fall asleep during your talk, Wigner said, don't worry—Professor Russell always falls asleep. If Pauli appears to be nodding, don't assume he agrees—he nods from palsy. (Pauli could be ruthless in dismissing work he considered shallow or flimsy: "*ganz falsch*," utterly false—or worse, "*not even* false.") Feynman prepared carefully. He collected his notes and put them into a brown envelope. He entered the seminar room early and covered the black-board with equations. While he was writing, he heard a soft voice behind him. It was Einstein. He was coming to the lecture and first he wondered whether the young man might direct him to the tea.

Afterward Feynman could remember almost nothing: just the trembling of his hand as he pulled his notes from the envelope and then a feeling that his mind put itself at ease by concentrating on the physics and forgetting the occasion and the personalities. Pauli did object, perhaps sensing that the use of advanced potentials merely invoked a sort of mathematical tau-tology. Then, politely, Pauli said, "Don't you agree, Professor Einstein?" Feynman heard that soft Germanic voice again—so pleasant, it seemed—saying no, the theory seemed possible, perhaps there was a conflict with the theory of gravitation, but after all the theory of gravitation was not so well established . . .

The Reasonable Man

He suffered spells of excessive rationality. When these struck it was not enough to make progress in his scientific work, nor to rectify his mother's checkbook, nor to recompute his own equivocal balance sheet (eighteen dollars for laundry, ten dollars to send home . . .), nor to lecture his friends, as they watched him repair his bicycle, on the silliness of believing in God or the supernatural. During one occurrence he wrote out an hourly schedule of his activities, both scholarly and recreational, "so as to efficiently distribute my time," he wrote home. When he finished, he recognized

that no matter how careful he was, he would have to leave some indeter-
minate gaps—"hours when I haven't marked down just what to do but I
do what I feel is most necessary then—or what I am most interested in—
whether it be W.'s problem or reading Kinetic Theory of Gases, etc." If
there is a disease whose symptom is the belief in the ability of logic to
control vagarious life, it afflicted Feynman, along with his chronic digestive
troubles. Even Arline Greenbaum, sensible as she was, could spark flights
of reason in him. He grew concerned about the potential for emotional
disputes between husbands and wives. Even his own parents fought. He
hated the battles and the anger. He did not see why two intelligent people,
in love with each other, willing to converse openly, should get caught in
arguments. He worked out a plan. Before revealing it to Arline, however,
he decided to lay it out for a physicist friend over a hamburger at a diner
on the Route 1 traffic circle. The plan was this. When Dick and Arline
disagreed intensely about a matter of consequence, they would set aside a
fixed time for discussion, perhaps one hour. If at the end of that time they
had not found a resolution, rather than continue fighting they would agree
to let one of them decide. Because Feynman was older and more experienced (he explained), he would be the one.

His friend looked at him and laughed. He knew Arline, and he knew
what would really happen. They would argue for an hour, Dick would give
up, and Arline would decide. Feynman's plan was a sobering example of
the theoretical mind at work.

Arline was visiting more and more often. They would have dinner with
the Wheelers and go for long walks in the rain. She had the rare ability to
embarrass him: she knew where his small vanities were, and she teased
him mercilessly whenever she caught him worrying about other people's
opinions—how things might *seem*. She sent him a box of pencils embla-
zoned, "Richard darling, I love you! Putsie," and caught him slicing off
the incriminating legend, for fear of inadvertently leaving one on Professor
Wigner's desk. "What do you care what other people think?" she said again
and again. She knew he prided himself on honesty and independence, and
she held him to his own high standards. It became a touchstone of their
relationship. She mailed him a penny postcard with a verse written across
it:

> If you don't like the things I do
> My friend, I say, Pecans to you!

> If I irate with pencils new
> My bosom pal, Pecans to you!
>
> . . .
>
> If convention's mask is borne in view
>
> . . .
>
> If deep inside sound notions brew
> And from without you take your cue
> My sorry friend, Pecans to you!

Her words struck home. Meanwhile she had nagging health worries: a lump seemed to come and go on her neck, and she developed uncomfortable, unexplained fevers. Her uncle, a physician, had her rub the lump with a nostrum called omega oil. (This style of treatment had had its heyday a hundred years before.)

The day after his presentation to the physics colloquium in February, Richard went up to Cambridge for a meeting of the American Physical Society, and she took the train from New York to Boston's South Station to join him. An old fraternity friend picked her up and they crossed the bridge to MIT, catching a ride on a horse-drawn junk wagon. They found Richard in the corridor of building 8, the physics building. He walked by in animated conversation with a professor. Arline made eye contact with him, but he did not acknowledge her. She realized that it would be better not to speak.

When Richard returned to the fraternity house that evening he found her in the living room. He was ebullient; he grabbed her and swung her around, dancing. "He certainly believes in physical society," one of the fraternity boys said. At Wheeler's prodding Feynman had presented their space-time electrodynamics a second time, to a broader audience. The talk went well. After having faced a public of Einstein, Pauli, von Neumann, and Wigner, he had little to fear from the American Physical Society rank and file. Still, he worried that he might have bored his listeners by sticking nervously to his prepared text. There were a few polite questions, and Wheeler helped answer them.

Feynman had enunciated a set of principles for a theory of interacting particles. He wrote them out as follows:

1 The acceleration of a point charge is due only to the sum of its interactions with other charged particles. . . . A charge does not act on itself.

2   The force of interaction which one charge exerts on a second is cal-
     culated by means of the Lorentz force formula, in which the fields
     are the fields generated by the first charge according to Maxwell's
     equations.

Phrasing the third principle was more difficult. He tried:

3   The fundamental equations are invariant with respect to a change of
     the sign of the time . . .

Then, more directly:

3   The fundamental (microscopic) phenomena in nature are symmetrical
     with respect to interchange of past and future.

Pauli, despite his skepticism, understood the power of the last principle.
He pointed out to Feynman and Wheeler that Einstein himself had argued
for an underlying symmetry of past and future in a little-known 1909 paper.
Wheeler needed little encouragement; he made an appointment to call at
the white clapboard house at 112 Mercer Street.

   Einstein received this pair of ambitious young physicists sympathetically,
as he did most scientists who visited in his last years. They were led into
his study. He sat facing them behind his desk. Feynman was struck by how
well the reality matched the legend: a soft, nice man wearing shoes without
socks and a sweater without a shirt. Einstein was well known to be unhappy
with the acausal paradoxes of quantum mechanics. He now spent much
of his time writing screeds on world government which, from a less revered
figure, would have been thought crackpot. His distaste for the new physics
was turning him into, as he would have it, "an obstinate heretic" and "a
sort of petrified object, rendered blind and deaf by the years." But the theory
Wheeler and Feynman described was not yet a quantum theory—so far,
it used only classical field equations, with none of the quantum-mechanical
amendments that they knew would ultimately be necessary—and Einstein
saw no paradox. He, too, he told them, had considered the problem of
retarded and advanced waves. He reminisced about the strange little paper
he had published in 1909, a manifesto of disagreement with a Swiss colleague, Walter Ritz. Ritz had declared that a proper field theory should
include only retarded solutions, that the time-backward advanced solutions
should simply be declared impermissible, innocent though the equations

On his bicycle in Far Rockaway.

Melville, Lucille, Richard, and Joan at the house they shared with Lucille's sister's family, at 14 New Broadway.

Richard and Arline: left, at
Presbyterian Sanatorium.

At Los Alamos: "*I opened the safes which contained behind them the entire secret of the atomic bomb…*"

Slouching beside J. Robert Oppenheimer at a Los Alamos meeting: "*He is by all odds the most brilliant young physicist here, and everyone knows this.*"

Awaiting the Trinity test: "*And we scientists are clever—too clever—are you not satisfied? Is four square miles in one bomb not enough? Men are still thinking. Just tell us how big you want it!*"

I. I. Rabi (left) and Hans Bethe: *Physicists are the Peter Pans of the human race, Rabi said.*

At the Shelter Island Conference, June 1947: Willis Lamb and John Wheeler, standing; Abraham Pais, Feynman, and Herman Feshbach, seated; Julian Schwinger, kneeling.

Julian Schwinger: "*It seems to be the spirit of Macaulay which takes over, for he speaks in splendid periods, the carefully architected sentences rolling on, with every subordinate clause duly closing.*"

Feynman and Hideki Yukawa in Kyōto, 1955: Feynman presented his theory of superfluidity, the strange, frictionless behavior of liquid helium—quantum mechanics writ large.

At Caltech, before a
slide from his original
presentation on anti-
particles traveling back-
ward through time.

Victor Weisskopf (left) and Freeman Dyson.

looked. Einstein, however, could see no reason to rule out advanced waves. He argued that the explanation for the arrow of time could not be found in the basic equations, which truly were reversible.

That was Feynman and Wheeler's view. By insisting on the symmetry of past and future, they made the combination of retarded and advanced potentials seem a necessity. In the end, there was an asymmetry in the universe of their theory—the role of ordinary retarded fields far outweighs the backward advanced fields—but that asymmetry does not lie in the equations. It comes about because of the disordered, mixed-up nature of the surrounding absorber. A tendency toward disorder is the most universal manifestation of time's arrow. A movie showing a drop of ink diffusing in a glass of water looks wrong when run backward. Yet a movie showing the microscopic motion of any one ink molecule would look the same backward or forward. The random motions of each ink molecule can be reversed, but the overall diffusion cannot be. The system is microscopically reversible, macroscopically irreversible. It is a matter of chaos and probability. It is not impossible for the ink molecules, randomly drifting about, someday to reorganize themselves into a droplet. It is just hopelessly improbable. In Feynman and Wheeler's universe, the same kind of improbability guaranteed the direction of time by ensuring disorder in the absorber. Feynman took pains to spell out the distinction in the twenty-two-page manuscript he wrote early in 1941:

> We must distinguish between two types of irreversibility. A sequence of natural phenomena will be said to be microscopically irreversible if the sequence of phenomena reversed in temporal order in every detail could not possibly occur in nature. If the original sequence and the reversed in time one have a vastly different order of probability of occurrence in the macroscopic sense, the phenomena are said to be macroscopically irreversible. . . . The present authors believe that all physical phenomena are microscopically reversible, and that, therefore, all apparently irreversible phenomena are solely macroscopically irreversible.

Even now the principle of reversibility seemed startling and dangerous, defying as it did the sense of one-way time that Newton had implanted in science. Feynman called his last statement to Wheeler's attention with a note: "Prof Wheeler," he wrote—and then self-consciously crossed out "Prof"—"This is a rather sweeping statement. Perhaps you don't agree with it. RPF."

Meanwhile Wheeler was searching the literature, and he found several obscure precedents for their absorber model. Einstein himself pointed out that H. Tetrode, a German physicist, had published a paper in *Zeitschrift für Physik* in 1922 proposing that all radiation be considered an interaction between a source and an absorber—no absorber, no radiation. Nor did Tetrode shrink from the tree-falls-in-the-forest consequences of the idea:

> The sun would not radiate if it were alone in space and no other bodies could absorb its radiation. . . . If for example I observed through my telescope yesterday evening that star . . . 100 light years away, then not only did I know that the light which it allowed to reach my eye was emitted 100 years ago, but also the star or individual atoms of it knew already 100 years ago that I, who then did not even exist, would view it yesterday evening at such and such a time.

For that matter, the invisible reddened whisper of radiation emitted by a distant (and in the twenties, unimagined) quasar not one hundred but ten billion years ago—radiation that passed unimpeded for most of the universe's lifetime until finally it struck a semiconducting receiver at the heart of a giant telescope—this, too, could not have been emitted without the cooperation of its absorber. Tetrode conceded, "On the last pages we have let our conjectures go rather far beyond what has mathematically been proven." Wheeler found another obscure but provocative remark in the literature, from Gilbert N. Lewis, a physical chemist who happened to have coined the word *photon*. Lewis, too, worried about the seeming failure of physics to recognize the symmetry between past and future implied by its own fundamental equations, and for him, too, the past-future symmetry suggested a source-absorber symmetry in the process of radiation.

> I am going to make the . . . assumption that an atom never emits light except to another atom. . . . it is as absurd to think of light emitted by one atom regardless of the existence of a receiving atom as it would be to think of an atom absorbing light without the existence of light to be absorbed. I propose to eliminate the idea of mere emission of light and substitute the idea of *transmission*, or a process of exchange of energy between two definite atoms. . . .

Feynman and Wheeler pushed on their theory. They tried to see how far they could broaden its implications. Many of their attempts led nowhere. They worked on the problem of gravity in hopes of reducing it to a similar interaction. They tried to construct a model in which space itself was eliminated: no coordinates and distances, no geometry or dimension; only the interactions themselves would matter. These were dead ends. As the theory developed, however, one feature gained paramount importance. It proved possible to compute particle interactions according to a principle of least action.

The approach was precisely the shortcut that Feynman had gone out of his way to disdain in his first theory course at MIT. For a ball arcing through the air, the principle of least action made it possible to sidestep the computation of a trajectory at successive instants of time. Instead one made use of the knowledge that the final path would be the one that minimized action, the difference between the ball's kinetic and potential energy. In the absorber theory, because the field was no longer an independent entity, the action of a particle suddenly became a quantity that made sense. It could be calculated directly from the particle's motion. And once again, as though by magic, particles chose the paths for which the action was smallest. The more Feynman worked with the least-action approach, the more he felt how different was the physical point of view. Traditionally one always thought in terms of the flow of time, represented by differential equations, which captured a change from instant to instant. Using the principle of least action instead, one developed a bird's-eye perspective, envisioning a particle's path as a whole, all time seen at once. "We have, instead," Feynman said later, "a thing that describes the character of the path throughout all of space and time. The behavior of nature is determined by saying her whole space-time path has a certain character." In college it had seemed too pat a device, too far abstracted from the true physics. Now it seemed extraordinarily beautiful and not so abstract after all. His conception of light was still in flux—still not quite a particle, not quite a wave, still pressing speculatively against the unresolved infinities of quantum mechanics. The notion had come far since Euclid wrote, as the first postulate of his *Optics*, "The rays emitted by the eye travel in a straight line."

The empty space of the physicist's imagination—the chalkboard on which every motion, every force, every interaction played itself out—had undergone a transformation in less than a generation. A ball pursued a trajectory through the everyday space of three dimensions. The particles of Feynman's reckoning forged paths through the four-dimensional space-time so indispensable to the theory of relativity, and through even more abstract spaces whose coordinate axes stood for quantities other than distance and time. In space-time even a motionless particle followed a trajectory, a line extending from past to future. For such a path Minkowski coined the phrase *world-line*—"an image, so to speak, of the everlasting career of the substantial point, a curve in the world. . . . The whole universe is seen to resolve itself into similar world-lines." Science-fiction writers had already begun to imagine the strange consequences of world-lines twisting back from the future into the past. No novelist was letting his fantasies roam as far as Wheeler was, however. One day he called Feynman on the hall telephone in the Graduate College. Later Feynman remembered the conversation this way:

—Feynman, I know why all the electrons have the same charge and the same mass.

—Why?

—Because they are all the same electron! Suppose that all the world-lines which we were ordinarily considering before in time and space—instead of only going up in time were a tremendous knot, and then, when we cut through the knot, by the plane corresponding to a fixed time, we would see many, many world-lines and that would represent many electrons, except for one thing. If in one section this is an ordinary electron world-line, in the section in which it reversed itself and is coming back from the future we have the wrong sign . . . and therefore, that part of a path would act like a positron.

The positron, the antiparticle twin of the electron, had been discovered (in cosmic-ray showers) and named (another modern -tron, short for positive electron) within the past decade. It was the first antiparticle, vindicating a prediction of Dirac's, based on little more than a faith in the loveliness of his equations. According to the Dirac wave equation, the energy of a particle amounted to this: $\pm\sqrt{\text{something}}$. Out of that plus-or-minus sign the positron was born. The positive solution was an electron. Dirac boldly resisted the temptation to dismiss the negative solution as a quirk of algebra. Like Wheeler in making his leap toward advanced waves, he followed a mirror-image change in sign to its natural conclusion.

Feynman considered the wild suggestion coming through the earpiece of his telephone—that all creation is a slice through the spaghetti path of a single electron—and offered the mildest of the many possible rebuttals. The forward and backward paths did not seem to match up. An embroidery needle pulling a single thread back and forth through a canvas must go back as many times as it goes forth.

—But, Professor, there aren't as many positrons as electrons.

—Well, maybe they are hidden in the protons or something.

Wheeler was still trying to make the electron the basis of all other particles. Feynman let it pass. The point about positrons, however, reverberated. In his first published paper two years before, on the scattering of cosmic radiation by stars, he had already made this connection, treating antiparticles as ordinary particles following reversed paths. In a Minkowskian universe, why shouldn't the reversal apply to time as well as to space?

Mr. X and the Nature of Time

Twenty years later, in 1963, the problem of time having given up none of its mystery, a group of twenty-two physicists, cosmologists, mathematicians, and others sat around a table at Cornell to discuss the matter. Was time a quantity entered in the account books of their equations to mark the amount of *before* and *after*? Or it was an all-enveloping flow, carrying everything with it like a constant river? In either case, what did it mean to say *now*? Einstein had worried about this, accepting the unwelcome possibility that the present belongs to our minds alone and that science cannot comprehend it. A philosopher, Adolph Grünbaum, argued that the usual notion of the forward flow of time was merely an illusion, a "pseudoconception." If it seemed to us as conscious entities that new events kept "coming into being," that was merely one of the quirky consequences of the existence of conscious entities—"organisms which *conceptually register* (ideationally represent)" them. Physicists need not worry about it unduly.

When Grünbaum finished his presentation, a participant with a loathing for what he viewed as philosophical and psychological vagueness began a hard cross-examination. (The published version of the discussion identified this interlocutor only as "Mr. X," which fooled no one; by now, Feynman hiding behind such a cloak made himself as conspicuous as an American secretary of state quoted as "a senior official aboard the secretary of state's plane.")

GRÜNBAUM: I want to say that there is a difference between a conscious thing and an unconscious thing.

X: What is that difference?

GRÜNBAUM: Well, I don't have more precise words in which to say this, but I would not be worried if a computer is unemployed. If a human being is unemployed, I would worry about the sorrows which that human being experiences in virtue of conceptualized self-awareness.

X: Are dogs conscious?

GRÜNBAUM: Well, yes. It is going to be a question of degree. But I wonder whether they have *conceptualized* awareness.

X: Are cockroaches conscious?

GRÜNBAUM: Well, I don't know about the nervous system of the cockroach.

X: Well, they don't suffer from unemployment.

It seemed to Feynman that a robust conception of "now" ought not to depend on murky notions of mentalism. The minds of humans are manifestations of physical law, too, he pointed out. Whatever hidden brain machinery created Grünbaum's coming into being must have to do with a correlation between events in two regions of space—the one inside the cranium and the other elsewhere "on the space-time diagram." In theory one should be able to create a feeling of nowness in a sufficiently elaborate machine, said Mr. X.

One's sense of the now feels subjective, arbitrary, open to differences of definition and interpretation, particularly in the age of relativity. "One can say easily enough that any particular value of t can be taken as now and that would not be wrong, but it does not correspond to experience," the physicist David Park has said. "If we attend only to what is happening around us and let ourselves live, our attention concentrates itself on one moment of time. Now is when we think what we think and do what we do." For similar reasons many philosophers wished to banish the concept. Feynman, staking out a characteristic position in such debates, rejected the idea that human consciousness was special. He and other rigorous scientists, their tolerance broadened by their experience with quantum-mechanical measurement problems, found that they could live with the imprecision— the possibility that the *nows* of different observers would differ in timing and duration. Technology offered ways of tightening the definition, at least for the sake of argument: less subjectivity arose in the now recorded by a camera shutter or a computing machine. Wheeler, also present at the Cornell meeting, proposed the example of a computer on an antiaircraft gun. Its now is the finite interval containing not just the immediate past, the few moments of data coming from the radar tracks, but the immediate future, the flight of the target plane as extrapolated from the data. Our memories, too, blend the immediate past with the anticipation of the soon to be, and a living amalgam of these—not some infinitesimal pointlike instant forever fleeing out of reach—is our now. Wheeler quoted the White Queen's remark to Alice: "It's a poor sort of memory that only works backwards."

The absorber theory of Wheeler and Feynman had by then lost the interest of an increasingly single-minded particle physics, but it held center stage in this eclectic gathering. It had been born of their concern with reversible and irreversible processes, and now it served as common ground for three different approaches to understanding time's flow, the arrow of time. As particle physicists had passed the absorber theory by, a new generation of cosmologists had taken it up. Their field had begun a transition from mere stargazing astronomy to an enterprise asking the grandest questions about the universe: whence and wherefore. It was beginning to stand out among the modern sciences as an enterprise not fully scientific, but an amalgam of philosophy, art, faith, and not a little hope. They had so few windows through the murky atmosphere—a few overworked glass contraptions on mountain tops, a few radio antennae—yet they believed they could peer far enough, or guess shrewdly enough, to uncover the origins of space and time. Already their space was not the flat, neutral stuff of their parents' pre-Einsteinian intuition, but an eerily plastic medium that somehow embodied both time and gravity. Some of them, but not all, believed that space was expanding at high speed and dragging its contents farther and farther apart, on account of an explosive big bang ten or fifteen billion years before. It no longer seemed safe to assume that the universe was the same everywhere, infinite, static, Euclidean, ageless, and homogeneous: world without end, amen. The strongest evidence for an expanding universe was still, in 1963, Edwin Hubble's 1929 discovery that other galaxies are streaming away from ours, and that the farther away they are, the faster they seem to be moving. Whether this expansion would continue forever or whether it would reverse itself was—and would remain—an open question. Perhaps the universe bloomed and collapsed again and again in a cycle that ran through eternity.

The issue seemed linked to the nature of time itself. Assumptions about time were built into the equations for the particle interactions that led to the creation and dissipation of light. If one thought about time as Wheeler and Feynman had, one could not escape a cosmic connection between these intimate interactions and the process of universal expansion. As Hermann Bondi said at the meeting's outset, "This process leads to the dark night sky, to the disequilibrium between matter and radiation, and to the fact that radiated energy is effectively lost . . . we accept a very close connection between cosmology and the basic structure of our physics." By their boldness in constructing a time-symmetrical theory of half advanced and half retarded waves, Wheeler and Feynman had been forced into boldness of a cosmological sort. If the equations were to balance properly, they had to make the mathematical assumption that all radiation was eventually absorbed somewhere. A beam of light heading forever into an eternal future, never to cross paths with a substance that would absorb it, would violate their assumption, so their theory mandated a certain kind of universe. If the universe were to expand forever, conceivably its matter might so thin out that light would not be absorbed.

Physicists had learned to distinguish three arrows of time. Feynman described them: the thermodynamic or "accidents of life" arrow; the radiation or "retarded or advanced" arrow; and the cosmological arrow. He suggested keeping in mind three physical pictures: a tank with blue water on one side and clear water on the other; an antenna with a charge moving toward it or away; and distant nebulas moving together or apart. The connections between these arrows were connections between the pictures. If a film showed the water getting more and more mixed, must it also show the radiation leaving the antenna and the nebulas drifting apart? Did one form of time govern the others? His listeners could only speculate, and speculate they did.

"It's a very interesting thing in physics," said Mr. X, "that the laws tell us about permissible universes, whereas we only have one universe to describe."

Least Action in Quantum Mechanics

Omega oil did nothing for Arline's lumps and fevers, and she was admitted to the hospital in Far Rockaway with what her doctor feared was typhoid. Feynman began to glimpse the special powerlessness that medical uncertainty can inflict on a scientific person. He had come to believe that the scientific way of thinking brought a measure of calmness and control in difficult situations—but not now. However remotely, medicine was a part of the domain of knowledge he considered his. It belonged to science. At one time his father had hopefully studied a kind of medicine. Lately Richard had been sitting in on a physiology course, learning some basic anatomy. He read up on typhoid fever in Princeton's library, and when he visited Arline in the hospital he started questioning the doctor. Had a Widal test been administered? Yes. The results? Negative. Then how could it be typhoid? Why were all of Arline's friends and relatives wearing gowns to protect against supposed bacteria that even a sensitive laboratory test could not detect? What did the mysterious lumps appearing and disappearing in her neck and armpit have to do with typhoid? The doctor resented his questions. Arline's parents pointed out that his status as fiancé did not entitle him to interfere in her medical care. He backed down. Arline seemed to recover.

With Wheeler, meanwhile, Feynman was trying to move their work a crucial step forward. So far, despite its modern, acausal flavor, it was a classical theory, not a quantum one. It treated objects as objects, not as probabilistic smudges. It treated energy as a continuous phenomenon, where quantum mechanics required discrete packets and indivisible jumps in well-defined circumstances. The problem of self-energy was as severe in classical electrodynamics as in quantum theory. Unwanted infinities predated the quantum. They appeared as soon as one faced the consequences of a pointlike electron. It was as simple as dividing by zero. Feynman had felt from the beginning that the natural route would be to start with the classical case and only then work toward a quantized electrodynamics. There were already standard recipes for translating classical models into their modern quantum cousins. One prescription was to take all the momentum variables and replace them with certain more complicated expressions. The problem was that in Wheeler and Feynman's theory there were no momentum variables. Feynman had eliminated them in creating his simplified framework based on the principle of least action.

Sometimes Wheeler told Feynman not to bother—that he had already solved the problem. Later in the spring of 1941 he went so far as to schedule a presentation of the quantized theory at the Princeton physics colloquium. Pauli, still dubious, buttonholed Feynman on his way into Palmer Library one day. He asked what Wheeler was planning to say. Feynman said he didn't know.

"Oh?" Pauli said. "The professor doesn't tell his assistant how he has it worked out? Maybe the professor hasn't got it worked out."

Pauli was right. Wheeler canceled the lecture. He lost none of his enthusiasm, however, and made plans for not one but a grand series of five papers. Feynman, meanwhile, had a doctoral thesis to prepare. He decided to approach the quantizing of his theory just as he had approached complicated problems at MIT, by working out cases that were stripped to their bare essentials. He tried calculating the interaction of a pair of harmonic oscillators, coupled, with a time delay—just a pair of idealized springs. One spring would shake, sending out a pure sine wave. The other would bounce back, and out of their interaction new wave forms would evolve. Feynman made some progress but could not understand the quantum version. He had gone too far in the direction of simplicity.

Conventional quantum mechanics went from present to future by the solving of differential equations—the so-called Hamiltonian method. Physicists spoke of "finding a Hamiltonian" for a system: if they could find one, then they could go ahead and calculate; if not, they were helpless. In Wheeler and Feynman's view of direct action at a distance, the Hamiltonian method had no place. That was because of the introduction of time delays. It was not enough merely to write down a complete description of the present: the positions, momentums, and other quantities. One never knew when some delayed effect would hurtle into the picture out of the past (or in the case of Wheeler and Feynman, out of the future). Because past and future interacted, the customary differential-equation point of view broke down. The alternative least-action or Lagrangian approach was no luxury. It was a necessity.

With all this on his mind, Feynman went to a beer party at the Nassau Tavern. He sat with a physicist lately arrived from Europe, Herbert Jehle, a former student of Schrödinger in Berlin, a Quaker, and a survivor of prison camps in both Germany and France. The American scientific world was absorbing such refugees rapidly now, and the turmoil of Europe seemed more palpable and near. Jehle asked Feynman what he was working on. Feynman explained and asked in turn whether Jehle knew of any application of the least-action principle in quantum mechanics.

Jehle certainly did. He pointed out that Feynman's own hero, Dirac, had published a paper on just that subject eight years before. The next day Jehle and Feynman looked at it together in the library. It was short. They found it, "The Lagrangian in Quantum Mechanics," in the bound volumes of *Physikalische Zeitschrift der Sowjetunion*, not the best-read of journals. Dirac had worked out the beginnings of a least-action approach in just the style Feynman was seeking, a way of treating the probability of a particle's entire path over time. Dirac considered only one detail, a piece of mathematics for carrying the wave function—the packet of quantum-mechanical knowledge—forward in time by an infinitesimal amount, a mere instant.

Infinitesimal time did not amount to much, but it was the starting point of the calculus. That limitation was not what troubled Feynman. As he looked over the few bound pages, he kept stopping at a single word: *analogue*. "A very simple quantum analogue," Dirac had written. " . . . They have their classical analogues. . . . It is now easy to see what the quantum analogue of all this must be." What kind of word was that, Feynman wondered, in a paper on physics? If two expressions were analogous, did it mean they were equal?

No, Jehle, said—surely Dirac had not meant that they were equal. Feynman found a blackboard and started working through the formulas. Jehle was right: they were not equal. So he tried adding a multiplication constant. Calculating more rapidly than Jehle could follow, he substituted terms, jumped from one equation to the next, and suddenly produced something extremely familiar: the Schrödinger equation. There was the link between Feynman's Lagrangian-style formulation and the standard wave function of quantum mechanics. A surprise—by *analogous* Dirac had simply meant *proportional*.

But now Jehle had produced a small notebook. He was rapidly copying from Feynman's blackboard work. He told Feynman that Dirac had meant no such thing. In his view Dirac's idea had been strictly metaphorical; the Englishman had not meant to suggest that the approach was useful. Jehle told Feynman he had made an important discovery. He was struck by the unabashed pragmatism in Feynman's handling of the mathematics, so different from Dirac's more detached, more aesthetic tone. "You Americans!" he said. "Always trying to find a use for something."

The Aura

This was Richard Feynman nearing the crest of his powers. At twenty-three he was a few years shy of the time when his vision would sweep hawklike across the breadth of physics, but there may now have been no physicist on earth who could match his exuberant command over the native materials of theoretical science. It was not just a facility at mathematics (though it had become clear to the senior physicists at Princeton that the mathematical machinery emerging in the Wheeler-Feynman collaboration was beyond Wheeler's own ability). Feynman seemed to possess a frightening ease with the substance behind the equations, like Einstein at the same age, like the

Soviet physicist Lev Landau—but few others. He was a sculptor who sleeps and dreams with the feeling of clay alive in his fingers. Graduate students and instructors found themselves wandering over to the afternoon tea at Fine Hall with Feynman on their minds. They anticipated his bantering with Tukey and the other mathematicians, his spinning of half-serious physical theories. Handed an idea, he always had a question that seemed to pierce toward the essence. Robert R. Wilson, an experimentalist who arrived at Princeton from the famous cauldron of Ernest Lawrence's Berkeley laboratory, talked casually with Feynman only a few times before making a mental note: Here is a great man.

The Feynman aura—as it had already become—was strictly local. Feynman had not yet finished his second year of graduate school. He remained ignorant of the basic literature and unwilling even to read through the papers of Dirac or Bohr. This was now deliberate. In preparing for his oral qualifying examination, a rite of passage for every graduate student, he chose not to study the outlines of known physics. Instead he went up to MIT, where he could be alone, and opened a fresh notebook. On the title page he wrote: NOTEBOOK OF THINGS I DON'T KNOW ABOUT. For the first but not the last time he reorganized his knowledge. He worked for weeks at disassembling each branch of physics, oiling the parts, and putting them back together, looking all the while for the raw edges and inconsistencies. He tried to find the essential kernels of each subject. When he was done he had a notebook of which he was especially proud. It was not much use in preparing for the examination, as it turned out. Feynman was asked which color was at the top of a rainbow; he almost got that wrong, reversing in his mind the curve of refraction index against wavelength. The mathematical physicist H. P. Robertson asked a clever question about relativity, involving the apparent path of the earth as viewed through a telescope from a distant star. Feynman did get that wrong, he realized later, but in the meantime he persuaded the professor that his answer was correct. Wheeler read a statement from a standard text on optics, that the light from a hundred atoms, randomly phased, would have fifty times the intensity of one atom, and asked for the derivation. Feynman saw that this was a trick. He replied that the textbook must be wrong, because by the same logic a pair of atoms would glow with the same intensity as one. All this was a formality. Princeton's senior physicists understood what they had in Feynman. In writing up course notes on nuclear physics, Feynman had been frustrated by a complicated formula of Wigner's for particles in the nucleus. He did not understand it. So he worked the problem out for himself, inventing a

diagram—a harbinger of things to come—that enabled him to keep a tally
of particle interactions, counting the neutrons and protons and arranging
them in a group-theoretical way according to pairs that were or were not
symmetrical. The diagram bore an odd resemblance to the diagrams he
invented for understanding the pathways of folded-paper flexagons. He did
not really understand why his scheme worked, but he was certain that it
did, and it proved to be a considerable simplification of Wigner's own
approach.

In high school he had not solved Euclidean geometry problems by
tracking proofs through a logical sequence, step by step. He had manipulated
the diagrams in his mind: he anchored some points and let others float,
imagined some lines as stiff rods and others as stretchable bands, and let
the shapes slide until he could see what the result must be. These mental
constructs flowed more freely than any real apparatus could. Now, having
assimilated a corpus of physical knowledge and mathematical technique,
Feynman worked the same way. The lines and vertices floating in the space
of his mind now stood for complex symbols and operators. They had a
recursive depth; he could focus on them and expand them into more
complex expressions, made up of more complex expressions still. He could
slide them and rearrange them, anchor fixed points and stretch the space
in which they were embedded. Some mental operations required shifts in
the frame of reference, reorientations in space and time. The perspective
would change from motionlessness to steady motion to acceleration. It was
said of Feynman that he had an extraordinary physical intuition, but that
alone did not account for his analytic power. He melded together a sense
of forces with his knowledge of the algebraic operations that represented
them. The calculus, the symbols, the operators had for him almost as
tangible a reality as the physical quantities on which they worked. Just as
some people see numerals in color in their mind's eye, Feynman associated
colors with the abstract variables of the formulas he understood so intimately. "As I'm talking," he once said, "I see vague pictures of Bessel
functions from Jahnke and Emde's book, with light tan j's, slightly violet-
bluish n's, and dark brown x's flying around. And I wonder what the hell
it must look like to the students."

In the past eight years neither Dirac nor any other physicist had been
able to follow up on the notion of a Lagrangian in quantum mechanics—
a way of expressing a particle's history in terms of the quantity of action.
Now Dirac's idea served as an explosive release in Feynman's imagination.
The uneasy elements of quantum mechanics broke loose and rearranged

themselves into a radically new formulation. Where Dirac had pointed the way to calculating how the wave function would evolve in an infinitesimal slice of time, Feynman needed to carry the wave function farther, through finite time. A considerable barrier separated the infinitesimal from the finite. Making use of Dirac's infinitesimal slice required a piling up of many steps—infinitely many of them. Each step required an integration, a summing of algebraic quantities. In Feynman's mind a sequence of multiplications and compounded integrals took form. He considered the coordinates that specify a particle's position. They churned through his compound integral. The quantity that emerged was, once again, a form of the action. To produce it, Feynman realized, he had to make a complex integral encompassing every possible coordinate through which a particle could move. The result was a kind of sum of probabilities—yet not quite probabilities, because quantum mechanics required a more abstract quantity called the probability amplitude. Feynman summed the contributions of every conceivable path from the starting position to the final position— though at first he saw more a haystack of coordinate positions than a set of distinct paths. Even so, he realized that he had burrowed back to first principles and found a new formulation of quantum mechanics. He could not see where it would lead. Already, however, his sense of paths in space-time seemed somehow cleaner—more direct. There seemed something quaint now about the peculiarly constrained oscillations of the post-ethereal field, the wavy inheritance of the 1920s.

The White Plague

Twentieth-century medicine was struggling for the scientific footing that physics began to achieve in the seventeenth century. Its practitioners wielded the authority granted to healers throughout human history; they spoke a specialized language and wore the mantle of professional schools and societies; but their knowledge was a pastiche of folk wisdom and quasi-scientific fads. Few medical researchers understood the rudiments of controlled statistical experimentation. Authorities argued for or against particular therapies roughly the way theologians argued for or against their theories, by employing a combination of personal experience, abstract reason, and aesthetic judgment. Mathematics played no role in a biologist's education. The human body was still largely a black box, its contents accessible only by means of the surgeon's knife or the crepuscular outlines of the early X rays. Researchers were stumbling toward the first rudimentary understanding of diet. The modern-sounding word *vitamin* had been coined and a few examples isolated in laboratories, but Feynman's father, Melville, having been diagnosed with chronic high blood pressure, was being slowly poisoned with an enriched, salty diet of eggs, milk, and cheese. Immunology and genetics were nothing but wells of ignorance. The prevailing theory of the mind was less a science than a collection of literary conceits blended with the therapeutic palliative of the confessional. Cancers, viruses, and diseases of the heart and brain resisted even the first glimmers of understanding. They would continue to mock medical science throughout the century.

Yet medicine was within reach of its first planetwide triumphs against bacterial epidemics, with the twin weapons of vaccination and antibiotic drugs. The year Feynman entered graduate school, Jonas Salk became a medical doctor; his assault on polio was just a few years away. Still, the habits of large clinical trials and statistical thinking had yet to become engrained in medical research. Alexander Fleming had noticed the antibacterial effect of the mold *Penicillium notatum* a decade before and then failed to take what a later era would consider the obvious next steps. He published his observation in a paper titled "A Medium for the Isolation of Pfeiffer's Bacillus." He tried rubbing his mold onto the open wounds of a few patients, with unclear results, but it never occurred to him to attempt a systematic study of its effects. A full decade passed, while biologists (and Fleming himself) dreamed futilely of a magic antibacterial agent that would save millions of lives, before finally two researchers happened upon his paper, extracted penicillin, and in 1940 crossed the line separating anecdote from science: they injected it into four sick mice, leaving another four untreated. In the context of 1930s medical science the lost decade was hardly noteworthy. Fleming's contemporaries did not deride him as a bungler. They hailed him as a hero and awarded him the Nobel Prize.

Tuberculosis—consumption, the wasting disease, scrofula, phthisis, the white plague—killed more people at its prime, in more parts of the globe, than any other disease. To novelists and poets it carried a romantic aura. It was a disease of pale aesthetes. It was a disease of rarefaction, of the body squandering itself. Its long, slow fevers gave the false impression of life intensified, the metabolism heightened, the processes of existence stimulated. Thomas Mann, allowing tuberculosis to inspire his most famous novel, associated the ruin and inflammation of the tubercles with sin, with

the Fall, with the creation of life itself from cool inorganic molecules—
"that pathologically luxuriant morbid growth, produced by the irritant of
some unknown infiltration . . . an intoxication, a heightening and unli-
censed accentuation of the physical state." He wrote those words in 1924,
when the *Magic Mountain* resort-style sanatoriums of Europe were already
dinosaurs of the past. To American public-health authorities faced with the
reality of the disease, even then tuberculosis was more simply a disease of
the poor.

Tuberculosis had infected Arline Greenbaum's lymphatic system, perhaps having been carried by unpasteurized milk. Swelling reappeared in
the lymph nodes on her neck and elsewhere, the lumps rubbery and pain-
less. She suffered fevers and fatigue. But an accurate diagnosis remained
beyond the abilities of her doctors. Arline did not strike them as the typical
tuberculosis victim; she was not poor enough or young enough. Nor was
lymphatic tuberculosis as common as tuberculosis that began in the lung
(it was twenty to thirty times rarer). When they abandoned the notion of
typhoid fever and considered the other standard possibilities, they focused
on cancerous outbreaks: lymphoma, lymphosarcoma, Hodgkin's disease.

Feynman was back in the library at Princeton, reading everything he
could find. One standard book listed the possibilities. First was local in-
fection. This was out of the question because the swellings were traveling
too far. Second was lymphatic tuberculosis. This was easy to diagnose, the
book said. Then came the cancers, and these, he read to his horror, were
almost invariably fatal. For a moment he mocked himself for jumping to
the most morbid possibility. Everyone who reads such catalogues must start
thinking about death, he thought. He went off to the Fine Hall tea, where
the conversation seemed unnaturally normal.

Those months in 1941 were a blur of visits to hospitals, symptoms
appearing and fading, consultations with more and more doctors. He hovered on the outside, hearing most news secondhand through Arline's parents. He and Arline promised each other that they would face whatever
came, bravely and honestly. Arline insisted, as she had when less was at
stake, that honesty was the bedrock of their love and that what she treasured
in Richard was his eagerness to confront the truth, his unwillingness to be
embarrassed or evasive. She said she did not want euphemisms or pretense
about her illness. Few patients did, but the weight of medical practice
opposed forthrightness in the face of terminal illness. Honest bad news was
considered antitherapeutic. Richard faced a dilemma, because the doctors

were finally settling on a grim diagnosis of Hodgkin's disease. There would be periods of remission, they said, but the course of the illness could not be reversed.

For Arline's benefit they proposed a camouflage diagnosis of "glandular fever." Richard refused to go along with it. He explained that he and Arline had a pact—no lies, not even white ones. How would he be able to face her with this biggest lie of all?

His parents, Arline's parents, and the doctors all urged him not to be so cruel as to tell a young woman she was dying. His sister, Joan, sobbing, told him he was stubborn and heartless. He broke down and bowed to tradition. In her room at Farmingdale Hospital, with her parents at her side, he confirmed that she had glandular fever. Meanwhile, he started carrying around a letter—a "goodbye love letter," as he called it—that he planned to give her when she discovered the truth. He was sure she would never forgive the unforgivable lie.

He did not have long to wait. Soon after Arline returned home from the hospital she crept to the top of the stairs and overheard her mother weeping with a neighbor down in the kitchen. When she confronted Richard—his letter snug in his pocket—he told her the truth, handed her the letter, and asked her to marry him.

Marriage was not so simple. It had not occurred to universities like Princeton to leave such matters to their students' discretion. The financial and emotional responsibilities were considered grave in the best of circumstances. He was supporting himself as a graduate student with fellowships—he was the Queen Junior Fellow and then the Charlotte Elizabeth Proctor Fellow, entitling him to earn two hundred dollars a year as a research assistant. When he told a university dean that his fiancée was dying and that he wanted to marry her, the dean refused to permit it and warned him that his fellowship would be revoked. There would be no compromise. He was dismayed at the response. He considered leaving graduate school for a while to find work. Before he made his decision, more news came from the hospital.

A test had found tuberculosis in Arline's lymph glands. She did not have Hodgkin's disease after all. Tuberculosis was not treatable—or rather it was treatable by any of dozens of equally ineffectual methods—but its onslaught was neither swift nor certain. Relief came over Richard in a flood. To his surprise the first note he heard in Arline's voice was disappointment. Now they would have no reason to marry immediately.

Preparing for War

As the spring of 1941 turned to summer, the prospect of war was everywhere. For scientists it seemed especially real. The fabric of their international community was already tearing. Refugees from Hitler's Europe had been establishing themselves in American universities for more than half a decade, often in roles of leadership. The latest refugees, like Herbert Jehle, had increasingly grim stories to tell, of concentration camps and terror. War work began to swallow up scientists long before the Japanese attack on Pearl Harbor. A Canadian colleague of Feynman's returned home to join the Royal Air Force. Others seemed to slip quietly away: the technologies of war were already drawing scientists into secret enterprises, as advisers, engineers, and members of technical subcommittees. It was going to be a physicists' war. When scientists were covertly informed about the Battle of Britain, the critical details included the detection of aircraft by reflected radio pulses—"radar" did not yet have a name. A few even heard about the breaking of codes by advanced mathematical techniques and electro-mechanical devices. Alert physicists knew from the published record that nuclear fission had been discovered at the Kaiser Wilhelm Institutes outside Berlin; that great energies could be released by a reaction that would proceed in a neutron-spawning chain; that any bomb, however, would require large quantities of a rare uranium isotope. How large? A number in the air at Princeton was 100 kilograms, more than the weight of a man. That seemed forbidding. Not so much as a grain of uranium 235 existed in pure form. The world's only experience in separating radioactive isotopes on a scale greater than the microscopic was in Norway—now a German colony—where a distilling plant tediously produced "heavy," deuterium-enriched, water. And uranium was not water.

Scientists picked up tidbits from casual conversation or found themselves fortuitously introduced into inner circles of secret activity. While Feynman remained mostly oblivious, his senior professor Eugene Wigner had for two years been a part of "the Hungarian conspiracy," with Leo Szilard and Edward Teller, conniving to alert Einstein and through him President Franklin D. Roosevelt to the possibility of a bomb. ("I never thought of that!" Einstein had told Wigner and Szilard.) Another Princeton instructor, Robert Wilson, had been drawn in by a sequence that began with a telegram from his old mentor at the Berkeley cyclotron, Ernest Lawrence. At MIT, under cover of a conventional scientific meeting, Wilson and several other physicists learned about the new Radiation Laboratory, already called the

Rad Lab, formed to turn the nascent British experience with radar into a technology that would guide ships, aim guns, hunt submarines, and altogether transform the nature of war. The idea was to beam radio waves in pulses so strong that targets would send back detectable echoes. Radar had begun at wavelengths of more than thirty feet, which meant fuzzy resolution and huge antennae. Clearly a practical radar would need wavelengths measured in inches, down toward the microwave region. The laboratory would have to invent a new electronics combining higher intensities, higher frequencies, and smaller hardware than anything in their experience. The British had invented a "magnetron" producing a microwave beam so concentrated that it could light cigarettes—enough to confound the Americans. ("It's simple—it's just a kind of whistle," I. I. Rabi told one of the first groups of physicists to gather uneasily around the British prototype. One of them snapped back, "Okay, Rabi, how does a whistle work?") These scientists acted long before the American public accepted the inevitability of the conflict. Wilson agreed to join the Rad Lab, though he had considered himself a pacifist at Berkeley. But when he tried to leave Princeton, Wigner and the department chairman, Smyth, decided it was time for another initiation. They told Wilson that Princeton would soon take on a project to create a nuclear reactor, and they told him why.

Fueling the prewar collaboration of scientists and weapons makers was a patriotic ethos that no subsequent war would command. It easily overcame Wilson's pacifism. Feynman himself visited an army recruitment office and offered to join the Signal Corps. When he was told he would have to start with unspecialized basic training—no promises—he backed down. That spring, in 1941, after three years of frustration, he finally got a job offer from Bell Laboratories in New York, and he wanted to accept. When his friend William Shockley showed him around, he was thrilled by the atmosphere of smart, practical science in action. From their windows the Bell researchers could see the George Washington Bridge going up across the Hudson River, and they had traced the curve of the first cable on the glass. As the bridge was hung from it, they were marking off the slight changes that transformed the curve from a catenary to a parabola. Feynman thought it was just the sort of clever thing he might have done. Still, when a recruiter from the Frankford Arsenal nearby in Philadelphia—an army general—visited Princeton seeking physicists, Feynman did not hesitate to turn down Bell Laboratories and sign up with the army for the summer. It was a chance to serve his country.

In one way or another, by the time the United States entered the war in December, one-fourth of the nation's seven-thousand-odd physicists had joined a diffuse but rapidly solidifying military-research establishment. A generation brought up with the understanding that science meant progress, the harnessing of knowledge and the empowerment of humanity, now found a broad national purpose. A partnership was already forming between the federal establishment and the leaders of scientific institutions. The government created in the summer of 1941 an Office of Scientific Research and Development, subsuming the National Defense Research Committee, charged with coordinating research in what MIT's president, Karl Compton, the epitome of the new partnership, called "the field of mechanisms, devices, instrumentalities and materials of warfare." Not just radar and explosives but calculating machines and battlefield medicines occupied the urgent war effort. An area like artillery was no longer a matter of haphazard trial-and-error lobbing of randomly designed shells. The nuclear physicist Hans Bethe had turned on his own initiative to a nascent theory of armor penetration; he also took on the issue of the supersonic shock waves that would shudder from the edge of a projectile. Less glamorously, Feynman spent his summer at the Frankford Arsenal working on a primitive sort of analog computer, a combination of gears and cams designed to aim artillery pieces. It all seemed mechanical and archaic—later he thought Bell Laboratories would have been a better choice after all.

Still, even in his college workshops, he had never confronted such an urgent blending of mathematics and metal. To aim a gun turret meant converting sines and tangents into steel gears. Suddenly trigonometry had engineering consequences: long before the tangent of a near-vertical turret diverged to infinity, the torque applied to the teeth of the gears would snap them off. Feynman found himself drawn to a mathematical approach he had never considered, the manipulation of functional roots. He divided a sine into five equal subfunctions, so that the function of the function of the function of the function of the function equaled the sine. And the gears could handle the load. Before the summer ended he was given a new problem as well: how to make a similar machine calculate a smooth curve— the path of an airplane, for example—from a sequence of positions coming in at regular intervals of a few seconds. Only later did he learn where this problem had arisen—from radar, the new technology from the MIT Radiation Laboratory.

After the summer he returned to Princeton, nothing remaining in his graduate education except the final task of writing his thesis. He worked slowly, trying out his least-action view of quantum mechanics on a variety of basic, illustrative problems. He considered the case of two particles or particle systems, A and B, which do not interact directly but through an intermediary system with wavelike behavior, a harmonic oscillator, O. A causes O to oscillate; O in turn acts on B. Complicated time delays enter the picture because, once O is set in motion, B will feel an influence that depends on A's behavior some time in the past—and vice versa. This case was a carefully reduced version of the familiar problem of two particles interacting through the mediation of the field. He asked himself in what circumstances the equations of motion could be derived from a principle of least action, strictly from the available information about the two particles A and B, completely disregarding O, the stand-in for the field. The least-action principle had come to seem like more than merely a useful shortcut. He now felt that it bore directly on the issues on which physics traditionally turned, such principles as the conservation of energy.

"This preoccupation with . . ." he wrote—then reconsidered.

"This desire for a principle of least action is besides the simplicity gained that, when the motions can be so represented, conservation of energy, momentum, etc. are guaranteed."

One morning Wilson came into his office and sat down. Something secret was going on, he said. He was not supposed to reveal the secret, but he needed Feynman and there was no other way. Furthermore, there were no rules about this secret. The military still did not take the physicists completely seriously. Physicists had decided on their own not to discuss certain matters, and now Wilson had decided to take it on himself to discuss one. It was time for Feynman's initiation.

There was a possibility of a nuclear bomb, Wilson said. British physicists had heard the message of Bohr and Wheeler about uranium 235 two years earlier and had arrived at a new estimate for the critical mass of material that would be needed. An expatriate German chemist on the British team, Franz Simon, had made the Atlantic crossing by "flying boat" with the latest news from their Birmingham laboratory. Perhaps a pound or two would be enough. Perhaps even less. The British were working hard on the problem of separating the uranium isotopes, winnowing the rare lighter isotope, uranium 235, from the far more common chaff, uranium 238. The two forms of uranium are chemically indistinguishable—a chemical reaction sees just one kind of atom. But the atoms of different isotopes have different masses, a fact that theorists could exploit in several plausible ways. Simon himself was investigating a scheme of slow gaseous diffusion through metal foil riddled with pinpoint holes; the uranium 238 molecules, ever so slightly heavier, would lag behind as the gas drifted through. Secret committees and directorates were forming around the uranium problem. The British had a code name: tube alloy, soon contracted to *tubealloy*. The Americans were building a nuclear reactor; other Princeton professors were involved. And Wilson said he had come up with an idea of his own. He had invented a device—so far existing only in his head—that he hoped would solve the separation problem much faster. Where Simon was thinking about holes in metal—one morning he had gone into his kitchen and attacked a wire strainer with a hammer—Wilson had in mind a combination of novel electronics and cyclotron technology.

He had persuaded Harry Smyth to let him assemble a team from among the instructors, graduate students, and engineers. A sort of countrywide "body shop" trading in the available technical talent was taking shape with the help of the National Defense Research Council; that would help him find some necessary staff. Graduate students were being pressed into service with the help of a simple expedient—Princeton called a halt to most degree work. Students were asked to choose from among three war-related projects: Wilson's; an effort to develop a new blast gauge for measuring explosive pressure; and a dully irrelevant-sounding investigation of the thermal properties of graphite. (Only later did it become clear that this meant the thermal-*neutron* properties of a material destined for nuclear reactors.) Wilson wanted to sign Feynman first. It occurred to him that Feynman's persistent skepticism, his unwillingness to accept any assertion on authority, would be useful. If there was any baloney or self-deception in the idea, he thought, Feynman would find it. He wanted Feynman in place when he presented the plan to the other graduate students.

To his dismay Feynman turned him down flat. He was too deep in his thesis; also, though he did not say so, the Frankford Arsenal had left him slightly disillusioned with war work. He said that he would keep the secret but that he wanted no part of it. Wilson asked him at least to come to the meeting.

Long afterward, after all the bomb makers had taken second looks back at their moments of decision, Feynman remembered the turmoil of that afternoon. He had not been able to go back to work. As he recalled it, he thought about the importance of the project; about Hitler; about saving the world. Elsewhere a few physicists already guessed, making delicate inferences from university rosters and published papers, that Germany was mounting no more than a cursory nuclear-weapons research project. Still, among the physicists who had disappeared from view was Werner Heisenberg. The threat seemed real enough. Later Feynman remembered the decisive physical act of opening his desk drawer and placing in it the loose sheets of his thesis.

The Manhattan Project

Chicago, Berkeley, Oak Ridge, Hanford: the first outposts of the Manhattan Project eventually became permanent capitals of a national nuclear establishment. To produce purified uranium and plutonium on a scale of mere pounds would require the rapid establishment of the largest single-purpose industrial enterprise ever. General Electric, Westinghouse, Du Pont, Allis-Chalmers, Chrysler, Union Carbide, and dozens of smaller companies combined in an effort that would see giant new factory towns rising from the earth. Yet in the first uncertain months after the attack on Pearl Harbor nothing in the modest scale of nuclear research even remotely foreshadowed the impending transformation of the nation's war-making capacity. Workshops were converted according to happenstance and convenience. At Princeton no more than a few thousand dollars was available for Wilson's project. To get help with the electronics he resorted to throwing a near tantrum in I. I. Rabi's office at the MIT Rad Lab. Including shop workers and technicians, his team grew to number about thirty. The experimental division amounted to one ungainly tube the length of an automobile, sprouting smaller tubes and electrical wiring. The theoretical division comprised, in its entirety, two cocky graduate students sitting side by side at roll-top desks in a small office.

They found they were able to bear the pressure of working on the nation's most fateful secret research project. The senior theoretician crumpled a piece of paper one day, passed it to his assistant, and ordered him to throw it in the wastebasket.

"Why don't *you?*" the assistant replied.

"My time is more valuable than yours," said Feynman. "I'm getting paid more than you." They measured the distances from scientist to wastebasket; multiplied by the wages; bantered about their relative value to nuclear science. The number-two man, Paul Olum, threw away the paper. Olum had considered himself the best undergraduate mathematician at Harvard. He arrived at Princeton in 1940 to be Wheeler's second research assistant. Wheeler introduced him to Feynman, and within a few weeks he was devastated. What's happening here? he thought. Is this the way physicists are, and I missed it? No physicist at Harvard was like this. Feynman, a cheerful, boyish presence spinning across the campus on his bicycle, scornful of the formalisms of modern advanced mathematics, was running mental circles around him. It wasn't that he was a brilliant calculator; Olum knew the tricks of that game. It was as if he were a man from Mars. Olum could not track his thinking. He had never known anyone so intuitively at ease with nature—and with nature's seemingly least accessible manifestations. He suspected that when Feynman wanted to know what an electron would do under given circumstances he merely asked himself, "If I were an electron, what would I do?"

Feynman found a vast difference between intuiting the behavior of electrons in rarefied theoretical contexts and predicting the behavior of a bulky jury-rigged assemblage of metal and glass tubing and electronics. He and Olum worked hastily. They could see from the start that Wilson's idea sat somewhere near the border between possible and hopeless—but on which side of the border? The calculations were awkward. Often they had to resort to guesswork and approximation, and it was hard to see which pieces of the work could accommodate guesses and which demanded rigorous exactitude. Feynman realized that he did not completely trust theoretical physics, now that its procedures were put to such an unforgiving test. Meanwhile the technicians moved forward; they could not afford to wait for the theorists' numbers. It was like a cartoon, Feynman thought; every time he looked around, the apparatus had sprouted another tube or a new set of dials.

Wilson called his machine an isotron (a near-meaningless name; his old mentor, Ernest Lawrence, was calling a competing device a calutron, *Cali*fornia + *tron*). Of all the separation schemes, Wilson's isotron owed the least to ordinary intuition about physical objects. It came the closest to treating atoms as denizens of a wavy electromagnetic world, rather than miniature balls to be pushed about or squeezed through holes. The isotron first vaporized and ionized chunks of uranium—heated them until they gave up an electron and thus became electrically charged. Then a magnetic field set them in motion. The stream of atoms passed through a hole that organized it into a tight beam. Then came the piece of wizardry that set the isotron apart from all the other separation schemes, the piece Feynman was struggling to evaluate.

A particularly jagged, sawtooth oscillation would be set up in the magnetic field. The voltage would swing sharply up and down, at radio wavelengths. Some of the uranium atoms would hit the field just as the energy fell to zero. Then some later atoms would enter the field as the energy rose, and they would accelerate enough to catch up with the first atoms. Then the energy would fall off again, so that the next atoms would travel more slowly. The goal was to make the beam break up into bunches, like traffic clumping on a highway. Wilson estimated that the bunches would be about a yard long. Most important, the uranium 235 and uranium 238 atoms, because of their differing masses, would accelerate differently in the magnetic field and would therefore bunch at different points. If the experimenters could get the timing right, Wilson thought, the bunches of each isotope should be distinct and separable. As they reached the end of the tube another precisely timed oscillating field, like a flag man at a detour, would deflect the bunches alternately left and right into waiting containers.

Complications appeared. As the ions' own momentum pushed them together, their tendency to repel one another came into play. Furthermore some atoms lost not one but two or more electrons when ionized, doubling or tripling their electric charge and sabotaging Feynman's calculations. When experimenters tried higher voltages than Feynman had initially calculated, they found that the bunches were springing back, the waves rebounding and forming secondary waves. It was with something like shock that Feynman realized that these secondary effects appeared in his equations, too—if only he could persuade himself to trust them. Nothing about the isotron project was simple. The physicists had to invent a way of feeding the machine with uranium powder instead of uranium wire, because the wire had a tendency to alloy with the electrodes, destroying them spectacularly. One of the experimenters found that, by setting a flame to the end of the uranium wire, he could create a shower of dazzling stars—an unusually expensive sparkler.

Meanwhile the project's worst enemy was proving to be its closest competitor, Lawrence, at Berkeley. He wanted to absorb the isotron into his own project, shutting down the Princeton group and taking on its staff and equipment for his calutron. The California-tron similarly used the new accelerator technology to create a beam of uranium ions but accelerated them instead around a three-foot racetrack. The heavier atoms swung farther out. The light atoms made the tight turn into a carefully positioned collector. Or so they would in theory. When General Leslie R. Groves, the new head of the Manhattan Project, first made the drive up the winding road from San Francisco Bay to Berkeley's Radiation Hill, he was appalled to find that the entire product of Lawrence's laboratory could barely be seen without the aid of a magnifying glass. Worse, the microgram samples were not even half pure. Even so, they outweighed the total output of the Princeton group. Feynman carried the isotron's flyspeck sample by the train to Columbia for analysis late in 1942; Princeton had no equipment capable of measuring the proportions of the isotopes in a tiny piece of uranium. Wearing his battered sheepskin coat, he had trouble finding anyone in the building who would take him seriously. He wandered around with his radioactive fragment until finally he saw a physicist he knew, Harold Urey, who took him in hand. Urey was a distinguished physicist who, as it happened, had delivered the first scientific lecture Feynman had ever heard, a public talk in Brooklyn on the subject of heavy water, sharing the bill with the wife of the Belgian balloonist Auguste Piccard. More recently Feynman had come to know Urey by attending meetings of the Manhattan Project's de facto steering committee. In that way he also met for the first time I. I. Rabi, Richard Tolman, and the physicist, so like Feynman and yet so unlike him, who would control his destiny for the next three years, J. Robert Oppenheimer.

Soon after Feynman's trip to Columbia bearing uranium, these men made their final decision on Princeton's adventure with the isotron. On the recommendation of Lawrence, nominally in charge of all electromagnetic separation research, they closed the Princeton project down. Operationally the calutron seemed a full year ahead, and money had to be committed as well to the more conventional diffusion approach, with pumps and pipes instead of magnets and fields, the atoms drifting in random trajectories, at ever-so-slightly different speeds, through many miles of metal barriers pricked with billions of microscopic holes. Wilson was stunned. He thought the committee was acting not just hastily but hysterically. To his senior colleagues it seemed that Wilson had lost to the personal strength and promotional skill of his former mentor Lawrence. Smyth and Wigner both felt privately that, given a fuller trial, the isotron might conceivably have shortened the war. "Lawrence's calutron simply used raw brute force to pry the beam a little way apart," a younger team member said. "Our method was *elegant.*" Blown up to the scale needed for mass production—thousands of giant machines—the isotron promised a yield many times greater. Feynman had produced detailed calculations for the design of a vast manufacturing plant, with isotrons working in a "cascade" of increasing purity. He took into account everything from wall-scrapings to uranium that would be lost in workers' clothing. He conceived arrays of several thousand machines—yet that proved a modest scale, in light of the later reality.

For Feynman one legacy of the Princeton effort was the friendship with Olum, a friendship, like many that followed, intellectually rich and emotionally unequal. Encounters with Feynman left marks on a series of young physicists and mathematicians, in the glare of a bright light, out-thought for the first time in their lives. They found different ways of adapting to this new circumstance. Some subordinated their own abilities to his and accepted his occasional bantering abuse in exchange for the surprising pleasure that came with his praise. Some found their self-image enough changed that they abandoned physics altogether. Olum himself eventually returned to mathematics, where he was more comfortable. He worked with Feynman throughout the war and then Feynman drifted away. They met only a few times in the next forty years. Olum thought of his old friend often, though. He was president of the University of Oregon when he heard of Feynman's death. He realized that the young genius he had met at Princeton had become a part of him, impossible to extricate. "My wife died three years ago, also of cancer," he said.

. . . I think about her a lot. I have to admit I have Dick's books and other things of Dick's. I have all of the Feynman lectures and other stuff. And there are things that have pictures of Dick on them. The article in *Science* about the *Challenger* episode. And also some of the recent books.

I get a terrible feeling every time I look at them. How could someone like Dick Feynman be dead? This great and wonderful mind. This extraordinary feeling for things and ability is in the ground and there's nothing there anymore.

It's an awful feeling. And I feel it—— A lot of people have died and I know about it. My parents are both dead and I had a younger brother who is dead. But I have this feeling about just two people. About my wife and about Dick.

I suppose, although this wasn't quite like childhood, it was graduate students together, and I do have more—— I don't know, romantic, or something, feelings about Dick, and I have trouble realizing that he's dead. He was such an extraordinarily special person in the universe.

Finishing Up

Absent from Princeton's nuclear effort was John Wheeler. He had already departed for Chicago, where Enrico Fermi and his team at the Metallurgical Laboratory—that enigmatic laboratory employing no metallurgists—were driving toward the first nuclear reactor. They intended to use less-than-bomb-grade uranium to produce slow fission. In the spring of 1942 Chicago was the place where it was easiest to gain a sense of what the future held. Wheeler knew how deeply his former student was mired in the isotope-separation work. In March he sent Feynman a message. It was time to finish his thesis, no matter how many questions remained open. Wigner—who was also more and more a part of the Chicago work—agreed that Feynman had accomplished enough for his degree.

Feynman heard the warning. He requested a short leave from the isotron project. Even now he did not feel quite ready to write, especially under such pressure. Later he remembered spending the first day of his leave lying on the grass, guiltily looking at the sky. Finally, writing with fountain pen in his fast adolescent scrawl, he filled sheaves of scratch paper—but paper was expensive, so he used the stationery of the *Lawrencian*, the Lawrence High School newspaper (Arline Greenbaum, editor in chief) or surplus order forms of G. B. Raymond & Company, sewer pipe, flue linings, etcetera, of Glendale, Long Island. He had now thoroughly assimilated Wheeler's revolutionary attitude, the stance that declared a break with the past. When the quantum mechanics of Max Planck was applied to the problem of light and the electromagnetic field, he wrote, "great difficulties have arisen which have not been surmounted satisfactorily." Other interactions, with more recently discovered particles, were creating similar difficulties, he pointed out: "Meson field theories have been set up in analogy to the electromagnetic field theory. But the analogy is unfortunately all too perfect; the infinite answers are all too prevalent and confusing." So he disposed of the field—at least the old idea of the field as a free medium for carrying waves. The field is a "derived concept," he wrote. "The field in actuality is entirely determined by the particles." The field is a mere "mathematical construction." Just as radically, he deprecated the wave function of Schrödinger, the now-orthodox means of describing the full state of a quantum-mechanical system at a given time. It was practically useless, after all, when the interaction of particles involved a time delay. "We can take the viewpoint, then, that the wave function is just a mathematical construction, useful under certain conditions"——no, "certain particular conditions . . . but not generally applicable."

He also took pains to leave his collaboration with Wheeler decisively behind. He wanted his thesis to be his own; he may already have sensed that the absorber theory in itself was leading toward a quirky dead end. It was his conception of the principle of least action that now consumed him. Wheeler-Feynman had been only a starting point, he wrote. It happened to provide most of the "illustrative examples" that would fill out the thesis. But he declared that his least-action method "is in fact independent of that theory, and is complete in itself."

When he was done, the first part of the thesis looked deceptively old-fashioned. It worked out some nearly textbook equations for the description of mechanical systems, such as springs, coupled together by means of another oscillator. Then this intermediate oscillator disappeared. A stroke of mathematical ingenuity eliminated it. A shorthand calculation appeared, very much like the classical Lagrangian. Soon the ground shifted, and the subject was quantum mechanics. The classical machinery of the first part turned into something quite modern. Where there had been two mechanical systems coupled by an oscillator, now there were two particles interacting through the medium of an oscillating field. The field, too, was now eliminated. A new quantum electrodynamics arose from a blank slate.

Feynman concluded with a blunt catalog of the flaws in his thesis. It was a theory untested by any connection to experiment. (He hoped to find an application to laboratory problems in the future.) The quantum mechanics remained nonrelativistic: a working version would have to take into account the distortions of Newtonian physics that occur near the speed of light. Above all he felt dissatisfied with the physical meaning of his equations. He felt they lacked a clear interpretation. Although few concepts in science seemed more frightening or abstruse than Schrödinger's wave function, in fact the wave function had achieved a kind of visualizability for physicists, if only as a sort of probabilistic smudge at the edge of consciousness. Feynman acknowledged that his scheme discarded even that fragment of a mental picture. Measurement was a problem: "In the mathematics we must describe the system for all times, and if a measurement is going to be made in the interval of interest, this fact must be put somehow into the equations from the start." Time was a problem: his approach required, as he said, "speaking of states of the system at times very far from the present." In the long run this would prove a virtue. For now it seemed to turn the method into a formalism with no ready physical interpretation. For Feynman, an unvisualizable formalism was anathema. The official thesis readers, Wheeler and Wigner, were unperturbed. In June Princeton awarded Feynman his doctoral degree. He attended the ceremony wearing the academic gown that had made him so uncomfortable three years before. He was proud in the presence of his parents. Fleetingly he was annoyed at sharing the platform with honorary-degree recipients; always pragmatic, he thought it was like giving an "honorary electrician's license" to people who had not done the work. He imagined being offered such an honor and told himself that he would turn it down.

Graduation removed one obstacle to marriage, but only one. According to medical and quasi-medical dogma, tuberculosis was a burden on love. "Should Consumptives Marry?" was the title of a chapter in Dr. Lawrence F. Flick's 1903 monograph, *Consumption a Curable and Preventable Disease*. Not without gravely weighing the "risks and burdens," he warned. And:

> The relationship between husband and wife is so intimate that even with great care there may be given opportunity in moments of forgetfulness for conveyance of the disease.

And:

> Many a young consumptive mother gets her shroud shortly after she has purchased the christening frock for her babe.

A 1937 *Manual of Tuberculosis for Nurses and Public Health Workers* declared that marriage should be forbidden:

> Marriage is apt to be a very expensive and dangerous luxury to those who are suffering, or have recently suffered, from tuberculosis of the lungs. . . .
> If the patient is a woman, she has not only to face the risk of infecting her husband and her children, but she must take into consideration the fact that pregnancy is liable to aggravate existing disease.

As late as 1952 an authoritative text cited Somerset Maugham's short story "Sanatorium," about a young couple in love who disregard the customary strictures.

They were both so young and brave that it was a great pity. . . . One could wish the novelist would rewrite the story with the boy and girl sensibly waiting for several years. . . . I am addicted to happy endings.

The textbook phrases gave no hint of the howling whirlpool of emotions that came when love and tuberculosis combined. Richard's parents dreaded his marriage to Arline. Lucille Feynman, especially, found the idea impossible to bear. Her dealings with her son became harsher as she realized how serious his intention was. In the late spring she sent him a cold, handwritten screed bristling with her fear for his health, her fear for his career, her worry about money, and, indirectly, her revulsion at the possibility of sexual relations. She held nothing in reserve.

"Your health is in danger, no I should say your life is in danger," she wrote. "It is only natural that when you are married you will see more of her." She worried about what other people would think (an enemy against which Richard and Arline were learning to circle the wagons). Tuberculosis carried a stigma, and the stigma would attach to Richard. "People dread T.B. When you have a wife in a T.B. sanatorium, no one knows it is not a real marriage. & I know the world considers such a man dangerous to associate with." She told Richard that he was not earning enough money, that he had been loyal enough already, and that Arline "should be satisfied with the status of 'engagement' instead of 'marriage,' because in such a marriage you are not getting *any* of the pleasures of marriage, but only the severe burden." She warned that she and Melville would not help the couple with money under any circumstances. She appealed to his patriotism, saying that the burden of a sick wife would compromise his ability to serve his country. She reminded him that his grandparents had fled European persecution and pogroms for a country whose freedom he took for granted. "Your marriage at this time, seems a selfish thing to do, just to please one person." She doubted that he sincerely wanted to marry Arline; she asked whether he was not merely trying to please her, "just as you used to occasionally eat spinach to please me." She said that she loved him and hated to see him make a noble but useless gesture. She said, "I was surprised to learn such a marriage is not unlawful. It ought to be."

Melville took a calmer tack. He asked Richard to get professional advice at Princeton, and Richard obeyed, consulting his department chairman, Smyth, and the university doctor. Smyth merely said he preferred to keep out of his staff's private affairs. He kept to that position even when Feynman went to the extreme of pointing out that he would be in contact both with a tubercular wife and with students. The doctor was concerned to make sure that Feynman understood the danger of pregnancy, and Feynman told him they did not intend to make love. (The doctor noted that tuberculosis was an infectious rather than a contagious disease, and Feynman, typically, pressed him on that point. He had a suspicion that the distinction was an artifact of unscientific medical jargon—that, if there was a difference at all, it was a difference of degree only.)

He told his father that he and Arline did not plan to marry any time within the next year. But just a few days later, having received his degree and his new status, he wrote back to his mother, proudly updating his letterhead by penning "PH.D." after the printed "RICHARD FEYNMAN." He tried to respond reasonably to each argument. Neither Smyth nor the university physician were concerned about any danger to his health, he said. If marriage to Arline would be a burden, it was a burden he coveted. He had realized one day, arranging Arline's transfer to the sanatorium nearby, that he was actually singing aloud with the sheer pleasure of planning their life together. As far as his duty to country was concerned, he would do whatever was necessary and go wherever he was sent. It was not that he wanted to be noble, he told his mother. Nor was it that he felt obliged to keep a promise he had made years before under different circumstances.

Marrying Arline was distinctly different from spinach. He did not like spinach. Anyway, he said, he had not eaten spinach out of love for his mother. "You misunderstood my motives as a small boy—I didn't want you angry at me."

He had made up his mind. He moved into a flat at 44 Washington Road immediately after graduation and for a while did not even tell his mother the address. He rapidly made the final arrangements—as Arline said, "in no time flat":

I guess maybe it is like rolling off of a log—my heart is filled again & I'm choked with emotions—and love is so good & powerful—it's worth preserving—I know nothing can separate us—we've stood the tests of time and our love is as glorious now as the day it was born—dearest riches have never made people great but love does it every day—we're not little people—we're giants . . . I know we both have a future ahead of us— with a world of happiness—now & forever.

With his parents frightened and unreconciled, he borrowed a station wagon from a Princeton friend, outfitted it with mattresses for the journey, and picked up Arline in Cedarhurst. She walked down her father's hand-poured concrete driveway wearing a white dress. They crossed New York Harbor on the Staten Island ferry—their honeymoon ship. They married in a city office on Staten Island, in the presence of neither family nor friends, their only witnesses two strangers called in from the next room. Fearful of contagion, Richard did not kiss her on the lips. After the ceremony he helped her slowly down the stairs, and onward they drove to Arline's new home, a charity hospital in Browns Mills, New Jersey.

LOS ALAMOS

÷

Feynman tinkered with radios again at the century's big event. Someone passed around dark welding glass for the eyes. Edward Teller put on sun lotion and gloves. The bomb makers were ordered to lie face down, their feet toward ground zero, twenty miles away, where their gadget sat atop a hundred-foot steel tower. The air was dense. On the way down from the hill three busloads of scientists had pulled over to wait while one man went into the bushes to be sick. A moist lightning storm had wracked the New Mexican desert. Feynman, the youngest of the group leaders, now grappled more and more urgently with a complicated ten-dial radio package mounted on an army weapons carrier. The radio was the only link to the observation plane, and it was not working.

He sweated. He turned the dials with nervous fingers. He knew what frequency he needed to find, but he asked again anyway. He had almost missed the bus after having flown back from New York when he received the urgent coded telegram, and he had not had time to learn what all those dials did. In frustration he tried rearranging the antenna. Still nothing—static and silence. Then, suddenly, music, the eerie, sweet sound of a Tchaikovsky waltz floating irrelevantly from the ether. It was a shortwave transmission on a nearby frequency, all the way from San Francisco. The signal gave Feynman a bench mark for his calibrations. He worked the dials again until he thought he had them right. He reset them to the airplane's wavelength one last time. Still nothing. He decided to trust his calibrations and walk away. Just then a raspy voice broke through the darkness. The radio had been working all along; the airplane had not been transmitting. Now Feynman's radio announced, "Minus thirty minutes."

Distant searchlights cut the sky, flashing back and forth between the clouds and the place Feynman knew the tower must be. He tried to see his flashlight through his welder's glass and decided, to hell with it, the glass was too dim. He looked at the people scattered about Campaña Hill, like a movie audience wearing 3-D glasses. A bunch of crazy optimists, he thought. What made them so sure there would be any light to filter? He went to the weapons carrier and sat in the front seat; he decided that the windshield would cut out enough of the dangerous ultraviolet. In the command center twenty-five miles away, Robert Oppenheimer, thin as a specter, wearing his tired hat, leaned against a wooden post and said aloud, "Lord, these affairs are hard on the heart," as though there had ever been such an affair.

At 5:29:45 A.M., July 16, 1945, just before dawn would have lighted the place called (already) the Jornada del Muerto, Journey of Death, instead came the flash of the atomic bomb. In the next instant Feynman realized that he was looking at a purple blotch on the floor of the weapons carrier. His scientific brain told his civilian brain to look up again. The earth was paper white, and everything on it seemed featureless and two-dimensional. The sky began to fade from silver to yellow to orange, the light bouncing off new-formed clouds in the lee of the shock wave. *Something creates clouds!* he thought. An experiment was in progress. He saw an unexpected glow from ionized air, the molecules stripped of electrons in the great heat. Around him witnesses were forming memories to last a lifetime. "And then, without a sound, the sun was shining; or so it looked," Otto Frisch recalled afterward. It was not the kind of light that could be assessed by human sense organs or scientific instruments. I. I. Rabi was not thinking in foot-candles when he wrote, "It blasted; it pounced; it bored its way into you. It was a vision which was seen with more than the eye." The light rose and fell across the bowl of desert in silence, no sound heard until the expanding shell of shocked air finally arrived one hundred seconds after the detonation.

Then came a crack like a rifle shot, startling a *New York Times* correspondent at Feynman's left. "What was that?" the correspondent cried, to the amusement of the physicists who heard him.

"That's the *thing*," Feynman yelled back. He looked like a boy, lanky and grinning, though he was now twenty-seven. A solid thunder echoed in the hills. It was felt as much as heard. The sound made it suddenly more real for Feynman; he registered the physics acoustically. Enrico Fermi, closer to the blast, barely heard it as he tore up a sheet of paper and calculated the explosive pressure by dropping the pieces, one by one, through the sudden wind.

The jubilation, the shouting, the dancing, the triumph of that day have been duly recorded. On the road back, another physicist thought Feynman was going to float through the roof of the bus. The bomb makers rejoiced and got drunk. They celebrated the thing, the device, the gadget. They were smart, can-do fellows. After two years in this brown desert they had converted some matter into energy. The theorists, especially, had now tested an abstract blackboard science against the ultimate. First an idea—now fire. It was alchemy at last, an alchemy that changed metals rarer than gold into elements more baneful than lead.

Accustomed to spending their days in a mostly mental world, the theorists had sweated over messy problems that they could touch and smell. Almost everyone was working in a new field, the theory of explosions, for example, or the theory of matter at extremely high temperatures. The practicality both sobered and thrilled them. The purest mathematicians had to soil their hands. Stanislaw Ulam lamented that until now he had always worked exclusively with symbols. Now he had been driven so low as to use actual numbers, and, even more humbling, they were numbers with decimal points. There was no choosing issues for their elegance or simplicity. These problems chose themselves—ticklish chemicals and exploding pipes. Feynman himself interrupted diffusion calculations to repair typewriters, interrupted typewriter repair to check the safety of accumulating masses of uranium, and invented new kinds of computing systems, part machine and part human, to solve equations that theoretically could not be solved at all. A pragmatic spirit had taken over the mesas of Los Alamos; no wonder the theorists were exhilarated.

Later they remembered having had doubts. Oppenheimer, urbane and self-torturing aficionado of Eastern mysticism, said that as the fireball stretched across three miles of sky (while Feynman was thinking, "Clouds!") he had thought of a passage from the *Bhagavad-Gita*, "Now I am become

Death, the destroyer of worlds." The test director, Kenneth Bainbridge, supposedly told him, "We are all sons of bitches now." Rabi, when the hot clouds dissipated, said he felt "a chill, which was not the morning cold; it was a chill that came to one when one thought, as for instance when I thought of my wooden house in Cambridge . . ." In the actuality of the event, relief and excitement drowned out most such thoughts. Feynman remembered only one man "moping"—his own recruiter to the Manhattan Project, Robert Wilson. Wilson surprised Feynman by saying, "It's a terrible thing that we made." For most the second thoughts did not come until later. On the scene the scientists, polyglot and unregulation though they seemed to the military staff, shared a patriotic intensity that faded from later accounts. Three weeks after the test, and three days after Hiroshima— on the day, as it happened, of Nagasaki—Feynman used a typewriter to set down his thoughts in a letter to his mother.

> We jumped up and down, we screamed, we ran around slapping each other on the backs, shaking hands, congratulating each other. . . . Everything was perfect but the aim—the next one would be aimed for Japan not New Mexico. . . . The fellows working for me all gathered in the hall with open mouths, while I told them. They were all proud as hell of what they had done. Maybe we can end the war soon.

The experiment code-named Trinity was the threshold event of an age. It permanently altered the psychology of our species. Its prelude was a proud mastery of science over nature—irreversible. Its sequel was violence and death on a horrible scale. In the minute that the new light spread across that sky, humans became fantastically powerful and fantastically vulnerable. A story told many times becomes a myth, and Trinity became the myth that illuminated the postwar world's anxiety about the human future and its reckless, short-term approach to life. The images of Trinity—the spindly hundred-foot tower waiting to be vaporized, the jackrabbits found shredded a half-mile from the blast, the desert sand fused to a bright jade-green glaze—came to presage the central horror of an age. We have hindsight. We know what followed: the blooding of the scientists, the loss of innocence—Hiroshima, Dr. Strangelove, throw weights, radwaste, Mutual Assured Destruction. The irony is built in. At first, though, ground zero stood for nothing but what it was, a mirrored surface, mildly radioactive, where earlier had stood a tower of steel. Richard Feynman, still not much more than a boy, wrote, "It is a wonderful sight from the air to see the green area with the crater at the center in the brown desert."

The Man Comes In with His Briefcase

Thirty months had passed since the closing of the isotron project at Princeton. Feynman and the rest of Wilson's team had been left in a tense limbo—not knowing. Wilson thought they were like professional soldiers awaiting their next orders. "We became then what I suppose is the worst of all possible things," he said later, "a research team without a problem, a group with lots of spirit and technique, but nothing to do." To pass the time he decided to invent some neutron-measuring equipment, sure to be needed before long. He meanwhile felt a dearth of hard information from Chicago, the project's temporary center, domain of Enrico Fermi and his atomic "pile" (the leather-jacketed physicist from Rome was using his freshly acquired Anglo-Saxon vocabulary to coin a blunt nuclear jargon). The pile—graphite bricks and uranium balls assembled into a lattice on a university squash court—was chain-reacting. Wilson sent Feynman as his emissary.

First came a briefing on the art of information gathering. He told Feynman to approach each department in turn and offer to lend expertise. "Have them describe to you in every detail the problem to such a point that you really could sit down and work on it without asking any more questions."

"That's not fair!" Feynman recalled saying.

"That's all right, that's what we're going to do, and that way you'll know everything."

Feynman took the train to Chicago early in 1943. It was his first trip west since the Century of Progress fair a decade before. He did gather information as efficiently as a spy. He got to know Teller and they talked often. He went from office to office learning about neutron cross sections and yields. He also left behind an impressed group of theorists. At one meeting he handed them a solution to an awkward class of integrals that had long stymied them. "We all came to meet this brash champion of analysis," recalled Philip Morrison. "He did not disappoint us; he explained on the spot how to gain a quick result that had evaded one of our clever calculators for a month." Feynman saw that the problem could be broken into two parts, such that part B could be looked up in a table of Bessel functions and part A could be derived using a clever trick, differentiation with respect to parameter on the integral side—something he had practiced as a teenager. Now the audience was new and the stakes were higher.

He was not the last prodigy to plant the kernel of a legend at the Metallurgical Laboratory. Five months after he passed through, Julian Schwinger arrived from Columbia, by way of Berkeley, where he had already collaborated with Oppenheimer, and the MIT Radiation Laboratory. Schwinger was Feynman's exact contemporary, and the contrast between these two New Yorkers was striking. Their paths had not yet crossed. Schwinger impressed the Chicago scientists with his pristine black Cadillac sedan and his meticulous attire. His tie never seemed to loosen through that hot summer. A colleague trying to take notes while he worked at the blackboard through the night found the process hectic. Schwinger, who was ambidextrous, seemed to have fashioned a two-handed blackboard technique that let him solve two equations at once.

Strange days for physicists reaching what should have been the intense prime of their creative careers. The war disrupted young scientists' lives with infinite gentleness compared with the disruption suffered by most draft-age men; still, Feynman could only wait uneasily for the course change war would entail. Almost as a lark he had accepted a long-distance job offer from the University of Wisconsin, as a visiting assistant professor on leave without pay. It gave him some feeling of security, though he hardly expected to become more than a professor on leave. Now, in Chicago, he decided at the last moment to take a side trip to Madison and spent a day walking about the campus almost incognito. In the end he introduced himself to a department secretary and met a few of his nominal colleagues before heading back.

He returned to Princeton with a little briefcase full of data. He briefed Wilson and the others: telling them how the bomb looked as of the winter of early 1943, how much uranium would be needed, how much energy would be produced. He was a twenty-four-year-old standing in shirtsleeves in a college classroom. Wisecracks and laughter echoed from the corridor. Feynman was not thinking about history, but Paul Olum was. "Someday when they make a moving picture of the dramatic moment at which the men of Princeton learn about the bomb, and the representative comes back from Chicago and presents the information, it will be a very serious situation, with everybody sitting in their suit coats and the man comes in with his briefcase," he told Feynman. "Real life is different than one imagines."

The army had made its unlikely choice of a civilian chief: a Jew, an aesthete, a mannered, acerbic, left-flirting, ultimately self-destructive scientist whose administrative experience had not extended beyond a California physics group. J. Robert Oppenheimer—Oppy, Oppie, Opje—held the respect of colleagues more for his quicksilver brilliance than for the depth of his work. He had no feeling for experimentation, and his style was unphysical; so, when he made mistakes, they were notoriously silly ones: "Oppenheimer's formula . . . is remarkably correct for him, apparently only the numerical factor is wrong," a theoretician once wrote acidly. In later physicist lingo a calculation's *Oppenheimer factors* were the missing π's, i's, and minus signs. His physics was, as the historian Richard Rhodes commented, "a physics of bank shots"—"It works the sides and the corners . . . but prefers not to drive relentlessly for the goal." No one understood the core problems of quantum electrodynamics and elementary particle physics better than he, but his personal work tended toward esoterica. As a result, though he became the single most influential behind-the-scenes voice in the awarding of Nobel Prizes in physics, he never received one himself. In science as in all things he had the kind of taste called exquisite. His suits were tailored with exaggerated shoulders and broad lapels. He cared about his martinis and black coffee and pipe tobacco. Presiding over a committee dinner at a steak house, he expected his companions to follow his lead in specifying rare meat; when one man tried to order well-done, Oppenheimer turned and said considerately, "Why don't you have fish?" His New York background was what Feynman's mother's family had striven toward and fallen back from; like Lucille Feynman he had grown up in comfortable circumstances in Manhattan and attended the Ethical Culture School. Then, where Feynman assimilated the new, pragmatic, American spirit in physics, Oppenheimer had gone abroad to Cambridge and Göttingen. He embraced the intellectual European style. He was not content to master only the modern languages. To physicists Oppenheimer's command of Sanskrit seemed a curiosity; to General Groves it was another sign of genius. And genius was what the general sought. Solid administrator that he was, he saw no value in a merely solid chief scientist. Much to the surprise of some, Groves's instincts proved correct. Oppenheimer's genius was in leadership after all. He bound Feynman to him in the winter of early 1943, as he bound so many junior colleagues, taking an intimate interest in their problems. He called long-distance from Chicago—Feynman had never had a long-distance telephone call from so far—to say that he had found a sanatorium for Arline in Albuquerque.

In the choice of a site for the atomic bomb project, the army's taste and Oppenheimer's coincided. Implausible though it may have seemed after-

ward, military planning favored desert isolation for security against enemy
attack as well as more reasonably for the quarantine of a talkative and
unpredictable scientific community. Oppenheimer had long before fallen
in love with New Mexico's unreal edges, the air clear as truth, the stunted
pines cleaving to canyon walls. He had made Western work shirts and belt
buckles part of his casual wear, and now he led Groves up the winding
trail to the high mesa where the Los Alamos Ranch School for boys looked
back across the wide desert to the Sangre de Cristo Mountains. Not everyone
shared their immediate sympathy with the landscape. Leo Szilard, the
Budapest native who first understood the energy-liberating chain reaction—
at other times so prescient about the bomb project—declared: "Nobody
could think straight in a place like that. Everybody who goes there will go
crazy."

The impatient Princeton group signed up en masse. Wilson rushed out
to see the site and rushed back to report on the mud and confusion, a
theater being built instead of a laboratory, water lines being mislaid. The
state of secrecy was such that Feynman already knew that Groves and
Oppenheimer were arguing over the state of secrecy. Cyclotron parts and
neutron-counting gear started heading out by rail in wooden crates from
the Princeton station. Princeton's carloads provided the new laboratory's
core equipment, followed eventually by a painstakingly dismantled cyclotron from Harvard and other generators and accelerators. Soon Los Alamos
was the best-equipped physics center in the world. The Princeton team
began leaving soon after the crates of gear. Richard and Arline went with
the first wave, on Sunday, March 28. Instructions were to buy tickets for
any destination but New Mexico. Feynman's contrariety warred for a moment with his common sense, and contrariety won out. He decided that,
if no one else was buying a New Mexico ticket, he would. The ticket seller
said, Aha—all these crates are for you?

The railroad provided a wheelchair and a private room for Arline. She
had begged Richard tearfully to pay the extra price for the room and hinted
that at last she might have a chance to be all that a wife should be to the
husband she loves. For both of them the move out West portended an
open-skied, open-ended future. It cut them off finally from their protective
institutions and their childhoods. Arline cried night after night from worry
and filled Richard with her dreams: curtains in their home, teas with his
students, chess before the fireplace, the Sunday comics in bed, camping
out in a tent, raising a son named Donald.

Chain Reactions

Fermi's pile of uranium and graphite, sawed and assembled by professional cabinetmakers in a University of Chicago racquets court, became the world's first critical mass of radioactive material on December 2, 1942. Amid the black graphite bricks, the world's first artificial chain reaction sustained itself for half an hour. It was a slow reaction, where a bomb would have to be a fast reaction—less than a millionth of a second. From the two-story-high ellipsoid of Chicago pile number one to the baseball-size sphere of plutonium that exploded at Trinity, there could be no smooth evolutionary path. To go from the big, slow pile to a small, fast bomb would require a leap. There were few plausible intermediate stages.

Yet one possibility was playing itself out in Feynman's mind the next April, as he sat in a car just outside the makeshift security gate on the Los Alamos mesa. Hydrogen atoms slowed neutrons, as Fermi had discovered ages ago. Water was cheaply bound hydrogen. Uranium dissolved in water could make a powerful compact reactor. Feynman waited while the military guards tried to straighten out a mistake about his pass. Left and right from the security gate stretched the beginnings of a barbed-wire fence. Behind it lay no laboratory, but a few ranch buildings and a handful of partially complete structures rose from the late-winter mud in what the army called modified mobilization style, namely fast-setting concrete foundations, wood frame, plain siding, asphalt roofs. The thirty-five-mile ride from Santa Fe had ended in a harrowing dirt road cut bluntly into the mesa walls. Feynman was not the only physicist who had never been farther west than Chicago. The recruiters had warned scientists that the army wanted isolation, but no one quite realized what isolation would mean. At first the only telephone link was a single line laid down by the Forest Service. To make a call one had to turn a crank on the side of the box.

As he sat waiting for the military police to approve his pass, Feynman was running through some calculations for the hypothetical in-between reactor that would be called a water boiler. Instead of blocks of uranium interspersed with graphite, this unit would use a uranium solution in water, uranium enriched with a high concentration of the 235 isotope. The hydrogen in the water would increase the effectiveness many times over. He was trying to figure out how much uranium would be needed. He worked on the water-boiler problem, picking it up and putting it down again over the next weeks, thinking about the detailed geometry of neutrons colliding in hydrogen. Then he tried something quirky. Perhaps the ideal arrangement of uranium, the one that would require the least material, would be different from the obvious uniform arrangement. He converted the equations into a form that would allow a shortcut solution in terms of a minimum principle, now his favorite technique. He worked out a theorem for the spatial distribution of fissionable material—and discovered that the difference would not matter in a reactor as small as this. When enriched uranium finally began to arrive, the water boiler took form as a one-foot sphere inside a three-foot cube of black beryllium oxide, sitting on a table behind a heavy concrete wall at the pine-shaded bottom of Omega Canyon, miles away from the main site. It served as the project's first large-scale experimental source of neutrons and the first real explosion hazard. For all the theorists, the elements of this first problem became leitmotivs of their time working on the bomb: the paths of neutrons, the mixing of esoteric metals, the radiation, the heat, the probabilities.

In the muddy weeks of April the population of scientists reached about thirty. They came and went through a temporary office in Santa Fe and disappeared from there into a void in the landscape. If they had seen their destination from the air, they would have understood that they were to be situated in a compound atop a flat finger of ancient lava, one of many radiating from the giant crater of a long-quiet volcano. Instead, their imagining of the place began with mysterious addresses: P. O. Box 1663 for mail, Special List B for driver's licenses. Not all the procedures devised in the name of security helped allay the suspicions of the local population. Any local policeman who pulled over Richard Feynman on the road north of Santa Fe would see the driver's license of a nameless *Engineer* identified only as *Number 185*, residing at *Special List B*, whose signature was, for some reason, *Not required*. The name Los Alamos meant hardly anything. A canyon? A boys' school? When scientists reached the site they would see, as likely as not, a former professor standing outdoors and peppering a military construction crew with unwanted instructions. If Oppenheimer happened to be there to greet them, he would say from beneath the already famous hat, "Welcome to Los Alamos and who the devil are you?" The first familiar face that Feynman saw belonged to his Princeton friend Olum—Olum was standing in the road with a clipboard, checking off each truckload of lumber as it arrived. At first Feynman slept in one of a row of beds lined up on the balcony of a school building. Food was still coming up from Santa Fe in the form of box lunches.

Amid the turmoil of construction, the concrete hardening in the open air, the noise of hand-held buzz saws everywhere, only the theorists had the equipment they needed to start work immediately—one blackboard on rollers. Their true ground-breaking ceremony came on April 15. Oppenheimer gathered them together, along with the first few experimentalists and chemists, to learn officially what they had been told in hushed tones. They were to build a bomb, a weapon, a working device that would concentrate the neutron-spraying phenomenon of radioactivity into a speck of space and time concentrated enough to force an explosion. As the lecture began, Feynman opened a notebook and wrote the cautionary words, "Talks are not necessarily on things we *should* discuss but things we have worked out." Much was known to the teams from Berkeley and Chicago, or so it seemed. The splitting of an ordinary uranium atom required a blow from a fast, high-energy neutron. Every atom was its own tiny bomb: it split with a jolt of energy and released more neutrons to trigger its neighbors. The neutrons tended to slow, however, dropping below the necessary threshold for further fission. The chain reaction would not sustain itself. However, the rarer isotope, uranium 235, would fission when struck by a slow neutron. If a mass of uranium were enriched with these more volatile atoms, neutrons would find more targets and chain reactions would live longer. Pure uranium 235—though it would not be available in any but microscopic quantities for months—would make an explosive reaction possible. Another way to encourage a chain reaction was to surround the radioactive mass with a shell of metal, a tamper, that would reflect neutrons back toward the center, intensifying their effects as the glass of a greenhouse intensifies its infrared warming. A lanky Oppenheimer aide, Robert Serber, described the different tamper possibilities to his audience of thirty-odd men radiating an almost palpable energy of nerves. Feynman wrote quickly. " . . . reflect neutrons . . . keep bomb in . . . *critical mass* . . . Non absorbing equiscattering factor 3 in mass . . . a *good* explosion . . ." He sketched some hasty diagrams. From nuclear physics the discussion was forced to turn to the older but messier subject of hydrodynamics. While the neutrons were doing their work, the bomb would heat and expand. In a crucial millisecond would come shock waves, pressure gradients, edge effects. These would be hard to calculate, and for a long time the theorists would be calculating blind.

Making a bomb was not like making a theory of quantum electrodynamics, where the ground had already been mined by the greatest scientists. Here the problems were fresh, close to the surface, and therefore—this surprised Feynman at first—easy. Beginning with the issues raised by the first indoctrination lectures, he produced a string of small triumphs, gratifying by contrast with the long periods of wandering in the dark of pure theory. There were compensating difficulties, however.

"Most of what was to be done was to be done for the first time," an anonymous ghostwriter of the bomb's official history wrote afterward. (The ghostwriter was Feynman, called to this unaccustomed service by his former department head, Harry Smyth.) Struggling to sum up the problems of theoretical science at Los Alamos, he added "untried," and then "with materials which were for a long time practically unavailable." *Materials*— he could not bring himself to write *uranium* or *plutonium* after the euphemistic years of *tubealloy* and *49*. The wait for tubealloy had been agonizing, for the theorists no less than the experimenters. More mundane materials could be requisitioned—at the laboratory's request Fort Knox delivered two hemispheres of pure gold, each the size of half a basketball. Feynman, giving Smyth a tour one day, pointed out that he was absently kicking one of them, now in use as a doorstop. A request for osmium, a dense nonradioactive metal, had to be denied when it became clear that the metallurgists had asked for more than the world's total supply. In the cases of uranium 235 and plutonium, the laboratory had to wait for the world's supply to be multiplied a millionfold.

For now the only knowledge of these materials came from experiments on quantities so tiny as to be invisible. The experiments were expensive and painstaking. Even getting an early measurement of plutonium's density challenged the team at Chicago. The first dot of plutonium did not arrive at Los Alamos until October 1943. Trials with more comfortable quantities would have to wait; in the event, just one full-size experiment would be possible. Most questions would have to be answered with pencil and paper. It soon became clear that theory at Los Alamos would be performed on a high wire without a net. The theoretical division was small, just thirty-five physicists and a computing staff, charged with providing analysis and prediction for all the much larger practical divisions: experimental, ordnance, weapons, and chemical and metallurgical. Analysis and prediction—what would happen if . . . ? Theorists at Los Alamos had dispensed with the luxury of contemplating simple mysteries—the way a single atom of hydrogen emits a single packet of light in such and such a color, or the way an idealized wave might travel through an idealized gas. The materials at hand were not idealized, and the theorists, no less than the experimenters, had to poke about in the rubble-strewn territory of nonlinear mathematics. Crucial decisions had to be made before the experimenters could conduct trials. Feynman, in his anonymous account, listed the main questions:

How big must the bombs be (the imploding sphere of plutonium or the gun device in the case of uranium)? What would be the critical mass and the critical radius for each material, the dimensions beyond which a chain reaction would sustain itself?

What materials would best serve as tamper, a surrounding liner that would reflect neutrons back into the bomb? The metallurgists had to begin the work of fabricating tamper long before a true test was possible.

How pure would the uranium have to be? On this calculation rested a decision to build or not build an enormous third stage in the isotope-separation complex at Oak Ridge.

How much heat, how much light, how much shock would a nuclear explosion create in the atmosphere?

The Battleship and the Mosquito Boat

They occupied a two-story green-painted box called T building (T for theoretical), which Oppenheimer made his headquarters and the laboratory's spiritual center. He placed Hans Bethe, Cornell's famous nuclear physicist, in charge. The corridors were narrow, the walls thin. As the scientists worked, they would hear from time to time Bethe's booming laughter. When they heard that laugh they suspected that Feynman was nearby.

Bethe and Feynman—strange pair, some of their colleagues thought, a pedantic-seeming German professor and a budding quicksilver genius. Someone coined the nicknames "Battleship" and "Mosquito Boat." Their collaborative method was for Bethe to plow solidly ahead, a determined giant, while Feynman buzzed back and forth across his bow, gesticulating, yelling in his scabrous New York accent, "You're crazy" and "That's nuts." Bethe would respond calmly in his slow professorial way, working his way through the problem analytically and explaining that he was not crazy, Feynman was crazy. Feynman would consider and pace back and forth, and finally through the partitions the other scientists would hear him shout back, "No, no, you're *wrong*." He was reckless where Bethe was careful, and he was just what Bethe was looking for, someone who would perform the severest and most imaginative criticism, who would find flaws before an idea went too far. Challenges and fresh insights came easily from Feynman. He did not wait, as Bethe did, to double-check every intuitive leap. His first idea did not always work. His cannier colleagues developed a rule of thumb: If Feynman says it three times, it's right.

Bethe was a natural choice as leader of the theoretical division. His sweeping three-article review of the state of nuclear physics in the thirties had established him as the authoritative theorist in that field. As Oppenheimer well knew, Bethe had not just organized the existing knowledge of the subject but had calculated or recalculated every line of theory himself. He had worked on probability theory, on the theory of shock waves, on the penetration of armor by artillery shells (this last paper, born of his eagerness in 1940 to make some contribution to the looming war, was immediately classified by the army so that Bethe himself, not yet an American citizen, could not see it again). His explanation in 1938 of the thermonuclear fires that light the sun would win him the Nobel Prize. Since arriving at Cornell in 1935 he had made it one of the new world centers in physics, as Oppenheimer and Ernest O. Lawrence had done for Berkeley.

Oppenheimer wanted him badly and strained to persuade him that the atomic bomb was practical enough to draw him from the MIT Radiation Laboratory, where he had begun to make a contribution in 1942. (When Bethe agreed, the news was sent to Oppenheimer by a prearranged code: a Western Union kiddiegram.) Bethe's friend Edward Teller had pressed hard for his participation. No one but Teller was now surprised when Oppenheimer appointed Bethe, the sturdy pragmatist, to head the theoretical division, to nurse the egos and the prodigies, to run the most eccentric, temperamental, insecure, volatile assortment of thinkers and calculators ever squeezed together in one place.

Bethe had learned his physics all across Europe: first at Munich, where he studied with Arnold Sommerfeld, a prodigious producer of future Nobel Prize winners, and then at Cambridge and Rome. At Cambridge, Dirac's lectures on the new quantum mechanics held center stage, but Bethe quit attending after discovering that Dirac, having perfected his formulation of the subject, was simply reading his book aloud. At Rome, where he was the first foreign student of physics in the university's history, the attraction was Fermi. For a short time they worked together closely, and Bethe acquired from him a style that he called "lightness of approach." His first great teacher, Sommerfeld, had always begun work on a problem by writing down a formalism selected from a heavy arsenal of mathematical equipment. He would work out the equations and only then translate the results into an understanding of the physics. By contrast, Fermi would begin by gently turning a problem over in his mind, by thinking about the forces at work, and only later sketching out the necessary equations. "Lightness" was a difficult attitude to sustain in a time of abstract, unvisualizable quantum mechanics. Bethe combined the physicality of Fermi's attitude with an almost compulsive interest in computing the actual numbers that an equation entailed. That was far from typical. Most physicists could happily string equations down a page, working out the algebra without keeping in mind a sense of real quantities, or ranges of quantities, that a symbol might represent. For Bethe a theory only mattered when he could get actual numbers out.

From Fermi's Rome, Bethe returned to a Germany whose scientific establishment was nearing the precipice. In his classroom at the ancient university of Tübingen, where he took an assistant professorship, he saw students wearing swastikas on arm bands. It was the autumn of 1932. That winter Hitler took power. In February the Reichstag burned. By spring the first of the Nazis' anti-Jewish ordinances entailed the immediate dismissal of one-fourth of the country's university physicists—non-Aryan civil servants. Bethe, his father a Prussian Protestant, did not consider himself a Jew, but because his mother was Jewish his status in Nazi Germany was clear. He was immediately shed from the faculty he had just entered. Across Europe the greatest intellectual migration in history was already beginning, and Bethe had little choice but to join it. Scientists in general had the advantage of working in a polyglot community, where international study and temporary overseas lectureships eased their emotional transition—from citizen to refugee. He reached the New World in 1935.

Feynman had known Bethe's name since he was an undergraduate— the Bethe Bible, the three famous review articles on nuclear physics, had provided the entire content of MIT's course. He had seen Bethe once from a distance at a scientific meeting. An ugly man, he had thought at first glance, awkward, with slightly squashed features on a strong frame, light brown hair bristling skyward above a broad brow. Feynman's first impression dissolved when they met up close in Santa Fe before heading up to Los Alamos for the first time. Bethe, thirty-seven years old, had the body of a mountain climber, and he spent as much time as possible hiking in the canyons or up to the peaks behind the laboratory. He radiated solidity and warmth. Soon after their arrival on the mesa, a statistical fluctuation in the comings and goings of the theorists left Bethe stripped of the people he needed to consult. Victor Weisskopf, his deputy, was away. Teller was away—but Teller, anyway, had immediately grown more aloof than useful;

not only had Oppenheimer passed him over in favor of Bethe, but Bethe had passed him over in favor of Weisskopf. So Bethe drifted into Feynman's office one day, and soon people down the corridor could hear his booming laugh.

Bethe left the initial lectures trying to work out a way of calculating the efficiency of a nuclear explosion. Serber had presented a formula for the simplest case, when the mass of uranium or plutonium was just above critical. For bombs, which would require masses substantially over critical, the problem was far more difficult. He and Feynman developed a method of classic elegance that became known as the Bethe-Feynman formula. The dangerous practicalities of nuclear physics brought other questions. A lump of uranium or plutonium, even smaller than critical mass, raised the possibility of a runaway chain reaction—predetonation. Chemical explosives were far more stable. Bethe assigned this problem to Feynman in the project's first months. Stray neutrons were always a presence, at some low level of probability—from cosmic rays, from spontaneous individual fissions, and from nuclear reactions caused by impurities. Cosmic rays alone sparked enough fission to make uranium 235 noticeably hotter in the high altitudes of Los Alamos than in sea-level laboratories. Without understanding predetonation, the scientists could not understand detonation itself, because they would not know how the bomb would behave during the split-second transition from subcritical to supercritical. Feynman spent a long time thinking about the properties of a chunk of matter in the peculiar condition of near-criticality, a form of matter that science had not had occasion to ponder before. He recognized that the essence of the problem was not its average behavior but its fluctuations: bursts of neutron activity here and there, spreading in chains before dying out.

Mathematics, in the form of probability theory, had barely begun to provide tools for handling such complex patterns; he discussed the problem with the Polish mathematician Stanislaw Ulam, and Ulam's approach to it helped midwife a new field of probability called branching-processes theory. Feynman himself worked out a theory of fluctuations building upward from the easier-to-calculate probabilities of short chain reactions: a neutron splits one atom; a newly liberated neutron finds another target; but then the chain breaks. Some measurable fluctuations—audible bursts of noise on a Geiger counter—could be traced back to an origin in a single fission event. Others were combinations of chains. As with so many other problems, Feynman took a geometrical approach, considering the probability that a burst in a certain unit volume would lead to a burst in another unit volume at a given time later. He arrived at a practical method that reliably computed the chances of any premature reaction taking hold. It was suitable even for the odd-shaped segments of uranium that would be blasted into one another in the Hiroshima bomb.

Bethe found in Feynman the perfect foil and goad. This young man was quick, fearless, and ambitious. He was not satisfied to take away one problem and work on it; he wanted to work on everything at once. Bethe decided to make him a group leader, a position otherwise reserved for prominent physicists like Teller, Weisskopf, Serber, and the head of the British contingent at Los Alamos, Rudolf Peierls. For his part Feynman, who had lived through twenty-five years and a full formal education without ever falling under the spell of a mentor, began to love Hans Bethe.

Diffusion

Feynman did some recruiting for the project. He had invited one of his MIT fraternity friends to join the secret work. He even tried to recruit his father. Melville's health had turned poor—his chronic high blood pressure affected him more and more—and Lucille wished he could afford to travel less. Richard wrote his mother that there might be a job available as a purchasing clerk. He wished, too, that Melville could see at close range the heady intellectual world toward which he had so long aimed his son. "He would be partly out of the rush, etc. of the business work, & would be among academic men to a great extent, which I'm sure he would enjoy . . . Purchasing these days is quite difficult, & everyone here is in a hell of a hurry for their stuff . . . it will be a damn important position in our project and scientific venture."

Nothing came of that suggestion. In the spring of 1944 Feynman came across a familiar name on a list of available physicists: T. A. Welton. He filled out a requisition. His college friend, working as an instructor at the University of Illinois, had been trying to remain a civilian by teaching military-related courses and had watched unhappily as the more distinguished members of his department disappeared to mysterious locales. Feynman's requisition rescued him. Welton, like so many physicists by then, had pieced together more than the army security officials liked to think possible. When he was invited to meet a stranger in a hotel room in Chicago, and then invited by the stranger to drop everything and move to

New Mexico, he understood that this was, as he said later, the classic impossible-to-refuse offer. The day he arrived, Feynman took him on a long hike down into a gorge that had lately been named Omega Canyon. He was able to startle Feynman with an affirmative answer to his first question, "Do you know what we're doing here?"

"Yes," Welton said. "You're making an atomic bomb."

Feynman recovered quickly. "Well," he asked, "did you know we're going to make it with a new element?" His friend admitted that the news of plutonium had not drifted as far as Illinois. While they walked—Welton's lungs desperately drawing in the underpressurized air of 7,000 feet above sea level—Feynman intoxicated him with a briefing. They talked about the bomb. There were now two designs. A uranium bomb would take the form of a gun, creating a critical mass by firing a uranium bullet at a uranium target. A plutonium bomb would use another audacious method. A hollow sphere would be blown inward on itself by the shock from explosives packed all around it. The hot plutonium atoms would be compressed not through one dimension, as in the gun, but through three dimensions. The implosion method, as it was accurately named, was starting to look better and better—in part because so many problems had plagued the alternatives. (Feynman did not mention his own initial reaction when implosion's inventor, Seth Neddermeyer, first reported experiments on explosives wrapped around steel pipes. He had raised his hand in the back row and announced, "It stinks.")

As Welton listened, trying to keep up along the narrow canyon walls, he understood that Feynman was also saying that he had worked hard to establish himself as a smart kid to be reckoned with—that a young researcher had to impress the senior people with his usefulness, that he, Feynman, had been through that process, and that he had succeeded. They talked only briefly about Arline. She was not well, spending most of her days in a wooden bed in the Presbyterian Sanatorium, a small, poorly staffed facility by the side of a highway in Albuquerque. Feynman, visiting her almost every weekend, hitchhiked or borrowed a car to head down the unpaved road toward Santa Fe on Friday afternoon or Saturday. Away from the laboratory he would turn his thoughts back to the pure theory of quantum mechanics. He used the long trip, and the hours when Arline slept, to push his thesis work further. Welton remembered how obstinately his friend had resisted the Lagrangian simplification of dynamical problems when they were a pair of precocious sophomores in MIT's theory course. He was amused and impressed to hear how far Feynman had taken the Lagrangian method in reformulating the most basic quantum mechanics. Feynman sketched out his idea of expressing quantum behavior as a sum of all the possible space-time trajectories a particle could take, and he told Welton frankly that he did not know how to apply it. He had a wonderful recipe that had not gelled.

Welton became the fourth physicist in the group Feynman headed, now formally known as T-4, Diffusion Problems. As a group leader Feynman was ebullient and original. He drove his team hard in pursuit of his latest unorthodox idea for solving whatever problem was at hand. Sometimes one of the scientists would object that a Feynman proposal was too complex or too bizarre. Feynman would insist that they try it out, computing in groups with their mechanical calculators, and he had enough unexpected successes this way to win their loyalty to the cause of wide-ranging experimentation. They all tried to innovate in his fashion—no idea too wild to be considered. He could be ruthless with work that did not meet his high standards. Even Welton experienced the humiliation of a Feynman rebuke—"definitely ungentle humor" to which "only a fool would have subjected himself twice." Still Feynman managed to build esprit. He had taught himself to flip a pencil in one motion from a table into his hand, and he taught the same trick to his group. One day, amid a typical swirl of rumors that military uniforms were going to be issued to scientists working in the technical area, Bethe walked in to talk about a calculation. Feynman said he thought they should integrate it by hand, and Bethe agreed. Feynman swiveled around and barked, "All right, pencils, calculate!"

A roomful of pencils flipped into the air in unison. "Present pencils!" Feynman shouted. "Integrate!" And Bethe laughed.

Diffusion, that faintly obscure and faintly pedestrian holdover from freshman physics, lay near the heart of the problems facing all the groups. Open a perfume bottle in a still room. How long before the scent reaches a set of nostrils six feet away, eight feet away, ten feet away? Does the temperature of the air matter? The density? The mass of the scent-bearing molecules? The shape of the room? The ordinary theory of molecular diffusion gave a means of answering most of these questions in the form of a standard differential equation (but not the last question—the geometry of the containing walls caused mathematical complications). The progress of a molecule depended on a herky-jerky sequence of accidents, collisions with other molecules. It was progress by wandering, each molecule's path the sum of many paths, of all possible directions and lengths. The same problem arose in different form as the flow of heat through a metal. And the central issues of Los Alamos, too, were problems of diffusion in a new guise.

The calculation of critical mass quickly became nothing more or less than a calculation of diffusion—the diffusion of neutrons through a strange, radioactive minefield, where now a collision might mean more than a glancing, billiard-ball change of direction. A neutron might be captured, absorbed. And it might trigger a fission event that would give birth to new neutrons. By definition, at critical mass the creation of neutrons would exactly balance the loss of neutrons through absorption or through leakage beyond the container boundaries. This was not a problem of arithmetic. It was a problem of understanding the macroscopic spreading of neutrons as built up from the microscopic individual wanderings.

For a spherical bomb the mathematics resembled another strange and beautiful diffusion problem, the problem of the sun's limb darkening. Why does the sun have a crisp edge? Not because it has a solid or liquid surface. On the contrary, the gaseous ball of the sun thins gradually; no boundary marks a division between sun and empty space. Yet we see a boundary. Energy diffuses outward from the roiling solar core toward the surface, particles scattering one another in tangled paths, until finally, as the hot gas thins, the likelihood of one more collision disappears. That creates the apparent edge, its sharpness more an artifact of the light than a physical reality. In the language of statistical mechanics, the mean free path—the average distance a particle travels between one collision and the next— becomes roughly as large as the radius of the sun. At that point photons have freed themselves from the pinball game of diffusion and can fly in a straight line until they scatter again, in the earth's atmosphere or in the sensitive retina of one's eye. The difference in brightness between the sun's center and its edge gave an indirect means of calculating the nature of the internal diffusion. Or should have—but the mechanics proved difficult until a brilliant young mathematician at MIT, Norbert Wiener, devised a useful method.

If the sun were a coolly radioactive metal ball a few inches across, with neutrons rattling about inside, it would start to look like a miniaturized version of the same problem. For a while this approach proved useful. Past a certain point, however, it broke down. Too many idealizing assumptions had to be made. In a real bomb, cobbled together from mostly purified uranium, surrounded by a shell of neutron-reflecting metal, the messy realities would defy the most advanced mathematics available. Neutrons would strike other neutrons with a wide range of possible energies. They might not scatter in every direction with equal probability. The bomb might not be a perfect sphere. The difference between these realities and the traditional oversimplifications arose in the first major problem assigned to Feynman's group. Bethe had told them to evaluate an idea of Teller's, the possibility of replacing pure uranium metal with uranium hydride, a compound of uranium and hydrogen. The hydride seemed to have advantages. For one, the neutron-slowing hydrogen would be built into the bomb material; less uranium would be needed. On the other hand, the substance was pyrophoric—it tended to burst spontaneously into flame. When the Los Alamos metallurgists got down to the work of making hydride chunks for testing, they set off as many as half a dozen small uranium fires a week.

The hydride problem had one virtue. It pushed the theorists past the limits of their methods of calculating critical masses. To make a sound judgment of Teller's idea they would have to invent new techniques. Before they considered the hydride, they had got by with methods based on an approximation of Fermi's. They been able to assume, among other things, that neutrons would travel at a single characteristic velocity. In pure metal, or in the slow reaction of the water boiler, that assumption seemed to work out well enough. But in the odd atomic landscape of the hydride, with its molecules of giant uranium atoms bonded to two or three tiny hydrogen atoms, neutrons would fly about at every conceivable velocity, from very fast to very slow. No one had yet invented a way of computing critical mass when the velocities spread over such a wide range. Feynman solved that problem with a pair of approximations that worked like pincers. The method produced outer bounds for the answer: one estimate known to be too large and another known to be too small. The experience of actual computation showed that this would suffice: the pair of approximations were so close together that they gave an answer as accurate as was needed. As he drove the men in his group toward a new understanding of criticality (poaching sneakily, it seemed to them, on the territory of Serber's group, T-2), he delivered up a series of insights that struck even Welton, who understood him best, as mystical. One day he declared that the whole problem would be solved if they could produce a table of so-called eigenvalues, characteristic values of energies, for the simplified model that T-2 had been using. That seemed an impossible leap, and the group said so, but they soon found that he was right again. For Teller's scheme, the new model was fatal. The hydride was a dead end. Pure uranium and plutonium proved far more efficient in propagating a chain reaction.

In this way, amid these clusters of scientists, the theory of diffusion underwent a kind of scrutiny with few precedents in the annals of science. Elegant textbook formulations were examined, improved, and then discarded altogether. In their place came pragmatic methodologies, gimmicks with patches. The textbook equations had exact solutions, at least for special cases. In the reality of Los Alamos, the special cases were useless. In Feynman's Los Alamos work, especially, an accommodation with uncertainty became a running theme. Few other scientists filled the foreground of their papers with such blunt acknowledgments of what was not known: "unfortunately cannot be expected to be as accurate"; "Unfortunately the figures contained herein cannot be considered as 'correct'"; "These methods are not exact." Every practical scientist learned early to include error ranges in their calculations; they learned to internalize the knowledge that three miles times 1.852 kilometers per mile equals five and a half kilometers, not 5.556 kilometers. Precision only dissipates, like energy in an engine governed by the second law of thermodynamics. Feynman often found himself not just accepting the process of approximation but manipulating it as a tool, employing it in the creation of theorems. Always he stressed ease of use: ". . . an interesting theorem was found to be extremely useful in obtaining approximate expressions . . . it does permit, in many cases, a simpler derivation or understanding . . ."; " . . . in all cases of interest thus far investigated . . . accuracy has been found ample . . . extremely simple for computation and, once mastered, quite simple to use in thinking about a wide variety of neutron problems." Theorems as theorems, or objects of mathematical beauty, had never been so unappealing as at Los Alamos. Theorems as tools had never been so valued. Again and again the theorists had to devise equations with no hope of exact solution, equations that sentenced them to countless hours of laborious computation with nothing at the end but an approximation. When they were done, the body of diffusion theory had become a hodgepodge. The state of knowledge was written in no one place, but it was more practical than ever before.

For Feynman, thinking in his spare time about the pure theory of particles and light, diffusion dovetailed peculiarly with quantum mechanics. The traditional diffusion equation bore a family resemblance to the standard Schrödinger equation; the crucial difference lay in a single exponent, where the quantum mechanical version was an imaginary factor, i. Lacking that i, diffusion was motion without inertia, motion without momentum. Individual molecules of perfume carry inertia, but their aggregate wafting through air, the sum of innumerable random collisions, does not. With the i, quantum mechanics could incorporate inertia, a particle's memory of its past velocity. The imaginary factor in the exponent mingled velocity and time in the necessary way. In a sense, quantum mechanics was diffusion in imaginary time.

The difficulties of calculating practical diffusion problems forced the Los Alamos theorists into an untraditional approach. Instead of solving neat differential equations, they had to break the physics into steps and solve the problem numerically, in small increments of time. The focus of attention was pushed back down to the microscopic level of individual neutrons following individual paths. Feynman's quantum mechanics was evolving along strikingly similar lines. His private work, like the diffusion work, embodied an abandonment of a too simple, too special differential approach; the emphasis on step-by-step computation; and above all the summing of paths and probabilities.

Computing by Brain

Walking around the hastily built wooden barracks that housed the soul of the atomic bomb project in 1943 and 1944, a scientist would see dozens of men laboring over computation. Everyone calculated. The theoretical department was home to some of the world's masters of mental arithmetic, a martial art shortly to go the way of jiujitsu. Any morning might find men such as Bethe, Fermi, and John von Neumann together in a single small room where they would spit out numbers in a rapid-fire calculation of pressure waves. Bethe's deputy, Weisskopf, specialized in a particularly oracular sort of guesswork; his office became known as the Cave of the Hot Winds, producing, on demand, unjustifiably accurate cross sections (shorthand for the characteristic probabilities of particle collisions in various substances and circumstances). The scientists computed everything from the shapes of explosions to the potency of Oppenheimer's cocktails, first with rough guesses and then, when necessary, with a precision that might take weeks. They estimated by seat of the pants, as a cook who wants one-third cup of wine might fill half a juice glass and correct with an extra splash. Anyone who calculated logarithms by mentally interpolating between the entries in a standard table—a technique that began to vanish thirty years later, when inexpensive electronic calculators made it obsolete—learned to estimate this way, using some unconscious feeling for the right curve. Feynman had a toolbox of such curves in his head, precalibrated. His Los Alamos colleagues were sometimes amused to hear him, when thinking out loud, howl a sort of whooping glissando when he meant, *this rises exponentially*; a different sound signified *arithmetically*. When he started managing groups of people who handled laborious computation, he developed a reputation for glancing over people's shoulders and stabbing his finger at each error: "That's wrong." His staff would ask why he was putting them to such labor if he already knew the answers. He told them he could spot wrong results even when he had no idea what was right— something about the smoothness of the numbers or the relationships between them. Yet unconscious estimating was not really his style. He liked to know what he was doing. He would rummage through his toolbox for an analytical gimmick, the right key or lock pick to slip open a complicated integral. Or he would try various simplifying assumptions: *Suppose we treat some quantity as infinitesimal.* He would allow an error and then measure the bounds of the error precisely.

It seemed to colleagues that some of his computation was a matter of conscious reputation building. One day Feynman, who had made a point of considering watches to be affectations, received a pocket watch from his father. He wore it proudly, and his friends began to needle him; they asked the time at every opportunity, until he began responding, with a glance at the watch: "Well, four hours and twenty minutes ago it was twelve before noon," or "In three hours and forty-nine minutes it will be two seventeen." Few caught on. He was doing no arithmetic at all. Rather, he had designed a simple parlor trick in the spirit of gauge theories to come. Each morning he would turn his watch to a fixed offset from the true time—three hours and forty-nine minutes fast one day; the next day four hours and twenty minutes slow. He had only to remember one number and read the other directly from the watch. (This was the same Feynman who, years later, trying to describe to a layman the intricate shiftings of time and orientation on which theoretical physics depended, said, "You know how it is with daylight saving time? Well, physics has a dozen kinds of daylight saving.")

When Bethe and Feynman went up against each other in games of calculating, they competed with special pleasure. Onlookers were often surprised, and not because the upstart Feynman bested his famous elder. On the contrary, more often the slow-speaking Bethe tended to outcompute Feynman. Early in the project they were working together on a formula that required the square of 48. Feynman reached across his desk for the Marchant mechanical calculator.

Bethe said, "It's twenty-three hundred."

Feynman started to punch the keys anyway. "You want to know exactly?" Bethe said. "It's twenty-three hundred and four. Don't you know how to take squares of numbers near fifty?" He explained the trick. Fifty squared is 2,500 (no thinking needed). For numbers a few more or less than 50, the approximate square is that many hundreds more or less than 2,500. Because 48 is 2 less than 50, 48 squared is 200 less than 2,500—thus 2,300. To make a final tiny correction to the precise answer, just take that difference again—2—and square it. Thus 2,304.

Feynman had internalized an apparatus for handling far more difficult calculations. But Bethe impressed him with a mastery of mental arithmetic that showed he had built up a huge repertoire of these easy tricks, enough to cover the whole landscape of small numbers. An intricate web of knowledge underlay the techniques. Bethe knew instinctively, as did Feynman, that the difference between two successive squares is always an odd number, the sum of the numbers being squared. That fact, and the fact that 50 is half of 100, gave rise to the squares-near-fifty trick. A few minutes later they needed the cube root of 2½. The mechanical calculators could not handle cube roots directly, but there was a look-up chart to help. Feynman barely had time to open the drawer and reach for the chart before he heard Bethe say, "That's 1.35." Like an alcoholic who plants bottles within arm's reach of every chair in the house, Bethe had stored away a device for anywhere he landed in the realm of numbers. He knew tables of logarithms and he could interpolate with unerring accuracy. Feynman's own mastery of calculating had taken a different path. He knew how to compute series and derive trigonometric functions, and how to visualize the relationships between them. He had mastered mental tricks covering the deeper landscape of algebraic analysis—differentiating and integrating equations of the kind that lurk dragonlike in the last chapters of calculus texts. He was continually put to the test. The theoretical division sometimes seemed like the information desk at a slightly exotic library. The phone would ring and a voice would ask, "What is the sum of the series $1 + (½)^4 + (⅓)^4 + (¼)^4 + \ldots$?"

"How accurate do you want it?" Feynman replied.

"One percent will be fine."

"Okay," Feynman said. "One point oh eight." He had simply added the first four terms in his head—that was enough for two decimal places.

Now the voice asked for an exact answer. "You don't need the exact answer," Feynman said.

"Yeah, but I know it can be done."

So Feynman told him. "All right. It's pi to the fourth over ninety."

He and Bethe both saw their talents as labor-saving devices. It was also a form of jousting. At lunch one day, feeling even more ebullient than usual, he challenged the table to a competition. He bet that he could solve any problem within sixty seconds, to within ten percent accuracy, that could be stated in ten seconds. Ten percent was a broad margin, and choosing a suitable problem was hard. Under pressure, his friends found themselves unable to stump him. The most challenging problem anyone could produce was: Find the tenth binomial coefficient in the expansion of $(1 + x)^{20}$. Feynman solved that just before the clock ran out. Then Paul Olum spoke up. He had jousted with Feynman before, and this time he was ready. He demanded the tangent of ten to the hundredth. The competition was over. Feynman would essentially have had to divide one by π and throw out the first one hundred digits of the result—which would mean knowing the one-hundredth decimal digit of π. Even Feynman could not produce that on short notice.

He integrated. He solved equations taking the spirit of infinite summation into more difficult realms. Some of these perilous, nontextbook, nonlinear equations could be integrated through just the right combination of mental gimmicks. Others could not be integrated exactly. One could plug in numbers, make estimates, calculate a little, make new estimates, extrapolate a little. One might visualize a polynomial expression to approximate the desired curve. Then one might try to see whether the leftover error could be managed. One day, making his rounds, Feynman found a man struggling with an especially complicated varietal, a nonlinear three-and-a-half-order equation. There was a business of integrating three times and figuring out a one-half derivative—and in the end Feynman invented a shortcut, a numerical method for taking three integrals at once and a half integral besides, all more accurately than had been thought possible. Similarly, working with Bethe, he invented a new and general method of solving third-order differential equations. Second order had been manageable for several centuries. Feynman's invention was precise and practical. It was also doomed to a quick obsolescence in an age of machine computation, as was, for that matter, the skill of mental arithmetic that did so much to establish Feynman's legend.

Computing by Machine

Not only the atomic era but also the computer era had its start in those years. Scattered about the nation's military and civilian laboratories, a few researchers focused exclusively on the means of calculating instead of the calculations themselves. At Los Alamos, in particular, the demand for numerical computation grew more intense than anywhere else on earth. The means were mechanical and now partly electronic, though the crucial technological key, the transistor, remained to be invented at the decade's end. Calculating technology became a hybrid with machine parts and human parts: people carrying cards from place to place served as the memories and logical-branching units of near computers that stretched across rows and columns of desks.

The bomb project could draw on the best technology available anywhere, but the best technology offered little to the working scientist. The manufacturers of such equipment—the International Business Machines Corporation already preeminent among them—considered the scientific market to be negligible. It could not imagine the vast clientele that would soon consume as much calculating capacity as could be created: for forecasting weather, designing engines, analyzing proteins, scheduling airplanes, and simulating everything from ecosystems to heart valves. Business was thought to be the sole potential consumer for business machines, and business meant accounting, which meant addition and subtraction. Multiplication seemed a luxury, although it might be necessary to multiply monthly sales by twelve. Division by machine was esoteric. Computation of mortgage payments and bond yields could be managed by humans with standard tables.

The workhorse of scientific calculating was the Marchant calculator, a clattering machine nearly as large as a typewriter, capable of adding, subtracting, multiplying, and with some difficulty dividing numbers of up to ten digits. (At first, to save money, the project ordered slower, eight-digit versions as well. They were rarely used.) In these machines a carriage spun around, propelled at first by a hand crank and later by an electric motor. Keys and levers pushed the carriage left or right. Counter and register dials displayed painted digits. There were rows and columns of keys for entering numbers, a plus bar and a minus bar, a multiplier key and a negative multiplier key, shift keys, and a key for stopping the machine when division went out of control, as it often did. Mechanical arithmetic was no simple affair. With all its buttons and linkages the Marchant was not quite as powerful as the giant Difference Engine and Analytical Engine, invented in England a century before by Charles Babbage in hopes of generating the printed tables of numbers on which navigators, astronomers, and mathematicians had to rely. Not only did Babbage solve the problem of carrying digits from one decimal place to the next; his machines actually used punched cards, borrowed from mechanical looms, to convey data and instructions. In the era of steam power, few of his contemporaries appreciated the point.

The Marchants took a hard pounding at Los Alamos. Metal parts wore thin and came out of alignment. The officially nonexistent laboratory was poorly suited to field-service visits by the manufacturer's repair crews, so standard procedure required the shipping of broken machines back to California. Eventually three or four machines were in the pipeline at any one time. Feynman, frustrated, turned to Nicholas Metropolis, a mustached Greek mathematician who later became an authority on computation and numerical methods, and said, "Let's learn about these damned things and not have to send them to Burbank." (Feynman grew a temporary mustache, too.) They spent hours taking apart new and old machines for comparative diagnosis; learned where the jams and slippages began; and hung out a shingle advertising, "Computers Repaired." Bethe was not amused at this waste of his theoreticians' time. He finally ordered a halt to the tinkering. Feynman complied, knowing that within weeks the shortage of machines would change Bethe's mind.

Escalation of the computation effort came in the fall of 1943 with an order to IBM for business machines to be delivered to an unknown location: three 601 multipliers, one 402 tabulator, one reproducer-summary punch, one verifier, one keypunch, one sorter, and one collator. Astronomers at Columbia had been experimenting with punch-card computing before the war. A multiplier, an appliance the size of a restaurant stove, could process calculations in large batches. Electrical probes found the holes in the cards, and operations could be configured by plugging groups of wires into a patchboard. Among the computation-minded at Los Alamos, the prospect of such machines caused excitement. Even before they arrived, one of the theorists, Stanley Frankel, set about devising improvements: for example, tripling the output by rearranging the plugs so that three sets of three- or four-digit numbers could be multiplied in a single pass. Having requisitioned the machines, the scientists now also requisitioned a maintenance man—an IBM employee who had been drafted into the army. They were gaining adroitness at military procurement. The crates arrived two days before the repairman; in those two days Feynman and his colleagues managed to get the machines unpacked and assembled, after a fashion, with the help of nothing but a set of wiring blueprints. So much more powerful were they that Feynman—sensitive to rhythms as always—rapidly discovered that he could program them to clatter out the cadence of well-known songs. The theorists began to organize something new in the annals of computation: a combination of the calculating machine and the factory assembly line. Even before the IBM machines arrived Feynman and Metropolis set up an array of people—mostly wives of scientists, working at three-eighths salary—who individually handled pieces of complex equations, one cubing a number and passing it on, another performing a subtraction, and so on. It was mass production married to numerical calculation. The banks of women wielding Marchants simulated the internal workings of a computer. As a later generation would discover, there was something mentally seductive in the act of breaking calculus into the algorithmic cogs needed for machine computation. It forced the mind back down into the essence of arithmetic. It also began a long transformation in the understanding of what kinds of equations were *solvable*. Stacks of punch cards could *solve* equations for a ball of fire rising through a suddenly turbulent atmosphere, by stepping through successive approximations for time 0:01, time 0:02, time 0:03 . . . though by the lights of traditional analysis those sharply nonlinear equations were unsolvable.

Of the many problems put to the Los Alamos computers, none better anticipated the coming age of massive scientific simulation than implosion itself: how to calculate the motion of an inward-flowing shock wave. An explosive charge wrapped around the bomb was to set the shock wave in motion, and the pressure would crush a nugget of plutonium into criticality. How should the bomb assembly be configured to assure a stable detonation? What kind of fireball would ensue? Such questions required a workable formula for the propagation of a spherical detonation wave in a compressible fluid, the "compressible fluid" in this case being the shotput-size piece of plutonium liquefied in the microseconds before it became a nuclear blast. The pressure would be more intense than at the earth's center. The temperature would reach 50 million degrees Centigrade. The theorists were on their own here; experimentalists could offer little more than good wishes. All during 1944 the computation effort grew. John von Neumann served as a traveling consultant with an eye on the postwar future. Von Neumann—mathematician, logician, game theorist (he was more and more a

fixture in the extraordinary Los Alamos poker game), and one of the fathers
of modern computing—talked with Feynman while they worked on the
IBM machines or walked though the canyons. He left Feynman with two
enduring memories. One was the notion that a scientist need not be responsible for the entire world, that social irresponsibility might be a reasonable stance. The other was a faint, early recognition of the mathematical
phenomena that would later be called chaos: a persistent, repeatable irreg-
ularity in certain equations as they prepared to run them through their
primitive computers. As a shock wave, for example, passed though a material, it left oscillations in its wake. Feynman thought at first that the
irregular wiggles must be numerical errors. Von Neumann told him that
the wiggles were actually features of interest.

Von Neumann also kept these new computer specialists up to date with
the other sites he visited. He brought news of an electromechanical Mark
I under construction at Harvard, a relay calculator at Bell Laboratories,
human neuronal research at the University of Illinois, and at the Aberdeen
Proving Ground in Maryland, where problems of ballistic trajectories mo-
tivated the calculators, a more radical device with a new kind of acronym:
ENIAC, for Electronic Numerical Integrator and Computer, a machine composed of eighteen thousand vacuum tubes. The tubes controlled binary on-
off flip-flops; in a bow to the past, the flip-flops were arranged in rings of
ten, to simulate the mechanical wheels used in decimal calculating machines. The ENIAC had too many tubes to survive. Von Neumann estimated:
"Each time it is turned on, it blows two tubes." The army stationed soldiers
carrying spare tubes in grocery baskets. The operators borrowed *mean free
path* terminology from the ricocheting particles of diffusion theory; the
computer's mean free path was its average time between failures.

Meanwhile, under the influence of this primal dissection of mathematics, Feynman retreated from pragmatic engineering long enough to put
together a public lecture on "Some Interesting Properties of Numbers." It
was a stunning exercise in arithmetic, logic, and—though he would never
have used the word—philosophy. He invited his distinguished audience
("all the mighty minds," he wrote his mother a few days later) to discard
all knowledge of mathematics and begin from first principles—specifically,
from a child's knowledge of counting in units. He defined addition, $a +
b$, as the operation of counting b units from a starting point, a. He defined
multiplication (counting b times). He defined exponentiation (multiplying
b times). He derived the simple laws of the kind $a + b = b + a$ and $(a
+ b) + c = a + (b + c)$, laws that were usually assumed unconsciously,

though quantum mechanics itself had shown how crucially some mathematical operations did depend on their ordering. Still taking nothing for granted, Feynman showed how pure logic made it necessary to conceive of inverse operations: subtraction, division, and the taking of logarithms. He could always ask a new question that perforce required a new arithmetical invention. Thus he broadened the class of objects represented by his letters a, b, and c and the class of rules by which he was manipulating them. By his original definition, negative numbers meant nothing. Fractions, fractional exponents, imaginary roots of negative numbers—these had no immediate connection to counting, but Feynman continued pulling them from his silvery logical engine. He turned to irrational numbers and complex numbers and complex powers of complex numbers—these came inexorably as soon as one from facing up to the question: What number, i, when multiplied by itself, equals negative one? He reminded his audience how to compute a logarithm from scratch and showed how the numbers converged as he took successive square roots of ten and thus, as an inevitable by-product, derived the "natural base" e, that ubiquitous fundamental constant. He was recapitulating centuries of mathematical history—yet not quite recapitulating, because only a modern shift of perspective made it possible to see the fabric whole. Having conceived of complex powers, he began to *compute* complex powers. He made a table of his results and showed how they oscillated, swinging from one to zero to negative one and back again in a wave that he drew for his audience, though they knew perfectly well what a sine wave looked like. He had arrived at trigonometric functions. Now he posed one more question, as fundamental as all the others, yet encompassing them all in the round recursive net he had been spinning for a mere hour: To what power must e be raised to reach i? (They already knew the answer, that e and i and π were conjoined as if by an invisible membrane, but as he told his mother, "I went pretty fast & didn't give them a hell of a lot of time to work out the reason for one fact before I was showing them another still more amazing.") He now repeated the assertion he had written elatedly in his notebook at the age of fourteen, that the oddly polyglot statement $e^{\pi i} + 1 = 0$ was the most remarkable formula in mathematics. Algebra and geometry, their distinct languages notwithstanding, were one and the same, a bit of child's arithmetic abstracted and generalized by a few minutes of the purest logic. "Well," he wrote, "all the mighty minds were mightily impressed by my little feats of arithmetic."

Indeed, if Feynman was, as his friend Welton thought, consciously trying to establish himself among these influential physicists, he was succeeding even more than he knew. As early as November 1943, seven months after the Los Alamos project began, Oppenheimer began trying to persuade his department at Berkeley to hire Feynman for after the war. He wrote to the department chairman, Birge:

> He is by all odds the most brilliant young physicist here, and everyone knows this. He is a man of thoroughly engaging character and personality, extremely clear, extremely normal in all respects, and an excellent teacher with a warm feeling for physics in all its aspects.

Oppenheimer warned that Feynman was sure to have other job offers, because "a not inconsiderable number of 'big shots' " had already noticed him. He quoted two of the big shots. Bethe, according to Oppenheimer, had said bluntly that he would sooner lose any two scientists than lose Feynman. And Wigner of Princeton had made what was, for a physicist's physicist in the 1940s, perhaps the ultimate tribute.

"He is a second Dirac," Wigner said, "only this time human."

Fenced In

Feynman celebrated his wedding anniversary by grilling steak outdoors at the Presbyterian Sanatorium in a small charcoal broiler that Arline had ordered from a catalog. She also got him a chef's hat, apron, and gloves. He wore them self-consciously, along with his new mustache, while she reveled in the domesticity of it all, until he could no longer stand the idea of people watching him from passing cars. She laughed, asking, as she so often did, why he cared what other people thought. Steak was an extravagance—eighty-four cents for two pounds. With it they ate watermelon, plums, and potato chips. The hospital lawn sloped down to Route 66, the cross-country highway, where the traffic roared by. Albuquerque was sweltering, and they were happy. Arline talked to her parents by long-distance telephone for seven minutes, another extravagance. After Richard left to hitchhike back north, a late-afternoon thundershower blackened the desert. Arline worried about him in the downpour. She still had not gotten used to the raw force of storms in the open West.

His near-weekly trips through the valley that lay between the Jemez and Sangre de Cristo mountains made him a rarity on the mesa. Few residents of that hermetic community had occasion to leave at all. Once, in a fanciful conversation about likely candidates to be a Nazi spy, one friend, Klaus Fuchs, a German turned Briton, suggested that it could only be Dick Feynman—who else had insinuated himself into so many different parts of the laboratory's work? Who else had a regular rendezvous in Albuquerque? In its unreal isolation, with its unusual populace, Los Alamos was growing into a parody of a municipality. It took its place in the mental geography of its residents as it was officially: not a village in the lee of the Jemez Mountains, not only a fenced-in circle of houses on dirt paths by a pond, with ducks, but also a fictitious abstraction, P. O. Box 1663, Santa Fe, New Mexico. To some it carried an ersatz resonance of a certain European stereotype of America, as one resident noted—"a pioneer people starting a new town, a self-contained town with no outside contacts, isolated in vast stretches of desert, and surrounded by Indians." Victor Weisskopf was elected mayor. Feynman was elected to a town council. The fence that marked the city line heightened a magic-mountain atmosphere—it kept the world apart. An elite society had assembled on this hill. Elite and yet polyglot—in this cauldron, as in the other wartime laboratories, a final valedictory was being written to the Protestant, gentlemanly, leisurely class structure of American science. Los Alamos did gather an aristocracy—"the most exclusive club in the world," one Oxonian said—yet the princely, exquisitely sensitive Oppenheimer made it into a democracy, where no invisible lines of rank or status were to impede the scientific discourse. The elected councils and committees furthered that impression. Graduate students were supposed to forget that they were talking to famous professors. Academic titles were mainly left behind with the business suits and neckties. It was a democracy by night, too, when inflamed parties brought together cuisines and cocktails of four continents, dramatic readings and political debates, waltzes and square dances (the same Oxonian, bemused amid the clash of cultures, asked, "What exactly is square about it—the people, the room, or the music?"), a Swede singing torch songs, an Englishman playing jazz piano, and Eastern Europeans playing Viennese string trios. Feynman played brassy drum duets with Nicholas Metropolis and organized conga lines. He had never been exposed to culture as such a flamboyant stew (certainly not when he was a student learning to disdain the packaged morsels that MIT handed to its would-be engineers). One party featured an original ballet, to modernistic-sounding music by Gershwin, titled *Sacre du Mesa*. At the end a clattering, flashing mechanical brain noisily revealed the sacred mystery of the mesa: $2 + 2 = 5$.

Los Alamos built its wall against the outside world and thrived within. Separately and privately Richard and Arline, too, sought what refuge they could. They made their secret lives. They built a fence of their own. None of his scientific friends knew that he called her Putzie and she called him Coach; that she noticed the muscles hardening in his legs from all his hiking; that the days of respite from her illness were growing rarer. She wrote him in code, playing to his love of unraveling puzzles; his father did this, too. Their letters caught the eye of the military censors at the laboratory's Intelligence Office. The censors alerted Feynman to regulation 4(e): *Codes, ciphers or any form of secret writing will not be used. Crosses, X's or other markings of a similar nature are equally objectionable.* Censorship had been designed delicately to accommodate a nonmilitary clientele, university people who still liked to imagine that they were volunteers in a project of scientific research in a nation where the privacy of the mail was sacred. The censors trod carefully. They tried to turn mail around the day they received it, and they agreed to allow correspondence in French, German, Italian, and Spanish. They felt entitled at least to ask Feynman for the key to the codes.

He said he did not have a key or want a key. Finally they agreed that if Arline would enclose a key for their benefit they would remove it before the envelope got to Feynman.

Inevitably, he then ran afoul of regulation 8(l), a delightfully (to Feynman) self-referential law requiring the censorship of *any information concerning these censorship regulations or any discourse on the subject of censorship.* He got the message to Arline nonetheless, and her acid sense of fun took over. She started sending letters with holes cut in them or blotches of ink covering words: "It's very difficult writing because I feel that the ———— is looking over my shoulder." He would respond with numerical fancies, pointing out how peculiarly the decimal expansion of 1/243 repeats itself: .004 115 226 337 448 . . . and his increasingly frustrated official audience would have to ensure that the string of digits was neither a cipher nor a technical secret. Feynman explained with subtle glee that this fact had the empty, tautological, zero-information-content quality of all mathematical truths. In one of her mail-order catalogs Arline found a kit for do-it-yourself jigsaw puzzles; the next letter from the Albuquerque sanatorium to Box 1663 came disassembled in a little sack. From another the censors deleted a suspicious-sounding shopping list. Richard and Arline talked about a booby-trapped letter that would begin, "I hope you remembered to open this letter carefully because I have included the Pepto Bismol powder . . ." Their letters were a lifeline. No wonder, under watchful eyes, the lovers found ways to make them private.

The censorship, like the high barbed-wire fence, reminded the mesa's more sensitive residents of their special status: watched, enclosed, restricted, isolated, surrounded, guarded. They understood that no other civilian post office box had all its mail opened and read. The fence was a double-edged symbol. Few scientists were so important as to merit armed soldiers patrolling their laboratory perimeters. They could not help feeling some pride. Feynman admonished his parents to maintain secrecy: "There are Captains in the Army who live up here who don't know what we're doing. (Even Majors.)" Much later, in a post–*Catch-22* world, the military trappings were remembered as irritants and targets of mockery. At the time it was not so simple. The men and women of Los Alamos resented the fence and respected the fence. Feynman explored most of its length. When he discovered holes, with well-beaten paths leading through, he pointed them out in a spirit of good citizenship, annoyed only that the guards responded so lackadaisically. ("I explained it to him & the officer in charge," he wrote Arline, "but I bet they don't do anything.") He never realized that the holes had semiofficial sanction. The security staff tolerated them—with Oppenheimer's connivance, it seemed—so that people from the local tribes could come to the laboratory's twelve-cent movies.

Feynman's exploring drew him to every secret and private place. He had a fidgety way of prying into things—the laboratory's new Coca-Cola dispenser, for example, a contraption that secured the bottles with a locked steel collar around their necks. This device replaced an older container, the most ancient prototype of the soda machine: customers would open the lid, take a bottle, and honorably drop their coin in a box. The new dispenser struck Feynman as a withdrawal of trust; thus he felt entitled to accept the technological challenge and finesse the mechanism. Was that right or wrong? He debated the moral principles with his friends. Meanwhile he found himself abstaining from liquor. He had got so drunk one night that he could tell it was ruining his drum playing and joke telling, although it did not stop him from running all over the base singing and beating pots and pans; finally he passed out, and Klaus Fuchs took him home. He decided to give up alcohol, along with tobacco, and wondered whether it was a sign of encroaching conventionality. Was he getting "moraller and moraller" as he got older? ("That's bad.")

His reputation as a skilled prier spread. One scientist left some belongings in a storeroom at Fuller Lodge and borrowed Feynman's fingers to pick the Yale lock. Paper clips, screwdriver, two minutes. Two men arrived, breathless from running up the stairs, and begged Feynman to crack a file cabinet holding a crucial document about a ski tow. Combination locks still seemed too hard. As a group leader he had been issued a special steel safe for sensitive material of his own, and he had not yet managed a way to break in. He would spin the dial from time to time. Occasionally it occurred to him that his interest in locks was turning into an obsession. Why? "Probably," he told Arline,

> because I like puzzles so much. Each lock is just like a puzzle you have to open without forcing it. But combination locks have me buffaloed.
> You do too, sometimes, but eventually I figure out you.

Locks mixed human logic and mechanical logic. The designer's strategy was constrained by the manufacturer's convenience or the limits of the metal, as it was in so many of the bomb project's puzzles. The official logic of a Los Alamos safe, as displayed in the dial's numbers and hatch marks, indicated a million different combinations—three numbers from 0 to 99. Some experimentation, though, showed Feynman that the markings disguised a considerable margin of error, plus or minus two, attributable to plain mechanical slackness; if the correct number was 23, anything from 21 to 25 would work as well. When he was searching combinations systematically, therefore, he needed only to try one number in every five—0, 5, 10, 15 . . . —to be sure of hitting the target. By thinking in terms of error ranges, instead of accepting the authority of the numerals on the dial, he brought a pragmatic physicist's intuition to bear. That one insight effectively reduced the total combinations from one million to a mere eight thousand, almost few enough to try, given a few hours.

An American folklore had developed about safes and the yeggs who cracked them. Through the cowboy era and the gangster era safes grew thicker and more elaborate—double walls of cast iron and manganese, triple side bolts and bottom bolts, curb tumblers and pressure handles—and the legend, too, grew thicker and more elaborate. The consummate safeman was thought to need sandpapered fingers and hypersensitive ears.

His essential skill: a feeling for the vibrations of tumblers lining up or falling into place. This was pure myth. It was true that once in a long while someone would open a safe by feel, but, the lore notwithstanding, the chief tools of successful safecrackers were crowbars and drills. Safes were cracked; holes were torn in their sides; handles and dials were torn off. When all else failed, safes were burned. The safeman used "soup"—nitroglycerin. The Los Alamos physicists had been conditioned by the myth, and when word started spreading that the laboratory had a skilled safecracker on its staff, most of them believed—and never stopped believing—that Feynman had mastered the art of listening to the tiny clicks.

To learn how to crack safes he had to find his way past the same myth. He read pulp memoirs of safemen to look for their secrets. They inspired him to dreams of glory: these authors boasted about opening bullion-filled safes underwater; he would write the book that would top them all. In its preface he would intone, *I opened the safes which contained behind them the entire secret of the atomic bomb: the schedules for the production of plutonium, the purification procedures, how much was going to be needed, how the atomic bomb worked, how the neutrons are generated . . . the whole schmeer.* Only gradually, as he looked for the nuggets of useful information, did he realize how mundane the business was. Because his repertoire would have to omit drills and nitroglycerin, it would have to make the most of such practical rules as he could find. Some he read; others he learned as he went along. Most were variations on a theme: People are predictable.

They tend to leave safes unlocked.

They tend to leave their combinations at factory settings such as 25-0-25.

They tend to write down the combinations, often on the edge of their desk drawers.

They tend to choose birthdays and other easily remembered numbers.

This last insight alone made an enormous difference. Of the 8,000 effective possible combinations, Feynman figured that only 162 worked as dates. The first number was a month from 1 to 12—given the margin of error, that meant he need try just three possibilities, 0, 5, and 10. For a day from 1 to 31 he needed six; for a year from 1900 to the present, just nine. He could try $3 \times 6 \times 9$ combinations in minutes. He also discovered that it took just a few inexplicable successes to make a safecracker's reputation.

By fiddling with his own safe he learned that when a door was open he could find the last number of a combination by turning the dial and feeling when the bolt came down. Given some time, he could find the second number that way, too. He made a habit of absently leaning against his colleagues' safes when he visited their offices, twirling the dials like the perpetual fidgeter he was, and thus he built up a master list of partial combinations. The remaining trial and error was so trivial that he found himself—for the sake of cultivating his legend—carrying tools as red herrings and pretending that safe jobs took longer than they really did.

The Last Springtime

Friday afternoon again. Gravel switchbacks wound perilously down the mesa. Across a desert spotted with pale green bristles, the Sangre de Cristo Mountains rose like luminous cutouts thirty miles to the east, as bright as if they were a few city blocks away. The air was clearer than any Feynman had seen. The scenery left emotional fingerprints on many of the Easterners and Europeans who lived in its spell for two years. When it snowed, the shades of whiteness seemed impossibly rich. Feynman reveled in the clouds skimming low across the valley, the mountains visible above and below the clouds at once, the velvet glow of cloud-diffused moonlight. The sight stirred something within the most rational of minds. He mocked himself for feeling it: *See, I'm getting an aesthetic sense.* The days blurred, especially now—no more banker's hours, not much theory to divert the mind. The pace of computation was hectic. Feynman's day began at 8:30 and ended fifteen hours later. Sometimes he could not leave the computing center at all. He worked through for thirty-one hours once and the next day found that an error minutes after he went to bed had stalled the whole team. The routine allowed just a few breaks: a hasty ride across the mesa to help put out a chemical fire; or one of those Los Alamos seminar-briefing-colloquium-town-meetings, where, slouching as far as his frame would permit, he would sit in the second row next to a detached-looking Oppenheimer; or a drive with his friend Fuchs to some Indian caves, where they could explore on hands and knees until dusk.

Still, each Friday or Saturday, if he could, Feynman left this place behind, making his way down the rutted road in Paul Olum's little Chevrolet coupe or sometimes now in Fuchs's blue Buick. He turned over and over in his mind some nagging puzzle and let his thoughts drift back to the hard quantum problems he had left behind at Princeton. He made a difficult mental transition to his weekend. The trips down from those heights marked off full weeks for him, empty ones for Arline. He was like a spy invented by a novelist: "not certain whether this time spent traveling between his two secret worlds was when he was truly himself, when he was able to hold the two in balance and know them to be separate from himself; or whether this was the one time he was nothing at all, a void traveling between two points." Later, when Fuchs, shockingly, turned out to have been a spy for the Soviet Union, Feynman thought it might not have been so strange after all that his friend had been able to hide his inner thoughts so well. He, too, had felt he was leading a double life. His anguish over Arline, so dominating his mind, stayed invisible to the colleagues who saw his aggressively carefree self. He would sit in a group and look at someone—even at Fuchs—and think, how easy it is to hide my thoughts from others.

A third springtime was coming to Los Alamos, and Feynman knew it would be the last. For a moment he thought he felt a break in the tension. He found a way to get the computation group running smoothly enough to allow him a few hours more sleep. He took a shower. For a half hour he read a book before falling asleep. It seemed, just for a moment, that the worst might be over. He wrote Arline:

> You are a strong and beautiful woman. You are not always as strong as other times but it rises & falls like the flow of a mountain stream. I feel I am a reservoir for your strength—without you I would be empty and weak . . . I find it much harder these days to write these things to you.

He never wrote without saying *I love you* or *I'm still loving you* or *I have a serious affliction: loving you forever.* The pace quickened again, and Feynman sometimes thought about long days he had worked for twenty dollars a week waiting on tables and helping in the kitchen of his aunt's summer hotel, the Arnold, on the beach at Far Rockaway. Wherever he went, his drumming could be heard through the walls, nervous or jaunty, a rapping that his staff had to enjoy or endure. It was not music. Feynman himself could barely endure the more standard tunes of his friend Julius Ashkin's recorder, "an infernally popular wooden tube," he called it, "for making noises bearing a one-one correspondence to black dots on a piece of paper—in imitation to music."

Stresses were tightening, too, between the security staff and the scientists, and Feynman had lost his eager spirit of cooperation. A colleague had been interrogated for more than an hour in a smoky room, questions fired by

men sitting in the dark, as in a melodramatic movie. "Don't get scared
tho," Feynman wrote Arline, "they haven't found out that I am a relativist
yet." Fear sometimes clutched Feynman now. His intestines suffered chron-
ically. He had a chest X ray: clear. Names rushed through his head: maybe
Donald; if a girl, maybe Matilda. Putzie wasn't drinking enough milk—
how could he help her build her strength at this distance? They were
spending $200 a month on the room and oxygen and $300 more on nurses,
and $300 was the shortfall between income and expenditures. His salary
as a Manhattan Project group leader: $380 a month. If they spent Arline's
savings, $3,300 plus a piano and a ring, they could cover ten more months.
Arline seemed to be wasting away.

Letters went back and forth almost daily. They wrote like a boy and
a girl without experience at the art of love letters. They catalogued the
everyday—how much sleep, how much money. Macy's sent Arline an
unexpected mail-order refund of forty-four cents: *I feel like a million-
aire . . . I.O.U. 22¢.* His sporadic bad digestion or swollen eyelid; her
waning or waxing strength, her coughed-up blood and her access to oxygen.
They used matching stationery. It was a mail-order project of Arline's—
soon most of her relatives and many of Richard's friends on the hill had
the same green or brown block letterhead from the Dollar Stationery Company. For herself she ordered both formal (Mrs. Richard P. Feynman) and
informal, with the same legend she had once caught Richard slicing from
her pencils:

RICHARD DARLING,

I LOVE YOU

PUTZIE

She decorated the envelopes with red hearts and silver stars. The army
decorated them with tape: OPENED BY U. S. ARMY EXAMINER.

They called each other "Dope" and then worried about whether they
had given offense. *You're never that—just silly & cute & lots of fun—you
know what I mean, don't you coach?* Alone in her cramped sanatorium
room, decorated with a few pictures and knickknacks received as wedding
gifts, Arline worried about Richard and other women. He was a popular
dancer at Los Alamos parties; he flirted intently with nurses, wives, and a
secretary of Oppenheimer's. All it took to set Arline's mind racing was an
offhand mention of the wife of a colleague. Or worse: the scientists were
in an uproar over the appearance of M.P.'s around a women's dormitory

(the army had discovered an active prostitution trade there), and for some reason Richard had been chosen to lead the protest. He reassured her continually. *Everything is under control—& I love you only.* She explained and reexplained the facts of their love like an incantation: he is tall, gentle, kind, strong; he supports her, but once in a while can lean on her, too; he must confide everything in her, as she has slowly learned to confide in him; we have to think in terms of us, always; she loves the way he stretches casually to open a high window beyond her reach, and she loves the way he talks babytalk with her.

Not until the beginning of this grim year did they make love. Their gingerly discussions had led nowhere. He was afraid of taking advantage, or afraid of harming her, or just afraid. Arline grasped ever more tightly her sense of romantic love. She read *Lady Chatterley's Lover* ("No!" she said. "Love me! Love me, and say you'll keep me. Say you'll keep me! Say you'll never let me go, to the world nor to anybody!") and a popular 1943 book, *Love in America.* "I do not know—although there are those who profess to know with mathematical accuracy—whether sex is all-important in the life of a man or a woman," the author wrote provocatively. Americans lag Europeans in such matters. "We have developed no concepts of love as an art or a rite. . . . We do not seem to realize that woman's love is not prompted by good deeds on a man's part or by Boy Scout conduct; that neither gratitude nor pity are love; that loving lies in demanding as well as in giving; that the woman who loves yearns to give and give again."

Arline herself finally made the decision and set aside a Sunday when she would allow no other visitors. She missed him spiritually and physically, she told him.

Darling I'm beginning to think that perhaps this restlessness I feel within myself is due to pent up emotions—I really think we'd both feel happier and better dear if we released our desires.

She wrote Richard a few days before to tell him it was time. She could not sleep. She clipped a phrase from a newspaper advertisement: "OUR MARRIAGE COMES FIRST." She reminded him of the future that waited for them: just a few more years in bed for her; then he would be a *renowned professor* (*physicist* still did not denote a profession with stature) and she a mother. She apologized, as she so often did, for being moody, for being difficult, for saying hurtful things, and for having to lean on him without respite. Her thoughts rambled.

. . . We have to fight hard—every inch of the way—we can't slip ever—
a slip costs too much. . . . I'll be all a women would be to you—I'll always
be your sweetheart & first love—besides a devoted wife—we'll be proud
parents too—we'll fight to make Donald real—I want him to be like
you. . . . I am proud of you always Richard—your a good husband, and
lover, & well, coach, I'll show you what I mean Sunday.

Your Putzie

False Hopes

Her health continued to fail. "Drink some milk!" Richard wrote in May.
Her weight had fallen to eighty-four pounds. She looked like a woman
starving.

You are a nice girl. Every time I think about you, I feel good. It must be
love. It sounds like a definition of love. It is love. I love you.
I'll see you in two days.

R. P. F.

More and more they talked of medical tests. They needed optimism. He
was near despair. *Time passes fast. Maybe we should start looking for another
doctor. . . . Why don't you drink an extra bottle of milk right now while
you are thinking of it.*

The scientific knowledge that empowered the physicists seemed to mean
nothing on the soft soil of medicine. With the final desperation of the
dying, Richard and Arline reached out for slender possibilities. He had
heard about a new drug, sulf-something—he was not sure—and had written
to researchers in the East, who told him apologetically that studies of
sulfabenamide were in the most preliminary stage. The discovery that
substances of the sulfonamide family retarded bacterial growth was not yet
a decade old. They were destined to prove poor substitutes for true antibiotics.

Now Richard was writing to faraway doctors again. It seemed that Arline
was pregnant. After ending the celibacy of their marriage, she had immediately missed her menstrual period. Was it possible? They were frightened and jubilant at once. Richard did not tell his parents, but he told his
sister, now a college student. Joan was dazzled at the prospect of becoming

an aunt. They talked about names and began making new plans. Yet to
Richard it still seemed that Arline was wasting away. He thought he saw
symptoms of starvation. Perhaps no rational observer could have construed
the cessation of menses at this stage of the disease as a sign of pregnancy,
but that was how they construed it. The alternative was so grim. Their
doctors saw little reason for hopefulness. The chief physician from the
sanatorium in Browns Mills, New Jersey, advised urgently that any pregnancy must be "interrupted"—"have it done by a specialist." Then a pregnancy test gave a negative result after all. They did not know what to think.
A doctor at Los Alamos told Richard that the tests were notoriously unreliable but that they could try again at an Albuquerque laboratory. He
thought the laboratory had the necessary rabbits for the Friedman test.

The same doctor said he had heard of a new substance made from mold
growths—"streptomicin"?—that seemed to cure tuberculosis in guinea
pigs. If it worked, the doctor thought it might soon become widely available.
Arline refused to believe the negative pregnancy result. She wrote cryptic
remarks about "P.S. 59-to-be." The same day a nurse wrote Feynman from
the sanatorium to say that Arline had been spitting blood. He opened his
encyclopedia yet again. Nothing. He drifted through the pages: *tuberculosis,
tuff, tularemia* . . . Tuff was a kind of volcanic rock; Tunicata an animal
group. He wrote Arline another letter. "Tumors you know about & Turkey,
the country, also." Some days she was now too weak even to write back.
He grasped his uncertainty. Not knowing was frustration, anguish, and
finally his only solace.

"*Keep hanging on,*" he wrote. "*Nothing is certain. We lead a charmed
life.*"

In the midst of their private turmoil came V-E day and then Richard's
twenty-seventh birthday. Arline had prepared another mail-order surprise:
the laboratory was flooded with newspapers—handed about and tacked to
walls—proclaiming with banner headlines, "Entire Nation Celebrates Birth
of R. P. Feynman!" The war in Europe, having provided so many of the
scientists with their moral purpose, had now ended. The bloody circle was
closing in the Pacific. They needed no threat of a German or Japanese
bomb to urge them onward. Uranium was arriving. There would be one
test—one last experiment.

At the Mayo Clinic in Minnesota another kind of experiment was under
way, the first clinical trial of streptomycin, a substance that had been discovered nearly two years before, in August 1943. The population participating in the trial: two patients. Both had been near death from tuberculosis

when the experiment began in the fall of 1944; both were improving rapidly. Even so, it was not until the next August that the Mayo trial had expanded to as many as thirty patients. The doctors could see lesions healing and lungs clearing. A year after that, the study of streptomycin as an antitubercular agent had become the most extensive research project ever devoted to a drug and a disease. Researchers were treating more than one thousand patients. In 1947 streptomycin was released to the public.

Streptomycin's discovery, like penicillin's a few years earlier, had been delayed by medicine's slow embrace of the scientific method. Physicians had just begun to comprehend the power of controlled experiments repeated thousands of times. The use of statistics to uncover any but the grossest phenomena remained alien. The doctor who first isolated the culture he named *Streptomyces griseus*, by cultivating some organisms swabbed from the throat of a chicken, had seen the same microbes in a soil sample in 1915 and had recognized even then that they had a tendency to kill disease-causing bacteria. A generation had to pass before medicine systematized its study of such microbes, by screening them, culturing them, and measuring their antibiotic strengths in carefully labeled rows of test tubes.

Nuclear Fear

In its infancy, too, was the branch of science that would have to devote itself to the safety, short-term and long-term, of humans in the presence of nuclear radiation. The sense of miasmic dread that would become part of the cultural response to radioactivity lay in the future. The Manhattan Project's researchers handled their heavy new substances with a breeziness that bordered on the cavalier. Workers handling plutonium were supposed to wear coveralls, gloves, and a respirator. Even so, some were overexposed. The prototype reactors leaked radioactive material. Scientists occasionally ignored or misread their radiation badges. Critical-mass experiments always flirted with danger, and by later standards the safety precautions were flimsy. Experimentalists assembled perfect shining cubes of uranium into near-critical masses by hand. One man, Harry Daghlian, working alone at night, let slip one cube too many, frantically grabbed at the mound to halt the chain reaction, saw the shimmering blue aura of ionization in the air, and died two weeks later of radiation poisoning. Later Louis Slotin used a screwdriver to prop up a radioactive block and lost his life when the screwdriver slipped. Like so many of these worldly scientists he had performed a faulty kind of risk assessment, unconsciously mis-multiplying a low probability of accident (one in a hundred? one in twenty?) by a high cost (nearly infinite).

To make measurements of a fast reaction, the experimenters designed a test nicknamed the dragon experiment after a coolly ominous comment of Feynman's that they would be "tickling the tail of a sleeping dragon." It required someone to drop a slug of uranium hydride through a closely machined ring of the same substance. Gravity would be the agent in achieving supercriticality, and gravity, it was hoped, would carry the slug on through to a safe ending. Feynman himself proposed a safer experiment that would have used an absorber made of boron to turn a supercritical material into a subcritical one. By measuring how rapidly the neutron multiplication died out, it would have been possible to calculate the multiplication rate that would have existed without boron. The arithmetical inference would have served as a shield. It was dubbed the Feynman experiment, and it was not carried out. Time was too short.

Los Alamos hardly posed the most serious new safety challenges, for all its subsequent visibility. These belonged to the vast new factory cities— Oak Ridge, Tennessee, and Hanford, Washington—where plants thrown up across thousands of acres now manufactured uranium and plutonium in bulk. Compounds and solutions of these substances were accumulating in metal barrels, glass bottles, and cardboard boxes piled on the cement floors of storerooms. Uranium was combined with oxygen or chlorine and either dissolved in water or kept dry. Workers moved these substances from centrifuges or drying furnaces into cans and hoppers. Much later, large epidemiological studies would overcome obstacles posed by government secrecy and disinformation to show that low-level radiation caused more harm than anyone had imagined. Yet the authorities at the processing plants were overlooking not only this possibility but also a more immediate and calculable threat: the possibility of a runaway, explosive chain reaction.

Feynman had seemed to be everywhere at once as the pace of work accelerated in 1944 and 1945. At Teller's request he gave a series of lectures on the central issues of bomb design and assembly: the critical-mass calculations for both metal and hydride; the differences between reactions in pile, water boiler, and gadget; how to compute the effects of various tamper materials in reflecting neutrons back into the reactions; how to convert the pure theoretical calculations into the practical realities of the gun method and the implosion method. He became responsible for calculating the way the efficiency of a uranium bomb would depend on the concentration of uranium 235 and for estimating safe amounts of radioactive materials under a variety of conditions. When Bethe had to assign theorists to G Division (Weapon Physics Division—G for gadget) he assigned Feynman to four different groups. Furthermore, he let Oppenheimer know that, as far as the implosion itself was concerned, "It is expected that a considerable fraction of the new work coming in will be carried out by group T-4 (Feynman)." Meanwhile, though Feynman was officially only a consultant to the group handling computation by IBM machines, Bethe decreed that Feynman would now have "complete authority."

At Oak Ridge, where the first batches of enriched uranium were accumulating, a few officials began to consider some of the problems that might arise. One letter that made its way to Los Alamos from Oak Ridge opened, "Dear Sir, At the present time no provisions have been made in the 9207 Area for stopping reactions resulting from the bringing together by accident of an unsafe quantity of material. . . ." Would it make sense, asked the writer—a plant superintendent with the Tennessee Eastman Corporation— to install some kind of advanced fire-extinguishing equipment, possibly using special chemicals? Oppenheimer recognized the peril waiting in such questions. He brought in Teller and Emilio Segrè, head of the experimental division's radioactivity group. Segrè paid an inspection visit, other theorists were assigned, and finally the problem was turned over to Feynman, with his expertise in critical-mass calculations.

As Segrè had discovered, the army's compartmentalization of information created a perilous combination of circumstances at Oak Ridge. Workers there did not know that the substance they were wheeling about in large bottles of greenish liquid was grist for a bomb. A few officials did know but assumed that they could ensure safety by never assembling any amount close to the critical mass estimated by the physicists. They lacked knowledge that had become second nature to the experts at Los Alamos: that the presence of hydrogen, as in water, slowed neutrons to dangerously effective speeds and so reduced the amount of uranium 235 needed to sustain a reaction. Segrè astounded his Oak Ridge hosts by telling them that their accumulating stores of wet uranium, edging closer to bomb-grade purity, were likely to explode.

Feynman began by retracing Segrè's steps and found that the problem was even worse than reported. In one place Segrè had been led into the same storeroom twice and had inadvertently noted two batches as though

they were accumulating in separate rooms. Through dozens of rooms in a
series of buildings Feynman saw drums with 300 gallons, 600 gallons, 3,000
gallons. He made drawings of their precise arrangements on floors of brick
or wood; calculated the mutual influence of solid pieces of uranium metal
stored in the same room; tracked the layouts of agitators, evaporators, and
centrifuges; and met with engineers to study blueprints for plants under
construction. He realized that the plant was headed toward a catastrophe.
At some point the buildup of uranium would cause a nuclear reaction that
would release heat and radioactivity at near-explosive speed. In answer to
the Eastman superintendent's question about extinguishing a reaction, he
wrote that dumping cadmium salts or boron into the uranium might help,
but that a supercritical reaction could run away too quickly to be halted by
chemicals. He considered seemingly remote contingencies: "During centrifuging some peculiar motion of the centrifuge might possibly gather metal
together in one lump, possibly near the center." The nightmare was that
two batches, individually safe, might accidentally be combined. He asked
what each possible stuck valve or missing supervisor might mean. In a few
places he found that the procedures were too conservative. He noted minute
details of the operations. "Is CT-1 empty when we drop from WK-1. . . ?
Is P-2 empty when solt'n is transfered . . . ? Supervisor OK's solution of P-
2's ppt. Under what circumstances?" Eventually, meeting with senior army
officers and company managers, he laid out a detailed program for ensuring
safety. He also invented a practical method—using, once again, a varia-
tional method to solve an otherwise unsolvable integral equation—that
would let engineers make a conservative approximation, on the spot, of the
safe levels of bomb material stored in various geometrical layouts. A few
people, long afterward, thought he had saved their lives.

Wielding the authority of Los Alamos was an instructive experience.
Feynman's first visit to Oak Ridge was his first ride on an airplane, and the
thrill was heightened by his special-priority military status on the flight,
with a satchel of secret documents actually strapped to his back under his
shirt. Oppenheimer had briefed his young protégé with care. Feynman
decided that the plant could not be operated safely by people kept ignorant
of the nature of their work, and he insisted that the army allow briefings
on basic nuclear physics. Oppenheimer had armed him with a means of
handling difficult negotiations:

"You should say: *Los Alamos cannot accept the responsibility for the safety
of the Oak Ridge plant unless——*"

". . . You mean me, little Richard, is going to go in there and say——"

". . . Yes, little Richard, you go in there and do that."

John von Neumann may have advised him during their thin-air walks that there could be honor in irresponsibility, but amid the barrels and carboys of the world's first nuclear hoards, responsibility caught up with him. Lives depended on his methods and judgments. What if his estimates were not conservative enough? The plant designers had taken his calculations as fact. He hovered outside himself, a young man watching, unsure and giddy, while someone carried off an impersonation of an older, more powerful man. As he said, recalling the feeling many years later, he had to grow up fast.

The possibility of death at Oak Ridge tormented him more urgently than the mass slaughter to come. Sometime that spring it struck him that the seedy El Fidel hotel, where he had nonchalantly roomed on his trips to Albuquerque, was a firetrap. He could not stay there any more.

I Will Bide My Time

Hitchhiking back one Sunday night, nearing the unpaved turnoff to Los Alamos, he saw the lights of a carnival shining from a few miles north in Espanola. Years had passed since he and Arline last went to a carnival, and he could not resist. He rode a rickety Ferris wheel and spun about in a machine that whirled metal chairs hanging on chains. He decided not to play the hoop-toss game, with unappealing Christ figures as prizes. He saw some children staring at an airplane device and bought them a ride. It all made him think sadly about Arline. Later he got a lift home with three women. "But they were kind of ugly," he wrote Arline, "so I remained faithful without even having the fun of exerting will power to do it."

A week later he rebuked her for some act of weakness and then, miserable, wrote the last letter she would read.

My Wife:

I am always too slow. . . . I understand at last how sick you are. I understand that this is not the time to ask you to make any effort to be less of a bother to others. . . . It is a time to comfort you as you wish to be comforted, not as I think you should wish to be comforted. It is a time to love you in any way that you wish. Whether it be by not seeing you or by holding your hand or whatever.

This time will pass—you will get better. You don't believe it, but I do. So I will bide my time & yell at you later and now I am your lover devoted to serving you in your hardest moments. . . .

I am sorry to have failed you, not to have provided the pillar you need to lean upon. Now, I am a man upon whom you can rely, have trust, faith, that I will not make you unhappy any longer when you are so sick. Use me as you will. I am your husband.

I adore a great and patient woman. Forgive me for my slowness to understand.

I am your husband. I love you.

He also wrote to his mother, breaking a long silence. One night he awoke at 3:45 A.M. and could not get back to sleep—he did not know why—so he washed socks until dawn.

His computing team had put everything aside to concentrate on one final problem: the likely energy of the device to be exploded a few weeks hence at Alamogordo in the first and only trial of the atomic bomb. The group's productivity had risen many times since he took over. He had invented a system for sending three problems through the machine simultaneously. In the annals of computing this was an ancestor to what would later be called parallel processing or pipelining. He made sure that the component operations of an ongoing computation were standardized, so that they could be used with only slight variations in different computations, and he had his team use color-coded cards, with a different color for each problem. The cards circled the room in a multicolored sequence, small batches occasionally having to pass other batches like impatient golfers playing through. He also invented an efficient technique for correcting errors without halting a run. Because a mistake only propagated a certain distance in each cycle, when an error was found it would have tainted only certain cards. Thus he was able to substitute small new card decks that eventually caught up with the main computation.

He was at work in the computing room when the call came from Albuquerque that Arline was dying. He had arranged to borrow Klaus Fuchs's car. When he reached her room she was still. Her eyes barely followed him as he moved. He sat with her for hours, aware of the minutes passing on her clock, aware of something momentous that he could not quite feel. He heard her breaths stop and start, heard her efforts to swallow, and tried to think about the science of it, the individual cells starved of air, the heart unable to pump. Finally he heard a last small breath, and a nurse came and said that Arline was dead. He leaned over to kiss her and made a mental note of the surprising scent of her hair, surprising because it was the same as always.

The nurse recorded the time of death, 9:21 P.M. He discovered, oddly, that the clock had halted at that moment—just the sort of mystical phenomenon that appealed to unscientific people. Then an explanation occurred to him. He knew the clock was fragile, because he had repaired it several times, and he decided that the nurse must have stopped it by picking it up to check the time in the dim light.

The next day he arranged an immediate cremation and collected her few possessions. He returned to Los Alamos late at night. A party was under way at the dormitory. He came in and sat down, looking shattered. His computing team, he found the next day, was deep in a computing run, not needing his help. He let his friends know that he wanted no special attention. In her papers he found a small spiral notebook she had used to log her medical condition. He carefully penned a final entry: "June 16— Death."

He returned to work, but soon Bethe ordered him home to Far Rockaway for a rest. (His family did not know he was coming until the telephone rang and a foreign-accented voice asked for him. Joan replied that her brother had not been home for years. The voice said, *When he comes in, tell him Johnny von Neumann called.*) There Richard stayed for several weeks, until a coded telegram arrived. He flew from New York Saturday night and reached Albuquerque at noon the next day, July 15. An army car met him and drove him directly to Bethe's house. Rose Bethe had made sandwiches. Feynman was barely in time to catch the bus to the observation site, a ridge overlooking the patch of New Mexican desert, the Jornada del Muerto, already called by its more modern name, ground zero.

We Scientists Are Clever

The test seared images into all their memories: for Bethe the perfect shade of ionized violet; for Weisskopf the eerie Tchaikovsky waltz and the unbidden memory of the halo in a medieval painting of Christ's ascension; for Otto Frisch the cloud rising on its tornado stem of dust; for Feynman the awareness of his "scientific brain" trying to calm his "befuddled one," and then the sound he felt in his bones; for so many of them the erect figure of Fermi, letting his bits of paper slip through the wind. Fermi measured the displacement, consulted a table he had prepared in his notebook, and estimated that the first atomic bomb had released the energy of 10,000 tons of TNT, somewhat more than the theorists had predicted and somewhat less than later measurements would suggest. Two days later, calculating that the ground radiation should have decayed sufficiently, he drove with Bethe and Weisskopf to inspect the glazed area that Feynman saw from an observation plane. The molten sand, the absent tower. Later a small monument marked the spot.

The aftermath changed them all. Everyone had played a part. If a man had merely calculated a numerical table of corrections for the effect of wind on the aerodynamically clumsy Nagasaki bomb, the memory would never leave him. No matter how innocent they remained through the days of Trinity and Hiroshima, those who had worked on the hill had knowledge that they could not keep from themselves. They knew they had been complicit in the final bringing of fire; Oppenheimer gave public lectures explaining that the legend of Prometheus had been fulfilled. They knew, despite their labors and ingenuity, how easy it had all been.

The official report on its development stated later that year that the bomb was a weapon "created not by the devilish inspiration of some warped genius but by the arduous labor of thousands of normal men and women working for the safety of their country." Yet they were not normal men and women. They were scientists, and some already sensed that a dark association like a smoke cloud would attach itself to the hitherto-innocent word *physicist*. (A draft of the same report had said, "The general attitude of Americans toward their scientists is a curious mixture of exaggerated admiration and amused contempt"—never again was it quite so amused.) Not long after writing his triumphant letter home, Feynman wrote some arithmetic on a yellow pad. He estimated that a Hiroshima bomb in mass production would cost as much as one B-29 superfortress bomber. Its destructive force surpassed the power of one thousand airplanes carrying ten-ton loads of conventional bombs. He understood the implications. "No monopoly," he wrote. "No defense." "No security until we have control on a world level."

Under the heading "SKILL & KNOWLEDGE" he concluded:

Most was known. . . . Other peoples are not being hindered in the development of the bomb by any secrets we are keeping. They might be helped a little by our mentioning which of two processes is found to be more efficient, & by our telling them what size parts to plan for—but soon they will be able to do to Columbus, Ohio, and *hundreds* of cities like it what we did to Hiroshima.

And we scientists are clever—too clever—are you not satisfied? Is four square miles in one bomb not enough? Men are still thinking. Just tell us how big you want it!

Many of the scientists found their magic mountain hard to leave. Lingering for months, they continued minor research that had acquired its own momentum, or skied near the Valle Grande, where they were intermittently aware that their tow rope had previously served to hoist the bomb up the tower at ground zero. Some joined the hydrogen bomb project that Teller would lead, and some remained at Los Alamos permanently, as the compound behind the fence grew into a major national laboratory and a central fixture of the American weapons-research establishment. The scientists who slowly dispersed began to realize how unlikely they were to work ever again in such a purposeful, collegial, and passionate scientific enterprise.

Nothing held Feynman to Los Alamos. He was joining Bethe's faculty at Cornell. Raymond Birge at Berkeley had angered Oppenheimer by delaying the job offer he had recommended. Oppenheimer wrote again: "It would seem to me that under these circumstances too much of courage was not required in making a commitment to a young scientist. . . . I perhaps presumed too much on the excellence of his reputation among those to whom he is known. . . . He is not only an extremely brilliant theorist, but a man of the greatest robustness, responsibility and warmth, a brilliant and lucid teacher . . . one of the most responsible men I have ever met. . . . We regard him as invaluable here; he has been given a responsibility and his work carries a weight far beyond his years. . . ." Birge finally came through with an offer to Feynman that summer, but too late.

When Arline was alive they had talked about moving to California for her health. Now Bethe easily swayed him.

Feynman became the first of the group leaders to leave, in October 1945. There were only a few reports to write up and some final safety tours of Oak Ridge and Hanford. It was on his last trip to Oak Ridge, as he walked past a shop window, that he happened to see a pretty dress. Before he could prevent it, a thought came. *Arline would like that.* For the first time since her death, he wept.

CORNELL

· ÷ ·

For physics as an enterprise within American culture there were two eras. One ended and the other began in the summer of the atomic bombs. Politicians, educators, newspaper editors, priests, and the scientists themselves began to understand the divide that had been crossed.

"Among the divinities of ancient Greece, there was a Titan named Prometheus," ran a typical essay in *The Christian Century* the next winter. "He stole fire from heaven and gave it to man. . . . For this act, Prometheus has been held in highest honor as a benefactor of humanity and the divine patron of science and learning." No more. Now, rather to the cleric essayist's delight, the atomic bomb had humbled Prometheus's heirs, the scientists. Their centuries of progress had decisively ended with their invention of a device of human self-destruction. Now it was time for Christian ministers to step in. Even the scientists, he said, "have for the first time in history turned aside from their vocation and become statesmen and evangelists, preaching the grim gospel of damnation unless men repent." Here he was alluding to J. Robert Oppenheimer, for Oppenheimer had already seen the aptness of the Promethean legend—who could have missed it?—and had begun to speak out both to the public and to scientists. What Oppenheimer preached, however, was more subtle than a gospel of damnation. He reminded listeners that the religious had long felt threatened by science, and now the only mildly God-fearing public had something real to fear. He suspected that atomic weapons would scare people more than any scientific development since Darwin's theory of evolution.

Already, in November 1945, with relieved soldiers and sailors streaming home from the Pacific Theater, before fallout shelters, nuclear proliferation, and ban the bomb had a chance to enter the language, Oppenheimer anticipated the time when celebration would give way to dread. "Atomic weapons are a peril which affect everyone in the world," he told his friends and colleagues of the past thirty months. His audience filled the largest assembly hall in Los Alamos, its movie theater. He knew that the newspapers and magazines glorifying the scientists' achievement would soon recognize how little real mystery there had been, how unremarkable, actually, were the problems of nuclear fission (if not implosion), how easy atomic bombs would be to make, and how affordable for many nations.

Prometheus was not the only mythic figure standing in for the scientist; the other was Faust. Lately the Faustian bargain for knowledge and power had not seemed so horrible as it had in medieval times. Knowledge meant washing machines and medicines, and the devil had softened into an amusing character for Saturday cartoons and Broadway musicals. But now the fires in two Japanese cities renewed a primal understanding that the devil was not so tame. It might mean something, after all, to sell him one's soul. Oppenheimer knew, partly from introspection, that the scientists had immediately begun to question their own motives. "It's a terrible thing that we made," Robert Wilson had said to Feynman, surprising him and pricking his ebullient bubble. Others were beginning to agree. Oppenheimer reminded them of what they were reminding themselves: that two years earlier a Nazi bomb had seemed possible and that the American victory had seemed far from inevitable. He acknowledged that these justifications had faded. Some people, he said, might have been driven by a less high-minded motivation, no more than curiosity and a sense of adventure, and he surprised some of them by saying, "and rightly so." He said it again: "And rightly so." Feynman had left Los Alamos several days before, so he did not hear, nor did he need to hear, Oppy's reminder of their shared credo, a credo now being welded to the most painful act of self-justification they had ever had to perform:

When you come right down to it the reason that we did this job is because it was an organic necessity. If you are a scientist you cannot stop such a thing. If you are a scientist you believe that it is good to find out how the world works; that it is good to find out what the realities are; that it is good to turn over to mankind at large the greatest possible power to control the world. . . . It is not possible to be a scientist unless you believe that the knowledge of the world, and the power which this gives, is a thing which is of intrinsic value to humanity, and that you are using it to help in the spread of knowledge, and are willing to take the consequences.

Thus spoke a bringer of fire.

The relations between Americans and their scientists had changed. It became an instant truism that science meant power. Science as an institution—"organized science"—ranked second only to the military as a guarantor of what was being called national security. President Harry S Truman told the Congress that fall that America's role in the world would depend directly on research coordinated by universities, industrial companies, and the government: "The events of the past few years are both proof and prophecy of what science can do." In short order the government established an Atomic Energy Commission, an Office of Naval Research, and a National Science Foundation. Permanent national laboratories with no precedent in the prewar world arose at Los Alamos; at Oak Ridge; at Argonne, south of Chicago; at Berkeley; and at Brookhaven, Long Island, on a six-thousand-acre former army site. Money flowed copiously. Before the war the government had paid for only a sixth of all scientific research. By the war's end the proportions had flipped: only a sixth was financed by all nongovernment sources combined. The government and the public gained a new sense of proprietorship over the whole scientific enterprise. As physicists began to speak out about world government and the international control of nuclear arms, so an army of clerics, foundation heads, and congressmen now made the mission and the morality of science a part of their lecture-circuit repertoire.

On the whole, the popular press lionized Oppenheimer and his colleagues. To have worked on the bomb gave a scientist a stature matched only by the Nobel Prize. By comparison it was nothing to have created radar at the MIT Radiation Laboratory, though by a plausible calculus radar had done more to win the war. The word *physicist* itself finally came into vogue. Einstein was now understood to be a physicist, not a mathematician. Even nonnuclear physicists acquired prestige by association. Soon Wilson, Feynman's recruiter, would look back wistfully to "the quiet times when physics was a pleasant, intellectual subject, not unlike the study of Medieval

French in its popular interest." The atomic scientists felt the guilt that flowed from the sudden deaths of at least one hundred thousand residents of Hiroshima and Nagasaki; meanwhile the scientists found themselves hailed as hero wizards, and this was a more complex role than many of them realized at first, containing as it did the seeds of darker relationships. In less than a decade Oppenheimer himself would lose his security clearance in the classic McCarthy-era auto-da-fé. The public would find that knowledge created by scientists was a commodity requiring special handling. It could be stamped CLASSIFIED or betrayed to foreign enemies. Knowledge was the grist of secrets and the currency of spies.

Theoretical physicists, too, had learned something about their kind of knowledge. Oppenheimer reminded them of it, in his November 1945 talk at Los Alamos. The nature of the work in theoretical physics before the war had forced a certain recognition on them, he said—the recognition that human language has limits, that people choose concepts that correspond only faintly to things in the real world, like the shadows of ghosts. Before the bomb work began, quantum mechanics had already altered the relations between science and common sense. We make models of experience, and we know that our models fail to meet the reality.

The University at Peace

Their remarkable change in status buffeted every American institution that made a home for physicists. At Cornell, President Edmund Ezra Day was one of the first to feel the force of the transition, in the stark contrast between two budget meetings with his physicists, one during and one after the war.

In the first, he sat down with his chief experimentalist, Robert F. Bacher, who was setting off on his leave of absence; ultimately Bacher led the bomb project's experimental physics division. Bacher pleaded for a cyclotron like those at Berkeley and Princeton. He pressed Day to find a way of providing operating costs that he said might amount to as much as a professor's salary, from four thousand to five thousand dollars a year.

In the second, two months after Hiroshima, Day's physicists told him that a far more powerful accelerator would be required, along with a new laboratory to house it. This time they asked for a capital expenditure of

$3,000,000 and an operating budget that would begin at $250,000. They suggested, furthermore, that without this commitment they would have to look elsewhere for a more propitious environment for nuclear science. The trustees had no obvious source of funds, but after a heated meeting with Day they voted unanimously to proceed. Day declared: "The problem is not to control nuclear forces but to control nuclear physicists. They are in tremendous demand, and at a frightful premium." Bacher himself, after returning to Cornell briefly, left for Washington to serve as the first scientist on the newly formed Atomic Energy Commission. Three years later Cornell had a new accelerator, a synchrotron. The trustees' leap of faith had been vindicated by generous funding from the Office of Naval Research. Three years after that, the synchrotron had passed into obsolescence and a new version was already under construction.

Feynman's first glimpse of the postwar university came in the dead of night before the start of classes in the fall of 1945. Ithaca was a village at the dimmest reaches of a New York City boy's sense of his state's geography, practically in Ohio. He made the journey by train, using the long hours to begin sketching out a basic graduate course he was supposed to teach in mathematical methods for physicists. He debarked with a single suitcase and a self-conscious sense of being, finally, a professor. He suppressed the urge to sling his bag over his shoulder as usual. Instead he let a porter guide him to the rear seat of a taxicab. He told the driver to take him to the biggest hotel in town.

In Ithaca, as in towns and cities across America that fall, the hotels and short-term apartments were booked. Housing was scarce. With demobilization college enrollments were exploding. Boom was in the air. Even sleepy Ithaca seemed like a Western town amid the gold rush. Cornell was building houses and barracks at emergency speed. The week before Feynman arrived, five new barracks burned down. He tried a second hotel. Then he realized he could not afford to wander by taxicab, so he checked his suitcase and began to walk, past darkened houses and dormitories. He realized he must have found Cornell. Huge raked piles of leaves dotted the campus, and they started to look like beds—if only he could find one out of the glare of the streetlights. Finally he spotted an open building with couches in the lobby and asked the janitor if he could spend the night on one. He explained awkwardly that he was a new professor.

The next morning he washed as well as he could in the public bathroom, checked in at the physics department, and made his way to a campus housing office in Willard Straight Hall, near the center of the sloping campus. There a clerk told him haughtily that the housing situation was so bad that last night a professor had had to sleep in the lobby. "Look, buddy," Feynman snapped back, "I'm that professor. Now do something for me." He was unpleasantly startled to realize that in a town Ithaca's size he could set off a rumor and circle back into its wake within a matter of hours. He also began to realize that he was going to have to readjust his internal clock. The war had left him with a sense of urgency about appointments and deadlines. Even as ten thousand undergraduates arrived, Cornell seemed slack. He was surprised to discover that the administration had scheduled a full week with nothing for him to do but explore the campus and prepare for classes. Speech patterns struck him as slow, with none of the *beep-beep-beep* nervousness he had got used to. People took time to talk about the weather.

His first months were lonely. None of his close colleagues had been in such a hurry to begin postwar life. Even Bethe did not leave Los Alamos for Cornell until December. The school year began late and stayed unsettled. Space ran short. Workers subdivided rooms in Rockefeller Hall. Closets became offices. Outside, three tennis courts gave way to hasty wooden barracks. Feynman soon shared his dingy Rockefeller office with a colleague from Los Alamos, Philip Morrison, who had carried the atomic bomb's plutonium core to Alamogordo in the back seat of an army sedan. Morrison had been lured by the sweet, serious Bethe, so full of integrity—and also by Feynman, though it now seemed, surprisingly, that Feynman was depressed and lonely. Bethe sensed this, too, but few others noticed. Later Bethe noted dryly, "Feynman depressed is just a little more cheerful than any other person when he is exuberant."

He spent time in the library reading the mildly bawdy *Arabian Nights* and staring hopefully at women. Unlike most of the Ivy League universities, Cornell had accepted women as undergraduates since its founding, after the Civil War, though they automatically matriculated in the College of Home Economics. He went to freshman dances and ate in the student cafeteria. He looked younger than his twenty-seven years, and he did not stand out amid all the returning servicemen. His dance partners looked askance at what sounded like a line—that he was a physicist just back from building the atomic bomb. He missed Arline. Even before leaving Los Alamos he had begun dating other women—especially beautiful women, in what some of his friends saw as a frenetic, razor-edge denial of grief.

A gulf had opened between Feynman and his mother. Lucille, after so adamantly opposing Richard's marriage, had written painfully on Arline's death:

> . . . now I want you to know that I'm proud and glad you married her & did what you could to make her short life happy. She worshipped you. Forgive me for not seeing things your way. I was frightened for you—for what you would have to bear. But you bore it so well. Now try to face life without her . . .

Begging him to come home, she promised him piles of rice and sugar buns and gave her word that no one would tell him to comb his hair. He did come, briefly, for a few days in July. Then, in August, the news of the atomic bomb broke over the household like a lightning storm. Friends and relatives called almost continuously. Lucille tried in vain to get through to Santa Fe by telephone. One cousin called from a wire-service office to read a comment of Oppenheimer's that had just come across the ticker. After 11 P.M. the phone rang and a voice said, "This is the Princeton *Triangle*. Is it true that your son R. P. Feynman had more gravy stains on his gown that any other man at the Graduate College in 1940?" It was another cousin.

"I have a sense of humor, too," Lucille wrote to Richard, "but I don't think this is a funny occasion."

> I felt thrilled & frightened at your part in this tremendous thing. No one can be really joyous. It is with horror that I listen to the death & destruction the bomb has caused. . . . I pray that this horrible destruction of man by man may be the climax of all such destruction. . . . No wonder I thought you were nervous. Who wouldn't be, playing around in such a dangerous place.

The combination of pride and terror—the scientists, too, were feeling it that night—stirred a remarkable memory. "It reminded me of the time I was playing bridge in the living room & my child prodigy had a little fire in a trash basket he was holding outside the window.

"By the way," she added, "I don't think you ever told me how you put it out."

Feynman did not stop at home on his way to Ithaca from New Mexico that fall. At some point Lucille began to realize how much damage had been done by her opposition to the marriage. Late one night, unable to sleep, she got out of bed and penned an anguished letter—a love letter from mother to son—beginning, "Richard, What has happened between you and your family? What has driven us apart? My heart yearns for you. . . . My heart is full to bursting & hot tears burn my eyes as I write."

She wrote about his childhood: how much he had been wanted and treasured; how she had read him beautiful stories; how Melville had made patterns for him from colored tiles; how they had tried to invest him with a sense of morality and duty. She reminded him of the pride they had felt in all his achievements, from high school through graduate school.

More times than I can enumerate here my heart has leaped for joy because of you. . . . And now—now—strange harvest that I reap. We are as far apart as the poles.

Without mentioning Arline, she said she felt a sense of shame. "The fault must be mine. Some where along the way I lost you." Other mothers, she said, had sons who loved them. Why not her? She closed with as impassioned a plea as any spurned lover could make.

I need *you*. I want you. I will *never* give you up. Not even death can break the bond between us. . . . Think of me sometimes & let me know that you are thinking of me. My darling, oh my darling, what more can I say to you. I adore you & *always* will.

He did go home for Christmas in 1945. Gradually the wound began to heal. In the meantime Feynman made some indirect efforts to find his way back into the unfinished theory that had occupied him at Princeton, but they did not lead to anything usable. The culmination of the driven, purposeful work of the past three years had left a void that he could not easily fill. He found it hard to concentrate on research. As spring came he would sit on the grass outdoors and worry about whether he had slipped past his best working years without achieving anything. He had built a reputation among senior physicists, but now, back in a world returning to normal, he realized that he had not done the normal work to go with the reputation. Since his two published papers in college—his squib on cosmic rays with Vallarta and his undergraduate thesis—his only journal publications had been accounts of the work with Wheeler on the absorber theory, already looking short-lived.

Phenomena Complex—Laws Simple

If Feynman was struggling to find his footing, Julian Schwinger was not. Since growing up at opposite ends of New York City, in neighborhoods that might as well have been a thousand miles apart, they had become competitors without either quite acknowledging it. Their routes into physics had remained utterly separate, as had their styles. Schwinger, with heavy owlish eyes and a mild stoop even in his twenties, took as great pains to achieve elegance as Feynman did to remain rough-hewn. He dressed carefully and expensively and drove a Cadillac. He worked nocturnally, usually sleeping until late afternoon. His lectures had already become famous for their seamlessness and uninterruptibility. He prided himself on speaking without notes. A young Englishman who heard him (and who considered Feynman's ebullience slightly tiring, by contrast) thought Schwinger became "a man possessed"—"It seems to be the spirit of Macaulay which takes over, for he speaks in splendid periods, the carefully architected sentences rolling on, with every subordinate clause duly closing." He liked to make his listeners think. He would never announce directly that he had married and taken a honeymoon, when he could say, "I abandoned my bachelor quarters and embarked on an accompanied, nostalgic trip around the country. . . ." His equations had something of the same style.

His patron had been I. I. Rabi, who never tired of describing their first encounter: Schwinger, a seventeen-year-old waiting quietly in his office, had finally piped up to settle an argument over a controversial foray into quantum-mechanical paradox just published by Einstein, Boris Podolsky, and Nathan Rosen. With the arrogance of a shy young man determined to plow his own course, Schwinger was already in administrative difficulties at City College because he rarely attended classes. Rabi helped him transfer to Columbia and then took devilish pleasure in encouraging his irate instructors to carry out their threats to flunk him. "Are you a mouse or a man? Give him an F," he told one dull chemistry professor; he judged correctly that the grade would come to haunt the professor more than it would the student. Even before Schwinger got his college diploma at the age of nineteen, Rabi was having him fill in as the lecturer in his quantum-mechanics course. Also before graduating, he completed the research that served as his doctoral dissertation. Fermi, Teller, and Bethe each knew him, knew his work, or had collaborated with him. Meanwhile Feynman, barely three months younger, was completing his sophomore year at MIT. Schwinger published a fecund series of research papers, mostly in the *Physical Review*, each highly polished, with a dozen different collaborators. By the time Feynman published his undergraduate thesis, Schwinger was in Berkeley as a National Research Council fellow, working directly with Oppenheimer.

With Rabi, he chose to avoid Los Alamos in favor of radar and the Radiation Laboratory. He never seemed to lose a stride. By the war's end Rabi had him replace Pauli as a special lecturer in charge of bringing the laboratory's scientists up to date with nonwar physics. For the atomic bomb scientists, isolated as they were behind their desert fence, the war brought a more total interruption of normal careers. Physicists Feynman's age were especially aware of it. They had just reached what should have been their crucial, productive years. Schwinger made one tour through Los Alamos in 1945 and met Feynman briefly for the first time. Feynman marveled at how much this contemporary had managed to publish. He had thought Schwinger was older. When he had long since forgotten the content of Schwinger's lecture to the Los Alamos theorists, he still remembered the style: the way Schwinger walked into the room, his head tilted, like a bull into the ring; the way he conspicuously set his notebook aside; the intim-idating perfection of his discourse.

Now Schwinger was at Harvard, where he was shortly to become a twenty-nine-year-old full professor. The Harvard committee had seriously considered only Bethe for the same opening and worried meanwhile whether Schwinger would be able to wake up to teach classes that met as early as noon. He managed, and his lectures on nuclear physics quickly became a draw for the entire Harvard and MIT physics community.

Feynman, meanwhile, poured energy into his more mundane course in the methods of mathematical physics. This was a standard course, taught in every physics department, though it occurred to Feynman that he had just lived through a momentous change in physicists' mathematical meth-ods. At Los Alamos mathematical methods had been put through a crucible:

refined, clarified, rewritten, reinvented. Feynman thought he knew what was useful and what was mere textbook knowledge taught because it had always been taught. He intended to emphasize nonlinearity more than was customary and to teach students the patchwork of gimmicks and tricks that he used himself to solve equations. Beginning with his jottings on the night train that brought him to Ithaca, he designed a new course from the bottom up.

On the first page of a cardboard notebook like the ones he had used in high school he began with first principles:

Phenomena complex—laws simple—
connection is math-phys—the solution of equ obtained from laws.

He was thinking about how to mold students in his own image. How did *he* solve problems?

Know what to leave out. . . . physical insight knowing what can be done by math.

He decided to give the students a blunt summary of what did and did not lie ahead.

Lots of tricks to introduce—no time for complete study or math rigor demonstration. Lots of work.

He crossed that out.

Really introduce each subject.

But after all it *would* be lots of work.

Lots of work—practice. Interested in more detail, read books, see me, practice more examples. If no go—OK we slow up. Hand in some problems so I can tell.

He would promise them important mathematical methods left out of ordinary courses, as well as methods that were altogether new. It would be practical, not perfect, mathematics.

Specify accuracy required. Let's go

He scanted some of the laborious traditional techniques, such as contour integration, because he had so often found—winning bets in the process—that he could handle most such integrals directly by frontal assault. Whether he would succeed in conveying such skills to his students was a question that worried some of his colleagues as they watched Feynman plow apart the mathematical-methods syllabus. Nevertheless, during the few years that he taught the course, it drew some of the younger members of the physics and mathematics faculty along with the captive graduate students. The coolest among them had to feel the jolt of an examination problem that began, "In an atom bomb in the form of a cylinder radius a, height 2π, the density of neutrons n . . ." The students found themselves in the grip of a theorist whose obsession with mathematical methods concerned the uneasy first principles of quantum mechanics. Again and again he showed his affinity with the purest core issues of the propagation of sound and light. He drove his students through calculations of the total intensity of radiation in all directions when emitted by a periodic source; through the reluctant visualization of vectors, matrices, and tensors; through the summations of infinite series that sometimes converged and sometimes failed to converge, running inconveniently off toward infinity.

Gradually he settled in at Cornell, though he still made no progress on his theoretical research. The atomic bomb was on his mind, and he went on the local radio to speak about it in unadorned language. *Announcer: Last week Dr. Feynman told you what one atom bomb did to Hiroshima, and what one bomb would do to Ithaca* . . . The interviewer asked about atomic-powered automobiles. Many listeners, he said, were awaiting the day when they could slip a spoonful of uranium into the tank and thumb their noses at the filling stations. Feynman said he doubted the practicality of that—"the rays emitted by the fission of the uranium in the engine would kill the driver." Still, he had spent time working out other applications of nuclear power. At Los Alamos he had invented a type of fast reactor for generating electric power and had patented it (in the government's behalf). He was also thinking about space travel. "Dear Sir," he wrote to a physicist colleague as 1945 came to a close, "I believe that interplanetary travel is now (with the release of atomic energy) a definite *possibility*." He had a radically quirky, almost flaky, proposal. Rocket propulsion would not be the answer, he said. It was fundamentally limited by the temperature and atomic weight of the propulsive gas, the temperature in turn being limited by the ability of metal to withstand heat. He predicted—anticipating the ungainly disposable boosters and giant fuel tanks that became the curse of space travel thirty years in the future—that the weight and bulk of fuel would exceed by too many times the weight and bulk of the vehicle.

Instead he proposed a form of jet propulsion, using air as the propellant. Jet technology had just now reached practicality in airplanes. Feynman's spacecraft would use the outer edges of the Earth's atmosphere as a sort of warm-up track and accelerate as it circled the earth. An atomic reactor would power the jet by heating the air that was sucked into the engine. Wings would be used first to provide lift and then, when the speed rose beyond five miles per second, "flying upside down to keep you from going off the earth, or rather out of the atmosphere." When the craft reached a useful escape velocity, it would fly off at a tangent toward its destination like a rock from a slingshot.

Yes, air resistance, heating the ship, would be a problem. But Feynman thought this could be overcome by delicately adjusting the altitude as the craft sped up—"if there is enough air to cause appreciable heating by friction there surely is enough to feed the jet engines." The engines would need impressive engineering to operate in such a wide range of air densities, he admitted. He did not address a problem of symmetry: how such a spacecraft would slow down on reaching an airless destination such as the moon. In any event he could not have anticipated the killing flaw in his idea: that people would lose faith in the innocence of nuclear reactors flying about overhead.

They All Seem Ashes

He visited Far Rockaway just before the fall semester began in 1946 and gave another talk on the atomic bomb at the local Temple Israel the day after Yom Kippur. The synagogue had a glamorous new rabbi, Judah Cahn, who delivered widely admired sermons on modern problems. Feynman's parents, despite their atheism, had started attending from time to time. Melville's health seemed slightly better. His uncontrollable high blood pressure had become a constant source of worry to the family, and in the preceding spring he had traveled out to the Mayo Clinic, in Minnesota, where he was enrolled in an early experiment on the effect of diet. He accepted a strict regimen of rice and fruit. It seemed to work. His blood pressure decreased. He returned home and occasionally sneaked out, in violation of doctors' orders, to play golf with friends. He was fifty-six years old. One day Feynman saw him at the table, staring at a salt shaker. Melville closed one eye, opened it, closed the other eye, and said he had a blind spot. A small blood vessel must have burst in his brain, he said.

The knowledge that sudden death might come at any time hung over the family. Melville and his son almost never wrote each other—Lucille handled the intrafamily correspondence—but when Richard first accepted the Cornell professorship he sent his father a letter expressing twenty-five years of love and gratitude, and Melville, moved, responded in kind. His chest was swelled with pride, he wrote (while Lucille complained that he was wasting paper by writing on only one side):

It is not so easy for a Dope of a father to write to a son who has already arrived to a state of learning and wisdom beyond his. . . . That was all right when you were small and I had a great advantage over you—but today it would be more equitable if I could bask in the sunlight of *your* knowledge, and sit by your side and learn from you some of the more wondrous secrets of nature that now are beyond my ken but are known to you.

On October 7 he collapsed from a stroke. He died the next day. Richard signed his second death certificate in two years. Melville Feynman had written him: "The dreams I have often had in my youth for *my* own development, I see coming true in your career. . . . I envy the life of culture you will have being constantly with so many other big men of equal culture."

The interment took place at Bayside Cemetery nearby in Queens, a vast rolling field of gravestones and monuments as far as the eye could see. Lucille's father had built a mausoleum there, a stone hut like a small bomb shelter. Midway through the ceremony Rabbi Cahn asked Richard, as eldest son, to say the Kaddish with him. Joan watched in anguish as her brother's face froze. He wanted no part of a mourners' prayer in praise of God.

He told the rabbi he did not understand the Hebrew. Cahn merely switched to English. Richard listened to the words and refused to repeat them. He did not believe in God; he knew that his father had not believed in God; and the hypocrisy seemed unbearable. His disbelief had nothing of indifference in it. It was a determined, coolly rational disbelief, a conviction that the myths of religion cheated knowledge. He stood there surrounded by stone and grass near the undersized sepulchral vaults, assembled one atop another, that held the bones of his grandparents. One shelf, too, held the remains of his infant brother, Henry, memorialized now for twenty-two years after his life of one month. On Feynman's face was a look of tension and determination and also, it seemed to Joan at that moment, utter isolation. Leaving his father's coffin, he exploded in a rage. Their mother broke down and wept.

At Cornell the next week he seemed unchanged. Just as at Los Alamos—it had been barely a year before—if he grieved, he showed no one. He was proudly rational as ever—"realistic," he told himself. Classes began. Cornell's 1946 fall-term enrollment was its largest ever, nearly double prewar levels. Feynman was already a draw for young physicists, and he lectured with absolute confidence. Then, a few nights into the term—it was October 17—he took a pen and paper, let realism slip away, and wrote one last letter to the only person who could help him now:

D'Arline,

I adore you, sweetheart.

I know how much you like to hear that—but I don't only write it because you like it—I write it because it makes me warm all over inside to write it to you.

It is such a terribly long time since I last wrote to you—almost two years but I know you'll excuse me because you understand how I am, stubborn and realistic; & I thought there was no sense to writing.

But now I know my darling wife that it is right to do what I have delayed in doing, and that I have done so much in the past. I want to tell you I love you. I want to love you. I always will love you.

I find it hard to understand in my mind what it means to love you after you are dead—but I still want to comfort and take care of you—and I want you to love me and care for me. I want to have problems to discuss with you—I want to do little projects with you. I never thought until just now that we can do that together. What should we do. We started to learn to make clothes together—or learn Chinese—or getting a movie projector. Can't I do something now. No. I am alone without you and you were the "idea-woman" and general instigator of all our wild adventures.

When you were sick you worried because you could not give me something that you wanted to & thought I needed. You needn't have worried. Just as I told you then there was no real need because I loved you in so many ways so much. And now it is clearly even more true—you can give me nothing now yet I love you so that you stand in my way of loving anyone else—but I want you to stand there. You, dead, are so much better than anyone else alive.

I know you will assure me that I am foolish & that you want me to have full happiness & don't want to be in my way. I'll bet you are suprised that I don't even have a girl friend (except you, sweetheart) after two years. But you can't help it, darling, nor can I—I don't understand it, for I have met many girls & very nice ones and I don't want to remain alone—but in two or three meetings they all seem ashes. You only are left to me. You are real.

My darling wife, I do adore you.

I love my wife. My wife is dead.

Rich.

P.S. Please excuse my not mailing this—but I don't know your new address.

That he had written such a letter to the woman he loved, two years after her death, could never become part of the iconography of Feynman, the collection of stories and images that was already beginning to follow him about. The letter went into an envelope, the envelope into a box. It was not read again until after his death. Nor did Feynman speak of his graveside outburst at the burial of his father, even to friends, although they would have recognized at least one of its potential morals, his unwillingness to submit to hypocrisy. The Feynman who could be wracked by strong emotion, the man stung by shyness, insecurity, anger, worry, or grief—no one got close enough any more to see him. His friends heard a certain kind of story instead, in which Feynman was an inadvertent boy hero, mastering a bureaucracy or a person or a situation by virtue of his naïveté, his good humor, his brashness, his commonsense cleverness (not brilliance), and his emperor's-new-clothes honesty. The stories were true, at least in spirit, though like all stories they were selectively incomplete. They were admired, polished, retold, and once in a while even relived.

Many of his friends at Los Alamos had already heard variations of a draft-examination story, in which he needled an army examiner who asked him to hold out his hands. Feynman held them out: one palm up, the other palm down. The examiner asked him to turn them over, and he did, providing a wise-guy lesson in symmetry: one palm down, the other palm up. Shortly after his first year at Cornell, Feynman got a chance to refine the story. The army was still drafting, and his educational deferments had run their course. The Selective Service scheduled a new physical examination. Feynman's version of the story, told countless times in the decades that followed, varied from the half serious to the strictly comic. The basic form went like this:

Stripped to his underwear, he goes from booth to booth, until—"Finally, we get to Booth No. 13, Psychiatrist."

Witch doctor. Baloney. Faker. Feynman held an extreme view of psychiatry. His mind was his bailiwick, and he liked to think himself in control. Sensitive psychiatrists might have noted his tendency to deny the occasional roiling undercurrents; the undercurrents and the denial were their bailiwick. He preferred to stress the unscientific hocus-pocus of their enterprise (conveniently shifting terminology, lack of reproducible experiments), as reflected in a movie he had seen recently, Alfred Hitchcock's *Spellbound*, in which "a woman" (Ingrid Bergman), "her hand is stuck and she can't play the piano . . . she used to be a great pianist. . . ." Certainly he never considered whether he (himself at that moment unable to work) might have had any but the most rational of reasons for feeling: "It's boring as hell. . . ." The woman ducks off-screen with her psychiatrist, comes back, sits down at the piano, and plays. "Well, this kind of baloney, you know, I can't stand it. I'm very anti. Okay?" Apart from everything else, psychiatrists are doctors, and Feynman has his reasons for holding doctors in contempt.

The psychiatrist looks at his file and says with a smile, *Hello, Dick! Where do you work?* ("Well, what the hell is he calling me *Dick* for? You know, he don't know me that well.")

Feynman says coldly, Schenectady. (This is temporarily true. He and Bethe are supplementing their Cornell salaries by working that summer at General Electric.)

Where at Schenectady, Dick?

Feynman tells him.

You like your work, Dick? "I couldn't like him less, you know? Like a guy bothering you in a bar."

Now a fourth question—*Do you think people talk about you?*—and

Feynman detects that this is the routine: three innocent questions and then down to business.

"So I say, Yeah . . ." At this point Feynman, relating the story, takes on a tone of misunderstood innocence. He is scrupulously honest. If only the psychiatrist would forget the formulas, forget the mumbo jumbo, and try to understand him. "I wasn't trying to fake it. . . . I meant in the sense that my mother talks to her friends. . . . I tried to explain—honest. . . ." The psychiatrist makes a note.

Do you think people stare at you? Feynman would say no—honest— but the psychiatrist adds, *For example, do you think that any of the fellows sitting on the benches are looking at us now.* Well, Feynman has sat on one of those benches, and there was not much else to look at. He does some mental arithmetic. "So I figure . . . there are about twelve guys in the thing and about three of them are looking—well, that's all they've got to do—so I say, to be conservative, 'Yeah, maybe two of them are looking at us.' "

He turns around to check, and sure enough. But the psychiatrist, "this nincompoop, this nincompoop . . . doesn't bother to turn around and find out if it's true or not." (No scientist he.)

Do you talk to yourself? "I admitted that I do. . . ." ("Incidentally, I didn't tell him something which I can tell you, which is I find myself sometimes talking to myself in quite an elaborate fashion . . . : 'The integral will be larger than this sum of the terms, so that would make the pressure higher, you see?' 'No, you're crazy.' 'No, I'm not! No, I'm not!' I say. I argue with myself . . . I have two voices that work back and forth.")

I see you lost a wife recently. Do you talk to her? (The resentment that this question must stir goes beyond the comic bounds of the anecdote.)

Do you hear voices in your head? "No," Feynman says. "Very rarely." He admits a few occasions. Sometimes, in fact, just as he was falling asleep, he would hear Edward Teller, with a distinctive Hungarian accent, in Chicago giving him his first briefing on the atomic bomb.

There was much more: an argument about the nature of insanity, an argument about the value of life—Feynman in both cases continuing to get under the examiner's skin. Feynman acknowledged that one of his mother's sisters was mentally ill. And then the punch line, more serious than Feynman's audiences tended to realize.

Well, Dick, I see you have a Ph.D. Where did you study?

MIT and Princeton. Where did *you* study?
Yale and London. And what did you study, Dick?
Physics. And what did you study?
Medicine.
And *this* is medicine?

The story never included several plausible points. Feynman never pleaded that, having contributed three years of wartime service in the Manhattan Project, he ought to be exempt from a further contribution. Nor did he mention how destructive it would have been to his career as a theoretical physicist if he had been conscripted now, at the age of twenty-eight. He had to walk a narrow line. There was nothing amusing or stylish in the summer of 1946 about evading the draft. For most people, to be declared mentally deficient by one's draft board was a more frightening possibility than army service—far more damaging to one's civilian prospects. So the Selective Service established few safeguards against fakery in the psychiatric examination. It did not expect to see records of a previous history of mental illness, for example; in any case private psychiatric treatment was far more unusual than it became in the next generation. Examiners felt they could rely on a subject's naïve self-description to answer their checklist questions. Feynman repeated his answers to a second psychiatrist. His ability to conjure the voice of Teller was recorded as *hypnagogic hallucinations.* It was noted that the subject had a *peculiar stare.* ("I think it was probably when I said, 'And *this* is medicine?' ") He was rejected.

It occurred to him that the Selective Service would examine its own files and discover a series of official letters requesting deferment so that Feynman could conduct essential research in physics during the war. More recent letters stated that he was performing an important service educating future physicists at Cornell. Might someone conclude that he was deliberately trying to deceive the examiners? To protect himself, he wrote a letter, carefully phrased, stating for the record that he believed no weight should be given to the finding of psychiatric deficiency. The Selective Service replied with a new draft card: 4F.

Around a Mental Block

Princeton was celebrating the bicentennial of its founding with a grand explosion of pomp that fall: parties, processions, and a series of formal conferences that drew scholars and dignitaries from long distances. Dirac had agreed to speak on elementary particles as part of a three-day session on the future of nuclear science. Feynman was invited to introduce his one-time hero and lead a discussion afterward.

He disliked Dirac's paper, a restatement of the now-familiar difficulties with quantum electrodynamics. It struck him as backward-looking in its Hamiltonian energy-centered emphasis—a dead end. He made so many nervous jokes that Niels Bohr, who was due to speak later in the day, stood up and criticized him for his lack of seriousness. Feynman made a heartfelt remark about the unsettled state of the theory. "We need an intuitive leap at the mathematical formalism, such as we had in the Dirac electron theory," he said. "We need a stroke of genius."

As the day wore on—Robert Wilson speaking about the high-energy scattering of protons, E. O. Lawrence lecturing on his California accelerators—Feynman looked out the window and saw Dirac lolling on a patch of grass and gazing at the sky. He had a question that he had wanted to ask Dirac since before the war. He wandered out and sat down. A remark in a 1933 paper of Dirac's had given Feynman a crucial clue toward his discovery of a quantum-mechanical version of the *action* in classical mechanics. "It is now easy to see what the quantum analogue of all this must be," Dirac had written, but neither he nor anyone else had pursued this clue until Feynman discovered that the "analogue" was, in fact, exactly proportional. There was a rigorous and potentially useful mathematical bond. Now he asked Dirac whether the great man had known all along that the two quantities were proportional.

"Are they?" Dirac said. Feynman said yes, they were. After a silence he walked away.

Feynman's reputation was traveling around the university circuit. Job offers floated his way. They seemed perversely inappropriate and did nothing to help his mood of frustration. Oppenheimer had invited him to California for the spring semester; now he turned the invitation down. Cornell promoted him to associate professor and raised his salary again. The chairman of the University of Pennsylvania's physics department needed a new chief theorist. Here Bethe stepped in paternalistically: he had no intention of letting go of Feynman, and he was sensitive to his protégé's mood. He thought it would be harmful for this suddenly unproductive twenty-eight-year-old to take on the psychological responsibility of a lead role in a university theory group. More than anything, he thought Feynman needed shelter. (He told the Pennsylvania administrator that Feynman was the second-best young physicist around: second to Schwinger.) For Feynman the most surprising—and oppressive—offer came from the Institute for Advanced Study, Einstein's institute in Princeton, in the spring. Oppenheimer had now been named as the institute's director, and he wanted Feynman. H. D. Smyth, Feynman's old chairman at Princeton, wanted him, too, and the two institutions had sounded him out about a special joint appointment. His anxiety about failing to live up to such expectations was reaching a peak. He experimented with various tactics to break his mental block. For a while he got up every morning at 8:30 and tried to work. Looking in the mirror one morning as he shaved, he told himself the Princeton offer was absurd—he could not possibly accept, and furthermore he could not accept the responsibility for their impression of him. He had never claimed to be an Einstein, he told himself. It was their mistake. For a moment he felt lighter. Some of his guilt seemed to lift away.

His old friend Wilson had just arrived to direct the nuclear laboratory. Along with Bethe, he caught Feynman's mood and invited him in for a talk. Don't worry so much, he told Feynman. *We* are responsible. We hire professors; we take the risks; as long as they teach their classes satisfactorily they fulfill their part of the bargain. It made Feynman think wistfully about the days before *the future of science* had begun to seem like his mission— the days before physicists changed the universe and became the most potent political force within American science, before institutions with fast-expanding budgets began chasing nuclear physicists like Hollywood stars. He remembered when physics had been a game, when he could look at the graceful narrowing curve in three dimensions that water makes as it streams from a tap, and he could take the time to understand why.

A few days later he was eating in the student cafeteria when someone tossed a dinner plate into the air—a Cornell cafeteria plate with the university seal imprinted on one rim—and in the instant of its flight he experienced what he long afterward considered an epiphany. As the plate spun, it wobbled. Because of the insignia he could see that the spin and the wobble were not quite in synchrony. Yet just in that instant it seemed to him—or was it his physicist's intuition?—that the two rotations were related. He had told himself he was going to *play*, so he tried to work the problem out on paper. It was surprisingly complicated, but he used a Lagrangian, least-action approach and found a two-to-one ratio in the relationship of wobble and spin. That was satisfyingly neat. Still, he wanted to understand the Newtonian forces directly, just as he had when he was a sophomore taking his first theory course and he provocatively refused to use the Lagrangian approach. He showed Bethe what he had discovered.

But what's the importance of that? Bethe asked.

It doesn't have any importance, he said. I don't *care* whether a thing has importance. Isn't it fun?

It's fun, Bethe agreed. Feynman told him that was all he was going to do from now on—have fun.

Sustaining that mood took deliberate effort, for in truth he had given up none of his ambition. If he was floundering, so were far more distinguished theoretical physicists, committed to resolving the flaws in quantum mechanics. He had not forgotten his painful disagreement with Dirac that fall—his conviction that Dirac had turned squarely back toward the past and that an alternative approach must surely be possible. Early in 1947 Feynman let his friend Welton know how grand his plans had become. (Welton was now working at the permanent plant at Oak Ridge; many years later he would finish his career there, still affected by the peculiar disappointment that hobbled so many others who had crossed Feynman's path at the wrong time.) Feynman said nothing about having fun. "I am engaged now in a general program of study—I want to understand (not just in a mathematical way) the ideas of all branches of theor. physics," he wrote. "As you know I am now struggling with the Dirac Equ." Despite what he told Bethe, he did make a connection between the axial wobble of a cafeteria plate and the abstract quantum-mechanical notion of spin that Dirac had so successfully incorporated in his electron.

Many years later Feynman and Dirac met one more time. They exchanged a few awkward words—a conversation so remarkable that a physicist within earshot immediately jotted down the Pinteresque dialogue he thought he heard drifting his way:

I am Feynman.

I am Dirac. (Silence.)

It must be wonderful to be the discoverer of that equation.

That was a long time ago. (Pause.) What are you working on?
Mesons.
Are you trying to discover an equation for them?
It is very hard.
One must try.

More than anyone else, Dirac had made the mere discovery of an equation into a thing to be admired. To aficionados the Dirac equation never did quite lose its rabbit-out-of-a-hat quality. It was relativistic—it survived without strain the manipulations required to accommodate near-light velocities. And it made spin a natural property of the electron. Understanding spin meant understanding the deceptive unreality of some of physics' new language. Spin was not yet as whimsical and abstract as some of the particle properties that followed it, properties called *color* and *flavor* in a half-witty, half-despairing acknowledgment of their unreality. It was still possible, barely, to understand spin literally: to view the electron as a little moon. But if the electron was also an infinitesimal point, it could hardly rotate in the classical fashion. And if the electron was also a smear of probabilities and a wave reverberating in a constraining chamber, how could these objects be said to *spin*? What sort of spin could come only in unit amounts or half-unit amounts (as quantum-mechanical spin did)? Physicists learned to think of spin not so much as a kind of rotation, but as a kind of symmetry, a way of stating mathematically that a system could undergo a certain rotation.

Spin was a problem for Feynman's theory as he had left it in his Princeton thesis. The quantity of action in ordinary mechanics contained no such property. And his theory would be useless if he could not apply it to a spinning, relativistic electron—the Dirac electron. Among the obstacles blocking his path, this was one of the heaviest. No wonder his eye might have been drawn to things that spun—a cafeteria plate, for example, wobbling in a split-second trajectory. His next step was peculiar and characteristic. He reduced the problem to a skeleton, a universe with just one dimension (or two: one space and one time). This universe was merely a line, and in it a particle could take just one kind of path, back and forth, reversing direction like a crazed insect. Feynman's goal was to begin with the method he had invented at Princeton—the summing of all possible paths a particle could take—and see whether he could derive, in this one-dimensional world, a one-dimensional Dirac equation. He jotted:

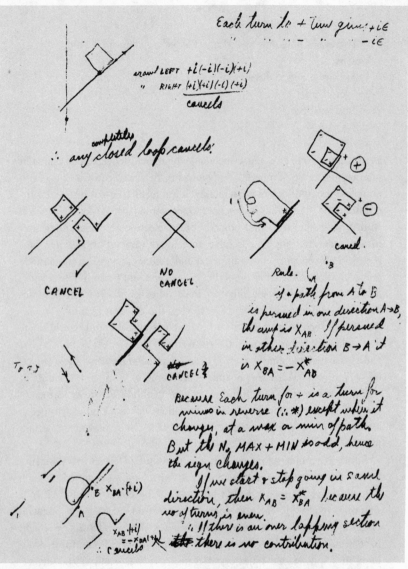

Feynman considered the path a particle would take in a one-dimensional universe—that is, a particle restricted to moving back and forth on a line, always at the speed of light. He diagrammed the back-and-forth motion by visualizing the space dimension horizontally and the time dimension vertically: the passage of time is represented as motion upward on the page. In this toy model, he found that he could derive a central equation of quantum mechanics by adding the contributions made by all the possible paths a particle could take.

Geometry of Dirac Equ. 1 dimension
Prob = squ. of sum of contrib. each path
Paths zig zag at light velocity.

And he added something new—a diagram, purely schematic, for keeping track of the zigs and zags. The horizontal dimension represented his one spatial dimension, and the vertical dimension represented time. He successfully negotiated the details of this one-dimensional shadow theory. The spin of his particles implied a phase, like the phase of a wave, and he made some assumptions, only partly arbitrary, about what would happen to the phase each time a particle zagged. Phase was crucial to the mathematics of summing the paths, because paths would either cancel or reinforce one another, depending on how their phases overlapped. Feynman did not attempt to publish this fragment of a theory, excited though he was by the progress. The challenge was to extend the theory to more dimensions—to let the space unfold—and this he could not do, though he spent long hours in the library, for once reading old mathematics.

Shrinking the Infinities

Feynman's frustration in these first postwar years mirrored a growing sense of impotence and defeat among established theoretical physicists. The feeling, at first private and then shared, remained invisible outside their small community. The contrast with the physicists' public glory could hardly have been greater.

The cause was abstruse. The single difficulty at the core of this anguish was a mathematical tendency of certain quantities to diverge as successive terms of an equation were computed—terms that should have been vanishing in importance. Physically it seemed that the closer one stood to an electron, the greater its charge and mass would appear. The result: the infinities with which Feynman had been struggling since Princeton. It meant that quantum mechanics produced good first approximations followed by a Sisyphean nightmare. The harder a physicist pushed, the less accurate his calculations became. Such quantities as the mass of the electron became—if the theory were taken to its limit—infinite. The horror of this was hard to comprehend, and no glimmer of it appeared in popular accounts of science at the time. Yet it was not merely a theoretical knot. A pragmatic physicist eventually had to face it. "Thinking I understand geometry," Feynman said later, "and wanting to fit the diagonal of a five foot square, I try to figure out how long it must be. Not being very expert I get infinity—useless. . . ."

> It is not philosophy we are after, but the behavior of real things. So in despair, I measure it directly—lo, it is near to seven feet—neither infinity nor zero. So, we have measured these things for which our theory gives such absurd answers. . . .

Experimental yardsticks for the electron were not so easy to come by, and it was a tribute to the original theory of Heisenberg, Schrödinger, and Dirac that first approximations matched any experimental results that the laboratories had produced so far. Better results were on the way, however.

Meanwhile, the scientists contemplating the state of theoretical physics descended into a distinct gloominess; in the aftermath of the bomb, their mood seemed postcoital.

"The last eighteen years"—the period, that is, since the quick birth of quantum mechanics—"have been the most sterile of the century," remarked I. I. Rabi to a colleague over lunch in that spring of 1947, though Rabi himself was thriving as head of a fruitful group at Columbia.

"Theoreticians were in disgrace"—so it seemed to one especially precocious student of physics, Murray Gell-Mann.

"The theory of elementary particles has reached an impasse," Victor Weisskopf wrote. Everyone had been struggling futilely, he said, especially since the war, and everyone had had enough of "knocking a sore head against the same old wall."

Merely a few dozen men in mathematical difficulty—or the generation's deepest crisis in theoretical physics. It was all the same. Weisskopf was preparing for an unusual gathering. A former president of the New York Academy of Sciences, Duncan MacInnes, had been nursing a conviction that modern-day conferences were growing too unwieldy. Hundreds of people would appear. Speakers were starting to cater to these diffuse audiences by delivering generalized and retrospective talks. As an experiment, MacInnes proposed an intimate meeting restricted to twenty or thirty invited guests, to take place in a relaxed, country-inn setting. With "Fundamental Problems of Quantum Mechanics" as a topic, he managed—though it took more than a year—to draw a select group in early June to an inn called the Ram's Head, just opening for the summer season on New York's Shelter Island, between the forks of eastern Long Island. Weisskopf was one of those charged with setting the agenda. Other participants were Oppenheimer, Bethe, Wheeler, Rabi, Teller, and several representatives of the younger generation, including Julian Schwinger and Richard Feynman.

So two dozen suit-jacketed physicists met on a Sunday afternoon on the East Side of New York and motored across Long Island in a rickety bus. Somewhere along the way a police escort picked them up, sirens wailing, and a banquet was arranged by a local chamber of commerce official who had been serving in the Pacific when, he felt, the atomic bomb saved his life. A ferry carried them across to Shelter Island, and to some of the physicists there was an air of unreality about it all. When they gathered for breakfast the next morning, they noticed the phrase "restricted clientele" on the menus and performed a quick head count: their group contained more Jews, they decided, than the inn's dining room had ever seen. One New York newspaper reporter had come along, and he telephoned his report to the *Herald Tribune*: "It is doubtful if there has ever been a conference quite like this one. . . . They roam through the corridors mumbling mathematical equations, eat their meals amid the fury of technical discussions. . . ." Island residents, he wrote,

are reasonably confused about this sudden descent of science among them. The principal theory is that the scientists are busy making another type of atomic bomb, and nothing could be farther from the truth. . . .

Quantum mechanics is the never-never land of science, a world in which matter and energy become confused and where all the verities of day-to-day life become meaningless. . . .

To those sensitive to small breezes, it was beginning to seem that two of the younger men in particular, Schwinger and Feynman, were engaged in a gestation of fresh ideas. Schwinger mostly kept his own counsel during these three days. Feynman tried his methods out on a few people; a young Dutch physicist, Abraham Pais, watched him derive results at lightning speed with the help of sketchy pictures that left Pais baffled. On the last morning, after some words by Oppenheimer, Feynman was asked to give the whole group an informal description of his work, and he did, happily. No one really understood, but he left the memory of—as one listener recorded in his diary—"a clear voice, great rush of words and illustrative gestures sometimes ebullient."

Above all, however, it was a conference dominated by news from experimenters, and particularly experimenters in the furnace Rabi was stoking at Columbia. The Columbia groups favored techniques that seemed homely and unspectacular in this era of the burgeoning particle accelerator, though their arsenal also included technologies fresh from the wartime Radiation Laboratory, magnetrons and microwaves. Willis Lamb had just shined a beam of microwaves onto a hot wisp of hydrogen blowing from an oven. He was trying to measure the precise energy levels of electrons in the hydrogen atom. He succeeded—the art of spectroscopy had never seen such precision—and he found a distinct gap between two energy levels that should have been identical. *Should have been*, that is, according to the clearest existing guide to hydrogen atoms and electrons, the theory of Dirac. That was in April. Lamb had gone to bed thinking about knobs and magnets and a bouncing spot of light from the galvanometer and the clear discrepancy between his experiment and Dirac's theory, and he had awakened the next day thinking (accurately, as it turned out): *Nobel Prize*. News of what would soon be called the Lamb shift had already reached most of the Shelter Island participants before Lamb made a detailed report the first day. The theorists present had often repeated the truism that progress in science comes when experiments contradict theory. Rarely had any of them seen such a clean example (more often it was theory that contradicted theory). To Schwinger, listening, the point was that the problem with quantum electrodynamics was neither infinite nor zero: it was a number, now standing before them, finite and small. The alumni of Los Alamos and the Radiation Laboratory knew that the task of theoretical physics was to justify such numbers. The rest of the conference fed off a nervous euphoria, as it seemed to Schwinger: "The facts were incredible—to be told that the sacred Dirac theory was breaking down all over the place." As the meeting adjourned, Schwinger left with Oppenheimer by seaplane.

Quantum electrodynamics was a "debacle," another physicist said. Harsh assessments of a theory accurate enough for all but this delicate experiment. But after all, the physicists had known that the theory was fatally pocked with infinities. The experiment gave them real numbers to calculate, numbers marking the exact not-quite-rightness of the world according to Dirac.

Dyson

That fall Freeman Dyson arrived at Cornell. Some of Cornell's mathematicians knew the work of a Briton by that name. It was hardly a common name, and mathematics was certainly known for its prodigies, but surely, they thought, this small, hawk-nosed twenty-three-year-old joining the physics department could not be the same man. Other graduate students found him genial but inscrutable. He would sleep late, bring his *New York Times* to the office, read it until lunch time, and spend the afternoon with his feet up and perhaps his eyes closed. Just occasionally he would wander into Bethe's office. What they did there, no one knew.

Indeed, Dyson was one of England's two or three most brilliant mathematical prodigies. He was the son of two supremely cultured members of the middle class who were late to marry and entering middle age when he was born. His father, George, composed, conducted, and taught music at a boys' college in the south. Eventually he became director of England's Royal College of Music. His mother, Mildred, trained as a lawyer, though she did not practice, and passed on to Freeman her deep love of literature, beginning with Chaucer and the poets of ancient Greece and Rome. As a six-year-old he would sit with encyclopedia volumes spread open before him and make long, engrossing calculations on sheets of paper. He was intensely self-possessed even then. His older sister once interrupted him to ask where their nanny was and heard him reply, "I expect her to be in the absolute elsewhere." He read a popular astronomy book, *The Splendour of the Heavens*, and the science fiction of Jules Verne, and when he was eight and nine wrote a science-fiction novel of his own, *Sir Phillip Roberts's Erolunar Collision*, with a maturely cadenced syntax and an adult sense of literary flow. His scientist hero has a knack for both arithmetic and spaceship design. Freeman, who did not favor short sentences, imagined a scientist comfortable with public acclaim, yet solitary in his work:

"I, Sir Phillip Roberts, and my friend, Major Forbes," he began, "have just unravelled an important secret of nature; that Eros, that minor planet that is so well-known on account of its occasional proximity with the Earth, Eros, will approach within 3,000,000 miles of the Earth in 10 years 287 days hence, instead of the usual 13,000,000 miles every 37 years; and, therefore it may, by some great chance fall upon the Earth. Therefore I advise you to calculate the details of this happening!" . . .

When the cheers were over, and everybody had gone home, it did not mean that the excitement was over; no, not at all; everybody was making the wildest calculations; some reasonable, some not; but Sir Phillip only wrote coolly in his study rather more than usual; nobody could tell what his thoughts were.

He read popular books about Einstein and relativity and, realizing that he needed to learn a more advanced mathematics than his school taught, sent away to scientific publishers for their catalogs. His mother finally felt that his interest in mathematics was turning into an obsession. He was fifteen and had just spent a Christmas vacation working methodically, from six each morning until ten each evening, through the seven hundred problems of H. T. H. Piaggio's *Differential Equations*. That same year, frustrated at learning that a classic book on number theory by I. M. Vinogradov existed only in Russian, he taught himself the language and wrote out a full translation in his careful hand. As Christmas vacation ended, his mother went for a walk with him and began a cautionary lecture with the words of the Latin playwright Terence: "I am human and I let nothing human be alien to me." She continued by telling him Goethe's version of the Faust story, parts one and two, rendering Faust's immersion in his books, his lust for knowledge and power, his sacrifice of the possibility of love, so powerfully that years later, when Dyson happened to see the film *Citizen Kane*, he realized that he was weeping with the recognition of his mother's Faust incarnate once again on the screen.

As the war began, Dyson entered Trinity College, Cambridge. At Cambridge he heard intimate lectures by England's greatest mathematicians, Hardy, Littlewood, and Besicovitch. In physics Dirac reigned. Dyson's war could hardly have been more different from Feynman's. The British war organization wasted his talents prodigiously, assigning him to the Royal Air Force bomber command in a Buckinghamshire forest, where he researched statistical studies that were doomed, when they countered the official wisdom, to be ignored. The futility of this work impressed him. He and others in the operational research section learned—contrary to the essential bomber command dogma—that the safety of bomber crews did not increase with experience; that escape hatches were too narrow for airmen to use in emergencies; that gun turrets slowed the aircraft and bloated the crew sizes without increasing the chances of surviving enemy fighters; and that the entire British strategic bombing campaign was a failure. Mathematics repeatedly belied anecdotal experience, particularly when the anecdotal experience was colored by a lore whose purpose was to keep young men flying.

Dyson saw the scattershot bomb patterns in postmission photographs, saw the Germans' ability to keep factories operating amid the rubble of civilian neighborhoods, worked through the firestorms of Hamburg in 1943 and Dresden in 1945, and felt himself descending into a moral hell. At Los Alamos a military bureaucracy worked more successfully than ever before or since with independent-minded scientists. The military bureaucracy of Dyson's experience embodied a routine of petty and not-so-petty dishonesty, and the scientists of the bomber command were unable to challenge it.

These were black days for the combination of science and machinery called technology. England, which had invented so much, had always been prone to misgivings. Machines disrupted traditional ways of living. In the workplace they seemed dehumanizing. At the turn of the century, amid the black soot clouds of the English industrial city, it was harder to romanticize the brutal new working conditions of the factory than the brutal old working conditions of the peasant farm. America, too, had its Luddites, but in the age of radio, telephone, and automobile few saw a malign influence in the progress that technology brought. For Americans the loathing of technology that would become a theme of late-twentieth-century life began with fears born amid the triumph of 1945. Among the books that had most influenced Dyson was a children's tale called *The Magic City*, written in 1910 by Edith Nesbit. Among its lessons was a bittersweet one about technology. Her hero—a boy named Philip—learns that in the magic city, when one asks for a machine, he must keep using it forever. Given a choice between a horse and a bicycle, Philip wisely chooses the horse, at a time when few in England or America were failing to trade their horses for bicycles, motorcars, or tractors. Dyson remembered *The Magic City* when he learned about the atomic bomb—remembered that new technology, once acquired, in always with us. But nothing is simple, and Dyson also took to heart a remark of D. H. Lawrence's about the welcome minimal purity of books, chairs, bottles, and an iron bedstead, all made by machines: "My wish for something to serve my purpose is perfectly fulfilled. . . . Wherefore I do honour to the machine and to its inventor." The news of Hiroshima came partly as a relief to Dyson. It released him from his own war. Yet he knew that the strategic bombing campaign had killed four times as many civilians as the atomic bombs. Years later, when Dyson had a young son, he woke the boy in the middle of the night because he—Freeman—had awakened from an unbearable nightmare. A plane had crashed to the ground in flames. People were nearby, and some ran into the fire to rescue the victims. Dyson, in his dream, could not move.

He sometimes struck people as shy or diffident, but his teachers in England had learned that he had enormous self-possession. As a high-school student he had worked on the problem of pure number theory known as partitions—a number's partitions being the ways it can be subdivided into sums of whole numbers: the partitions of 4 are $1 + 1 + 1 + 1$, $1 + 1 + 2$, $1 + 3$, $2 + 2$, and 4. The number of partitions rises fairly rapidly—14 has 135 partitions—and the question of just how rapidly has all the hallmarks of classic number theory. It is easy to state. A child can work out the first few cases. And from its contemplation arises a glorious world with the intricacy and beauty of origami. Dyson followed a path trod earlier by the Indian prodigy Srinivasa Ramanujan at the beginning of the century. By his sophomore year at Cambridge he arrived at a set of con-jectures about partitions that he could not prove. Instead of setting them aside, he made a virtue of his failure. He published them as only his second paper. "Professor Littlewood," he wrote of one of his famous professors, "when he makes use of an algebraic identity, always saves himself the trouble of proving it; he maintains that an identity, if true, can be verified in a few lines by anybody obtuse enough to feel the need of verification. My ob-ject . . . is to confute this assertion. . . ." Dyson promised to state a series of interesting identities that he could not prove. He would also, he boasted, "indulge in some even vaguer guesses concerning the existence of identities which I am not only unable to prove but unable to state. . . . Needless to say, I strongly recommend my readers to supply the missing proofs, or, even better, the missing identities." Routine mathematical discourse was not for him.

One day an assistant of Dirac's told Dyson, "I am leaving physics for mathematics; I find physics messy, unrigorous, elusive." Dyson replied, "I am leaving mathematics for physics for exactly the same reasons." He felt that mathematics was an interesting game but not so interesting as the real world. The United States seemed the only possible place to pursue physics now. He had never heard of Cornell, but he was advised that Bethe would be the best person in the world to work with, and Bethe was at Cornell.

He went with the attitude of an explorer to a strange land, eager to expose himself to the flora and fauna and the possibly dangerous inhabitants. He played his first game of poker. He experienced the American form of "picnic," which surprisingly involved the frying of steak on an open-air grill. He ventured forth on automobile excursions. "We go through some wild country," he wrote his parents shortly after his arrival—the wild country in this case being the stretch of exurban New York lying between

Ithaca and Rochester. He traveled with a theoretician called Richard Feynman: "the first example I have met of that rare species, the native American scientist."

> He has developed a private version of the quantum theory . . . ; in general he is always sizzling with new ideas, most of which are more spectacular than helpful, and hardly any of which get very far before some newer inspiration eclipses them. . . . when he bursts into the room with his latest brain-wave and proceeds to expound it with the most lavish sound effects and waving about of the arms, life at least is not dull.

Although Dyson was nominally a mere graduate student, for his first assignment Bethe had handed him a live problem: a version of the Lamb shift, fresh from Shelter Island. Bethe himself had already made the first fast break in the theoretical problem posed by Lamb's experiment. On the train ride home, using a scrap of paper, he made a fast, intuitive calculation that soon made a dozen of his colleagues say, *if only I had* . . . He telephoned Feynman when the train reached Schenectady, and he made sure his preliminary draft was in the hands of Oppenheimer and the other Shelter Island alumni within a week. It was a blunt Los Alamos–style estimate, ignoring the effects of relativity and evading the infinities by arbitrarily cutting them off. Bethe's breakthrough was sure to be superseded by a more rigorous treatment of the kind Schwinger was known to have in the works. But it gave the right number, almost exactly, and it lent weight to the conviction that a proper quantum electrodynamics would account for the new, precise experiments.

The existing theory "explained" the existence of different energy levels in the atom. It gave physicists their only workable means of calculating them. The different energies arose from different combinations of crucial quantum numbers, the angular momentum of the electron orbiting the nucleus, and the angular momentum of the electron spinning around itself. A certain symmetry built into the equation made it natural for a pair of the resulting energy levels to coincide exactly. But they did not coincide in Willis Lamb's laboratory, so something must be missing and, as Bethe surmised, that something was the theorists' old bugbear, the self-interaction of the electron.

This extra energy or mass was created by the snake-swallowing-its-tail interplay of the electron with its own field. This quantity had been a tolerable nuisance when it was theoretically infinite and experimentally

negligible. Now it was theoretically infinite and experimentally real. Bethe
had in mind a suggestion that the Dutch physicist Hendrik Kramers had
made at Shelter Island: that the "observed" mass of the electron, the mass
the theorists tended to think of as a fundamental quantity, should be thought
of as a combination of two other quantities, the self-energy and an "intrinsic" mass. These masses, intrinsic and observed, also known as "bare"
and "dressed," made an odd couple. The intrinsic mass could never be
measured directly, and the observed mass could not be computed from first
principles. Kramers proposed a method by which the theorists would pluck
a number from experimental measurements and correct it, or "renormalize"
it. This Bethe did, crudely but effectively. Meanwhile, as the mass went,
so went the charge—this formerly irreducible quantity, too, had to be
renormalized. Renormalization was a process of adjusting terms of the
equation to turn infinite quantities into finite ones. It was almost like looking at a huge object through an adjustable lens, and turning a knob to
bring it down to size, all the while watching the effect of the knob turning on other objects, one of which was the knob itself. It required great
care.

From one perspective, renormalization amounted to subtracting infinities
from infinities, with a silent prayer. Ordinarily such an operation could be
meaningless: *infinity* (the number of integers, 0, 1, 2, 3, . . .) minus *infinity* (the number of even integers, 0, 2, 4, . . .) equals *infinity* (the
remaining, odd integers, 1, 3, 5, . . .), and all three of those infinities are
the same, unlike, for example, the distinctly greater infinity representing
the number of real numbers. The theorists implicitly hoped that when they
wrote *infinity − infinity = zero* nature would miraculously make it so,
for once. That their hope was granted said something important about the
world. For a while it was not clear just what.

Bethe assigned Dyson a stripped-down, toy version of the Lamb shift,
asking him to calculate the Lamb shift for an electron with no spin. It was
a way for Dyson to find a quick way into a problem of the most timely
importance and for Bethe to continue his own prodding. Dyson could see
that the calculation Bethe had published was both a swindle and a piece
of genius, a bad approximation that somehow coughed up the right answer.
More and more, too, Dyson talked with Feynman, who gradually began
to come into clearer focus for him. He watched this wild American dash
from the dinner table at the Bethes' to play with their five-year-old son,
Henry. Feynman did have an extraordinary affinity for his friends' children.

He would entertain them with gibberish, or with juggling tricks, or with what sounded to Dyson like a one-man percussion band. He could enthrall them merely by borrowing someone's eyeglasses and slowly putting them on, taking them off, and putting them on. Or he would engage them in conversation. He once asked Henry Bethe, "Did you know there are twice as many numbers as numbers?"

"No, there are not!" Henry said.

Feynman said he could prove it. "Name a number."

"One million."

Feynman said, "Two million."

"Twenty-seven!"

Feynman said, "Fifty-four," and kept on countering with the number that was twice Henry's, until suddenly Henry saw the point. It was his first real encounter with infinity.

For a while, because Feynman did not seem to take his work seriously, neither did Dyson. Dyson wrote his parents that Feynman was "half genius and half buffoon" (a description he later regretted). Just a few days later Dyson heard an account from Weisskopf, visiting Cornell, of Schwinger's progress at Harvard. He sensed a connection with the very different notions he was hearing from Feynman. He had begun to see a method beneath Feynman's flash and wildness. The next time he wrote his parents, he said:

> Feynman is a man whose ideas are as difficult to make contact with as Bethe's are easy; for this reason I have so far learnt much more from Bethe, but I think if I stayed here much longer I should begin to find that it was Feynman with whom I was working more.

A *Half-Assedly Thought-Out Pictorial Semi-Vision Thing*

By the physicists' own lights their difficulties were mathematical: infinities, divergences, unruly formalisms. But another obstacle lay in the background, rarely surfacing in the standard published or unpublished rhetoric: the impossibility of visualization. How was one to perceive the atom, or the electron in the act of emitting light? What mental picture could guide the scientist? The first quantum paradoxes had so shattered physicists' classical intuitions that by the 1940s they rarely discussed visualization. It seemed a psychological issue, not a scientific one.

The atom of Niels Bohr, a miniature solar system, had become an embarrassingly false image. In 1923, on the tenth anniversary of Bohr's conception, the German quantum physicist Max Born hailed it: "the thought that the laws of the macrocosmos in the small reflect the terrestrial world obviously exercises a great magic on mankind's mind"—but already he and his colleagues could see the picture fading into anachronism. It survived in the language of *angular momentum* and *spin*—as well as in the standard high-school physics and chemistry curriculums—but there was no longer anything plausible in the picture of electrons orbiting a nucleus. Instead there were waves with modes of resonance, particles that smeared out probabilistically, operators and matrices, malleable spaces with extra dimensions, and physicists who forswore the idea of visualization altogether. Bohr himself had set the tone. In accepting the Nobel Prize for his atomic model, he said it was time to give up the hope of explanations in terms of analogies with everyday experience. "We are therefore obliged to be modest in our demands and content ourselves with concepts which are formal in the sense that they do not provide a visual picture of the sort one is accustomed to require. . . ." This progress had not been altogether free of tension. "The more I reflect on the physical portion of Schrödinger's theory, the more disgusting I find it," was Heisenberg's 1926 comment to Pauli. "Just imagine the rotating electron whose charge is distributed over the entire space with axes in 4 or 5 dimensions. What Schrödinger writes on the visualizability of his theory . . . I consider trash." As much as physicists valued the conceptualizing skill they called intuition, as much as they spoke of a difference between physical understanding and formal understanding, they had nevertheless learned to mistrust any picture of subatomic reality that resembled everyday experience. No more baseballs, artillery shells, or planetoids for the quantum theorists; no more idle wheels or wavy waves. Feynman's father had asked him, in the story he told so many times: "I understand that when an atom makes a transition from one state to another, it emits a particle of light called a photon. . . . Is the photon in the atom ahead of time? . . . Well, where does it come from, then? How does it come out?" No one had a mental image for this, the radiation of light, the interaction of matter with the electromagnetic field: the defining event of quantum electrodynamics.

Where this image should have been, instead there was a void, as frothy and alive with possibility as the unquiet vacuum of the new physics. Unable to let their minds fix on even a provisional picture of quantum events, some physicists turned to a new kind of philosophizing, characterized by paradoxical thought experiments and arguments about *reality, consciousness, causality,* and *measurement.* Such arguments grew to form an indispensable part of the late twentieth century's intellectual atmosphere; they trailed the rest of physics as a cloud of smoke and flotsam trails a convoy. They were provocative and irresolvable. The paper of Einstein, Podolsky, and Rosen in 1935—the paper that provided the seventeen-year-old Schwinger with his first opportunity to impress Rabi—became an enduring example. It posed the case of two quantum systems—atoms, perhaps—linked by a particle interaction in their past but now separated by a great distance. The authors showed that the plain act of measuring one atom of this pair would affect what one could measure about the other atom, and the effect would be instantaneous—faster than light and thus retroactive, as it were. Einstein considered this a damning commentary on the laws of quantum mechanics. Bohr and younger theorists maintained a more sanguine attitude, noting that Einstein himself had already placed *past* and *distance* into the class of concepts about which one could no longer speak with comfortable, classical certainty. In the same vein was Schrödinger's famous cat: a poor hypothetical animal sitting in a box with a vial of poisonous gas attached to a detector, its fate thus linked to that same quantum-mechanical event, the emission of a photon from an atom. Schrödinger's point was that, while physicists now glibly calculated such events as probabilities—half yes and half no, perhaps—they still could not visualize a cat as anything but alive or dead.

Physicists made a nervous truce with their own inability to construct unambiguous mental models for events in the very small world. When they used such words as *wave* or *particle*—and they had to use both—there was a silent, disclaiming asterisk, as if to say: *not really.* As a consequence, they recognized that their profession's relationship to reality had changed. Gone was the luxury of supposing that a single reality existed, that the human mind had reasonably clear access to it, and that the scientist could explain it. It was clear now that the scientist's work product—a theory, a model—interpreted experience and construed experience in a way that was always provisional. Scientists relied on such models as intensely as someone crossing a darkened room relies on a conjured visual memory. Still, physicists now began to say explicitly that they were creating a lan-

guage—as though they were more like literary critics than investigators.
"It is wrong to think that the task of physics is to find out how nature is,"
said Bohr. "Physics concerns only what we can *say* about nature." This
had always been true. Never before, though, had nature so pointedly rubbed
physicists' noses in it.

Yet in the long run most physicists could not eschew visualization. They
found that they needed imagery. A certain kind of pragmatic, working
theorist valued a style of thinking based on a kind of seeing and feeling.
That was what *physical intuition* meant. Feynman said to Dyson, and Dyson
agreed, that Einstein's great work had sprung from physical intuition and
that when Einstein stopped creating it was because "he stopped thinking
in concrete physical images and became a manipulator of equations." Intuition was not just visual but also auditory and kinesthetic. Those who
watched Feynman in moments of intense concentration came away with a
strong, even disturbing sense of the physicality of the process, as though
his brain did not stop with the gray matter but extended through every
muscle in his body. A Cornell dormitory neighbor opened Feynman's door
to find him rolling about on the floor beside his bed as he worked on a
problem. When he was not rolling about, he was at least murmuring
rhythmically or drumming with his fingertips. In part the process of scientific visualization is a process of putting oneself *in* nature: in an imagined
beam of light, in a relativistic electron. As the historian of science Gerald
Holton put it, "there is a mutual mapping of the mind . . . and of the laws
of nature." For Feynman it was a nature whose elements interacted with
palpable, variegated, fluttering rhythms.

He thought about it himself. Once—uninterested though he was in
fiction or poetry—he carefully copied out a verse fragment by Vladimir Nabokov: "Space is a swarming in the eyes; and time a singing in the
ears."

"Visualization—you keep repeating that," he said to another historian,
Silvan S. Schweber, who was trying to interview him.

What I am really trying to do is bring birth to clarity, which is really a
half-assedly thought-out pictorial semi-vision thing. I would see the jiggle-
jiggle-jiggle or the wiggle of the path. Even now when I talk about the
influence functional, I see the coupling and I take this turn—like as if
there was a big bag of stuff—and try to collect it away and to push it. It's
all visual. It's hard to explain.

"In some ways you see the answer——?" asked Schweber.

——the character of the answer, absolutely. An inspired method of pic-
turing, I guess. Ordinarily I try to get the pictures clearer, but in the end
the mathematics can take over and be more efficient in communicating
the idea of the picture.

In certain particular problems that I have done it was necessary to
continue the development of the picture as the method before the mathematics could be really done.

The field itself presented the ultimate challenge. Feynman once told
students, "I have no picture of this electromagnetic field that is in any sense
accurate." In seeking to analyze his own way of visualizing the unvisualizable he had learned an odd lesson. The mathematical symbols he used
every day had become entangled with his physical sensations of motion,
pressure, acceleration . . . Somehow he invested the abstract symbols with
physical meaning, even as he gained control over his raw physical intuition
by applying his knowledge of how the symbols could be manipulated.

When I start describing the magnetic field moving through space, I speak
of the *E*- and *B*- fields and wave my arms and you may imagine that I
can see them. I'll tell you what I see. I see some kind of vague, shadowy,
wiggling lines . . . and perhaps some of the lines have arrows on them—
an arrow here or there which disappears when I look too closely. . . . I
have a terrible confusion between the symbols I use to describe the objects
and the objects themselves.

Yet he could not retreat into the mathematics alone. Mathematically the
field was an array of numbers associated with every point in space. That,
he told his students, he could not imagine at all.

Visualization did not have to mean diagrams. A complex, half-con-
scious, kinesthetic intuition about physics did not necessarily lend itself to
translation into the form of a stick-figure drawing. Nor did a diagram
necessarily express a physical picture. It could merely be a chart or a memory
aid. At any rate diagrams had been rare in the literature of quantum physics.
One typical example used a ladder of horizontal lines to represent the
notion of energy levels in the atom:

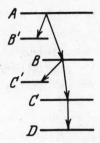

The "quantum jump" visualized as a sort of ladder.

The quantum jump from one level down to another accompanied the emission of a photon; the absorption of a photon would bring a jump upward. No depiction of the photons appeared in these diagrams; nor in another, more awkward schematic for the same process.

Feynman never used such diagrams, but he often filled his note pages with drawings of a different sort, recalling the space-time paths that had been so crucial a feature of his Princeton work with Wheeler. He drew the paths of electrons as straight lines, moving across the page to represent motion through space and up the page to represent progress through time. At first he, too, left the emission of a photon out of his pictures: that event would appear as the deflection of an electron from one path to another. The absence of photons did reflect an implicit choice from among the available pictorial landscapes: Feynman was still thinking mainly in terms of electrons interacting with the electromagnetic field as a field, rather than with the field as incarnated in the form of particles, photons.

In mid-1947 friends of Feynman persuaded him—threats and cajoling were required—to write for publication the theoretical ideas they kept hearing him explain. When he finally did, he used no diagrams. The result was partly a reworking of his thesis, but it also showed the maturing and broadening of his command of the issues of quantum electrodynamics. He expressed the tenets of his new vision with an unabashed plainness. For some physicists this would be the most influential set of ideas Feynman ever published.

He said he had developed an alternative formulation of quantum mechanics to add to the pair of formulations produced two decades before by Schrödinger and Heisenberg. He defined the notion of a *probability amplitude for a space-time path*. In the classical world one could merely add probabilities: a batter's on-base percentage is the 30 percent probability of a base hit plus the 10 percent probability of a base on balls plus the 5 percent probability of an error . . . In the quantum world probabilities were expressed as complex numbers, numbers with both a quantity and a phase, and these so-called *amplitudes* were squared to produce a probability. This was the mathematical procedure necessary to capture the wavelike aspects of particle behavior. Waves interfered with one another. They could enhance one another or cancel one another, depending on whether they were in or out of phase. Light could combine with light to produce darkness, alternating with bands of brightness, just as water waves combining in a lake could produce doubly deep troughs and high crests.

Feynman described for his readers what they already knew as the canonical thought experiment of quantum mechanics, the so-called two-slit experiment. For Niels Bohr it had illustrated the inescapable paradox of the wave-particle duality. A beam of electrons (for example) passes through two slits in a screen. A detector on the far side records their arrival. If the detector is sensitive enough, it will record individual events, like bullets striking; it might be designed to click as a Geiger counter clicks. But a peculiar spatial pattern emerges: the probabilities of electrons arriving at different places vary in the distinct manner of diffraction, precisely as though waves were passing through the slit and interfering with one another. Particles or waves? Sealing the paradox, quantum mechanically, is a conclusion that one cannot escape: that each electron "sees," or "knows about," or somehow *goes through* both slits. Classically a particle would have to go through one slit or the other. Yet in this experiment, if the slits are alternately closed, so that one electron must go through A and the next through B, the interference pattern vanishes. If one tries to glimpse the particle as it passes through one slit or the other, perhaps by placing a detector at a slit, again one finds that the mere presence of the detector destroys the pattern.

Probability amplitudes were normally associated with the likelihood of a particle's arriving at a certain place at a certain time. Feynman said he would associate the probability amplitude "with an entire motion of a particle"—with a path. He stated the central principle of his quantum mechanics: *The probability of an event which can happen in several different ways is the absolute square of a sum of complex contributions, one from each alternative way.* These complex numbers, these amplitudes, were written in terms of the classical action; he showed how to calculate the

The Two-Slit Experiment

The central mystery of quantum mechnics—the one to which all others could ultimately be reduced.

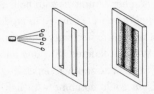

A gun (obeying the classical laws) sprays bullets toward a target. First they must pass through a screen with two slits. The pattern they make shows how their *probability* of arrival varies from place to place. They are likeliest to strike directly behind one of the slits. The pattern happens to be simply the sum of the patterns for each slit considered separately: if half the bullets were fired with only the left slit open and then half were fired with just the right slit open, the result would be the same.

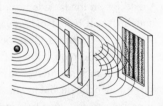

With waves, however, the result is very different, because of *interference*. If the slits were opened one at a time, the pattern would resemble the pattern for bullets: two distinct peaks. But when the slits are open at the same time, the waves pass through both slits at once and interfere with each other: where they are in phase they reinforce each other; where they are out of phase they cancel each other out.

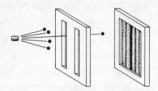

Now the quantum paradox: Particles, like bullets, strike the target one at a time. Yet, like waves, they create an interference pattern. If each particle passes individually through one slit, with what does it "interfere"? Although each electron arrives at the target at a single place and a single time, it seems that each has passed through—or somehow felt the presence of—both slits at once.

action for each path as a certain integral. And he established that this peculiar approach was mathematically equivalent to the standard Schrödinger wave function, so different in spirit.

The Physical Review had printed nothing by Feynman since his undergraduate thesis almost a decade before. To his dismay, the editors now rejected this paper. Bethe helped him rewrite it, showing him how to spell out for the reader what was old and what was new, and he tried the more retrospective journal *Reviews of Modern Physics*, where finally it appeared the next spring under the title "Space-Time Approach to Non-Relativistic Quantum Mechanics." He plainly admitted that his reformulation of quantum mechanics contained nothing new in the way of results, and he stated even more plainly where he thought the merit lay: "There is a pleasure in recognizing old things from a new point of view. Also, there are problems for which the new point of view offers a distinct advantage." (For example, when two particles interacted, it became possible to avoid the laborious bookkeeping of two different coordinate systems.) His readers—and at first they were few—found no fancy mathematics, just this shift of vision, a bit of physical intuition laid atop a foundation of clean, classical mechanics.

Few immediately recognized the power of Feynman's vision. One who did was the Polish mathematician Mark Kac, who heard Feynman describe his path integrals at Cornell and immediately recognized a kinship with a problem in probability theory. He had been trying to extend the work of Norbert Wiener on Brownian motion, the herky-jerky random motion in the diffusion processes that so dominated Feynman's theoretical work at Los Alamos. Wiener, too, had created integrals that summed many possible paths a particle could take, but with a crucial difference in the handling of time. Within days of Feynman's talk, Kac had created a new formula, the Feynman-Kac Formula, that became one of the most ubiquitous of mathematical tools, linking the applications of probability and quantum mechanics. He later felt that he was better known as the K in F-K than for anything else in his career.

Even to physicists well accustomed to theoretical constructions with awkward philosophical implications, Feynman's summings of paths—path integrals—seemed bizarre. They conjured a universe where no potential goes uncounted; where nothing is latent, everything alive; where every possibility makes itself felt in the outcome. He had expressed his conception to Dyson:

The electron does anything it likes. It just goes in any direction at any speed, forward or backward in time, however it likes, and then you add up the amplitudes and it gives you the wave function.

Dyson gleefully retorted that he was crazy. Still, Feynman had caught the intuitive essence of the two-slit experiment, where an electron seems aware of every possibility.

Feynman's path-integral view of nature, his vision of a "sum over histories," was also the principle of least action, the principle of least time, reborn. Feynman felt that he had uncovered the deep laws that gave rise to the centuries-old principles of mechanics and optics discovered by Christiaan Huygens, Pierre de Fermat, and Joseph-Louis Lagrange. How does a thrown ball know to find the particular arc whose path minimizes action? How does a ray of light know to find the path that minimizes time? Feynman answered these questions with images that served not only for the novel mysteries of quantum mechanics but for the treacherously innocent exercises posed for any beginning physics student. Light seems to angle neatly as it passes from air to water. It seems to bounce like a billiard ball off the surface of a mirror. It seems to travel in straight lines. These paths—the paths of least time—are special because they tend to be where the contributions of nearby paths are most closely in phase and most reinforce one another. Far from the path of least time—at the distant edge of a mirror, for example—paths tend to cancel one another out. Yet light does take every possible path, Feynman showed. The seemingly irrelevant paths are always lurking in the background, making their contributions, ready to make their presence felt in such phenomena as mirages and diffraction gratings.

Optics students learned alternative explanations for such phenomena in terms of waves like those undulating through water and air. Feynman was—with finality—eliminating the wave viewpoint altogether. Waviness was built into the phases carried by amplitudes, like little clocks. Once, with Wheeler, he had dreamed of eliminating the field itself. That idea had proved fanciful. The field had lodged itself deeply in the consciousness of physicists. It was indispensable and it was multiplying—a new particle, such as the meson, meant a new field, like a new plastic overlay, of which the particle was a quantized manifestation. Still, Feynman's theory retained the mark of its original scaffolding, though the scaffolding was long discarded. The actors were, more clearly than ever, particles. That became an attractive feature for physicists seeking help in visualization, in an ex-

perimental world dominated more and more by the cloud trails, the nomenclature, the behaviorism of particles.

Schwinger's Glory

Feynman's path integrals belonged to a loose kit of ideas and methods, a
private physics that he had assembled but not organized. Much relied on
guesswork or, as he said, "semi-empirical shenanigans." It was all hodgepodge and purpose-driven, and he could barely communicate it, let alone
prove it, even to his most sympathetic listeners, Bethe and Dyson. In the
fall of 1947 he attended a formal lecture by Bethe on his approach to the
Lamb shift. When Bethe concluded by stressing the need for a more reliable
way of making the theory finite, a way that would observe the requirements
of relativity, Feynman realized that he could compute the necessary correction. He promised Bethe an answer by the next morning.

By morning he realized that he did not know enough about Bethe's
calculation of the electron's self-energy to translate his correction into the
normal language of physics. They stood together at the blackboard for a
while, Bethe explaining his calculation, Feynman trying to translate his
technique, and the best answer they could reach diverged not modestly,
like Bethe's, but horrendously. Feynman, thinking about the problem physically, was sure it should not diverge at all.

In the days that followed, he taught himself about self-energy all over
again. When he reexpressed his equations in terms of the observed,
"dressed" mass of the electron instead of the theoretical, "bare" mass, the
correction came out just as he had thought, converging to a finite answer.
Meanwhile, glowing news of Schwinger's progress was reaching Ithaca from
Cambridge via Weisskopf and Bethe. When Feynman heard late in the fall
that Schwinger had worked out a calculation for the magnetic moment of
the electron—another tiny experimental anomaly newly found in Rabi's
laboratory—he solved the problem, too. Schwinger's elaborate piece of
calculating gave leading physicists a conviction that theory was once again
on the march. "God is great!" Rabi wrote Bethe with characteristic wryness,
and Bethe replied: "It is certainly wonderful how these experiments of yours
have given a completely new slant to a theory and how the theory has
blossomed out in a relatively short time. It is as exciting as in the early
days of quantum mechanics."

Feynman felt increasingly competitive about Schwinger, and increasingly frustrated. He had his quantum electrodynamics, he believed, and what he now thought of as "the Schwinger-Weisskopf-Bethe camp" had another. In January the American Physical Society met in New York, and Schwinger was the star. His program was not complete, but he had integrated the new idea of renormalization into the standard quantum mechanics in a way that let him demonstrate a series of impressive derivations. He showed how the anomalous magnetic moment, like the Lamb shift, came from the electron's interaction with its own field. His lecture drew a crowd that packed the hall. Too many physicists were forced to stand out in the corridors to hear the bursts of applause (and the embarrassed laughter that came when Schwinger finally said, "It is quite clear that . . ."). Hasty arrangements were made for Schwinger to repeat the lecture later the same day in Columbia's McMillin Theater. Dyson attended. Oppenheimer smoked his pipe conspicuously in the front row. Feynman rose during the question period to say that he, too, had reached these results and that, in fact, he could offer a small correction. Immediately he regretted it. He thought he must have sounded like a little boy piping up with "I did it too, Daddy." Few people that winter realized the depths of the rivalry he felt, but he made a bitter remark to a girlfriend, who understood the drift of his disappointment if not the exact circumstances.

"I'm so sorry that your long worked-on experiment was more or less stolen by someone else," she wrote back. "I know it just makes you feel sick. But Dick dear, how could life or things be interesting if there was not competition?" She wondered, why couldn't he and his competitor combine their ideas and work together?

Schwinger and Feynman were not alone in trying to produce the calculations—the explanation—required by the immediate experiments on the Lamb shift and the electron's magnetic moment. Other theorists followed the lead provided by Bethe's back-of-the-envelope approach. They saw no need to create a monumental new quantum electrodynamics, when they might generate the right numbers merely by patching the technique of renormalization onto the existing physics. Independently, two pairs of scientists succeeded in this, producing solutions that went beyond Bethe's in that they took into account the way masses fattened at relativistic speeds. Before publishing, one team, Weisskopf and a graduate student, Bruce French, committed a fatal act of indecision by consulting both Schwinger and Feynman. Engrossed in their more ambitious programs, Schwinger and Feynman each warned Weisskopf off, saying that he was in error by a small factor. Weisskopf decided it was inconceivable that these brilliant upstarts could both be wrong, independently, and delayed his manuscript. Months passed before Feynman called apologetically to say that Weisskopf's answer had been correct.

For Feynman's own developing theory, a breakthrough came when he confronted the ticklish area of antimatter. The first antiparticle, the negative electron, or positron, had been born less than two decades earlier as a minus sign in Dirac's equations—a consequence of a symmetry between positive and negative energy. Dirac had been forced to conceive of holes in a sea of energy, noting in 1931 that "a hole, if there were one, would be a new kind of particle, unknown to experimental physics." Unknown for the next few months—then Carl Anderson, at Caltech, found the trail of one in a cloud chamber built to detect cosmic rays. It looked like an electron, but it swerved up through a magnetic field when it should have swerved down.

The vivid photographs, along with the lively name coined by a journal editor against Anderson's will, gave the positron a legitimacy that theorists found hard to ignore. The collision of an electron with its antimatter cousin released energy in the form of gamma rays. Alternatively, in Dirac's picture of the vacuum as a lively sea populated by occasional holes, or bubbles, one could say that the electron fell into a hole and filled it, so that both the hole and the electron would disappear. As experimentalists continued to study their cosmic-ray photographs, they also found the reverse process: a gamma ray, nothing more than a high-frequency particle of light, could spontaneously produce a pair of particles, one electron and one positron.

Dirac's picture had difficulties. As elsewhere in his physics, unwanted infinities arose. The simplest description of the vacuum, empty space at absolute zero, seemed to require infinite energy and infinite charge. And from the practical perspective of anyone trying to write proper equations, the infinitude of presumed particles caused infernal complications. Feynman, seeking a way out, turned again to the forward- and backward-flowing version of time in his work with Wheeler at Princeton. Once again he proposed a space-time picture in which the positron was a time-reversed electron. The geometry of this vision could hardly have been simpler, but it was so unfamiliar that Feynman strained for metaphors:

"Suppose a black thread be immersed in a cube of collodion, which is then hardened," he wrote. "Imagine the thread, although not necessarily quite straight, runs from top to bottom. The cube is now sliced horizontally into thin square layers, which are put together to form successive frames of a motion picture." Each slice, each cross-section, would show a dot,

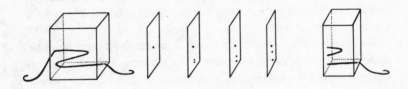

and the dot would move about to reveal the path of the thread, instant by instant. Now suppose, he said, the thread doubled back on itself, "somewhat like the letter N." To the observer, seeing the successive slices but not the thread's entirety, the effect would resemble the production of a particle-antiparticle pair:

> In successive frames first there would be just one dot but suddenly two new ones would appear when the frames come from layers cutting the thread through the reversed section. They would all three move about for a while then two would come together and annihilate, leaving only a single dot in the final frames.

The usual equations of electron motion covered this model, he said, though it did require "a more tortuous path in space and time than one is used to considering." He remained dissatisfied with the analogy of the thread and kept looking for more intuitive ways to express his view, capturing as it did the essence of the distinction between seeing paths in time-bound slices and seeing them whole. A Cornell student who had served as a wartime bombardier had a suggestion, and the bombardier metaphor, the one Feynman finally published, became famous.

> A bombardier watching a single road through the bomb-sight of a low flying plane suddenly sees three roads, the confusion only resolving itself when two of them move together and disappear and he realizes he has only passed over a long reverse switchback of a single road. The reversed section represents the positron in analogy, which is first created along with an electron and then moves about and annihilates another electron.

That was the broad picture. His path-integral method suited the model well: he knew from his old work with Wheeler that the summing of the phases of nearby paths would apply to "negative time" as well. He also found a shortcut past complications that had arisen because of the Pauli exclusion principle, the essential law of quantum mechanics that forbade two electrons from inhabiting the same quantum state. He granted himself a bizarre dispensation from the exclusion principle on the basis that, where earlier calculations had seen two particles, there was actually just one, taking a zigzag back and forth through a slice of time. "Usual theory says no, because then at time between t_y, t_x can't have 2 electrons in same state," he jotted in a note to himself. "We say it is same electron so Pauli exclusion doesn't operate." It sounded like something from the science fiction of time travel—hardly a notion designed for ready acceptance. He knew well that he was proposing a radical departure from the commonsense experience of time. He was violating the everyday intuition that the future does not yet exist and that the past has passed. All he could say was that time in physics had already departed from time in psychology—that nothing in the microscopic laws of physics seemed to mandate a distinction between past and future, and that Einstein had already ruined the notion of absolute time, independent of the observer. Yet Einstein had not imagined a particle's history reversing course and swerving back against the current. Feynman could only resort to an argument from utility: "It may prove useful in physics," he wrote, "to consider events in all of time at once and to imagine that we at each instant are only aware of those that lie behind us."

My Machines Came from Too Far Away

Schwinger and Feynman were both looking ahead to the inevitable sequel to the elite Shelter Island meeting. A new gathering was planned for late March at a resort in the Pocono Mountains of Pennsylvania: again the setting was to be pastoral, the roster intimate, the agenda profound. Success had enhanced the already high-status guest list. Fermi, Bethe, Rabi, Teller, Wheeler, and von Neumann were returning, along with Oppenheimer as chairman, and now they would be joined by two giants of prewar physics, Dirac and Bohr.

They gathered on March 30, 1948, in a lounge under a tarnished green clock tower with a view over a golf course and fifty miles of rolling woodlands. The presentations opened with the latest news of particle tracks in cosmic-ray showers and in the accelerator at Berkeley. With its sixteen-foot magnet the Berkeley synchrotron promised to push protons up to energies of 350 million electron volts by fall, enough to re-create copious bursts of the new elementary (so it seemed) particle called the meson, the cosmic-ray particle of most current topical interest. Instead of waiting for the cosmos to send samples down into their cloud chambers, experimenters would finally be able to make their own.

There had been a problem with the cosmic-ray data, an enormous discrepancy between the expected and the observed strengths of the mesons' interactions with other particles. At Shelter Island a young physicist, Robert Marshak, had proposed a solution requiring more courage and ingenuity in 1947 than such solutions would need in decades to come: namely, that there must be a second species of particle mixed in with the first. Not one meson but two—it seemed so obvious once someone dared break the ice. Feynman gleefully said they would have to call the new particle a marshak. Abetted by technology, the roster of elementary particles was climbing toward double digits. As the Pocono meeting opened, experimentalists warmed up the audience by showing pictures of an increasingly characteristic sort. Particles made impressive chicken-claw tracks in the photographs. No one could see fields, or matrices, or operators, but the geometry of particle scattering could not have been more vivid.

The next morning Schwinger took the floor. He began to present for the first time a complete theory of quantum electrodynamics that, as he stressed at the outset, met the dual criteria of "relativistic invariance" and "gauge invariance." It was a theory, that is, whose calculations looked the same no matter what velocity or phase its particles chose. These invariances assured that the theory would be unchanged by the arbitrary perspective of the observer, just as the time from sunrise to sunset does not depend on whether one has set one's clock forward to daylight saving time. The theory would have to make sure that calculations never tied themselves to a particular reference system, or "gauge." Schwinger told his listeners that he would consider a quantized electromagnetic field in which "each small volume of space is now to be handled as a particle"—a particle with more mathematical power and less visual presence than those of the previous day. He introduced a difficult new notation and set about to derive a sampling of specific results for such "applications" as the interaction of an electron with its own field. If his distinguished listeners found themselves in darkness, they were nevertheless not so easily cowed as Schwinger's customary audiences, and the usual express train found itself halted by interruptions. Bohr himself broke in with a question—Schwinger hated this and cut him off abruptly. Finally he managed to move forward, promising that all would be made clear in due course. As always, he made a point of lecturing without notes, and nearly all of his presentation was formal, deriving one equation after another. His talk became a marathon, lasting late into the afternoon. Bethe noticed that the formal mathematics silenced the critics, who raised questions only when Schwinger tried to express plainly physical ideas. He mentioned this to Feynman, suggesting that he, too, take a mathematical approach to his presentation. Fermi, glancing about at his famous colleagues, noticed with some satisfaction that one by one they had let their attention drift away. Only he and Bethe managed to stay with Schwinger to the end, he thought.

Then it was Feynman's turn. He was uneasy. It seemed to him that Schwinger's talk, though a bravura performance, had not gone well (but he was wrong—everyone, and crucially Oppenheimer, had been impressed). Bethe's warning made him reverse his planned presentation. He had meant to stay as closely as possible to physical ideas. He did have a mathematical formalism, as private though not as intricate as Schwinger's, and he could show how to derive his rules and methods from the formalism, but he could not justify the mathematics itself. He had reached it by trial and error. He knew it was correct, because he had tried it now on so many problems, including all of Schwinger's, and it worked, but he could not prove that it worked and he could not connect it to the old quantum mechanics. Nevertheless he took Bethe's advice and began with equations, saying, "This is a mathematical formula which I will now show you produces all the results of quantum mechanics."

He had always told his friends that once he started talking about physics he did not care who his audience was. One of his favorite stories was about Bohr, who had singled him out at Los Alamos as a young man unafraid to dispute his elders. Bohr had consulted Feynman privately there from time to time, often through his physicist son, Aage. Still, he had never fully warmed to Feynman, with his overeager, American, working-class style. Now Bohr waited, at the end of a long day, in this formidable audience of twenty-six men. Not even at Princeton, when he lectured to Einstein and Pauli, had Feynman stood before such a concentration of the great minds of his science. He had created a new quantum mechanics almost without reading the old, but he had made two exceptions: he had learned from the work of Dirac and Fermi, both now seated before him. His teachers

Wheeler and Bethe were there. So were Oppenheimer, who had built one bomb, and Teller, who was building the next. They had known him as a promising, fearless young light. His thirtieth birthday was seven weeks away.

Schwinger himself was hearing Feynman's theory for the first time. He thought it intellectually repulsive, though he did not say so (and afterward they cordially compared techniques and found themselves in nearly perfect agreement). He could see that Feynman was offering a patchwork of guesses and intuition. It struck him as engineering, all I-beams and T-beams. Bethe interrupted once, sensing that the audience was numbed with detail, and tried to return Feynman to fundamentals. Feynman explained his path integrals, an alien idea, and his positrons moving backward in time, even more disturbing. Teller caught the apparent infringement of the exclusion principle and refused to accept Feynman's unrigorous justification. It struck Feynman that everyone had a favorite principle or theorem and he was violating them all. When Dirac asked, "Is it unitary?" Feynman did not even know what he meant. Dirac explained: the matrix that carries one from the past to the future had to maintain an exact bookkeeping of total probability. But Feynman had no such matrix. The essence of his approach was a view of past and future together, with the freedom to go forward or backward in time at will. He was getting almost nothing across. Finally, as he sketched diagrams on the blackboard—schematic trajectories of particles—and tried to show his method of summing the amplitudes for different paths, Bohr rose to object. Had Feynman ignored the central lesson of two decades of quantum mechanics? It was obvious, Bohr said, that such trajectories violated the uncertainty principle. He stepped to the blackboard, gestured Feynman aside, and began to explain. Wheeler, taking notes, quickly jotted, "Bohr Has Raised The Question As To Whether This Point Of View Has Not The Same Physical Content As The Theory Of Dirac, But Differs In A Manner Of Speaking Of Things Which Are Not Well-Defined Physically." Bohr continued for long minutes. That was when Feynman knew he had failed. At the time, he was in anguish. Later he said simply: "I had too much stuff. My machines came from too far away."

There Was Also Presented (by Feynman) . . .

Wheeler had arranged as rapid a news service as the available technology permitted. On his first day back in Princeton he pressed his graduate students into service as scribes. They reproduced his notes page by page onto mimeograph blanks and printed dozens of copies, turning their forearms magenta. For months this samizdat document served as the only available introduction to the new Schwingerian covariant quantum electrodynamics. Only a few pages were devoted to Feynman, with his "alternative formulation" and curious diagrams. Dyson read the Wheeler notes avidly. Bethe had tried to get him an invitation to Pocono ("you can imagine that I was highly pleased and flattered," Dyson wrote his parents), but Oppenheimer refused to consider someone whose current caste was student.

Feynman himself was assigned the task of writing a nontechnical account of the Pocono meeting for a new trade journal for physicists, *Physics Today*—anonymously, he hoped. He explained renormalization à la Schwinger, concluding:

> A major portion of the conference was spent in hearing and discussing these results of Schwinger. (((One conferee put it: "We did not have time to discuss a great deal, for we had to take time out to learn some physics." He was referring to this work of Schwinger.)))
>
> There was also presented (by Feynman) a theory in which the equations of electrodynamics are artificially altered so that all quantities including the inertia of the electron turn out finite. The results of this theory are in essential agreement with those of Schwinger, but they are not as complete.

In the same runner-up vein Feynman was asked to help select a winner for a new prize the National Academy of Sciences was awarding for "an outstanding contribution to our knowledge of the nature of light." When Schwinger saw Feynman's name on the list of judges, he inferred correctly that the prize was meant for him. What was quantum electrodynamics about, if not light, in all its many dresses?

No one had been more definitively impressed by Schwinger, and unimpressed by Feynman, than Oppenheimer. Awaiting him back in Princeton was a startling confirmation of Schwinger's theory, in the form of a letter from a Japanese theorist, Shin'ichirō Tomonaga, whose claim to glory began with the words: "I have taken the liberty of sending you copies of several papers and notes . . ."

Japan's physicists had just begun making significant contributions to the international community in the 1930s—it had been Hideki Yukawa at Kyōto University who first proposed that a massive, short-lived, undiscovered particle might act as a "carrier" of the nuclear force, binding protons

together in the atom's core—when the war isolated them utterly. Even with
the war's end, channels to occupied Japan opened slowly. News of the
Lamb shift reached Kyōto and Tokyo not through American physicists and
not through journals, but from a squib in a newsmagazine.

Tomonaga, a native of Tokyo and a graduate of Kyōto University, a
classmate and friend of Yukawa, had been deeply influenced by Dirac; he
belonged to a small group that translated Dirac's famous textbook into
Japanese. In 1937 he traveled to Germany to study with Heisenberg; returning at the war's onset in 1939, he stopped briefly in New York to visit
the World's Fair. He worked out what he called a "super many time" theory,
in which every point in the field had its own clock—a workable notion,
he found, despite the seeming absurdity of trying to manipulate infinitely
many time variables. In his thoughts on physics he traversed much of the
ground covered by his European and American counterparts, but with a
far greater sense of solitude, hardly diminished by his time in Germany.
He recorded a dark mood in his diary from time to time:

> After supper I took up my physics again, but at last I gave up. Ill-starred
> work indeed! . . . Recently I have felt very sad without any reason, so I
> went to a film. . . . Returning home I read a book on physics. I don't
> understand it very well. . . . Why isn't nature clearer and more directly
> comprehensible? . . . As I went on with the calculation, I found the in-
> tegral diverged—was infinite. After lunch I went for a walk. The air was
> astringently cold. . . . All of us stand on the dividing line from which the
> future is invisible. We need not be too anxious about the results, even
> though they may turn out quite different from what you expect. . . .

His occasional emotional desolation paled in light of what faced him in
the months after the surrender, when shortages of food and housing over-
shadowed all else in Japan. He made a home and an office in a battered
Quonset hut on the Tokyo University grounds. He furnished it with mats.

Although Oppenheimer knew nothing of Tomonaga's personal circumstances, he knew what he and his Los Alamos compatriots had wrought
on Japan, and he also wished to preserve the internationalism of physics
in the face of what suddenly seemed an American hegemony. He could
hardly have been better placed to appreciate Tomonaga's letter—clear evidence that a Japanese physicist had not just matched the essentials of
Schwinger's work but had anticipated it. Tomonaga had not published, and
he had not created the entire Schwingerian tapestry, but he had been first.

Oppenheimer immediately gave Tomonaga his imprimatur in a letter to each of the Pocono participants. "Just because we were able to hear Schwinger's beautiful report," he wrote, "we may better be able to appreciate this independent development." For Dyson, working in Pocono's aftermath to understand the new theories, the revelation of Tomonaga's papers lay in what seemed a simple beauty. He thought that he now understood Schwinger and that not all Schwinger's complications were necessary. Graduate students poring over the Pocono notes already suspected this, despite the acclaim their elders were awarding. Later Dyson quoted "an unkind critic" as having said, "Other people publish to show how to do it, but Julian Schwinger publishes to show you that only he can do it." He seemed to strive for an exceptional ratio of equations to text, and the prose posed serious challenges to the *Physical Review*'s typesetters.

Schwinger occasionally heard what sounded like carping amid the applause: comments to the effect that he was a soulless Paganini, all flash and technique instead of music; that he was more mathematician than physicist; that he too carefully smoothed the rough edges. "I gather I stand accused," Schwinger said later, "of presenting a finished elaborate mathematical formalism from which had been excised all the physical insights that provide signposts to its construction."

He had removed the signposts. He never liked to show the rough pathways of his thinking, any more than he liked to let his audiences see notes when he lectured. Yet all his mathematical power could not have produced his joining of relativity and quantum electrodynamics if he had lacked the intuition of a physicist. Beneath the formalism lay a profound—and historically minded—conviction about the nature of particles and fields. To Schwinger renormalization was not just a mathematical trick. Rather it marked a mutation in physicists' understanding of what a particle was. His central physical insight, had he expressed it in the compromised language of everyday speech, might have sounded like this:

Are we talking about particles or are we talking about waves? Until now, everyone has thought that their equations—the Dirac equation, for example, which is supposed to describe the hydrogen atom—referred directly to the physical particles. Now, in a field theory, we recognize that the equations refer to a sublevel. Experimentally we are concerned with particles, yet the old equations describe fields. When you talk about fields, you presume that you can describe, and somehow experience, exactly what goes on at every point in space at every time; when you talk about particles, you merely sample the field with measurements at occasional instants.

A particle is a cohesive thing. We know we have a particle only when the same thing stays there as time goes on. The very language of particles implies phenomena with continuity over space and time. Yet if you make measurements at only disconnected instants, how do you know there is a particle? Experiments probe the field only crudely—they look at large spaces over long times.

The essence of renormalization is to make the transition from one level of description to the next. When you begin with field equations, you operate at a level when particles are not there from the start. It is when you solve the field equations that you see the emergence of particles. But the properties—the mass and the charge—that you ascribe to a particle are not those inherent in the original equations.

Other people say, "Oh, the equations have divergences, you have to cancel them out." That is only the form, not the essence of renormalization. The essence lies in recognizing that the theories of Maxwell and Dirac are not about electrons, positrons, and photons but about a deeper level.

Cross-Country with Freeman Dyson

Feynman had a tendency to vanish with the end of the school year, leaving behind a vacuum populated by uncorrected papers, ungraded tests, unwritten letters of recommendation. Often Bethe covered for his lapses in the paperwork of teaching. Still, June might bring a tirade from Lloyd Smith, the department chairman:

> Your sudden departure from Ithaca without completing the grades in your courses, especially those involving seniors who may thus be prevented from graduating, has caused the Department considerable embarrassment. I have begun to be somewhat apprehensive over what would appear to be a feeling of indifference concerning the obligations and responsibilities to the University . . .

Feynman would jot some grades—round numbers, none higher than 85—and then start doodling equations.

This June found him at the wheel of his secondhand Oldsmobile, rushing

across the country at a constant 65 miles per hour. In the passenger seat
Freeman Dyson eyed the scenery and occasionally wished Feynman would
slow down. Feynman thought Dyson was a bit dignified. Dyson liked the
role of foreign observer of the American scene: here was his chance to play
Tocqueville peering at the wild West from the vantage point of Route
66. Missouri, the Mississippi River (thick and reddish-brown, just as he
had imagined it), Kansas, Oklahoma—none of this struck him as very
Western, actually. In fact it looked not unlike his rural corner of New York.
He had decided that modern America resembled Victorian England, particularly in the attention devoted to furnishing middle-class homes and
women. His destination was Ann Arbor, Michigan, where he intended to
pursue Schwinger, who was presenting his work in a series of summer-
school lectures. Feynman, meanwhile, was heading for Albuquerque to
resolve an entanglement with a woman he had known at Los Alamos. (She
was Rose McSherry, a secretary whom he dated after Arline's death. Another of Feynman's current attachments was needling him by calling
McSherry his "movie queen." Dyson's guess was that he would marry
her.)

Dyson realized that he was not taking the direct route to Ann Arbor,
but he relished the chance to spend time with Feynman. No one interested
him as much. In the months since Pocono, he had begun to think that
his mission might be to find a synthesis of the difficult new theories of
quantum electrodynamics—rival theories, as he saw it, though to most of
the community the rivalry seemed lopsided. He had heard Feynman's
theory in informal blackboard sessions, and it still troubled him that Feynman was, as it seemed, merely writing down answers instead of solving
equations in the normal manner. He wanted to understand more.

They drove, sometimes stopping for hitchhikers, more often maintaining
a determined pace, and Feynman confided more in Dyson than he had
done with any friend in his adult life. He startled Dyson with a grim outlook
on the future. He felt certain that the world had seen only the beginning
of nuclear war. The memory of Trinity, sheer ebullient joy at the time,
haunted him now. Philip Morrison, his Cornell colleague, had published
an admonitory description of an atomic blast on East 20th Street in Manhattan—Morrison had witnessed the Hiroshima aftermath and wrote this
account in a horrifyingly vivid past tense—and Feynman could not meet
his mother at a midtown restaurant without thinking about the radius of
destruction. He could not shake a feeling that normal people, without the
burden of his accursed knowledge, were living a pitiful illusion, like ants

tunneling and building before the giant's boot comes down. This was a classic danger sign—the feeling of being the only sane man, the only man who truly sees—but Dyson suddenly felt that Feynman was as sane as anyone he knew. This was not the jester he had first described to his parents. Dyson wrote later: "As we drove through Cleveland and St. Louis, he was measuring in his mind's eye distances from ground zero, ranges of lethal radiation and blast and fire damage. . . . I felt as if I were taking a ride with Lot through Sodom and Gomorrah."

As they drew closer to Albuquerque, Feynman was also thinking about Arline. Sometimes it occurred to him that her death might have left him with a feeling of impermanence. Spring flooding in the Oklahoma prairie closed the highway. Dyson had never seen rain fall in such dense curtains— nature as raw as these plainspoken Americans, he thought. The car radio reported people trapped in cars, drowned or rescued by boats. They pulled off the road in a town called Vinita and found lodging in a hotel of the kind Feynman knew all too well from his weekend trips to visit Arline: an "office" on the second floor, a sign reading, "This hotel is under new management, so if you're drunk you came to the wrong place," a hanging cloth covering the doorway to the room he shared with Dyson for fifty cents apiece. That night he told Dyson more about Arline than ever before. Neither of them forgot it.

They talked about their aspirations for science. Feynman cared far less than Dyson about his still-patchwork scheme for renormalizing quantum electrodynamics. It was his sum-over-histories theory of physics that claimed his passion. As Dyson saw, it was a grand vision and a unifying vision— too ambitious, he thought. Too many physicists had already stumbled in pursuit of this grail, including Einstein, notoriously. Dyson—more than anyone who heard Feynman at Pocono or attended his occasional seminars at Cornell, more even than Bethe—was beginning to see just how far Feynman sought to reach. He was not ready to concede that his friend could out-Einstein Einstein. He admired Feynman's gall, the largeness of his dream, the implicit attempt to unify realms of physics that were more distant from one another than anything in human experience. On the largest scale, the scale of solar systems and galactic clusters, gravity reigned. On the smallest scale, particles still awaiting discovery bound the atom's nucleus with unimaginably strong forces. Dyson considered it enough to walk the "middle ground," the realm that after all encompassed everything in between: the furniture of everyday life, the foundations underlying chemistry and biology. The middle ground, where quantum theory ruled, extended to all phenomena that could be seen and studied without the help of either a mammoth telescope or a behemoth particle accelerator. Yet Feynman wanted more.

It was essential to his view of things that it must be universal. It must describe everything that happens in nature. You could not imagine the sum-over-histories picture being true for a part of nature and untrue for another part. You could not imagine it being true for electrons and untrue for gravity. It was a unifying principle that would either explain everything or explain nothing.

Many years later each man recalled their night in Vinita, Dyson showing how unshakably he revered his friend still, Feynman showing how he could use storytelling as a strategy—a dagger and a cloak. Dyson wrote:

In that little room, with the rain drumming on the dirty window panes, we talked the night through. Dick talked of his dead wife, of the joy he had had in nursing her and making her last days tolerable, of the tricks they had played together on the Los Alamos security people, of her jokes and her courage. He talked of death with an easy familiarity which can come only to one who has lived with spirit unbroken through the worst that death can do. Ingmar Bergman in his film *The Seventh Seal* created the character of the juggler Jof, always joking and playing the fool, seeing visions and dreams that nobody else believes in, surviving at the end when death carries the rest away. Dick and Jof have a great deal in common.

And Feynman:

The room was fairly clean, it had a sink; it wasn't so bad. We get ready for bed.

He says, "I've got to pee."

"The bathroom is down the hall."

We hear girls giggling and walking back and forth in the hall outside, and he's nervous. He doesn't want to go out there.

"That's all right; just pee in the sink," I say.

"But that's unsanitary."

"Naw, it's okay; you just turn the water on."

"I can't pee in the sink," he says.

We're both tired, so we lie down. It's so hot that we don't use any covers, and my friend can't get to sleep because of the noises in the place. I kind of fall asleep a little bit.

A little later I hear a creaking of the floor nearby, and I open one eye slightly. There he is, in the dark, quietly stepping over to the sink.

And Dyson:

That stormy night in our little room in Vinita, Dick and I were not looking thirty years ahead. I knew only that somewhere hidden in Dick's ideas was the key to a theory of quantum electrodynamics simpler and more physical than Julian Schwinger's elaborate construction. Dick knew only that he had larger aims in view than tidying up Schwinger's equations. So the argument did not come to an end, but left us each going his own way.

They reached Albuquerque, Dyson seeing for the first time the deceptively clear air and the red desert beneath still snowy peaks. Feynman bore into town at 70 miles per hour and was immediately arrested for a rapid sequence of traffic violations. The justice of the peace announced that the fine he handed down was a personal record. They parted—Feynman to find Rose McSherry (marriage was impossible, as it happened, in part because she was determinedly Roman Catholic and he could not be), Dyson to find a bus back toward Ann Arbor and Schwinger.

Oppenheimer's Surrender

With Bethe's blessing Dyson moved to the Institute for Advanced Study in Princeton in the fall of 1948. Oppenheimer had taken over as director the year before. Dyson was eager to impress him, and he immediately sensed he was not alone. "On Wednesday Oppenheimer returns," he wrote his parents. "The atmosphere at the Institute during these last days has been rather like the first scene in 'Murder in the Cathedral' with the women of Canterbury awaiting the return of their archbishop."

He did not wait for Oppenheimer's blessing, however, before mailing off to the *Physical Review* a manuscript representing a cathartic outpouring of work during the last days of the summer. He proudly told his parents that the concentration had nearly killed him. Inspiration came most snappily on the fifty-hour bus ride east to Princeton, he told colleagues. (When Oppenheimer heard this he retorted with a sarcastic allusion to the lightning-from-the-blue legend of Fermat's last theorem: "There wasn't enough room in the margin to write down the proof.") Dyson had found the mathematical common ground he was sure must exist. He, too, created and reshaped terminology to suit his purpose. His chief insight was to focus on a so-called scattering matrix, or S matrix, a mustering of all the probabilities associated with the different routes from an initial state to a given end point. He now advertised "a unified development of the subject"—more reliable than Feynman and more usable than Schwinger. His father said that Feynman-Schwinger-Dyson reminded him of a clause in the Athanasian Creed: "There is the Father incomprehensible, and the Son incomprehensible, and the Holy Ghost incomprehensible, yet there are not three incomprehensibles but one incomprehensible."

It occurred to Dyson that he was rushing into print with accounts of theories not yet published by their inventors and that the inventors themselves might take offense. He visited Bethe, temporarily in New York visiting Columbia, and they took a long walk in Riverside Park as the sun set over the Hudson River. Bethe warned him that there could be problems. Dyson said it was Schwinger's and Feynman's own fault that they had not published "any moderately intelligible account": Schwinger, he suspected, was polishing obsessively, while Feynman simply couldn't be bothered with paperwork. It was irresponsible. They were retarding the development of science. By publicizing their work Dyson was performing a service to humanity, he argued. He and Bethe ended up agreeing that Feynman would not mind but that Schwinger might, and that it would be poor tactics for an ambitious young physicist to irritate Schwinger. "So the result of all this," Dyson wrote his parents,

> is that I am reversing the tactics of Mark Antony, and saying very loud at various points in my paper, "I come to praise Schwinger, not to bury him." I only hope he won't see through it.

Still, he made his judgment clear. The distinctions he drew and the characterizations he set down soon became the community's conventional wisdom: that Schwinger's and Tomonaga's approach was the same, while

Feynman's differed profoundly; and that Feynman's method was original and intuitive, while Schwinger's was formal and laborious.

Dyson well understood that he was reaching out to an audience that wanted tools. When he showed a Schwinger formula with commutators threatening to subdivide like branches on a tree and remarked that "their evaluation gives rise to long and rather difficult analysis," he knew that his readers would not suspect him of overstating the difficulty. Ease of use was the Feynman virtue he stressed. To "write down the matrix elements" for a certain event, he explained, one need only take a certain set of products, replace them by sums of matrix elements from another equation, reassemble the various terms in a certain form, and undertake a certain type of substitution. Or, he said, one could simply draw a graph.

The simplest Dyson graph.

Graph was the mathematician's word for a network of points joined by lines. Dyson showed that there was a graph for every matrix and a matrix for every graph—the graphs provided a means of cataloging these otherwise-misplaceable arrays of probabilities. So alien did this conceit seem that Dyson left it to his readers to draw the graphs in their minds. The journal editors made room for just one figure. Dyson called the solid lines, with an implicit direction, electron lines. The directionless dotted lines were photon lines. Feynman, he mentioned, had something more in mind than the mere bookkeeping of matrices: "a picture of the physical process." For Feynman the points represented the actual creation or annihilation of particles; the lines represented paths of electrons and photons, not through a measurable real space but through the history from one quantum event to another.

Oppenheimer depressed Dyson with a coolness bordering on animosity. It was the last response he had expected: a defeatist Oppenheimer, a lethargic Oppenheimer, an Oppenheimer hostile to new ideas and unwilling to listen. He had been in Europe, where he had summarized the present state of the theory at two international conferences. It was "Schwinger's theory" and "Schwinger's program." There were developments "the first largely, the second almost wholly, due to Schwinger." In passing, there were "Feynman's algorithms"—an exotically disdainful phrase.

Dyson decided that there would be no prize for timidity and—still in his first weeks at the institute—sent Oppenheimer by interoffice mail an aggressive manifesto. He argued that the new quantum electrodynamics promised to be more powerful, more self-consistent, and more broadly applicable than Oppenheimer seemed to think. He did not mince words.

From Mr. F. J. Dyson.
Dear Dr. Oppenheimer:
As I disagree rather strongly with the point of view expressed in your Solvay Report (not so much with what you say as with what you do not say) . . .
I. . . . I am convinced that the Feynman theory is considerably easier to use, understand, and teach.
II. Therefore I believe that a correct theory, even if radically different from our present ideas, will contain more of Feynman than of Heisenberg-Pauli.
. . .
V. I do not see any reason for supposing the Feynman method to be less applicable to meson theory than to electrodynamics. . . .
VI. Whatever the truth of the foregoing assertions may be, we have now a theory of nuclear fields which can be developed to the point where it can be compared with experiment, and this is a challenge to be accepted with enthusiasm.

Enthusiasm was not immediately forthcoming, but Oppenheimer did set up a series of forums to let Dyson make his case. They became an occasion. Bethe came down from New York to listen and lend moral support. As the seminars went on, Oppenheimer was a dramatically nerve-tightening presence. He interrupted continually, criticizing, jabbing, pouncing on errors. To Dyson he seemed uncontrollably nervous—always chain-smoking and fidgeting in his chair. Feynman himself was following Dyson's progress by long-distance as he continued his own work. Dyson visited him at Cornell one weekend and watched, amazed, as he rattled off two new fundamental calculations in a matter of hours. Then Feynman fired off a hasty letter: "Dear Freeman: I hope you did not go bragging about how fast I could compute the scattering of light by a potential because on looking over the calculations last night I discovered the entire effect is zero. I am sure some smart fellow like Oppenheimer would know such a thing right off."

In the end Bethe turned Oppenheimer around. He cast his vote explicitly with the Feynman theory and let the audience know that he felt Dyson had more to say. He took Oppenheimer aside privately, and the mood shifted. By January, the war had been won. At the American Physical Society meeting Dyson found himself almost as much a hero as Schwinger had been the year before. Sitting in the audience with Feynman beside him, he listened as a speaker talked admiringly of "the beautiful theory of Feynman-Dyson." Feynman said loudly, "Well, Doc, you're in." Dyson had not even got a doctoral degree. He went on an excited lecture tour and told his parents that he was a certified big shot. The reward that lasted, however, was a handwritten note that had appeared in his mailbox in the dying days of the fall, saying simply, *"Nolo contendere. R. O."*

Dyson Graphs, Feynman Diagrams

It was the affair of Case and Slotnick at the same January meeting that brought home to Feynman the full power of his machinery. He heard a buzz in the corridor after an early session. Apparently Oppenheimer had devastated a physicist named Murray Slotnick, who had presented a paper on meson dynamics. A new set of particles, a new set of fields: would the new renormalization methods apply? With physicists looking inward to the higher-energy particles implicated in the forces binding the nucleus, meson theories were now rising to the fore. The flora and fauna of meson theories did seem to resemble quantum electrodynamics, but there were important differences—chief among them: the counterpart of the photon was the meson, but mesons had mass. Feynman had not learned any of the language or the special techniques of this fast-growing field. Experiments were delivering data on the scattering of electrons by neutrons. Infinities again seemed to plague many plausible theories. Slotnick investigated two species of theory, one with "pseudoscalar coupling" and one with "pseudovector coupling." The first gave finite answers; the second diverged to infinity.

So Slotnick reported. When he finished Oppenheimer rose and asked, "What about Case's theorem?"

Slotnick had never heard of Case's theorem—and could not have, since Kenneth Case, a postdoctoral fellow at Oppenheimer's institute, had not yet publicized it. As Oppenheimer now revealed, Case's theorem proved that the two types of coupling would have to give the same result. Case was going to demonstrate this the next day. For Slotnick, the assault was unanswerable.

Feynman had not studied meson theories, but he scrambled for a briefing and went back to his hotel room to begin calculating. No, the two couplings were not the same. The next morning he buttonholed Slotnick to check his answer. Slotnick was nonplussed. He had just spent six intensive months on this calculation; what was Feynman talking about? Feynman took out a piece of paper with a formula written on it.

"What's that Q in there?" Slotnick asked.

Feynman said that was the momentum transfer, a quantity that varied according to how widely the electron was deflected.

Another shock for Slotnick: here was a complication that he had not dared to confront in a half-year of work. The special case of no deflection had been challenge enough.

This was no problem, Feynman said. He set Q equal to zero, simplified his equation, and found that indeed his night's work agreed with Slotnick. He tried not to gloat, but he was afire. He had completed in hours a superior version of a calculation on which another physicist had staked a major piece of his career. He knew he now had to publish. He possessed a crossbow in a world of sticks and clubs.

He went off to Case's lecture. At the end he leapt up with the question he had ready: "What about Slotnick's calculation?"

Schwinger, meanwhile, found the spotlight sliding away. Dyson's paper carried a sting—Dyson, who had seemed such an eager student the summer before. Now this strange wave of Dyson-Feynman publicity. As Schwinger said later with his incomparably sardonic obliqueness, "There were visions at large, being proclaimed in a manner somewhat akin to that of the Apostles, who used Greek logic to bring the Hebrew god to the Gentiles."

Feynman now presented his own logic in his own voice. He and Dyson appeared at a third and last small gathering of physicists, this time at Oldstone-on-the-Hudson, New York, the final panel of the triptych that had begun at Shelter Island two years earlier. He published an extended set of papers—they would stretch over three years and one hundred thousand words—that defined the start of the modern era for the next generation of physicists. After his path-integrals paper came, in the *Physical Review*, "A Relativistic Cut-Off for Classical Electrodynamics," "Relativistic Cut-Off for Quantum Electrodynamics," "The Theory of Positrons," "Space-

Time Approach to Quantum Electrodynamics," "Mathematical Formulation of the Quantum Theory of Electromagnetic Interaction," and "An Operator Calculus Having Applications in Quantum Electrodynamics." As they appeared, the younger theorists who devoured them realized that Dyson had given only a bare summary of Feynman's vision. They felt invigorated by his images—beginning with the unforgettable bombardier metaphor in the positron paper—and by his way of insisting on the plainest statements of physical principles in physical language:

The rest mass particles have is simply the work done in separating them against their mutual attraction after they are created. . . .

How would such a path appear to someone whose future gradually became past through a moving present? He would first see . . .

No aspiring physicist could read these papers without thinking about what space was, what time was, what energy was. Feynman was helping physics live up to the special promise it made to its devotees: that this most fundamental of disciplines would bring them face to face with the primeval questions. Above all, however, to young physicists the diagrams spoke loudest.

Feynman had told Dyson, with a slight edge, that he had not bothered to read his papers. "Feynman and I really understand each other," Dyson wrote home cheerily. "I know that he is the one person in the world who has nothing to learn from what I have written; and he doesn't mind telling me so." Feynman's students, however, sometimes noticed what seemed to them an undercurrent of anger in the pointed comments he would make about Dyson. He had started hearing about Dyson's graphs—irritating. Why *graphs*? he asked Dyson. Was that the mathematician in him, putting on airs?

Feynman's space-time method had other antecedents besides Dyson's graphs, as it happened. A 1943 German textbook by Gregor Wentzel contained a parallel depiction of a particle exchange process in beta decay. A Swiss student of Wentzel's, Ernst Stückelberg, had developed a diagrammatic approach that even embraced the conception of time-reversed positrons; parts of this he published, in French, and parts were returned as unpublishable. (Wentzel himself was the unimpressed referee.) Their diagrams showed glimmerings of the style of visualization that Feynman now brought to fruition. His own full-dress version finally appeared in a paper he sent off in late spring 1949. "The fundamental interaction"— an image that would burn itself into the brains of the next generation of

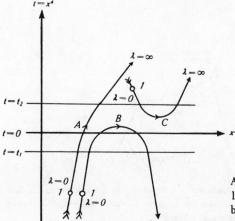

A diagram from a little-known 1941 paper of Ernst Stückelberg, showing a version of time reversal in particle trajectories.

field theorists—showed two electrons interacting by exchanging a single photon.

He drew electrons as solid lines with arrows. For photons he used wavy lines without arrows: no directionality needed because the photon's antiparticle is itself. "The fundamental interaction" reinterpreted the basic textbook process of electromagnetic repulsion. Two negative charges, electrons, repel. A standard picture, showing lines of force or merely two balls pressing apart from each other, would beg the question of how an entity feels the force of another entity at a distance. It would imply that force can be transmitted instantly, when in truth, as Feynman's diagrams automatically made explicit, whatever carries force can move only as fast as light. In the case of electromagnetism, it is light—in the form of fugitive "virtual" particles that flash into existence just long enough to help quantum theorists balance their books.

These were space-time diagrams, of course, representing time as one direction on the page. Typically the past sat at the bottom and the future at the top; one way to read the diagram would be to cover it with a sheet of paper, pull the paper slowly upward, and watch the history unfold. An electron changes course as it emits a photon. Another electron changes course when it absorbs the photon. Yet even the idea that the earlier event is *emission* and that the later is *absorption* represented a prejudice about time. It was built into the language. Feynman stressed how free his approach was from customary intuitions: these events are interchangeable.

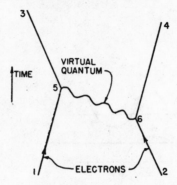

The Feynman diagram: "The fundamental interaction." It is a space-time diagram: the progress of time is shown upward on the page. If one covers it with a sheet of paper and then draws the paper slowly upward:

- A pair of electrons—their paths shown as solid lines—move toward each other.

- When (6) is reached, a virtual photon is emitted by the right-hand electron (wiggly line), and the electron is deflected outward.

- At (5) the photon is reabsorbed by the other electron, and it, too, is deflected outward.

Thus this diagram depicts the ordinary force of repulsion between two electrons as a force carried by a quantum of light. Because it is a *virtual* particle, coming into existence for a mere ghostly instant, it can temporarily violate the laws that govern the system as a whole—the exclusion principle or the conservation of energy, for example. And Feynman noted that it is arbitrary to think of the photon as being emitted in one place and absorbed in the other: one can say just as correctly that it is emitted at (5), travels *backward* in time, and is then (earlier) absorbed at (6).

The diagram is an aid to visualization. But it serves physicists mainly as a book-keeping device. Each diagram is associated with a complex number, an *amplitude* that is squared to produce a probability for the process shown.

In fact each diagram represented not a particular path, with specified times and places, but a sum of all such paths. There were other simple diagrams. He represented the self-energy of an electron—its interaction with itself—by showing a photon line returning to the same electron that spawned it. There was a grammar of permissible diagrams, corresponding, as Dyson had emphasized, to the permissible mathematical operations. Still, the diagrams could grow arbitrarily complicated, virtual particles

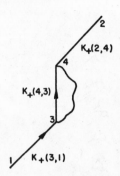

Self-interaction. It is necessary to sum the amplitudes corresponding to many Feynman diagrams—to add the contributions for every way an event can occur. The continual possibility of virtual particles materializing and vanishing causes increasing complexity. Here an electron interacts with itself, in effect—the self-energy problem that first troubled Feynman in his work with Wheeler. It emits and absorbs its own virtual photon.

appearing and disappearing in an intricate, recursive mesh. Feynman's first H-shaped diagram for interacting electrons was the only such diagram with one virtual photon. Drawing all the possible diagrams with two virtual photons showed how quickly the permutations grew. Each made a contribution to the final computation, and more complicated diagrams became enormously difficult to calculate. Fortunately the greater the complication the less the probability and the smaller, therefore, the effect on the answer. Even so, physicists would shortly find themselves agonizing over pages of diagrams resembling catalogs of knots. They found it was worth the effort; each diagram could replace an effective lifetime of Schwingerian algebra.

Feynman diagrams seemed to depict particles, and they had sprung from a mind focused on a particle-centered style of visualization, but the theory they anchored—quantum field theory—gave center stage to the field. In a sense the paths of the diagrams, and the paths summed in the path integrals, were the paths of the field itself. Feynman read the *Physical Review* more avidly than ever in the past, watching for citations. For a while it was all Schwinger—a paper would be pages of glyphs and would culminate in a neat expression that Feynman felt he could simply have written down as a starting point. He was sure this could not last. It did not. Feynman's method, Feynman's rules, began to take over. In the summer of 1950 a paper appeared with small "Feynman diagrams" on the first page—"following the simplified methods introduced by Feynman." A month later came another: "a technique due to Feynman. . . . The calculation of matrix elements can be simplified greatly by use of the Feynman-Dyson methods." The unreasonable power of the diagrams in the hands of students frustrated some of the elders, who felt that physicists were waving a sword that they did not understand. As the flood of papers began to cite Feynman,

Making the increasingly precise calculations for which quantum electrodynamics became famous required formidable exercises in combinatorics.

Schwinger went into what he described as his retreat. "Like the silicon chip of more recent years, the Feynman diagram was bringing computation to the masses," he said. Later, people who overlooked the note of hoi polloi quoted this remark as though Schwinger had intended a tribute. He had not. This was "pedagogy, not physics."

Yes, one can analyze experience into individual pieces of topology. But eventually one has to put it all together again. And then the piecemeal approach loses some of its attraction.

Schwinger's students at Harvard were put at a competitive disadvantage, or so it seemed to their fellows elsewhere, who suspected them of surreptitiously using the diagrams anyway. This was sometimes true. (They revered him, though—his night-owl ways, his Cadillac, his theatrically impeccable lecture performances. They emulated his way of saying, "We can effectively regard . . ." and they tried to construct the perfect Schwinger sentence: one graduate student, Jeremy Bernstein, liked a prototype that began, "Although 'one' is not perfectly 'zero,' we can effectively regard . . ." They also worried about Schwinger's ability to materialize silently beside them at the lunch table; a group of his graduate students protected themselves with a conversational convention in which *Schwinger* meant *Feynman* and *Feynman* meant *Schwinger*.)

Murray Gell-Mann later spent a semester staying in Schwinger's house in Cambridge and loved to say afterward that he had searched everywhere for the Feynman diagrams. He had not found any, but one room had been locked . . .

Away to a Fabulous Land

Bethe worried that Feynman was growing restless after four years at Cornell. There were entanglements with women: Feynman pursued them and dropped them, or tried to, with increasingly public frustration—so it seemed even to undergraduates, who knew him as the least professorial of professors, likely to be found beating a rhythm on a dormitory bench or lying supine and greasy beneath his Oldsmobile. He had never settled into any house or apartment. One year he lived as faculty guest in a student residence. Often he would stay nights or weeks with married friends until these arrangements became sexually volatile. He seemed to think that Cornell was alternately too large and too small—an isolated village with only a diffuse interest in science outside the confines of its physics department. Furthermore, Hans Bethe would always be the great man of physics at Cornell.

An old Los Alamos acquaintance, Robert Bacher, after serving on the new Atomic Energy Commission, was moving to Caltech, where he was charged with rebuilding an obsolete-looking physics program. He was swimming in a lake during a summer vacation in northern Michigan when Feynman's name came into his head. He rushed back to shore, tracked Feynman down by telephone, and within a few days had him there visiting.

Feynman agreed to consider Pasadena, but he was also thinking about possibilities even more faraway, exotic, and warm. South America was on his mind. He had gone so far as to study Spanish. Pan American Airways had opened the continent to American tourists on a large scale, jumping from New York to Rio de Janeiro in thirty-four hours for roughly the price of the fortnight-long ocean voyage, and the popular magazines were filling with sensual images: palms and plantations, hot beaches and gaudy nights. Carmen Miranda and bananas still dominated the travel writing. There was a new note, too, of the apocalyptic fear that had dogged Feynman: the Soviet Union had demonstrated its first working atomic bomb in September 1949, and worries about nuclear war were entering the national conscious- ness and spurring a panicky civil defense movement. Emigrations to South America became an odd symptom. One of Feynman's girlfriends told him seriously that he might be safer there. John Wheeler said—by way of imploring Feynman to join work on a thermonuclear bomb—that he was estimating "at least a 40 percent chance of war by September."

When a Brazilian physicist visiting Princeton, Jayme Tiomno, heard that Feynman was flirting with Spanish, he had suggested a switch to Portuguese and invited him to visit the new Centro Brasiliero de Pesquisas Físicas in Rio for several weeks in the summer of 1949. Feynman accepted, applied for a passport, and left the continental United States for the first time. He did learn enough Portuguese to teach physicists and beseech women in their native language. (By the end of the summer he had per- suaded one of them, a Copacabana resident named Clotilde, who called him *meu Ricardinho* in her mellifluous Portuguese, to come live with him in Ithaca—briefly.) Late the next winter he impulsively asked the centro to hire him permanently. Meanwhile he was negotiating seriously with Bacher. He had endured one too many days kneeling in cold slush as he tried to wrap chains around his tires. Caltech appealed to him. It reminded him of the other Tech, such a pure haven for the technically minded. Four years at a liberal-arts university had not softened his outlook. He was tired of "all the ins and outs of the small town and the bad weather," he wrote Bacher, and added, "The theoretical broadening which comes from having many humanities subjects on the campus is offset by the general dopiness of the people who study these things and by the Department of Home Economics." He warned Bacher about one of his weaknesses: he did not like having graduate students. At Cornell "poor Bethe" had ended up covering for him again and again.

I do not like to suggest a problem and suggest a method for its solution and feel responsible after the student is unable to work out the problem by the suggested method by the time his wife is going to have a baby so that he cannot get a job. What happens is that I find that I do not suggest any method that I do not know will work and the only way I know it works is by having tried it out at home previously, so I find the old saying that "A Ph.D. thesis is research done by a professor under particularly trying circumstances" is for me the dead truth.

He had a sabbatical year coming. He was going to make his escape, one way or another.

Once (and it was not yesterday), a diligent student of field theory wrote later at Niels Bohr's institute in Copenhagen, *there lived a very young mole and a very young crow who, having heard of the fabulous land called Quefithe, decided to visit it. Before starting out, they went to the wise owl and asked what Quefithe was like.*

Owl's description of Quefithe was quite confusing. He said that in Quefithe everything was both up and down. Physicists need more than ideas and methods. They need a version of history, too, a narrative cabinet for ordering their bits of knowledge. So they create a legend of search and discovery on the fly; they turn hearsay and supposition into instant lore. They discover that it is hard to teach a pure concept without clothing it in at least a fragment of narrative: who discovered it; what problem needed solving; what path led from not knowing to knowing. Some physicists learn that there is such a thing as physicists' history, necessary and convenient but often different from real history. The fable of Quefithe—"*quantum field theory*"—with a Schwinger mole and a Feynman crow, an owl resembling Bohr, and a fox like Dyson, lovingly satirized a story that had entered the community's store of self-knowledge as rapidly as the path integrals and Feynman diagrams: *If you knew where you were, there was no way of knowing where you were going and conversely, if you knew where you were going, there was no way of knowing where you were. . . .*

Clearly, if they were ever going to learn anything about Quefithe, they had to see it for themselves. And that is what they did.

After a few years had passed, the mole came back. He said that Quefithe consisted of lots of tunnels. One entered a hole and wandered through a maze, tunnels splitting and rejoining, until one found the next hole and got out. Quefithe sounded like a place only a mole would like, and nobody wanted to hear more about it.

Not much later the crow landed, flapping its wings and crowing excitedly. Quefithe was amazing, it said. The most beautiful landscape with high mountains, perilous passes and deep valleys. The valley floors were teeming with little moles who were scurrying down rutted paths. The crow sounded like he had taken too many bubble baths, and many who heard him shook their heads. The frogs kept on croaking "It is not rigorous, it is not rigorous!" . . . But there was something about crow's enthusiasm that was infectious.

The most puzzling thing about it all was that the mole's description of Quefithe sounded nothing like the crow's description. Some even doubted that the mole and the crow had ever gotten to the mythical land. Only the fox, who was by nature very curious, kept running back and forth between the mole and the crow and asking questions, until he was sure that he understood them both. Nowadays, anybody can get to Quefithe—even snails.

CALTECH

· ÷ ·

The California Institute of Technology had entered the 1920s with an engineering building, a physics building, a chemistry laboratory, an auditorium, and an orange grove on a dusty, underirrigated thirty acres a few minutes east of the thriving civic center of Pasadena, a town of new money in search of monuments. The scent of orange and rose floated from the gardens of porticoed homes often described as mansions, built in a relaxed Spanish and Italianate style that was coming to be thought of as Californian. Walls were a pale stucco, roofs a red tile. "Pasadena is ten miles from Los Angeles as the Rolls-Royces fly," one commentator said in 1932. "It is one of the prettiest towns in America, and probably the richest." Albert Einstein wintered there for three years, posing for pictures on a bicycle to the delight of the institute administrators, attending, as Will Rogers said, "every luncheon, every dinner, every movie opening, every marriage and two-thirds of the divorces," before he finally decided Princeton suited him better. Even as the Depression began to reverse Pasadena's fortunes, Caltech's rose on every new tide in science. A new Caltech laboratory polished the giant lens for the great telescope under way at Palomar Mountain. Caltech made itself the American center of systematic earthquake science; one of its young graduates, Charles Richter, devised the ubiquitous measurement scale that carries his name. The school moved quickly into aeronautic science, and a group of enthusiastic amateurs firing off rockets in the hills about the Rose Bowl became, by 1944, the Jet Propulsion Laboratory. Foundations and industrialists were eager to look beyond their usual East Coast funding targets. A cornflakes manufacturer paid for a building that became the Kellogg Radiation Laboratory, and its reigning expert, Charles Lauritsen, made it a national center for fundamental nuclear physics. Lauritsen spent much of the thirties investigating the nuclear properties of the light elements—hydrogen and deuterium, helium, lithium, up through carbon and beyond—filling in details of energy levels and spin with a patched-together arsenal of equipment.

He was still working in Kellogg in the winter of 1951, when oracular messages started coming in by ham radio. A blind operator in Brazil would establish a link every week or so with a student at Caltech. Lauritsen would receive terse predictions: *Could it be that nitrogen has two levels very close together at the lowest state, not just a single level?* He would check these, and often they would prove correct. His Brazilian informant apparently had a theory . . .

In Chicago, Fermi, too, heard from Feynman—a long "Dear Fermi" letter just before Christmas from the Miramar Palace Hotel in Copacabana. Feynman, following the thread he had picked up in the episode of Case *v.* Slotnick, was working on meson theory. It was messy—divergences everywhere—but he had reached a hodgepodge of conclusions. "I should like to make some comments at the risk of saying what is obvious to everybody in the U.S.," he wrote Fermi. Mesons are pseudoscalar . . . Yukawa's theory is wrong. He had heard some experimental news via the ham-radio link—"I am not entirely in the dark in Brazil." He had some predictions that he wanted checked. His approach to these particles, so essential to the binding of the atomic nucleus, centered increasingly on an even more abstract variant of spin: yet another quantum number called isotopic spin. So did Fermi's approach, as it turned out. Feynman was duplicating some of the Chicago work. In their ways they were trying to take the measure of a theory that resembled quantum electrodynamics yet resisted the lion tamers' favorite whips, renormalization, perturbation theory. "Don't believe any calculation in meson theory which uses a Feynman diagram!" Feynman wrote Fermi. Meanwhile, as they pushed more energetically inside the atom, they were watching the breakup of the prewar particle picture. With each new particle, the dream of a manageable number of building blocks faded. In this continually subdividing world, what was truly elementary?

What was made of what? "Principles," Feynman had written in the tiny address book he carried with him. "You can't say A is made of B or vice versa. All mass is interaction." That did not solve the problem, though. Cloud chamber photographs showed new kinds of forks and kinks in the trajectories—new mesons, it seemed, before anyone had understood the old. Fermi set the tone for the coming proliferation of particles with a declaration in the *Physical Review*.

In recent years several new particles have been discovered which are currently assumed to be "elementary," that is, essentially structureless. The probability that all such particles should be really elementary becomes less and less as their number increases.

It is by no means certain that nucleons, mesons, electrons, neutrinos are all elementary particles. . . .

Feynman had made his escape shortly after arriving in Pasadena. He accepted Caltech's offer of an immediate sabbatical year and fled to the most exotic place he could find. The State Department subsidized his salary. For the first time since Far Rockaway he could spend days at the beach, where he looked over the crowds in sandals and bathing suits and gazed at the endless waves and sky. He had never before seen a beach where mountains loomed just behind. At night the Serra da Carioca were black humps in the moonlight. Royal palms like dressed-up telephone poles—taller by far than the palms of Pasadena—lined the coast and the broad avenues of Rio. Feynman went down to the sea for inspiration. Fermi teased him: "I wish I could also refresh my ideas by swimming off Copacabana." Feynman liked the idea of helping build a new seat of physics at the Centro Brasiliero de Pesquisas Físicas. Fifteen years before, physics had hardly existed in Brazil or elsewhere in South America. A few lesser German and Italian physicists had grafted branches in the middle 1930s, and within a decade their students' students were creating new facilities with the support of industry and government agencies.

Feynman taught basic electromagnetism to students at the University of Brazil in Rio, who disappointed him by meekly refusing to ask questions. Their style seemed rote and hidebound after freewheeling Americans. European influence had dominated the construction of a curriculum. The nascent graduate programs did not have the luxury of a liberal mix of confident instructors. Memorization replaced understanding, or so it seemed to Feynman, and he began to proselytize the Brazilian educational establishment. Students learned names and abstract formulations, he said. Brazilian students could recite Brewster's Law: "Light impinging on a material of index n is 100 percent polarized with the electric field perpendicular to the plane of incidence if the tangent . . ." But when he asked what would happen if they looked out at the sunlight reflecting off the bay and held up a piece of polarized film and turned the film this way and that, he got blank stares. They could define "triboluminescence"—light emitted by crystals under mechanical pressure—and it made Feynman wish the professors would just send them into a dark room with a pair of pliers and a sugar cube or a Life Saver to see the faint blue flash, as he had when he was a child. "Have you got science? No! You have only told what a word means in terms of other words. You haven't told them anything about nature—*what* crystals produce light when you crush them, *why* they produce light. . . ." An examination question would read, "What are the four types of telescope?" (Newtonian, Cassegrainian, . . .) Students could answer, and yet, Feynman said, the real telescope was lost: the instrument that helped begin the scientific revolution, that showed humanity the humbling vastness of the stars.

Words about words: Feynman despised this kind of knowledge more intently than ever, and when he returned to the United States he found out again how much it was a part of American education, a mind-set showing itself not just in the habits of students but in quiz shows, popular what-should-you-know books, and textbook design. He wanted everyone to share his strenuous approach to knowledge. He would sit idly at a café table and cock his ear to listen to the sound sugar made as it struck the surface of his iced tea, something between a hiss and a rustle, and his temper would flare if anyone asked what the phenomenon was called—even if someone merely asked for an explanation. He respected only the not-knowing, first-principles approach: try sugar in water, try sugar in warm tea, try tea already saturated with sugar, try salt . . . see when the whoosh becomes a fizz. Trial and error, discovery, free inquiry.

He resented more than just the hollowness of standardized knowledge. Rote learning drained away all that he valued in science: the inventive soul, the habit of seeking better ways to do anything. His kind of knowledge—knowledge by doing—"gives a feeling of stability and reality about the world," he said, "and drives out many fears and superstitions." He was thinking now about what science meant and what knowledge meant. He told the Brazilians:

Science is a way to teach how something gets to be known, what is not known, to what extent things *are* known (for nothing is known absolutely), how to handle doubt and uncertainty, what the rules of evidence are, how to think about things so that judgments can be made, how to distinguish truth from fraud, and from show.

Telescopes, Newtonian or Cassegrainian, had flaws and limitations to go with their wondrous history. An effective scientist—even a theorist— needed to know about both.

Faker from Copacabana

Feynman told people that he had been born tone-deaf and that he disliked most music, despite the conventional observation that mathematical and musical aptitude run side by side. Classical music—music in the European tradition—he found not just dull but positively unpleasant. Above all it was the experience of listening that he could not stand.

Those who worked near him over the years knew nevertheless about the toneless music that seemed constantly to well up through his nerve endings, that clattered and pounded through their shared office walls. He drummed unconsciously as he calculated, and he drummed to attract a crowd at parties. Philip Morrison, who shared an office with him at Cornell, would say half seriously that Feynman was drawn to drumming because it was a noisy, staccato activity, because he had long fingers, and because it went with being a magician. But Morrison also noticed how freakish Western classical music had become by the twentieth century in one respect: of all the world's musical traditions, the West's had most decisively cast out improvisation. In Bach's era mastery of the keyboard still meant combining composer, performer, and improviser in one person. Even a century later, performers felt free to experiment with improvising cadenzas mid-concerto, and Franz Liszt toward the end of the nineteenth century gave concertgoers a taste of the athletic thrill of hearing music made up on the spot as fast as a pianist could play, hearing impromptu variations and embellishments along with the false steps and blind alleys from which the performer-composer would have to extricate himself like Houdini. Improvisation meant audible risk and wrong notes. In modern practice an orchestra or string quartet that plays a half-dozen wrong notes in an hour is judged incompetent.

Having resisted the MIT version of Western culture for engineers, having rejected the liberal arts version of culture at Cornell, Feynman finally began his own process of acculturation in Brazil. Travel for most Americans, physicists included, still began with the capitals of Europe, where Feynman never ventured until he was thirty-two and a conference brought him to Paris. In the streets of Rio he discovered a taste for the Third World and especially for the music, the slang, and the art that was not codified in books or taught in school—at least not American schools. For the rest of his life he preferred traveling to Latin American and Asia. He soon became one of the first American physicists to tour Japan and there, too, headed quickly for the countryside.

In Rio Feynman found a living musical tradition—rhythm-centered, improvisational, and hotly dynamic. The word *samba* was nowhere to be found in his *Encyclopaedia Britannica*, but the sound rattled through his windows high above the beach, all brass, bells, and percussion. Brazilian samba was an African-Latin slum-and-ballroom hybrid, played in the streets and nightclubs by members of clubs facetiously called "schools." Feynman became a sambista. He joined a local school, *Os Farçantes de Copacabana*, or, roughly, the Copacabana Burlesquers—though Feynman preferred to translate *farçantes* as "fakers." There were trumpets and ukuleles, rasps and shakers, snare drums and bass drums. He tried the *pandeiro*, a tambourine that was played with the precision and variety of a drum, and he settled on the *frigideira*, a metal plate that sent a light, fast tinkle in and around the main samba rhythms, the mood shifting from explosive abstract jazz to shameless pop schmaltz. At first he had trouble mastering the fluid wrist torques of the local players, but eventually he showed enough competence to win assignments on paid private jobs. He thought he played with a foreign accent that the other musicians found esoteric and charming. He played in beach contests and impromptu traffic-stopping street parades. The climactic event in the yearly samba calendar was Rio's *carneval* in February, the raucous flesh-celebrating festival that fills the nighttime streets with Cariocas half naked or in costume. In the 1952 *carneval*, amid the crepe paper and outsized jewelry, with revelers hanging from streetcars whose bells regurgitated the samba beat, a photographer for a local version of *Paris Match* snapped a carousing American physicist dressed as Mephistopheles.

As hard as he threw himself into life in Rio, he was lonely there. His ham-radio link was not enough to keep in touch with the fast-changing edge of postwar physics. He heard from hardly anyone, not even Bethe. That winter he drank heavily—enough to frighten himself one day into swearing off alcohol one more time, for good—and picked up women on the beach or in nightclubs. He haunted the Miramar Hotel's outdoor patio bar, where he socialized with an ever-changing group of expatriate Americans and Englishmen. He took out Pan American stewardesses, who stayed on the Miramar's fourth floor between flights. And in an act of rash abandon he proposed marriage, by mail, to a woman he had dated at Cornell.

Alas, the Love of Women!

The popular anthropologist Margaret Mead had recently reported what so many popular magazines were already noticing: that the courtship rituals of American culture were in ferment. Mead examined billboard advertisements and motion-picture plots and declared, "The old certainties of the past are gone, and everywhere there are signs of an attempt to build a new tradition . . ."

> In every pair of lovers the two are likely to find themselves wondering what the next steps are in a ballet between the sexes that no longer follows traditional lines, a ballet in which each couple must make up their steps as they go along. When he is insistent, should she yield, and how much? When she is demanding, should he resist, and how firmly?

Sometimes Feynman looked at his own mating habits with a similar detachment. Since Arline's death he had pursued women with a single-mindedness that violated most of the public, if not the private, scruples associated with the sexual ballet. He dated undergraduates, paid prostitutes in whorehouses, taught himself (as he saw it) how to beat bar girls at their own game, and slept with the young wives of several of his friends among the physics graduate students. He told colleagues that he had worked out a kind of all's-fair approach to sexual morality and argued that he was using women as they sought to use him. Love seemed mostly a myth—a species of self-delusion, or rationalization, or a gambit employed by women in search of husbands. What he had felt with Arline he seemed to have placed on a shelf out of the way.

Women told him that they loved him for his mind, for his looks, for the way he danced, for the way he did try to listen to them and understand them. They loved the company of his intellectual friends. They understood that work came first with him, and they loved that about him, although Rose McSherry, the New Mexican woman he courted intensely by mail at the height of his work on quantum electrodynamics, resented it when he returned from the Pocono conference and wrote her that work would always be his "first love." She would never marry a man to slave for him, she said. Sometimes she worried that he thought of women as mere recreation. She wished she could feel that he did his work because of her and for her. So many women wanted to be his muse.

The changing rules caught Feynman's lovers in a bind. The language of illicit sex relied on awkward euphemisms and old-fashioned labels, *spooning* and *jilting*, *heels* and *tramps*, defining their roles and leaving them at a disadvantage. In his first summer at Cornell, a woman he had met in Schenectady let him know as indirectly as possible that she was pregnant and then that the pregnancy was over. "I have been quite indisposed—something unusual for me—but I think you have undoubtedly guessed the reason." As she wrote, she knew that he was renewing a fling with his "Rose of Sharon." She knew she was supposed to hate him, but she preferred not to think of men as "heels." She assured him that she was not "in love."

> I almost envy you the wonderful and supreme happiness that you must have enjoyed before your wife passed away. Such happiness comes to so few people—I wonder—can it happen twice in one's lifetime?

She did offer him a warning, saying sarcastically that she was *sure* he would recognize a bit of Byron:

> Alas, the love of women! it is known
>   To be a lovely and a fearful thing; . . .
>   And their revenge is as the tiger's spring,
> Deadly, and quick, and crushing; yet, as real
>   Torture is theirs—what they inflict they feel.
> They are right; for man, to man so oft unjust,
>   Is always so to women . . .

In a postscript, she corrected his spelling of her name.

Women were expected to contend in the work force—another trend accelerated by the war—but they also stood in the centerpiece of a cozy domestic vision of family life. The professions, and particularly the sciences, remained in the rear guard. The new *Physics Today* summed up the difficulties from the sober perspective of someone who had spent more than a decade teaching physics to undergraduates at Bryn Mawr, where a local ditty asked,

> Tell me what it is like to be teaching these girls?
> Do you find that they have any brains?
> Do they take themselves seriously (may I ask) or do you?

The editors were determined to keep the tone lighthearted. The author argued, not without sympathy, that the single most grievous obstacle to the success of women as physicists was their own "tendency to defer to the superior male." Meanwhile employers continued to assume that women's eventual priority would be marriage and children. In the *Physical Review* women almost never appeared as authors.

In their wholly male world, physicists were even less likely than other American men to look for intellectual partnership in their sexual relationships. Some did, nevertheless. In the European tradition, where the professoriat implied a certain social class and cultural grounding, wives had tended to share their husbands' class and culture: Hans Bethe married the daughter of a theoretical physicist. In the American social stew, where science had become an upward pathway for children of the immigrant poor, whatever husbands and wives might be assumed to share, it was not necessarily a background in the academy. Feynman, alone anyway in the distant reaches of much of his work, seemed to date only women of obvious beauty, often blondes, sometimes heavily made-up and provocatively dressed—or so it seemed to some of the women he did not date. He hardly seemed interested in professional companionship from the women he chased, try though they might to offer it. "I'm learning more everyday about physics and realizing that there is just reams more to learn," one of his lovers wrote. "Somehow the field of physics has a fatal fascination for me." She suspected, though, that he had already moved on to someone else. She and all her successors shared an unforgivable handicap, and some of them guessed it: They were not Arline Greenbaum, Feynman's Juliet, the one perfect love, the girl who had died before the mundane, domestic, day-to-day, year-to-year realities of ordinary life could have time to add a tempering color and tone to the romantic ideal.

Every so often Feynman would feel the urge to bring a measure of rationality to his relations with women. He loved to work out the rules, to find the systems. He tired of the susurrus of promises, flattery, cajoling. He hated having to apologize. He turned Arline's favorite principle to a new purpose: "It seems to me that you go to lots of trouble to be sure the girl doesn't think ill of you," he wrote in a note to himself after one emotionally messy encounter.

WHAT DO YOU CARE WHAT SHE *THINKS*? It is all right to care whether you hurt her or not—just do your best, (if you insist) on trying not to—then if the fact is that you are O.K., don't bother to try to argue otherwise or try to get her to tell you you are wonderful. . . . Further, if you are selfish & look only to your physical pleasure—don't try to convince yourself otherwise—or rather—don't try to explain it to her or convince her otherwise.

In his favorite bar story he gradually deduces the procedural machinery of a bar: women flirt with the customers, the customers buy them drinks, the women move on. "How is it possible," he would say, "that an intelligent guy can be such a goddamn fool when he gets into a bar?" He is such a neophyte in a bar, such a naïve outside-the-experience anthropologist, that even his education in how to order a Black and White with water on the side holds interest. He watches as bar girls goad him to buy champagne cocktails. In retaliation he learns a new set of procedures. The main rule is to treat the women with disrespect. It is psychological warfare. "You are worse than a whore," he tells someone whom he has bought sandwiches and coffee for $1.10. His reward: she sleeps with him and repays him for the sandwiches, too. *All's fair.*

Feynman told these very stories to the women he dated. Despite their too-good-to-be-true quality, they were convincing and funny. No one ever caught him in a lie. Like many people who discover that storytelling is a talent—that they can hold an audience, focus a roomful of eyes—he honed his repertoire, never caring whether the crowd included people who had heard a story before. Nor, mostly, did they care. With his stories, his laughter, his dancing, his ability when alone with another person to concentrate his attention absolutely, he was intensely attractive to women. This despite the central coldness he held so close—this noetic Casanova. They suffered, sometimes, enormous pain. A second woman told him euphemistically that she had had an abortion: "The whole thing is horrible, cruel and wretched, and happens about once in two million. . . . I'm sure you never dreamt that any harm would come of such a sudden urge (shall we say, the 'shortest part' of an urge) but as I mentioned before the innocent have to pay, etc. etc." Later she asked him to forgive the mean things she had said.

They almost always did forgive him. They loved to recite his virtues. A catalog that one woman set down on paper:

1.	Handsome (could be)	6.	a drummer (whow!)
2.	clever (he thinks)	7.	personality plus (oh boy!)
3.	tall (very)	8.	smart (putting it mild)
4.	well dressed (trim)	9.	conversation (good)
5.	a dancer (From a whore in Mexico City)	10.	sweet (sometimes)

On professional trips overseas he seduced women so regularly that his hosts knew he expected them to make introductions. In London he would meet Pauline or Betty, in Paris Isabelle or Marina, in Amsterdam Marika or Genny. He would see a woman for days and then file her farewell letter with the others:

My love for you is so great that I'm sure it would have brought us both a wealth of happiness . . . please always remember, when in the evening of your life . . . that somewhere in the world there is me and that I love you. For I shall always remember you because you are the only person that I have felt at complete ease and sympathy with.

There were so many attitudes a woman could assume for a short-term love affair. His lovers would warn him jovially not to break too many hearts, or they would wish him luck with all his projects "be they blonde or mathematical—or physical!" They would hint that they might appear on his doorstep—that his "*sorcière*" might not know the way to the moon and stars but could find the USA—or implore, "concerning your work hurry up to find an atomic broom which could fly from Europe to California in a couple of hours." They would accuse him of preferring his own company—of a "Narcissus-of-the-mind complex." They would wonder aloud what home really meant to him—was he not a little lonely, after all?

He was. His friends refused to understand why he finally chose to settle

down with Mary Louise Bell of Neodesha, Kansas, who had met him in
a Cornell cafeteria and pursued him—they said cattily—all the way to
Pasadena and finally accepted his proposal by mail from Rio de Janeiro.
They considered her a platinum blonde ("the girl with the cellophane hair"
was one unkind nickname that floated behind Feynman's back) who wore
white high heels and tight white shorts to picnics. They thought she was
older than he was (the age difference was actually just a few months). Even
before they married, they quarreled by mail about how much they should
spend on interior furnishings and how he looked in old clothes. She made
clear that she did not usually think scientists were much fun. She had
studied the history of Mexican art and textiles—that was exotic enough to
interest him. While he was in Brazil, she taught courses at Michigan State
University in the History of Furniture and Institutional Interiors, mainly
to men pursuing careers in hotel or restaurant management. "The pattern
is that the girl who teaches this course usually marries one of those characters," she told him.

They married as soon as he returned from Brazil, in June 1952, and
they honeymooned in Mexico and Guatemala, where they ran up and
down Mayan pyramids. He made her laugh, but he also frightened her
with what she decided was a violent temper. She did not know what to
think when, riding down a Mexican highway, she complained that the car's
sun flap was annoying her, and he pulled out a screwdriver and repaired
it, with both hands off the wheel. She gave his friends the impression that
she did not altogether appreciate him. She wanted him to dress better; they
discovered that they could tell whether she was near by looking to see
whether he was wearing a necktie. She nagged him, they thought. She
liked to tell people that he was not "evolved" to the point of appreciating
music and that sometimes she thought she was married to an uneducated
man with a Ph.D.

They moved from Feynman's bungalow apartment near campus to a
larger place in Altadena, just across Pasadena's northern border. She resisted
socializing with other physicists. Once he missed a chance to catch Niels
Bohr while he was in Pasadena briefly; as he and Mary Lou were sitting
down to dinner, she said that she probably should have told him, but
someone had invited them over that evening to see an old bore. Politically
she was an extreme conservative, unlike most of Feynman's colleagues,
and as the Oppenheimer security hearings began, she irritated Feynman
by saying, "Where there's smoke there's fire." He, too, voted Republican,
at least for a while. Divorce was inevitable—Feynman realized early that

they should not have children, he confided in his sister—but it was nearly four years before they finally separated.

By agreement he confessed to Extreme Cruelty—

has wilfully, wrongfully, and without provocation, justification or excuse whatsoever inflicted grievous physical and mental suffering . . . ; plaintiff has suffered great physical pain and grievous mental suffering, and has suffered physical nervous shock to the extent that further married life between plaintiff and defendant has been rendered impossible.

He agreed to a circumscribed alimony, a total of ten thousand dollars over the next three years. She kept their 1950 Oldsmobile and all their household furniture. He kept their 1951 Lincoln Cosmopolitan, his scientific books, "All Drums and Percussion Instruments," and a set of dishes that his mother had given him. The divorce had a fleeting life in the national press—not because Feynman was a celebrity, but because columnists and cartoonists could not overlook the nature of the extreme cruelty: Prof Plays Bongos, Does Calculus in Bed. "The drums made terrific noise," his wife had testified. And: "He begins working calculus problems in his head as soon as he awakens. . . . He did calculus while driving his car, while sitting in the living room and while lying in bed at night."

One day near Thanksgiving 1954, as Southern California's winter neared with no discernible change of season, the smog had rolled up from Los Angeles toward the northern hills that cradled Pasadena, and for a moment their shared discontents had become too much. Feynman wrote to Bethe begging for his old job back. His eyes smarted from the smog; Mary Lou was complaining that she could not see the beautiful colors of the trees. He said he would take any salary—he surrendered unconditionally.

Soon afterward, someone rushed up to him with news of a discovery by Walter Baade, an astronomer at Mount Wilson Observatory up in the San Gabriel Mountains, demonstrating that the stars of the distant universe were several times older than anyone had established before. Caltech in the fifties was becoming an international center of cosmological discovery. The same day, a young microbiologist told him of a discovery he had made, confirming the fundamental irreducibility of the DNA molecule as bacteria divide and divide again. With Linus Pauling and Max Delbrück on hand, Caltech had some of the leading lights of molecular genetics as the field was undergoing its sensational birth. Meanwhile, although Bethe had been thrilled by Feynman's letter, he had to tell him that the most Cornell could offer on the spot was a temporary appointment.

Feynman changed his mind again. That same fall, Enrico Fermi died, and the University of Chicago decided to do whatever was necessary to hire Feynman. Its dean of the faculty, Walter Bartkey, and a younger physicist, Marvin Goldberger, later to become president of Caltech, traveled westward on the Super Chief—Bartkey was afraid to fly—and took a taxi directly from the railway station to Feynman's house. He refused to consider their proposition, and he begged them not even to tell him how much money they were offering. He was worried, he said, that Mary Lou would hear the amount and insist on moving. He had decided. He was going to stay at Caltech.

Onward with Physics

Where next, in the newly illuminated quantum world?

Feynman had reached maturity at a moment when the community of theoretical physicists shared a great unsolved problem, such a weighty knot that the enterprise could scarcely move forward until it was untied or cut. Now that quantum electrodynamics had been solved, no single problem seemed as universally compelling. Most theoretical physicists turned convoy fashion toward the smaller atomic distances and smaller time scales at which new particles appeared. They were driven in part by the logic of the past century's history: each new step inward toward the atom's core had brought not just new revelations but also a new simplification. The periodic table of elements had once served as a powerful unifying scheme; now it seemed more like a taxonomical catalog, itself unified by the deeper principles revealed by the quest inside the atom. A rhetoric was appearing in popular writing about physics by physicists and journalists: catchwords were *fundamental* and *constituents of matter* and *building blocks of nature* and *innermost sanctum of matter*. The phrases were seductive. Other kinds of science sought laws of nature, but a kind of priority seemed to belong to the search for elementary units.

The prestige of particle physics also rose with a flood tide of military support. Most plainly, the weapons laboratories prospered and such agencies as the Office of Naval Research financed specific military research projects.

A host of applied sciences, from electronics to cryptography, benefited from the concrete interest of military program officers. Academic scientists could immediately see the potential danger of allowing the armed forces to direct the course of scientific research. "When science is allowed to exist merely from the crumbs that fall from the table of a weapons development program," said Caltech's new president, Lee DuBridge, "then science is headed into the stifling atmosphere of 'mobilized secrecy' and it is surely doomed— even though the crumbs themselves should provide more than adequate nourishment." Yet the military also recognized this. One of the many legacies of the Manhattan Project was that generals and admirals now believed the scientists' dogma: that researchers left alone to follow their instincts will lay golden eggs. The bomb had been born of the esoteric fancies of the mandarins—that was clear. Now *pure* physicists wished to conduct *basic* research into forces and particles even stranger than those powering atomic bombs; the public and the government supported them enthusiastically. At institutions like DuBridge's Caltech, even the theoretical programs of research on particle physics flourished by accepting enormous government grants to which the professors applied in groups. The grants paid for salaries, graduate students, office expenses, and university overhead. The military actively encouraged, when it did not finance directly, the giant cyclotrons, betatrons, synchrotrons, and synchrocyclotrons, any one of which consumed more steel and electricity than a prewar experimentalist could have imagined. These were not so much crumbs from the weapons-development table as they were blank checks from officials persuaded that physics worked miracles. Who could say what was impossible? Free energy? Time travel? Antigravity? In 1954 the secretary of the army invited Feynman to serve as a paid consultant on an army scientific advisory panel, and he agreed, traveling to Washington for several days in November. At a cocktail party after one session, a general confided that what the army really needed was a tank that could use sand as fuel.

Earlier that year Feynman had picked up the telephone in Pasadena to hear the chairman of the AEC, Admiral Lewis L. Strauss, say that he had won his first major prize, the Albert Einstein Award: fifteen thousand dollars and a gold medal. He was the third winner, after Kurt Gödel and Julian Schwinger. Strauss informed him of the award (Feynman amused him by saying, "Hot dog!"). The public announcement came from Oppenheimer as director of the Institute for Advanced Study. Only gradually did it occur to Feynman that this was the same Strauss who was in the process of permanently removing Oppenheimer from public life. Strauss had carried out President Dwight D. Eisenhower's order to strip Oppenheimer of his security clearance, after a letter to J. Edgar Hoover accused him, in the fashion of the time, of being a "hardened Communist" who was probably "functioning as an espionage agent." The AEC began four weeks of hearings in April. Many physicists publicly defended the man they had so admired over the past decade. The famous, damaging exception was Teller, who complained that Oppenheimer had not supported his hydrogen bomb project and testified, choosing his words carefully, "I feel that I would like to see the vital interests of this country in hands which I understand better, and therefore trust more." Under the circumstances Feynman did not relish the prospect of accepting the award from Strauss. But Rabi, who was visiting Caltech, advised him to go ahead. "You should never turn a man's generosity as a sword against him," he recalled Rabi saying. "Any virtue that a man has, even if he has many vices, should not be used as a tool against him."

In the frightened climate, atomic scientists developed an invisible trail of agents, questioning their friends and childhood neighbors, painstakingly uncovering the obvious, trying to tune in to a hearsay of who liked whom, who resented whom, who might be likely to inform on whom. Feynman's own file at the FBI grew bulky. His Los Alamos friend Klaus Fuchs had been imprisoned in 1950 for spying for the Soviet Union. Fortunately for Feynman, the bureau did not realize how often Fuchs had lent Feynman his car. It was noted that Feynman had once made a speech at Temple Israel in Far Rockaway, "at which time he had spoken on brotherhood." He was described as a shy, retiring, introverted type of individual. Neighbors vouched for his loyalty and doubted that he had participated in the high school's Young People's Socialist League, which an investigating agent described as "a militant, pro-communistic group of students." Bethe was pestered by an officer of the Department of Commerce for information regarding Feynman's "loyalty." Finally he replied curtly, "Professor Feynman is one of the leading theoretical physicists of the world. His loyalty to the United States is unquestioned. Any further elaboration would be an insult to Dr. Feynman."

On one occasion the bureau discovered a "contact by OPPENHEIMER with one 'FINEMAN' (phonetic)" and surmised "that this 'FINEMAN' is in fact subject RICHARD FEYNMAN." Officials discussed the possibility of turning him into a confidential informant against Oppenheimer. They authorized a discreet approach and then placed Feynman on the "no contact" list when he refused to be interviewed by the bureau about anything at all. Agents interviewed his Los Alamos colleagues, who generally described him as a "prodigy" of "excellent character." Yet it was learned that he sometimes boasted of having "out-foxed" the Selective Service psychiatrists to obtain a 4-F classification. One colleague considered him a "screwball." Another felt that his interest in "jazz" was not in keeping with the usual demeanor of a physics professor. Yet he had voted for Eisenhower, according to informants, registered Independent (not to be confused with Independent Progressive), and "had no respect whatsoever for the Russians." The bureau carefully copied out newspaper accounts of his divorce. And one oddity had to be reported:

> FEYNMAN has developed a fair degree of skill opening sample tumbler and Yale type locks with hairpins, bits of wire, etc. . . . FEYNMAN has been trying to learn the workings of safe locks and has expressed an ambition to be able to open a safe.

In this first report the agent tried diligently to understand the exculpatory opinion of the informant that "this was not indicative of any criminal tendencies on the part of Feynman but was merely one of the works of a brilliant mathematical mind challenged by a device considered practically impossible of solution by an ordinary individual." Nevertheless, the suggestive combination of *opened safes containing atomic secrets* and *socialized with Klaus Fuchs* proved irresistible to the anonymous authors of memorandums, special inquiries, and secret airtels that swelled Feynman's file for years to come.

The bureau monitored one other incident with particular interest. The Soviet Academy of Sciences invited Feynman to a conference in Moscow, where he would have had a chance to meet the great Lev Landau and other Russian physicists. Nuclear physics, particularly in its sensitive guises, was not on the agenda. Still, the cream of Soviet physics was engaged in a weapons program quickly catching up with the Americans'. That year the Russians exploded an advanced, portable thermonuclear bomb over Siberia. (One of its principal architects, the future dissident Andrei Sakharov, watched from a platform on the snowy steppe, miles from ground zero. Having read an American primer called the black book, he decided it would be safe to remove his dark goggles.) Feynman accepted the invitation enthusiastically, the Soviet Academy having offered to cover his travel expenses. Then he had second thoughts. He wrote a careful letter to the AEC to ask for the government's advice. "I thought you would be interested," he said, "because I was connected to the Los Alamos project during the war, so the danger that I might not be able to return, or the attitude of public opinion must be considered." After a delay, officials at both the commission and the State Department replied, asking him to turn the Soviets down. His presence might be exploited for "propaganda gains." Feynman acquiesced. He wrote the head of the Soviet Academy that "circumstances have arisen which make it impossible for me to attend." The government also forced Freeman Dyson to withdraw, warning him that under the McCarran Immigration Act he might not be allowed back into the United States. Dyson did not surrender so quietly, however. He told newspaper reporters, "This is a clear case in which the law has been proved stupid."

In their basic, nonweapons research, Russian physicists eagerly pursued the latest developments in the United States and Europe. Yet a faint difference in outlook between East and West was already unfolding. The triumph of the atomic bomb had been an American triumph, had won the American war, and had not ingrained itself so firmly into the Soviet psyche (obsessed though policymakers were with the arms race). Although an international-class synchrocyclotron went up in Dubno, money was not so readily available for giant particle accelerators of the kind now under construction in the United States. And the most influential single figure in Soviet physics was Landau, famous for the catholicity of his interests across the whole breadth of phenomena that could be called theoretical physics. He had devoted his greatest work not to elementary particles but to condensed matter: the dynamics of fluids, transitions between one phase of matter and another, turbulence, plasmas, sound dispersion, and low-temperature physics. Fundamental though all these subjects were, in the United States their status was beginning to dim slightly next to the glamour of particle physics. Not so in the Soviet Union, where physicists were particularly eager in 1955 to meet Feynman. For his first major work since quantum electrodynamics, he had turned away from particle physics after all and chosen instead a subject close to Landau's heart: a theory of superfluidity, the frictionless motion of liquid helium cooled to near absolute zero.

A Quantum Liquid

By then science-fiction writers had learned an interesting rule: not to let their imaginations run too freely, too widely. It was often better to be conservative. To create a strange new world, they had only to alter one or two features of the usual reality and let the manifold unexpected implications play themselves out. Nature, too, seemed capable of adjusting a single rule and thereby creating the most bizarre phenomena.

Superfluid helium showed what happens when a liquid can flow with no friction—not just low friction, but zero friction. Resting in a beaker, the liquid spontaneously glides in a thin film up and over the walls, apparently in defiance of gravity. It passes through cracks or holes so microscopically small that even a gas would not fit through. No matter how perfectly a pair of glass plates are polished to a smooth surface, and no matter how hard they are pressed together, superfluid helium will still flow freely between them. The liquid conducts heat far better than any ordinary substance, and no amount of cooling will freeze it into a solid.

When Feynman talked about fluid flow, he knew he was returning to a childlike, elemental fascination with the world as it is. The pleasure of watching water in bathtubs or mud puddles on the sidewalk, of trying to dam a curbside rivulet after a rainstorm, of contemplating the movement in waterfalls and whirlpools—that was what made every child a physicist, he felt. In trying to understand superfluidity, he began once again with first principles. What was a fluid? A substance, liquid or gas, that cannot withstand a shear stress, but moves under the force. The tendency of a fluid to resist the shear is its viscosity, its internal friction—honey being more viscous than water, water more than air. Nineteenth-century physicists creating the first effective equations for fluid flow found viscosity especially troublesome, so uncomputable were its consequences. For the sake of simplicity, they often created models that ignored viscosity—and for that John von Neumann later mocked them. Modelers always tried to omit unnecessary complication—that was one thing. But classical fluid dynamicists had omitted what seemed an essential, defining quality. Sarcastically von Neumann called them theorists of "dry water." Superfluid helium, Feynman said, resembled that impossible idealization, fluid without viscosity. It was dry water.

Superfluidity had an equally bizarre twin, superconductivity, the flow of electricity with no dissipation or resistance. Both were phenomena of low-temperature experimentation. Superconductivity had been discovered in 1911; superfluidity not until 1938, because of the difficulties of watching the behavior of a liquid inside a pinhead-size container in a supercooled cryostat. Esoteric though they were, by the fifties this pair of phenomena had become crown jewels of the side of theoretical physics not devoted to elementary particles. Little progress had been made in understanding the perpetual-motion machinery that seemed to be at work. It seemed to Feynman that they were like "two cities under siege . . . completely surrounded by knowledge although they themselves remained isolated and unassailable." Besides Landau, the chief contributor to the theorizing on superfluidity was Lars Onsager, the distinguished Yale chemist whose notoriously difficult courses in statistical mechanics were sometimes called (in allusion to Onsager's accent) Norwegian I and Norwegian II.

Nature had exhibited another kind of perpetual motion, familiar to quantum physicists: motion at the level of electrons in the atom. No friction or dissipation slowed electrons. Only in the interactions of crowds of atoms did the energy drain of friction arise. Were these super phenomena somehow escaping the incoherent tumult of classical matter? Was this a case of quantum mechanics writ large? Could the whole apparatus of wave functions, energy levels, and quantum states translate itself onto macroscopic scales? The most basic clue that this was indeed large-scale quantum behavior came from the apparent unwillingness of helium to freeze into hard crystals at any temperature. Classically, absolute zero was often described as the temperature at which all motion ceases. Quantum mechanically, there is no such temperature. Atomic motion never does cease. That precise a zero would violate the uncertainty principle.

Landau and others had set the stage with a handful of useful conceptions of liquid helium. One powerful idea, which continued to dominate all kinds of solid-state physics, was the notion of new entities—"quasiparticles" or "elementary excitations"—collective motions that traveled through matter and interacted with one another as if they were particles. Quantum sound waves, now called phonons, were one example. Another: liquid helium seemed to contain units of rotational motion christened rotons. Feynman tried to work out the implications of these ideas. He also explored the notion that liquid helium acts as though it were (here, as elsewhere, the old-fashioned *is* had to be permanently replaced by the provisional *acts as though it were*) a mixture of two coexisting substances, a normal liquid and a pure superfluid.

One of the strangest of all the liquid-helium manifestations demonstrated how the mixture would work. A circular tube like a bicycle tire was packed with powder and then filled with liquid helium. It was set spinning and then abruptly halted. The powder would halt the flow of any normal liquid. But the superfluid component of liquid helium would continue to flow, around and around, passing through the microscopic interstices in the powder, in effect ignoring the presence of another, normal liquid. Students could sense the flow by feeling the tire's resistance to torque, as a spinning gyroscope resists sidelong pressure. And, once set in motion, the superflow would persist as long as the universe itself.

At a meeting in New York of the American Physical Society in 1955 Feynman startled a Yale group, students of Onsager, who described a new experiment they were conducting with rotating buckets. (In the low-temperature business "buckets" tended to be glass tubes the size of a thimble.) Feynman rose and said that a rotating bucket of superfluid would be filled with peculiar vortices, whirlpools hanging down like strings. The speakers had no idea what he was talking about. This peculiar image was the essence of his visualization of the atomic behavior of liquid helium. He had tried to picture how individual atoms would move together within the fluid; he calculated the forces between them as directly as he could, with tools dating back to his undergraduate research with John Slater. He saw that rotational motions would arise, just as Landau had suggested, and he applied the quantum-mechanical restriction that such motion would have to come in indivisible units. For a while he struggled to find the right image for an elementary excitation in a superfluid. He considered an atom in a cage, oscillating. A pair of atoms revolving one around the other. A tiny rotating ring of atoms. The challenge was to drive toward a solution of a many-particle problem in quantum mechanics without being able to begin with a formal, mathematical line of reasoning. It was a challenge in pure visualization.

He lay awake in bed one night trying to imagine how rotation could arise at all. He imagined a liquid divided by a thin sheet, an imaginary impermeable membrane. On one side the liquid was motionless; on the other side it flowed. He knew how to write the old-fashioned Schrödinger wave function for both sides. Then he imagined the sheet disappearing. How could he make the wave functions join? He thought about the different phases combining. He imagined a kind of surface tension, energy proportional to the surface area of his sheet. He considered what would happen when an individual atom moved across the boundary—at what point in the rising and falling wave of energy the surface tension would fall to zero and the atom would be able to move freely. He was starting to see a surface divided into strips of glue, where the atoms could not mix, and other narrow strips where atoms would be able to change places. He calculated how little energy it would take to distort the wave function until the atoms would be held back, and realized that the strips of free motion would be no more than the width of a single atom. Then he realized that he was seeing lines, vortex lines around which the atoms circulated in rings. The rings of atoms were like rings of children waiting to use a playground slide. As each child descended—the wave function changing from plus to minus—another would slip into position at the top. But the fluid version was more than just a two-dimensional ring. It also wound back on itself through the third dimension—like a smoke ring, Feynman concluded, twenty years after he had led an investigation of smoke-ring dynamics in his high school physics club. These quantum smoke rings, or vortex lines, would circle about the tiniest conceivable hole, just one atom's width across.

In a succession of articles spanning five years he worked out the consequences of his view of the interplay of energy and motion in this quantum fluid. The vortex lines were the fundamental units, the indivisible quanta of the system. They set limits on the ways in which energy could be exchanged within the fluid. In a thin enough tube, or a slow enough flow, the lines would not be able to form, and the flow would just coast, unchanging, losing no energy, and thus absolutely free of resistance. He showed when vortex lines would arise and when they would vanish. He showed when they would begin to entangle one another and ball up, creating another unexpected phenomenon that no one had yet seen in the laboratory: superfluid turbulence. Caltech hired low-temperature experimental specialists, and Feynman worked with them closely. He learned all the details of the apparatus, vacuum pumps for cooling by lowering the vapor pressure; rubber O-rings for ensuring tight seals. Before long, word was spreading of an experiment that struck physicists as "typical Feynman." Tiny wings, airfoils, were attached to a thin quartz fiber hanging down through a tube. The superfluid was pulled through vertically. A normal fluid would have spun the wings like a tiny propeller, but the superfluid refused to cause twisting. Instead it slipped frictionlessly past. In their search for lighter and lighter airfoils, the experimenters finally killed some local flies, or so they claimed, and the investigation became known as the flies'-wings experiment.

Physicists who had worked in the area of condensed matter for longer than Feynman—and who would remain there after Feynman had once again departed—were struck by his method as much as by his success. He used none of the technical apparatus for which he was now famous: no Feynman diagrams, no path integrals. Instead he began with mental pictures: this electron pushes that one; this ion rebounds like a ball on a spring. He reminded colleagues of an artist who can capture the image of a human face with three or four minimal and expressive lines. Yet he did not always succeed. As he worked on superfluidity, he also struggled with superconductivity, and here, for once, he failed. (Yet he came close. At one point, about to leave on a trip, he wrote a single page of notes, beginning, "Possibly I understand the main origin of superconductivity." He was focusing on a particular kind of phonon interaction and on one of the experimental signatures of superconductivity, a transition in a substance's specific heat. He could see, as he jotted to himself, that there was "something still a little haywire," but he thought he would be able to work out the difficulties. He signed the page: "*In case I don't return.* R. P. Feynman.") Three younger physicists, intensely aware of Feynman's competitive presence—John Bardeen, Leon Cooper, and Robert Schrieffer—invented a successful theory in 1957. The year before, Schrieffer had listened intently as Feynman delivered a pellucid talk on the two phenomena: the problem he had solved, and the problem that had defeated him. Schrieffer had never heard a scientist outline in such loving detail a sequence leading to failure. Feynman was uncompromisingly frank about each false step, each faulty approximation, each flawed visualization.

No tricks or fancy calculations would suffice, Feynman said. The only way to solve the problem would be to guess the outline, the shape, the quality of the answer.

We have no excuse that there are not enough experiments, it has nothing to do with experiments. Our situation is unlike the field, say, of mesons, where we say, perhaps there aren't yet enough clues for even a human mind to figure out what is the pattern. We should not even have to look at the experiments. . . . It is like looking in the back of the book for the answer . . . The only reason that we cannot do this problem of superconductivity is that we haven't got enough imagination.

It fell to Schrieffer to transcribe Feynman's talk for journal publication. He did not quite know what to do with the incomplete sentences and the frank confessions. He had never read a journal article so obviously spoken aloud. So he edited it. But Feynman made him change it all back.

New Particles, New Language

In the mere half-decade since the triumph of the new quantum electrodynamics the culture of high-energy physics had made and remade itself again and again. The language, the interests, and the machinery seemed to undergo a new transformation monthly. Experimentalists and theorists assembled yearly for meetings called Rochester conferences (after their initial site, Rochester, New York), descendants of the already mythic-seeming Shelter Island–Pocono–Oldstone meetings, but far larger and better financed, scores and then hundreds of participants. By the first of these meetings, at the close of 1950, quantum electrodynamics itself was already passé; it was so perfect experimentally and so far from the frontier of new forces and particles. That year had seen a kind of milestone, the discovery of a new particle not in cosmic rays but in an experimentalist's accelerator. This was a neutral pi meson, or pion—"neutral" because it carried no charge. Actually, the experimenters did not so much detect the neutral pion as the pair of gamma rays into which it immediately decayed. This particle's ephemerality made it less consequential in the everyday world of tables and chairs, chemistry and biology, than on this exciting frontier: it typically vanished after a lifetime of a tenth of a millionth of a billionth of a second. This qualified as a short time by 1950 standards. Yet standards were changing. Within a few years particle tabulations would list this fleeting entity in the category of STABLE. And meanwhile the legions of cosmic-ray explorers, many of them British, hoisting their photographic plates skyward with balloons, would find their specialty declining as spectacularly as it had risen. "Gentlemen, we have been invaded," one of their leaders declared. "The accelerators are here."

Of necessity physicists dispensed with their earlier squeamishness about the prospect of adding yet another particle to the already rich stew. On the contrary, an experimentalist could hardly aspire to more than the creation and discovery of a new particle. What it meant to measure these particles had also changed dramatically since the days when electrons had held center stage. Inferring the mass of a particle from the arcing traces left in a cloud chamber by its second- and third-generation decay products was not so simple. An enormous range of error had to be tolerated. It had become a serious and worthwhile intellectual challenge merely to identify the particles, to name them, to write down the rules of which particles could decay into which other particles. These rules were pithy new equations: $\pi^- + p \rightarrow \pi^0 + n$, a pion with negative charge and a proton produce a neutral pion and a neutron. Never mind assessing the mass; it was hard enough to identify the objects of study. Declaring the existence or nonexistence of a certain particle became a delicate rite imbued with as much anticipation and judgment as declaring a rain delay at a baseball game.

This was the experimenters' art, but, as the accelerator era began, Feynman took a special interest in the methodologies and pitfalls. He was influenced by Bethe, who always wanted to ground his theory in his own intuitions about the numbers, and by Fermi, the field's last great combination of experimenter and theorist. Bethe spent time working out formulas for the probabilities of various wrong curvatures in cloud-chamber photographs. One experimentalist, Marcel Schein, set off a typical commotion with the announcement that he had discovered a new particle in cyclotron experiments. Bethe was suspicious. The energies seemed far too low to produce the kind of particle Schein described. Feynman forever remembered the confrontation between the two men, their faces eerily illuminated by the glow from the light table used to view the photographic plates. Bethe looked at one plate and said that the gas of the cloud chamber seemed to be swirling, distorting the curvatures. In the next plate, and the next, and the next, he saw different sources of potential error. Finally they came to a clean-looking photograph, and Bethe mentioned the statistical likelihood of errors. Schein said that Bethe's own formula predicted only a one-in-five chance of error. Yes, Bethe replied, and we've already looked at five plates. To Feynman, looking on, it seemed like classic self-deception: a researcher believes in the result he is seeking, and he starts to overweight the favorable evidence and underweight the possible counterexamples. Schein finally said in frustration: You have a different theory for every case, while I have a single hypothesis that explains all at once. Bethe replied, Yes, and the difference is that each of my many explanations is right and your one explanation is wrong.

A few years later Feynman happened to be visiting Berkeley when experimenters excitedly thought they had discovered an antiproton—a particle clearly destined to be found at higher energies, but impossible, Feynman thought, at the mere hundreds of millions of electron volts available that year. As Bethe had, he went into a dark room to examine the photographs, a dozen questionable images and one that seemed absolutely perfect, the cornerstone of the discovery, with its track curving backward just as the antiparticle must.

There must be matter somewhere in the vacuum chamber, Feynman said.

Absolutely not, the experimenters told him—just thin glass walls on either side.

Feynman asked what held the upper and lower plates together. They said there were four small bolts.

He looked again at the white arc curving through the magnetic field. Then he jabbed his pencil down onto the table, inches away from the edge of the photograph. Right here, he said, must be one of those bolts.

The blueprint, retrieved from the files and laid out over the photograph, showed that his pencil had found the exact spot. An ordinary proton had struck the bolt and scattered backward into the picture. Later, experimenters at Caltech felt that Feynman's very presence exerted a sort of moral pressure on their findings and methods. He was mercilessly skeptical. He loved to talk about the famous oil-drop experiment of Caltech's first great physicist Robert Millikan, which revealed the indivisible unit charge of the electron by isolating it in tiny, floating oil drops. The experiment was right but some of the numbers were wrong—and the record of subsequent experimenters stood as a permanent embarrassment to physics. They did not cluster around the correct result; rather, they slowly closed in on it. Millikan's error exerted a psychological pull, like a distant magnet forcing their observations off center. If a Caltech experimenter told Feynman about a result reached after a complex process of correcting data, Feynman was sure to ask how the experimenter had decided when to stop correcting, and whether that decision had been made before the experimenter could see what effect it would have on the outcome. It was all too easy to fall into the trap of correcting until the answer looked right. To avoid it required an intimate acquaintanceship with the rules of the scientist's game. It also required not just honesty, but a sense that honesty required exertion.

As the particle era unfolded, however, it made other demands of top theorists—whose ranks, meanwhile, were expanding. They had to demonstrate new kinds of flair in sorting through the relations between particles. They competed to invent abstract concepts to help organize the information arriving from accelerators. A new quantum number like isotopic spin—a quantity that seemed to be conserved through many kinds of interactions—implied new incarnations of symmetry. This notion increasingly dominated the physicists' discourse. Symmetry for physicists was not far removed from symmetry for children with paper and scissors: the idea that something remains the same when something else changes. Mirror symmetry is the sameness that remains after a reflection of left and right. Rotational symmetry is the sameness that remains when a system turns on an axis. Isotopic spin symmetry, as it happened, was the sameness that existed between the two components of the nucleus, the proton and the neutron, two particles whose relationship had been oddly close, one carrying charge and the other neutral, their masses nearly but not exactly identical. The new way to understand these particles was this: They were two states of a single entity, now called a nucleon. They differed only in their isotopic spin. One was "up," the other "down."

Theorists of the new generation had not only to master the quantum electrodynamics set forth by Feynman and Dyson. They also had to arm themselves with a rococo repertoire of methods suited to the new territory. Physicists had long utilized exotic variations of the idea of *space*—imaginary spaces in which the axes might represent quantities other than physical distance. "Momentum space," for example, allowed them to plot and to visualize a particle's momentum as though it were merely another spatial variable. One grew comfortable with such spaces, and now they were multiplying. Isotopic spin space became essential to understanding the strong forces acting on nucleons.

Other concepts, too, had to become second nature. Symmetries suggested that various particles must come in families: pairs, or triplets, or (as physicists now said) multiplets. Physicists experimented with what they called "selection rules"—rules about what must or must not happen in particle collisions on account of the conservation of quantities like charge. A physicist Feynman's age, Abraham Pais, guessed at a rule called "associated production"—certain collisions must produce new particles in groups, preserving some putative new quantum number, the nature of which was unknown. Feynman had had a similar idea in Brazil but had not liked it enough to pursue it diligently. For a few years associated production became an important catchphrase. Experimenters looked for examples or counterexamples. In the longer term its main contribution to physics was that its popularity rankled a younger theorist, Murray Gell-Mann. He thought Pais was wrong, and he was jealous.

Murray

At fourteen he had been declared "Most Studious" and "wonder boy" by his classmates at Columbia Grammar, a private school on the Upper West Side of New York, and that was the last they saw of him, for he was already a senior, and he started at Yale that fall. Gell-Mann's surname was subtly difficult to pronounce. It was wrong to unstress the second syllable, as if the name were Gelman, although Murray's older brother, Benedict, had chosen that simpler spelling. Many people leaned the other way, toward a pedantic, European style of pronunciation, the accent on the second syllable and the *a* broad: *gel*-MAHN. This, too, was wrong. Later, when he had secretaries, they sometimes upbraided malefactors: "He's not German, you know." Of course the *g* was hard, despite the unconscious tug of the soft *g* in the word *gel*. Natives of New York and other regions that distinguish between the *a*'s of *man* and *mat* suspected rightly that the second, flatter *a* must be better for Gell-Mann. It was safest to stress the two syllables almost equally. By then anyone who knew anything about Gell-Mann knew that his own pronunciation of names in any language was impeccable. Supposedly he would lecture visitors from Strasbourg or Pago Pago about the niceties of their own Alsatian or Samoan dialects. He was so insistent about differentiating the pronunciations of Colombia and Columbia that colleagues suspected him of straining to bring references to the country into conversations about the university. From the beginning most physicists simply referred to him as Murray. There was never any doubt which Murray they meant. Feynman, preparing for a cameo performance as a tribal chieftain in a Caltech production of *South Pacific*, taught himself a few words of Samoan and then resignedly told a friend, "The only person who will know I'm pronouncing this wrong is Murray."

Gell-Mann attended Columbia Grammar on full scholarship. His father, born in Austria, had learned to speak a perfectly unaccented English and so, in the early 1920s, decided to start a language school for immigrants. It was the closest to success that he came, as his son saw it. The school moved several times—once, as Murray recalled, because his mother was afraid that his brother would catch whooping cough from someone in the building—and went out of business a few years later. It was his brother, nine years older and so adored by his parents, who taught him to read and to take pleasure in language, science, and art. Benedict was a bird-watcher and nature lover before nature became a practical field of interest; dropping out of college at the height of the Depression, he stunned his parents and left a complicated impression on his younger brother.

Murray did not find his way immediately to physics, talented as he was in so many subjects. When he applied to the Ivy League graduate schools, he was widely disappointed: Yale would take him only in mathematics, Harvard would take him only if he paid full fare, and Princeton would not take him at all. So he made a half-hearted application to MIT and heard directly back from Victor Weisskopf, whom he had not heard of. Gell-Mann decided to accept Weisskopf's offer, though grudgingly. MIT seemed so lumpen. The joke he told ever after was that the alternatives did not commute: he could try MIT first and suicide second, whereas the other ordering would not work. He reached MIT in 1948, close to his nineteenth birthday, just in time to watch the exploding competition in quantum electrodynamics from the vantage point of an office near Weisskopf's. When Weisskopf advised him that the future belonged to Feynman, he studied the available preprints. Feynman's struck him as a cuckoo private language, though correct; Schwinger's version struck him as hollow and pompous; Dyson's as crude and sloppy. He was already inclined toward scabrous assessments of his famous fellow physicists, though for now he kept them mostly private.

His own work was not quite living up to his severe expectations, though he was finally beginning to impress other physicists. After a year at the Institute for Advanced Study he joined Fermi's group at Chicago. He was in time to join the tumultuous effort to find the right concepts, the right ordering principles, the right quantum numbers for understanding the many new particles. There was confusion and there were regularities—coincidences in the experimental plots of particle masses and lifetimes. There were mesons that seemed to exist, and mesons that seemed plausible but absent. There were even more mysterious particles called V-particles. The problem with these enormously massive items was that particle accelerators produced them copiously, with relative ease, yet they did not decay with corresponding ease. They lingered for as long as a billionth of a second. Pais's approach to associated production had reached toward the heart of some of the regularities in need of explanation. It contained the crucial idea of another hidden symmetry. It was also reaching a peak of popularity: in the summer of 1953 Pais created such a stir at an international conference in Japan that *Time* magazine called him at his hotel. His roommate answered the phone—it happened to be Feynman, attending the same conference to present his liquid helium results. Feynman felt a flicker of envy when he realized that *Time* had no interest in him. Gell-Mann, in Chicago, felt even more, particularly since he now saw a far more powerful answer.

Physicists had learned to speak comfortably about four fundamental forces: gravity; electromagnetism, which dominated all chemical and electrical processes; the strong force binding the atom's nucleus; and the weak force, at work in the slow processes of radioactive decay. The quick ap-

pearance and slow disappearance of V-particles suggested that their creation
relied on strong forces and that weak forces came into play as they decayed.
Gell-Mann proposed a new fundamental quantity, which for a while he
called y. This y was like a new form of charge. Charge is conserved in
particle events—the total going in equals the total coming out. Gell-Mann
supposed that y is conserved, too—but not always. The algebraic logic of
Gell-Mann's scheme decreed that strong interactions would conserve y,
and so would electromagnetic interactions, but weak interactions would
not. They would break the symmetry. Thus strong interactions would create
a pair of particles whose y had to cancel each other (1 and -1, for example).
Such a particle, having flown away from its sibling, could not decay through
a strong interaction because there was no longer a canceling y. That gave
the slower weak interaction time to take over.

Artificial though it was, Gell-Mann's y qualified as not just a description
but an explanation. As he conceived his framework, it was an organizing
principle. It gave him a way of seeing families of particles, and its logic
was so compelling that the families had obvious missing members. He was
able to predict—and did predict, in papers he began publishing in August
1953—specific new particles not yet discovered, as well as specific particles
that he insisted could not be discovered. His timing was perfect. Experimenters bore out each of his positive predictions (and failed to contradict
the negative ones). But this was only part of Gell-Mann's triumph. He also
injected a piece of his fascination with language into the temporarily befuddled business of physics nomenclature. He decided to call his quantity
y "strangeness" and the families of V-like particles "strange." A Japanese
physicist, Kazuhiko Nishijima, who had independently hit upon the same
scheme just months after Gell-Mann, chose the considerably less friendly
name "η-charge." Amid all the -*ons* and Greek-lettered particles, *strange*
sounded whimsical and unorthodox. The editors of the *Physical Review*
would not allow "Strange Particles" in Gell-Mann's title, insisting instead
on "New Unstable Particles." Pais did not like it either. He pleaded with
the audience at a Rochester conference to avoid loaded terms like "strange."
Why should a broad-minded theorist consider one particle stranger than
another? The quirkiness of the word had a distancing effect: perhaps this
new construct was not quite as *real* as charge. But Gell-Mann's command
of language had an unstoppable force. Strangeness was only the beginning.

The winter Fermi died, just before Christmas 1954, Gell-Mann wrote
to the one physicist who seemed to him utterly genuine, free of phoniness,
the one who did not worship formalism and superficialities, whose own

Playing the bongos: "*On the infrequent occasions when I have been called upon in a formal place to play the bongo drums, the introducer never seems to find it necessary to mention that I also do theoretical physics.*"

Talking with a student as Murray Gell-Mann looks on: "*Murray's mask was a man of great culture... Dick's mask was Mr. Natural—just a little boy from the country that could see through things the city slickers can't.*"

With his hero, Paul A. M. Dirac, in Warsaw, 1962.

With Carl Feynman, three years old, facing photographers on the morning of the Nobel Prize: *"Listen, buddy, if I could tell you in a minute what I did, it wouldn't be worth the Nobel Prize."*

 WESTERN UNION

O CBU432 WUDO 103 SWB B 19 54 PD INTL FR CD STOCKHOLM VIA RCA
21 1155
PROFESSOR RICHARD FEYNMAN PHYSICS DEPT
  CALIFORNIA INSTITUTE OF TECHNOLOGY PASADENA (CALIF)
ROYAL ACADEMS OF SCIENCES TODAY AWARDED YOU AND TOMONAGA AND
SCHWINGER JOINTLY THE 1965 NOBEL PRIZE FOR PHYSICS FOR YOUR
FUNDAMENTAL WORK IN QUANTUM ELECTRODYNAMICS WITH DEEP PLOUGHING
CONSEQUENCES FOR THE PHYSICS OF ELEMENTARY PARTICLES STOP PRIZE
MONEY EACH ONE THIRD STOP OUR WARM CONGRATULATIONS STOP LETTER
WILL FOLLOW

ERIK RUNDBERG THE PERMANENT SECRETARY

 WESTERN UNION
SENDING BLANK

| CALL LETTERS | FBC | CHARGE TO | Physics/Box xx 20,000 |
| Pd RCA Cable | | | |

ERIK RUNDBERG
ROYAL ACADEMS OF SCIENCES
STOCKHOLM, SWEDEN

YOUR CABLEGRAM HAS MADE ME VERY HAPPY!

RICHARD P. FEYNMAN

Send the above message, subject to the terms on back hereof, which are hereby agreed to
PLEASE TYPE OR WRITE PLAINLY WITHIN BORDER—DO NOT FOLD
1269—(R 4-55)

Celebrating the Nobel Prize in Stockholm, 1965, with Gweneth Feynman (above) and a princess (below).

With Schwinger: "*I thought you would be happy that I beat Schwinger out at last,*" Feynman wrote his mother after winning one award, "*but it turns out he got the thing 3 yrs ago. Of course, he only got 1/2 a medal, so I guess you'll be happy. You always compare me with Schwinger.*"

Shin'ichirō Tomonaga, whose work in an isolated Japan paralleled the new theories of Feynman and Schwinger: "*Why isn't nature clearer and more directly comprehensible?*"

With Carl and Michelle
(right), and on a desert
camping trip.

Standing at a Caltech blackboard and playing a chieftain in a student production of *South Pacific*.

At the February 10, 1986, hearing of the presidential commission on the space shuttle accident: "*I took this stuff that I got out of your seal and I put it in ice water, and I discovered that when you put some pressure on it for a while and then undo it it doesn't stretch back. It stays the same dimension. In other words, for a few seconds at least and more seconds than that, there is no resilience in this particular material when it is at a temperature of 32 degrees. I believe that has some significance for our problem.*"

work was always sure to be interesting and real. Some of Feynman's colleagues were already beginning to think that he had drifted away from the mainstream of particle physics, but it did not seem that way to Gell-Mann. On the contrary, he knew from their few conversations that Feynman was thinking about all the outstanding problems, all the time. Feynman responded in a friendly way. Gell-Mann visited Caltech to give a talk on his current work. The two men met privately and spoke for hours. Gell-Mann described work he had done extending Feynman's quantum electrodynamics at short distances. Feynman said he knew of the work and admired it enormously—that in fact it was the only such work he had seen that he had not already done himself. He had pursued Gell-Mann's line of thinking and generalized it further—he showed what he meant—and Gell-Mann said he thought that was wonderful.

By the beginning of the new year Caltech had made Gell-Mann an offer and Gell-Mann had accepted. He moved into an office just upstairs from Feynman's. Caltech had now placed together in one building the two leading minds of their generation. To the close-knit, international community of physicists—a small world, no matter how rapidly it was growing—the collaborations and the rivalries between these men gained an epic quality. They were together, working or feuding, leaving their imprint on every area they cared to touch, for the rest of Feynman's life. They gave their colleagues a long time to muse on how strikingly different were the ways in which a giant intellect might choose to reveal itself, even in the person of a modern theoretical physicist.

In Search of Genius

In the spring of 1955 the man most plainly and universally identified with the word *genius* died at Princeton Hospital. Most of his body was cremated, the ashes scattered, but not the brain. The hospital's pathologist, Dr. Thomas S. Harvey, removed this last remnant to a jar of formaldehyde.

Harvey weighed it. A mediocre two and two-thirds pounds. One more negative datum to sabotage the notion that the brain's size might account for the difference between ordinary and extraordinary mental ability—a notion that various nineteenth-century researchers had labored futilely to establish (claiming along the way to have demonstrated the superiority of men over women, white men over black men, and Germans over Frenchmen). The brain of the great mathematician Carl Friedrich Gauss had been turned over to such scientists. It disappointed them. Now, with Einstein's cerebrum on their hands, researchers proposed more subtle ways of searching for the secret of genius: measuring the density of surrounding blood vessels, the percentage of glial cells, the degree of neuronal branching. Decades passed. Microscope sections and photographic slides of Einstein's brain circulated among a tight circle of anatomically minded psychologists, called neuropsychologists, unable to let go the idea that a detectable sign of the qualities that made Einstein famous might remain somewhere in these fragmentary trophies. By the 1980s this most famous of brains had been whittled down to small gray shreds preserved in the office of a pathologist retired to Wichita, Kansas—a sodden testament to the elusiveness of the quality called genius.

Eventually the findings were inconclusive, though that did not make them unpublishable. (One researcher counted a large excess of branching cells in the parietal sector called Brodmann area 39.) Those searching for genius's corporeal basis had little enough material from which to work. "Is there a neurological substrate for talent?" asked the editors of one neuropsychology volume. "Of course, as neuropsychologists we hypothesize that there must be such a substrate and would hardly think to relegate talent somehow to 'mind.' What evidence currently exists would be the results of the work on Einstein's brain . . ."—the brain that created the post-Newtonian universe, that released the pins binding us to absolute space and time, that visualized (in its parietal lobe?) a plastic fourth dimension, that banished the ether, that refused to believe God played dice, that piloted such a kindly, forgetful form about the shaded streets of Princeton. There was only one Einstein. For schoolchildren and neuropsychologists alike, he stood as an icon of intellectual power. He seemed—but was this true?—to have possessed a rare and distinct quality, genius as an essence, not a mere statistical extremum on a supposed bell-curve of intelligence. This was the conundrum of genius. Was genius truly special? Or was it a matter of degree—a miler breaking 3:50 rather than 4:10? (A shifting bell-curve, too: yesterday's record-setter, today's also-ran.) Meanwhile, no one had thought to dissect the brains of Niels Bohr, Paul A. M. Dirac, Enrico Fermi; Sigmund Freud, Pablo Picasso, Virginia Woolf; Jascha Heifetz, Isadora Duncan, Babe Ruth; or any of the other exceptional, creative, intuitive souls to whom the word was so often and so lubricously applied.

What a strange and bewildering literature grew up around the term *genius*—defining it, analyzing it, categorizing it, rationalizing and reifying it. Commentators have contrasted it with such qualities as (mere) talent, intellect, imagination, originality, industriousness, sweep of mind and elegance of style; or have shown how genius is composed of these in various combinations. Psychologists and philosophers, musicologists and art critics, historians of science and scientists themselves have all stepped into this quagmire, a capacious one. Their several centuries of labor have produced no consensus on any of the necessary questions. Is there such a quality? If so, where does it come from? (A glial surplus in Brodmann area 39? A doting, faintly unsuccessful father who channels his intellectual ambition into his son? A frightful early encounter with the unknown, such as death of a sibling?) When otherwise sober scientists speak of the genius as magician, wizard, or superhuman, are they merely indulging in a flight of literary fancy? When people speak of the borderline between genius and madness, why is it so evident what they mean? And a question that has barely been asked (the where-are-the-.400-hitters question): Why, as the pool of available humans has risen from one hundred million to one billion to five billion, has the production of geniuses—Shakespeares, Newtons, Mozarts, Einsteins—seemingly choked off to nothing, genius itself coming to seem like the property of the past?

"Enlightened, penetrating, and capacious minds," as William Duff chose to put it two hundred years ago, speaking of such exemplars as Homer, Quintilian, and Michelangelo in one of a string of influential essays by mid-eighteenth-century Englishmen that gave birth to the modern meaning of the word *genius*. Earlier, it had meant spirit, the magical spirit of a jinni or more often the spirit of a nation. Duff and his contemporaries wished to identify genius with the godlike power of invention, of creation, of making what never was before, and to do so they had to create a psychology of imagination: imagination with a "RAMBLING and VOLATILE power"; imagination "perpetually attempting to soar" and "apt to deviate into the mazes of error."

Imagination is that faculty whereby the mind not only reflects on its own operations, but which assembles the various ideas conveyed to the understanding by the canal of sensation, and treasured up in the repository of the memory, compounding or disjoining them at pleasure; and which, by its plastic power of inventing new associations of ideas, and of combining them with infinite variety, is enabled to present a creation of its own, and to exhibit scenes and objects which never existed in nature.

These were qualities that remained two centuries later at the center of cognitive scientists' efforts to understand creativity: the mind's capacity for self-reflection, self-reference, self-comprehension; the dynamical and fluid creation of concepts and associations. The early essayists on genius, writing with a proper earnestness, attempting to reduce and regularize a phenomenon with (they admitted) an odor of the inexplicable, nevertheless saw that genius allowed a certain recklessness, even a lack of craftsmanship. Genius seemed natural, unlearned, uncultivated. Shakespeare was—"in point of genius," Alexander Gerard wrote in 1774—Milton's superior, despite a "defective" handling of poetic details. The torrent of analyses and polemics on genius that appeared in those years introduced a rhetoric of ranking and comparing that became a standard method of the literature. Homer versus Virgil, Milton versus Virgil, Shakespeare versus Milton. The results—a sort of tennis ladder for the genius league—did not always wear well with the passage of time. Newton versus Bacon? In Gerard's view Newton's discoveries amounted to a filling in of a framework developed with more profound originality by Bacon—"who, without any assistance, sketched out the whole design." Still, there were those bits of Newtonian mathematics to consider. On reflection Gerard chose to leave for posterity "a question of very difficult solution, which of the two had the greatest genius."

He and his contemporary essayists had a purpose. By understanding genius, rationalizing it, celebrating it, and teasing out its mechanisms, perhaps they could make the process of discovery and invention less accidental. In later times that motivation has not disappeared. More overtly than ever, the nature of genius—genius as the engine of scientific discovery—has become an issue bound up with the economic fortunes of nations. Amid the vast modern network of universities, corporate laboratories, and national science foundations has arisen an awareness that the best financed and best organized of research enterprises have not learned to engender, perhaps not even to recognize, world-turning originality.

Genius, Gerard summed up in 1774, "is confessed to be a subject of capital importance, without the knowledge of which a regular method of invention cannot be established, and useful discoveries must continue to be made, as they have generally been made hitherto, merely by chance." Hitherto, as well. In our time he continues to be echoed by historians of science frustrated by the sheer ineffability of it all. But they keep trying to replace awe with understanding. J. D. Bernal said in 1939:

It is one of the hopes of the science of science that, by careful analysis of past discovery, we shall find a way of separating the effects of good organization from those of pure luck, and enabling us to operate on calculated risks rather than blind chance.

Yet how could anyone rationalize a quality as fleeting and accident-prone as a genius's inspiration: Archimedes and his bath, Newton and his apple? People love stories about geniuses as alien heroes, possessing a quality beyond human understanding, and scientists may be the world's happiest consumers of such stories. A modern example:

A physicist studying quantum field theory with Murray Gell-Mann at the California Institute of Technology in the 1950s, before standard texts have become available, discovers unpublished lecture notes by Richard Feynman, circulating samizdat style. He asks Gell-Mann about them. Gell-Mann says no, Dick's methods are not the same as the methods used here. The student asks, well, what are Feynman's methods? Gell-Mann leans coyly against the blackboard and says, Dick's method is this. You write down the problem. You think very hard. (He shuts his eyes and presses his knuckles parodically to his forehead.) Then you write down the answer.

The same story appeared over and over again. It was an old genre. From an 1851 tract titled *Genius and Industry*:

(A professor from the University of Cambridge calls upon a genius of mathematics working in Manchester as a lowly clerk.) " . . . from Geometry to Logarithms, and to the Differential and Integral Calculus; and thence again to questions the most foreign and profound: at last, a question was proposed to the poor clerk—a question which weeks had been required to solve. Upon a simple slip of paper it was answered immediately. 'But how,' said the Professor, 'do you work this? show me the rule! . . . The answer is correct but you have reached it by a different way.'

" 'I have worked it,' said the clerk, 'from a rule in my own mind. I cannot show you the law—I never saw it myself; the law is in my mind.'

" 'Ah!' said the Professor, 'if you talk of a law within your mind, I have done; I cannot follow you there.' "

Magicians again. As Mark Kac said: ". . . The working of their minds is for all intents and purposes incomprehensible. Even after we understand what they have done, the process by which they have done it is completely dark." The notion places a few individuals at the margin of their community—the impractical margin, since the stock in trade of the scientist is the method that *can* be transferred from one practitioner to the next.

If the most distinguished physicists and mathematicians believe in the genius as magician, it is partly for psychological protection. A merely excellent scientist could suffer an unpleasant shock when he discussed his work with Feynman. It happened again and again: physicists would wait for an opportunity to get Feynman's judgment of a result on which they had staked weeks or months of their career. Typically Feynman would refuse to allow them to give a full explanation. He said it spoiled his fun. He would let them describe just the outline of the problem before he would jump up and say, *Oh, I know that . . .* and scrawl on the blackboard not his visitor's result, A, but a harder, more general theorem, X. So A (about to be mailed, perhaps, to the *Physical Review*) was merely a special case. This could cause pain. Sometimes it was not clear whether Feynman's lightning answers came from instantaneous calculation or from a storehouse of previously worked-out—and unpublished—knowledge. The astrophysicist Willy Fowler proposed at a Caltech seminar in the 1960s that quasars—mysterious blazing radiation sources lately discovered in the distant sky—were supermassive stars, and Feynman immediately rose, astonishingly, to say that such objects would be gravitationally unstable. Furthermore, he said that the instability followed from general relativity. The claim required a calculation of the subtle countervailing effects of stellar forces and relativistic gravity. Fowler thought he was talking through his hat. A colleague later discovered that Feynman had done a hundred pages of work on the problem years before. The Chicago astrophysicist Subrahmanyan Chandrasekhar independently produced Feynman's result—it was part of the work for which he won a Nobel Prize twenty years later. Feynman himself never bothered to publish. Someone with a new idea always risked finding, as one colleague said, "that Feynman had signed the guest book and already left."

A great physicist who accumulated knowledge without taking the trouble to publish could be a genuine danger to his colleagues. At best it was unnerving to learn that one's potentially career-advancing discovery had been, to Feynman, below the threshold of publishability. At worst it undermined one's confidence in the landscape of the known and not known. There was an uneasy subtext to the genus of story prompted by this habit. It was said of Lars Onsager, for example, that a visitor would ask him about a new result; sitting in his office chair he would say, *I believe that is correct*; then he would bend forward diffidently to open a file drawer, glance sidelong at a long-buried page of notes, and say, *Yes, I thought so; that is correct.* This was not always precisely what the visitor had hoped to hear.

A person with a mysterious storehouse of unwritten knowledge was a wizard. So was a person with the power to tease from nature its hidden secrets—a scientist, that is. The modern scientist's view of his quest harkened back to something ancient and cabalistic: laws, rules, symmetries hidden just beneath the visible surface. Sometimes this view of the search for knowledge became overwhelming, even oppressive. John Maynard Keynes, facing a small audience in a darkened room at Cambridge a few years before his death, spoke of Newton as "this strange spirit, who was tempted by the Devil to believe . . . that he could reach *all* the secrets of God and Nature by the pure power of mind—Copernicus and Faustus in one."

> Why do I call him a magician? Because he looked on the whole universe and all that is in it *as a riddle*, as a secret which could be read by applying pure thought to certain evidence, certain mystic clues which God had laid about the world to allow a sort of philosopher's treasure hunt to this esoteric brotherhood. . . . He *did* read the riddle of the heavens. And he believed that by the same powers of his introspective imagination he would read the riddle of the Godhead, the riddle of past and future events divinely foreordained, the riddle of the elements and their constitution. . . .

In his audience, intently absorbing these words, aware of the cold and the gloom and the seeming exhaustion of the speaker, was the young Freeman Dyson. Dyson came to accept much of Keynes's view of genius, winnowing away the seeming mysticism. He made the case for magicians in the calmest, most rational way. No "magical mumbo-jumbo," he wrote. "I am suggesting that anyone who is transcendentally great as a scientist is likely also to have personal qualities that ordinary people would consider in some sense superhuman." The greatest scientists are deliverers and destroyers, he said. Those are myths, of course, but myths are part of the reality of the scientific enterprise.

When Keynes, in that Cambridge gloom, described Newton as a wizard, he was actually pressing back to a moderate view of genius—for after the eighteenth century's sober tracts had come a wild turning. Where the first writers on genius had noticed in Homer and Shakespeare a forgivable disregard for the niceties of prosody, the romantics of the late nineteenth century saw powerful, liberating heroes, throwing off shackles, defying God and convention. They also saw a bent of mind that could turn fully pathological. Genius was linked with insanity—*was* insanity. That feeling of divine inspiration, the breath of revelation seemingly from without, actually came from within, where melancholy and madness twisted the brain. The roots of this idea were old. "Oh! how near are genius and madness!" Denis Diderot had written. ". . . Men imprison them and chain them, or raise statues to them." It was a side effect of the change in focus from God-centeredness to human-centeredness. The very notion of revelation, in the absence of a Revealer, became disturbing, particularly to those who experienced it: ". . . something profoundly convulsive and disturbing suddenly becomes visible and audible with indescribable definiteness and exactness," Friedrich Nietzsche wrote. "One hears—one does not seek; one takes—one does not ask who gives: a thought flashes out like lightning. . . ." Genius now suggested Charles-Pierre Baudelaire or Ludwig van Beethoven, flying off the tracks of normality. *Crooked roads*, William Blake had said: "Improvement makes strait roads; but the crooked roads without Improvement are roads of Genius."

An 1891 treatise on genius by Cesare Lombroso listed some associated symptoms. *Degeneration. Rickets. Pallor. Emaciation. Lefthandedness.* A sense of the mind as a cauldron in tumult was emerging in European culture, along with an often contradictory hodgepodge of psychic terminology, all awaiting the genius of Freud to provide a structure and a coherent jargon. In the meantime: *Misoneism. Vagabondage. Unconsciousness.* More presumed clues to genius. *Hyperesthesia. Amnesia. Originality. Fondness for special words.* "Between the physiology of the man of genius, therefore, and the pathology of the insane," Lombroso concluded, "there are many points of coincidence. . . ." The genius, disturbed as he is, makes errors and wrong turns that the ordinary person avoids. Still, these madmen, "despising and overcoming obstacles which would have dismayed the cool and deliberate mind—hasten by whole centuries the unfolding of truth."

The notion never vanished; in fact it softened into a cliché. Geniuses display an undeniable obsessiveness resembling, at times, monomania. Geniuses of certain kinds—mathematicians, chess players, computer programmers—seem, if not mad, at least lacking in the social skills most easily identified with sanity. Nevertheless, the lunatic-genius-wizard did not play as well in America, notwithstanding the relatively unbuttoned examples of writers like Whitman and Melville. There was a reason. American genius as the nineteenth century neared its end was not busy making culture, playing with words, creating music and art, or otherwise impressing the academy. It was busy sending its output to the patent office. Alexander Graham Bell was a genius. Eli Whitney and Samuel Morse were geniuses. Let European romantics celebrate the genius as erotic hero (Don Juan) or the genius as martyr (Werther). Let them bend their definitions to accommodate the genius composers who succeeded Mozart, with their increasingly direct pipelines to the emotions. In America what newspapers already called the machine age was under way. The consummate genius, the genius who defined the word for the next generation, was Thomas Alva Edison.

By his own description he was no wizard, this Wizard of Menlo Park. Anyone who knew anything about Edison knew that his genius was ninety-nine percent perspiration. The stories that defined his style were not about inspiration in the mode of the Newtonian apple. They spoke of exhaustive, laborious trial and error: every conceivable lamp filament, from human hair to bamboo fiber. "I speak without exaggeration," Edison declared (certainly exaggerating), "when I say that I have constructed three thousand different theories in connection with the electric light, each one of them reasonable and apparently likely to be true." He added that he had methodically disproved 2,998 of them by experiment. He claimed to have carried out fifty thousand individual experiments on a particular type of battery. He had a classic American education: three months in a Michigan public school. The essential creativity that led him to the phonograph, the electric light, and more than a thousand other patented inventions was deliberately played down by those who built and those who absorbed his legend. Perhaps understandably so—for after centuries in which a rationalizing science had systematically drained magic from the world, the machine-shop inventions of Edison and other heroes were now loosing a magic with a frightening, transforming power. This magic buried itself in the walls of houses or beamed itself invisibly through the air.

"Mr. Edison is not a wizard," reported a 1917 biography.

Like all people who have prodigiously assisted civilization, his processes are clear, logical and normal.

Wizardry is the expression of superhuman gifts and, as such, is an impossible thing. . . .

And yet, Mr. Edison can bid the voices of the dead to speak, and command men in their tombs to pass before our eyes.

"Edison was not a wizard," announced a 1933 magazine article. "If he had what seems suspiciously like a magic touch, it was because he was markedly in harmony with his environment. . . ." And there the explication of Edisonian genius came more or less to an end. All that remained was to ask—but few did—one of those impossible late-night *what if* questions: What if Edison had never lived? What if this self-schooled, indefatigable mind with its knack for conceiving images of new devices, methods, processes had not been there when the flood began to break? The question answers itself, for it was a flood that Edison rode. Electricity had burst upon a world nearing the limits of merely mechanical ingenuity. The ability to understand and control currents of electrons had suddenly made possible a vast taxonomy of new machines—telegraphs, dynamos, lights, telephones, motors, heaters, devices to sew, grind, saw, toast, iron, and suck up dirt, all waiting at the misty edge of potentiality. No sooner had Hans Christian Oersted noticed, in 1820, that a current could move a compass needle than inventors—not just Samuel Morse but André-Marie Ampère and a half-dozen others—conceived of telegraphy. Even more people invented generators, and by the time enough pieces of technology had accumulated to make television possible, no one inventor could plausibly serve as its Edison.

The demystification of genius in the age of inventors shaped the scientific culture—with its plainspoken positivism, its experiment-oriented technical schools—that nurtured Feynman and his contemporaries in the twenties and thirties, even as the pendulum swung again toward the more mysterious, more intuitive, and conspicuously less practical image of Einstein. Edison may have changed the world, after all, but Einstein seemed to have reinvented it whole, by means of a single, incomprehensible act of visualization. He saw how the universe must be and announced that it was so. Not since Newton . . .

By then the profession of science was expanding rapidly, counting not hundreds but tens of thousands of practitioners. Clearly most of their work, most of science, was ordinary—as Freeman Dyson put it, a business of "honest craftsmen," "solid work," "collaborative efforts where to be reliable is more important than to be original." In modern times it became almost impossible to talk about the processes of scientific change without echoing Thomas S. Kuhn, whose *Structure of Scientific Revolutions* so thoroughly changed the discourse of historians of science. Kuhn distinguished between normal science—problem solving, the fleshing out of existing frameworks, the unsurprising craft that occupies virtually all working researchers—and revolutions, the vertiginous intellectual upheavals through which knowledge lurches genuinely forward. Nothing in Kuhn's scheme required individual genius to turn the crank of revolutions. Still, it was Einstein's relativity, Heisenberg's uncertainty, Wegener's continental drift. The new mythology of revolutions dovetailed neatly with the older mythology of genius—minds breaking with the standard methods and seeing the world new. Dyson's kind of genius destroyed and delivered. Schwinger's quantum electrodynamics and Feynman's may have been mathematically the same, but one was conservative and the other revolutionary. One extended an existing line of thought. The other broke with the past decisively enough to mystify its intended audience. One represented an ending: a mathematical style doomed to grow fatally overcomplex. The other, for those willing to follow Feynman into a new style of visualization, served as a beginning. Feynman's style was risky, even megalomaniacal. Reflecting later on what had happened, Dyson saw his own goals, like Schwinger's, as conservative ("I accepted the orthodox view . . . I was looking for a neat set of equations . . .") and Feynman's as visionary: "He was searching for general principles that would be flexible enough so that he could adapt them to anything in the universe."

Other ways of seeking the source of scientific creativity had appeared. It seemed a long way from such an inspirational, how-to view of discovery to the view of neuropsychologists looking for a *substrate*, refusing to speak merely about "mind." Why had *mind* become such a contemptible word to neuropsychologists? Because they saw the term as a soft escape route, a deus ex machina for a scientist short on explanations. Feynman himself learned about neurons; he taught himself some brain anatomy when trying to understand color vision; but usually he considered *mind* to be the level worth studying. Mind must be a sort of dynamical pattern, not so much founded in a neurological substrate as floating above it, independent of it. "So what is this mind of ours?" he remarked. "What are these atoms with consciousness?"

Last week's potatoes! They can now *remember* what was going on in my mind a year ago—a mind which has long ago been replaced. . . . The atoms come into my brain, dance a dance, and then go out—there are always new atoms, but always doing the same dance, remembering what the dance was yesterday.

Genius was not a word in his customary vocabulary. Like many physicists he was wary of the term. Among scientists it became a kind of style violation, a faux pas suggesting greenhorn credulity, to use the word *genius* about a living colleague. Popular usage had cheapened the word. Almost anyone could be a genius for the duration of a magazine article. Briefly Stephen Hawking, a British cosmologist esteemed but not revered by his peers, developed a reputation among some nonscientists as Einstein's heir to the mantle. For Hawking, who suffered from a progressively degenerative muscular disease, the image of genius was heightened by the drama of a formidable intelligence fighting to express itself within a withering body. Still, in terms of raw brilliance and hard accomplishment, a few score of his professional colleagues felt that he was no more a genius than they.

In part, scientists avoided the word because they did not believe in the concept. In part, the same scientists avoided it because they believed all too well, like Jews afraid to speak the name of Yahweh. It was generally safe to say only that Einstein had been a genius; after Einstein, perhaps Bohr, who had served as a guiding father figure during the formative era of quantum mechanics; after Bohr perhaps Dirac, perhaps Fermi, perhaps Bethe . . . All these seemed to deserve the term. Yet Bethe, with no obvious embarrassment or false modesty, would quote Mark Kac's faintly oxymoronic assessment that Bethe's genius was "ordinary," by contrast to Feynman's: "An ordinary genius is a fellow that you and I would be just as good as, if we were only many times better." *You and I would be just as good* . . . Much of what passes for genius is mere excellence, the difference a matter of degree. A colleague of Fermi's said: "Knowing what Fermi could do did not make me humble. You just realize that some people are smarter than you are, that's all. You can't run as fast as some people or do mathematics as fast as Fermi."

In the domains of criticism that fell under the spell of structuralism and then deconstructionism, even this unmagical view of genius became suspect. Literary and music theory, and the history of science as well, lost interest not only in the old-fashioned sports-fan approach—Homer versus Virgil—but also in the very idea of genius itself as a quality in the possession of certain historical figures. Perhaps genius was an artifact of a culture's psychology, a symptom of a particular form of hero worship. Reputations of greatness come and go, after all, propped up by the sociopolitical needs of an empowered sector of the community and then slapped away by a restructuring of the historical context. The music of Mozart strikes certain ears as evidence of genius, but it was not always so—critics of another time considered it prissy and bewigged—nor will it always be. In the modern style, to ask about his genius is to ask the wrong question. Even to ask why he was "better" than, say, Antonio Salieri would be the crudest of gaffes. A modern music theorist might, in his secret heart, carry an undeconstructed torch for Mozart, might feel the old damnably ineffable rapture; still he understands that *genius* is a relic of outmoded romanticism. Mozart's listeners are as inextricable a part of the magic as the observer is a part of the quantum-mechanical equation. Their interests and desires help form the context without which the music is no more than an abstract sequence of notes—or so the argument goes. Mozart's genius, if it existed at all, was not a substance, not even a quality of mind, but a byplay, a give and take within a cultural context.

How strange, then, that coolly rational scientists should be the last serious scholars to believe not just in genius but in geniuses; to maintain a mental pantheon of heroes; and to bow, with Mark Kac and Freeman Dyson, before the magicians.

"Genius is the fire that lights itself," someone had said. Originality; imagination; the self-driving ability to set one's mind free from the worn channels of tradition. Those who tried to take Feynman's measure always came back to originality. "He was the most original mind of his generation," declared Dyson. The generation coming up behind him, with the advantage of hindsight, still found nothing predictable in the paths of his thinking. If anything he seemed perversely and dangerously bent on disregarding standard methods. "I think if he had not been so quick people would have treated him as a brilliant quasi-crank, because he did spend a substantial amount of time going down what later turned out to be dead ends," said Sidney Coleman, a theorist who first knew Feynman at Caltech in the fifties.

There are lots of people who are too original for their own good, and had Feynman not been as smart as he was, I think he would have been too original for his own good.

There was always an element of showboating in his character. He was like the guy that climbs Mont Blanc barefoot just to show that it can be done. A lot of things he did were to show, you didn't have to do it that way, you can do it this other way. And this other way, in fact, was not as good as the first way, but it showed he was different.

Feynman continued to refuse to read the current literature, and he chided graduate students who would begin their work on a problem in the normal way, by checking what had already been done. That way, he told them, they would give up chances to find something original. Coleman said:

> I suspect that Einstein had some of the same character. I'm sure Dick thought of that as a virtue, as noble. I don't think it's so. I think it's kidding yourself. Those other guys are not all a collection of yo-yos. Sometimes it would be better to take the recent machinery they have built and not try to rebuild it, like reinventing the wheel.
>
> I know people who are in fact very original and not cranky but have not done as good physics as they could have done because they were more concerned at a certain juncture with being original than with being right. Dick could get away with a lot because he was so goddamn smart. He really *could* climb Mont Blanc barefoot.

Coleman chose not to study with Feynman directly. Watching Feynman work, he said, was like going to the Chinese opera.

> When he was doing work he was doing it in a way that was just—— absolutely out of the grasp of understanding. You didn't know where it was going, where it had gone so far, where to push it, what was the next step. With Dick the next step would somehow come out of—— divine revelation.

So many of his witnesses observed the utter freedom of his flights of thought, yet when Feynman talked about his own methods he emphasized not freedom but constraints. The kind of imagination that takes blank paper, blank staves, or a blank canvas and fills it with something wholly new, wholly free—that, Feynman contended, was not the scientist's imagination. Nor could one measure imagination as certain psychologists try to do, by displaying a picture and asking what will happen next. For Feynman the essence of the scientific imagination was a powerful and almost painful rule. What scientists create must match reality. It must match what is already known. Scientific creativity, he said, is imagination in a straitjacket. "The whole question of imagination in science is often misunderstood by people in other disciplines," he said. "They overlook the fact that whatever we are *allowed* to imagine in science must be *consistent with everything else we know.* . . ." This is a conservative principle, implying that the existing framework of science is fundamentally sound, already a fair mirror of reality. Scientists, like the freer-seeming arts, feel the pressure to innovate, but in science the act of making something new contains the seeds of paradox. Innovation comes not through daring steps into unknown space,

> not just some happy thoughts which we are free to make as we wish, but ideas which must be consistent with all the laws of physics we know. We can't allow ourselves to seriously imagine things which are obviously in contradiction to the known laws of nature. And so our kind of imagination is quite a difficult game.

Creative artists in modern times have labored under the terrible weight of the demand for novelty. Mozart's contemporaries expected him to work within a fixed, shared framework, not to break the bonds of convention. The standard forms of the sonata, symphony, and opera were established before his birth and hardly changed in his lifetime; the rules of harmonic progression made a cage as unyielding as the sonnet did for Shakespeare. As unyielding and as liberating—for later critics found the creators' genius in the counterpoint of structure and freedom, rigor and inventiveness.

For the creative mind of the old school, inventing by pressing against constraints that seem ironclad, subtly bending a rod here or slipping a lock there, science has become the last refuge. The forms and constraints of scientific practice are held in place not just by the grounding in experiment but by the customs of a community more homogeneous and rule-bound than any community of artists. Scientists still speak unashamedly of *reality*, even in the quantum era, of objective truth, of a world independent of human construction, and they sometimes seem the last members of the intellectual universe to do so. Reality hobbles their imaginations. So does the ever more intricate assemblage of theorems, technologies, laboratory results, and mathematical formalisms that make up the body of known science. How, then, can the genius make a revolution? Feynman said, "Our imagination is stretched to the utmost, not, as in fiction, to imagine things which are not really there, but just to comprehend those things which *are* there."

It was the problem he faced in the gloomiest days of 1946, when he was trying to find his way out of the mire that quantum mechanics had become. "We know so very much," he wrote to his friend Welton, "and then subsume it into so very few equations that we can say we know very little (except these equations) . . . Then we think we have *the* physical picture with which to interpret the equations." The freedom he found then was a freedom not from the equations but from the physical picture. He refused to let the form of the mathematics lock him into any one route to visualization: "There are so very few equations that I have found that many physical pictures can give the same equations. So I am spending my time in study—in seeing how many new viewpoints I can take of what is known." By then Welton had mastered the field theory that was becoming standard, and he was surprised to discover that his old friend had not. Feynman seemed to hoard shadow pools of ignorance, seemed to protect himself from the light like a waking man who closes his eyes to preserve a fleeting image left over from a dream. He said later, "Maybe that's why young people make success. They don't know enough. Because when you know enough it's obvious that every idea that you have is no good." Welton, too, was persuaded that if Feynman had known more, he could not have innovated so well.

"Would I had phrases that are not known, utterances that are strange, in new language that has not been used, free from repetition, not an utterance which has grown stale, which men of old have spoken." An Egyptian scribe fixed those words in stone at the very dawn of recorded utterance—already jaded, a millennium before Homer. Modern critics speak of the burden of the past and the anxiety of influence, and surely the need to innovate is an ancient part of the artist's psyche, but novelty was never as crucial to the artist as it became in the twentieth century. The useful lifetime of a new form or genre was never so short. Artists never before felt so much pressure to violate such young traditions.

Meanwhile, before their eyes, the world has grown too vast and multifarious for the towering genius of the old kind. Artists struggle to keep their heads above the tide. Norman Mailer, publishing yet another novel doomed to fall short of ambitions formed in an earlier time, notices: "There are no large people any more. I've been studying Picasso lately and look at who his contemporaries were: Freud, Einstein." He saw the change in his own lifetime without understanding it. (Few of those looking for genius understood where it had gone.) He appeared on a literary scene so narrow that conventional first novels by writers like James Jones made them appear plausible successors to Faulkner and Hemingway. He slowly sank into a thicket of hundreds of equally talented, original, and hard-driving novelists, each just as likely to be tagged as a budding genius. In a world into which Amis, Beckett, Cheever, Drabble, Ellison, Fuentes, Grass, Heller, Ishiguro,

Jones, Kazantzakis, Lessing, Nabokov, Oates, Pym, Queneau, Roth, Solzhenitsyn, Theroux, Updike, Vargas Llosa, Waugh, Xue, Yates, and Zoshchenko—or any other two dozen fiction writers—had never been born, Mailer and any other potential genius would have had a better chance of towering. In a less crowded field, among shorter yardsticks, a novelist would not just have seemed bigger. He would have been bigger. Like species competing in ecological niches, he would have had a broader, richer space to explore and occupy. Instead the giants force one another into specialized corners of the intellectual landscape. They choose among domestic, suburban, rural, urban, demimondaine, Third World, realist, postrealist, semirealist, antirealist, surrealist, decadent, ultraist, expressionist, impressionist, naturalist, existentialist, metaphysical, romance, romanticist, neoromanticist, Marxist, picaresque, detective, comic, satiric, and countless other fictional modes—as sea squirts, hagfish, jellyfish, sharks, dolphins, whales, oysters, crabs, lobsters, and countless hordes of marine species subdivide the life-supporting possibilities of an ocean that was once, for billions of years, dominated quite happily by blue-green algae.

"Giants have not ceded to mere mortals," the evolutionary theorist Stephen Jay Gould wrote in an iconoclastic 1983 essay. "Rather, the boundaries . . . have been restricted and the edges smoothed." He was not talking about algae, artists, or paleontologists but about baseball players. Where are the .400 hitters? Why have they vanished into the mythic past, when technical skills, physical conditioning, and the population on which organized baseball draws have all improved? His answer: Baseball's giants have dwindled into a more uniform landscape. Standards have risen. The distance between the best and worst batters, and between the best and worst pitchers, has fallen. Gould showed by statistical analysis that the extinction of the .400 hitter was only the more visible side of a general softening of extremes: the .100 hitter has faded as well. The best and worst all come closer to the average. Few fans like to imagine that Ted Williams would recede toward the mean in the modern major leagues, or that the overweight, hard-drinking Babe Ruth would fail to dominate the scientifically engineered physiques of his later competitors, or that dozens of today's nameless young base-stealers could outrun Ty Cobb, but it is inevitably so. Enthusiasts of track and field cannot entertain the baseball fan's nostalgia; their statistics measure athlete against nature instead of athlete against athlete, and the lesson from decade to decade is clear. There is such a thing as progress. Nostalgia conceals it while magnifying the geniuses of the past.

A nostalgic music lover will put on a scratchy 78 of Lauritz Melchior and sigh that there are no Wagnerian tenors any more. Yet in reality musical athletes have probably fared no worse than any other kind.

Is it only nostalgia that makes genius seem to belong to the past? Giants did walk the earth—Shakespeare, Newton, Michelangelo, DiMaggio— and in their shadows the poets, scientists, artists, and baseball players of today crouch like pygmies. No one will ever again create a *King Lear* or hit safely in fifty-six consecutive games, it seems. Yet the raw material of genius—whatever combination of native talent and cultural opportunity that might be—can scarcely have disappeared. On a planet of five billion people, parcels of genes with Einsteinian potential must appear from time to time, and presumably more often than ever before. Some of those parcels must be as well nurtured as Einstein's, in a world richer and better educated than ever before. Of course genius is exceptional and statistics-defying. Still, the modern would-be Mozart must contend with certain statistics: that the entire educated population of eighteenth-century Vienna would fit into a large New York apartment block; that in a given year the United States Copyright Office registers close to two hundred thousand "works of the performing arts," from advertising jingles to epic tone poems. Composers and painters now awake into a universe with a nearly infinite range of genres to choose from and rebel against. Mozart did not have to choose an audience or a style. His community was in place. Are the latter-day Mozarts not being born, or are they all around, bumping shoulders with one another, scrabbling for cultural scraps, struggling to be newer than new, their stature inevitably shrinking all the while?

The miler who triumphs in the Olympic Games, who places himself momentarily at the top of the pyramid of all milers, leads a thousand next-best competitors by mere seconds. The gap between best and second-best, or even best and tenth-best, is so slight that a gust of wind or a different running shoe might have accounted for the margin of victory. Where the measuring scale becomes multidimensional and nonlinear, human abilities more readily slide off the scale. The ability to reason, to compute, to manipulate the symbols and rules of logic—this unnatural talent, too, must lie at the very margin, where small differences in raw talent have enormous consequences, where a merely good physicist must stand in awe of Dyson and where Dyson, in turn, stands in awe of Feynman. Merely to divide 158 by 192 presses most human minds to the limit of exertion. To master— as modern particle physicists must—the machinery of group theory and current algebra, of perturbative expansions and non-Abelian gauge theories, of spin statistics and Yang-Mills, is to sustain in one's mind a fantastic house of cards, at once steely and delicate. To manipulate that framework, and to innovate within it, requires a mental power that nature did not demand of scientists in past centuries. More physicists than ever rise to meet this cerebral challenge. Still, some of them, worrying that the Einsteins and Feynmans are nowhere to be seen, suspect that the geniuses have fled to microbiology or computer science—forgetting momentarily that the individual microbiologists and computer scientists they meet seem no brainier, on the whole, than physicists and mathematicians.

Geniuses change history. That is part of their mythology, and it is the final test, presumably more reliable than the trail of anecdote and peer admiration that brilliant scientists leave behind. Yet the history of science is a history not of individual discovery but of multiple, overlapping, co-incidental discovery. All researchers know this in their hearts. It is why they rush to publish any new finding, aware that competitors cannot be far behind. As the sociologist Robert K. Merton has found, the literature of science is strewn with might-have-been genius derailed or forestalled—"those countless footnotes . . . that announce with chagrin: 'Since completing this experiment, I find that Woodworth (or Bell or Minot, as the case may be) had arrived at this same conclusion last year, and that Jones did so fully sixty years ago.'" The power of genius may lie, as Merton suggests, in the ability of one person to accomplish what otherwise might have taken dozens. Or perhaps it lies—especially in this exploding, multifarious, information-rich age—in one person's ability to see his science whole, to assemble, as Newton did, a vast unifying tapestry of knowledge. Feynman himself, as he entered his forties, prepared to undertake this very enterprise: a mustering and a reformulating of all that was known about physics.

Scientists still ask the *what if* questions. What if Edison had not invented the electric light—how much longer would it have taken? What if Heisenberg had not invented the S matrix? What if Fleming had not discovered penicillin? Or (the king of such questions) what if Einstein had not invented general relativity? "I always find questions like that . . . odd," Feynman wrote to a correspondent who posed one. Science tends to be created as it is needed.

"We are not that much smarter than each other," he said.

Weak Interactions

By the late 1950s and early 1960s, as the discovery of new particles became more commonplace, physicists found it harder to guess what might and might not be possible. The word *zoo* entered their vocabulary, and their scientific intuition sometimes seemed colored by a kind of aesthetic queasiness. Weisskopf declared at one meeting that it would be a shame if anyone found a particle with double charge. Oppenheimer added that he personally would hate to see a strongly interacting particle with spin greater than one-half. Both men were quickly disappointed. Nature was not so fastidious.

The methods assembled under the label of field theory just a few years before—direct computation of particle interactions, in the face of those still-troubling infinities—fell out of favor with many. The success of quantum electrodynamics did not extend easily to other particle realms. Of the four fundamental forces—electromagnetism; gravity; the strong force binding the atomic nucleus; and the weak force at work in radioactive beta decay and in strange-particle decays—renormalization seemed to work only for electromagnetism. With electromagnetism, the first, simplest Feynman diagrams told most of the story. Mathematically the relative weakness of the force expressed itself in the diminishing importance of more complicated diagrams (for the same reason that the later terms in a series like $1 + n + n^2 + \ldots$ vanish if n is 1/100). With the strong force, the forest of Feynman diagrams made an unendingly large contribution to any calculation. That made real calculations impossible. So where the more esoteric forces were concerned, it seemed impossible to match the success of quantum electrodynamics in making amazingly precise dynamical predictions. Instead, symmetries, conservation laws, and quantum numbers provided abstract principles by which physicists could at least organize the experimentalists' data. They looked for patterns, organized taxonomies, filled in holes. A diverging branch of mathematical physicists continued to pursue field theory, but most theorists now found it profitable to sift through particle data— the data now arriving in huge volume—looking for general principles. Searching for symmetries meant not tying oneself to the microscopic dynamics of particle behavior. It came to seem almost immoral, or at least silly, for a theorist to write down a specific dynamic or scale.

The understanding of symmetry also became an understanding of symmetry's imperfections, for, as symmetry laws came to dominate, they also began to break down. One of the most obvious of all symmetries led the way: the symmetry of left and right. Humans seem mostly symmetrical, but not perfectly so. The symmetry is "broken," as a modern physicist would say, by an off-center heart and liver and by more subtle or superficial differences. We learn to break the symmetry ourselves by internalizing an awareness of the difference between left and right, although sometimes this is not so easy. Feynman himself confessed to a group gathered around the coffee pot in a Caltech laboratory that even now he had to look for the mole on the back of his left hand when he wanted to be sure. As early as his MIT fraternity days he had puzzled over the classic teaser of mirror symmetry: why does a mirror seem to invert left and right but not top and bottom? That is, why are the letters of a book backward but not upside down, and why would Feynman's double behind the mirror appear to have a mole on his right hand? Was it possible, he liked to ask, to give a *symmetrical* explanation of what a mirror does—an explanation that treats up-and-down no differently from left-and-right? Many logicians and scientists had debated this conundrum. There were many explanations, some of them correct. Feynman's was a model of clarity.

Imagine yourself standing before the mirror, he suggested, with one hand pointing east and the other west. Wave the east hand. The mirror image waves its east hand. Its head is up. Its west hand lies to the west. Its feet are down. "Everything's really all right," Feynman said. The problem is on the axis running *through* the mirror. Your nose and the back of your head are reversed: if your nose points north, your double's nose points south. The problem now is psychological. We think of our image as another person. We cannot imagine ourselves "squashed" back to front, so we imagine ourselves turned left and right, as if we had walked around a pane of glass to face the other way. It is in this psychological turnabout that left and right are switched. It is the same with a book. If the letters are reversed left and right, it is because we turned the book about a vertical axis to face the mirror. We could just as easily turn the book from bottom to top instead, in which case the letters will appear upside down.

Our own asymmetries—our blemishes, hearts, handednesses—arise from contingent choices nature made in the process of building up complicated organisms. A preference for right or left appears in biology all the way down to the level of organic molecules, which can be right- or left-handed. Sugar molecules have this intrinsic corkscrew property. Chemists can make them with either handedness, but bacteria digest only "right-handed" sugar, and that is the kind that sugar beets produce. Earthly sugar beets, that is—for evolution might just as well have chosen a left-handed pathway, just as the industrial revolution might have settled on left-threaded rather than right-threaded screws.

On still smaller scales, at the level of elementary particle interactions, physicists assumed that nature would not distinguish between right and left. It seemed inconceivable that the laws of physics would change with mirror reflection, any more than they change when an experiment is conducted at a different place or a different time. How could anything so featureless as a particle embody the handedness of a corkscrew or a golf club? Right-left symmetry had been built into quantum mechanics in the form of a quantity called parity. If a given event conserved parity, as most physicists consciously or unconsciously assumed it must, then its outcome did not depend on any left-right orientation. Conversely, if nature did have some kind of handedness built into its guts, then an experimenter might be able to find events that did not conserve parity. When Murray Gell-Mann was a graduate student at MIT, a standard problem in one course was to derive the conservation of parity by mathematical logic, transforming coordinates from left-handed to right-handed. Gell-Mann spent a long weekend transforming coordinates back and forth without proving anything. He recalled telling the instructor that the problem was wrong: that the conservation of parity was a physical fact that depended on the structure of a particular theory, not on any inescapable mathematical truth.

Parity became an issue in theorists' unease about the liveliest experimental problem coming out of the accelerators in 1956: the problem of the theta and the tau, two strange particles (strange in Gell-Mann's sense). It was typical of the difficulties physicists were having in making taxonomical sense of the jumble of accelerator data. When the theta decayed, a pair of pions appeared. When the tau decayed, it turned into three pions. In other ways, however, the tau and the theta were beginning to look suspiciously similar. Data from cosmic rays and then accelerators made their masses and lifetimes seem indistinguishable. One experimenter had plotted thirteen data points in 1953. By the time the 1956 Rochester conference convened, he had more than six hundred data points, and the theorists were trying to face the obvious: perhaps the tau and the theta were one and the same. The problem was parity. A pair of pions had even parity. A trio of pions had odd parity. Assuming that a particle's decay conserved parity, a physicist had to believe that the tau and the theta were different. Intuitions were severely tested. Sometime after the Rochester conference ended, Abraham Pais wrote a note to himself: "Be it recorded here that on the train back from Rochester to New York, Professor Yang and the writer each bet Professor Wheeler one dollar that the theta- and tau-meson are distinct particles; and that Professor Wheeler has since collected two dollars."

Everyone was making bets. An experimenter asked Feynman what odds he would give against an experiment testing for the unthinkable, parity violation, and Feynman was proud later that he had offered a mere fifty to one. He actually raised the question at Rochester, saying that his roommate there, an experimenter named Martin Block, had wondered why parity could not be violated. (Afterward Gell-Mann teased him mercilessly for not having asked the question in his own name.) Someone had joked nervously about considering even wild possibilities with open minds, and the official note taker recorded:

Pursuing the open mind approach, *Feynman* brought up a question of Block's: Could it be that the [theta] and [tau] are different parity states of the same particle which has no definite parity, *i.e.*, that parity is not conserved. That is, does nature have a way of defining right- or left-handedness uniquely?

Two young physicists, Chen Ning Yang and Tsung Dao Lee, said they had begun looking into this question without reaching any firm conclusions. So desperately did the participants dislike the idea of parity violation that one scientist proposed yet another unknown particle, this time one that departed the scene with no mass, no charge, and no momentum—just carrying off "some strange space-time transformation properties" like a sanitation worker carting away trash. Gell-Mann rose to suggest that they keep their minds open to the possibility of other, less radical solutions. Discussion continued until, as the note taker put it, "The chairman"—Oppenheimer—"felt that the moment had come to close our minds."

But in Feynman's tentative question the answer had emerged. Lee and Yang undertook an investigation of the evidence. For electromagnetic interactions and strong interactions, the rule of parity conservation had a real experimental and theoretical foundation. Without parity conservation, a well-entrenched framework would be torn apart. But that did not seem to be true for weak interactions. They went through an authoritative text on beta decay, recomputing formulas. They examined the recent experimental literature of strange particles. By the summer of 1956 they realized that, as far as the weak force was concerned, parity conservation was a free-floating assumption, bound neither to any experimental result nor to any theoretical rationale. Furthermore, it occurred to them that Gell-Mann's conception of strangeness offered a precedent: a symmetry that held for the strong force and broke down for the weak. They quickly published a paper formally raising the possibility that parity might not be conserved by weak interactions and proposing experiments to test the question. By the end of the year, a team led by their Columbia colleague Chien Shiung Wu had set one of them up, a delicate matter of monitoring the decay of a radioactive isotope of cobalt in a magnetic field at a temperature close to absolute zero. Given an *up* and *down* defined by the alignment of the magnetic coil, the decaying cobalt would either spit out electrons symmetrically to the left and right or would reveal a preference. In Europe, awaiting the results, Pauli joined the wagerers: he wrote Weisskopf, "I do *not* believe that the Lord is a weak left-hander, and I am ready to bet a very large sum that the experiments will give symmetric results." Within ten days he knew he was wrong, and within a year Yang and Lee had received one of the quickest Nobel Prizes ever awarded. Although physicists still did not understand it, they appreciated the import of the discovery that nature distinguished right from left in its very core. Other symmetries were immediately implicated— the correspondence between matter and antimatter, and the reversibility of time (if the film of an experiment were run backwards, for example, it might look physically correct except that right would be left and left would be right). As one scientist put it, "We are no longer trying to handle screws in the dark with heavy gloves. We are being handed the screws neatly aligned on a tray, with a little searchlight on each that indicates the direction of its head."

Feynman made an odd presence at the high-energy physicists' meetings. He was older than the bright young scientists of Gell-Mann's generation, younger than the Nobel-wielding senators of Oppenheimer's. He neither withdrew from the discussions nor dominated them. He showed a piercing interest in the topical issues—as with his initial prodding on the question of parity—but struck younger physicists as detached from the newest ideas, particularly in contrast to Gell-Mann. At the 1957 Rochester conference it occurred to at least one participant that Feynman himself should have applied his theoretical talents to the question he had raised a year earlier, instead of leaving the plum to Yang and Lee. (The same participant noticed a revisionists' purgatory in the making: theorists from Dirac to Gell-Mann "busy explaining that they personally had never thought parity was anything special," and experimenters recalling that they had always meant to get around to an experiment like Wu's.) Publicly, Feynman was as serene as ever. Privately, he agonized over his inability to find the right problem. He wanted to stay clear of the pack. He knew he was not keeping up with even the published work of Gell-Mann and other high-energy physicists, yet he could not bear to sit down with the journals or preprints that arrived daily on his desk and piled up on his shelves and merely *read* them. Every arriving paper was like a detective novel with the last chapter printed first. He wanted to read just enough to understand the problem; then he wanted to solve it his own way. Almost alone among physicists, he refused to referee papers for journals. He could not bear to rework a problem from start to finish along someone else's track. (He also knew that when he broke his own rule he could be devastatingly cruel. He summarized one text by writing, "Mr. Beard is very courageous when he gives freely so many references to other books, because if a student ever did look at another book, I am sure he would not return again to continue reading Beard," and then urged the editor to keep his review confidential—"for Mr. Beard and I are good personal friends.") His persistently iconoclastic approach to other people's work offended even theorists whom he meant to compliment. He would admire what they considered a peripheral finding, or insist on what struck them as a cockeyed or baroque alternative viewpoint. Some theorists strived to collaborate with colleagues and to set a tone and an agenda for whole groups. Gell-Mann was one. Feynman seemed to lack that ambition—though a generation of physicists now breathed Feynman diagrams. Still, he was frustrated.

He sometimes confided in his sister, Joan, who had begun a career in science herself, getting a doctorate in solid-state physics at Syracuse University. She was still living in Syracuse, and Feynman visited her when he went to Rochester. He complained to her that he could not work. She reminded him of all the recent ideas that he had shared with her and then refused to pursue long enough to write a paper. *You've done it again and again*, she said. *You told me that Block might be right. And you don't do a damn thing about it. You should write it up, for crying out loud, when you have something like this.* She also reminded him that he had mentioned an idea for a universal theory of weak interactions—tying together beta decay and the strange-particle decays based on the weak force—and urged him, finally, to see where it would lead.

In its classic form, beta decay turns a neutron into a proton, throwing off an electron and another particle, a neutrino—massless, chargeless, and hard to detect. Charge is conserved: the neutron has none; the proton carries + 1 and the electron − 1. Analogously, in the meson family, a pion could decay into a muon (like a heavy electron) and a neutrino. A good theory would predict the rates of decay in such processes, as well as the energies of the outgoing particles. There were complications. The spins of the particles had to be reconciled, and for the massless neutrinos, especially, problems of handedness arose in calculating the appropriate spins. So the new understanding of parity violation immediately changed the weak-interaction landscape—for Feynman, for Gell-Mann, and for others.

In sorting the various kinds of particle interactions, theorists had created a classification scheme with five distinct transformations of one wave function into another. In one sense it was a classification of the characteristic algebraic techniques; in another, it was a classification of the types of virtual particles that arose in the interactions, according to their possible spins and parities. As shorthand, physicists used the labels S, T, V, A, and P, for *scalar, tensor, vector, axial vector,* and *pseudoscalar.* The different kinds of weak interaction had evident similarities, but this classification scheme posed a problem. As Lee pointed out at the 1957 Rochester meeting, most experiments on beta decay had demonstrated S and T interactions, while the new parity-violation experiments tended to suggest that meson decay involved V and A. Under the circumstances, the same physical laws could hardly be at work.

In reading Lee and Yang's preprint for the meeting—Joan had ordered him, for once, to sit down like a student and go through it step by step—Feynman saw an alternative way of formulating the violation of parity. Lee and Yang described a restriction on the spin of the neutrino. He liked the idea enough to mention it from the audience during five minutes cadged from another speaker. He went far back into the origins of quantum mechanics—back not only to the Dirac equation itself but beyond, to the Klein-Gordon equation that he and Welton had manufactured when they were MIT undergraduates. Using path integrals, he moved forward again, deriving—or "discovering"—an equation slightly different from Dirac's. It was simpler: a two-component equation, where Dirac's had four components. "Now I asked this question," Feynman said:

Suppose that historically [my equation] had been discovered before the Dirac equation? It has absolutely the same consequences as the Dirac equation. It can be used with diagrams the same way.

The diagrams for beta decay, of course, added a neutrino field interacting with the electron field. When Feynman made the necessary change to his equation, he found:

> Of course I can't do that because this term is parity unsymmetric. *But—*
> —beta decay is not parity symmetric, so it's now possible.

There were two difficulties. One was that he came out with the opposite sign for the spin: his neutrino would have to spin in the opposite direction from Lee and Yang's prediction. The other was that the coupling in his formulation would have to be V and A, instead of the S and T that everyone knew was correct.

Gell-Mann, meanwhile, had also thought about the problem of creating a theory for weak interactions. Nor were Feynman and Gell-Mann alone: Robert Marshak, who had put forward the original two-meson idea at the Shelter Island conference in 1947, was also leaning toward V and A with a younger physicist, E. C. G. Sudarshan. That summer, with Feynman traveling in Brazil, Marshak and Sudarshan met with Gell-Mann in California and described their approach.

Feynman returned at the end of the summer determined, for once, to catch up with the experimental situation and follow his weak-interaction idea through to the end. He visited Wu's laboratory at Columbia, and he asked Caltech experimenters to bring him up to date. The data seemed a shambles—contradictions everywhere. One of the Caltech physicists said that Gell-Mann even thought the crucial coupling could be V rather than S. That, as Feynman often recalled afterward, released a trigger in his mind.

> I flew out of the chair at that moment and said, "Then I understand everything. I understand everything and I'll explain it to you tomorrow morning."
> They thought when I said that, I'm making a joke. . . . But I didn't make a joke. The release from the tyranny of thinking it was S and T was all I needed, because I had a theory in which if V and A were possible, V and A were right, because it was a neat thing and it was pretty.

Within days he had drafted a paper. Gell-Mann, however, decided that he should write a paper, too. As he saw it, he had his own reasons for focusing on V and A. He wanted the theory to be universal. Electromagnetism depended on vector coupling, and the strange particles favored V and A. He was unhappy that Feynman seemed to be thoughtlessly dismissing his ideas.

Before the tension between them rose higher, their department head, Robert Bacher, stepped in and asked them to write a joint paper. He preferred not to see rival versions of the same discovery coming out of Caltech's physics group. Colleagues strained to overhear Feynman and Gell-Mann in the corridors or at a cafeteria table, engrossed in their oral collaboration. They stimulated each other despite the characteristic differences in their language: Feynman offering what sounded like *you take this and it zaps through here and you come out and pull this together like that*, Gell-Mann responding with *you substitute there and there and integrate like so*. . . . Their article reached the *Physical Review* in September, days before Marshak presented his and Sudarshan's similar theory at a conference in Padua, Italy. Feynman and Gell-Mann's theory went further in several influential respects. It proposed a bold extension of the underlying principles beyond beta decay to other classes of particle interactions; it would be years before experiment fully caught up, showing how prescient the two men had been. It also introduced the idea that a new kind of current—analogous to electrical current, a measure of the flow of charge—should be conserved; new extensions of the concept of current became a central tool of high-energy physics.

Feynman tended to recall that they had written the paper together. Gell-Mann sometimes disdained it, complaining particularly about the two-component formalism—a ghastly notation, he felt. It did bear Feynman's stamp. He was applying a formulation of quantum electrodynamics that went back to his first paper on path integrals in 1948; Gell-Mann allowed him to remark fondly, "One of the authors has always had a predilection for this equation." Yet it could hardly have been Feynman who wrote that their approach to parity violation "has a certain amount of theoretical *raison d'être*." Evident, too, was Gell-Mann's drive to make the theory as unifying and forward-looking as possible. The discovery was esoteric compared to other milestones of modern physics. If Feynman, Gell-Mann, Marshak, or Sudarshan had not made it in 1957, others would have soon after. Yet to Feynman it was as pure an achievement as any in his career: the unveiling of a law of nature. His model had always been Dirac's magical discovery of an equation for the electron. In a sense Feynman had discovered an equation for the neutrino. "There was a moment when I knew how nature worked," he said. "It had elegance and beauty. The goddamn thing was gleaming." To other physicists, "Theory of the Fermi Interaction," barely six pages long, shone like a beacon in the literature. It seemed to announce the beginning of a powerful collaboration between two great and complementary minds. They took a distinctive kind of theoretical high ground, repeatedly speaking of universality, of simplicity, of the preservation of symmetries, of broad future applications. They worked from general principles rather than particular calculations of dynamics. They made clear predictions about new kinds of particle decay. They listed specific experiments that contradicted their theory and declared that the experiments must therefore be wrong. Nothing could have more strikingly declared the supremacy of the theorists.

Toward a Domestic Life

The two-piece "bikini" bathing suit, named after the tiny Pacific atoll that was blasted by atomic and hydrogen bombs through the forties and fifties, had not yet taken over the beaches of the United States in 1958, but Feynman saw one, blue, on the sand of Genève-Plage, and laid his beach towel down nearby. He was visiting Geneva for a United Nations conference on the peaceful uses of atomic energy. He was preparing to give a summary talk in his own name and Gell-Mann's, telling the assembly:

> We are well aware of the fragility and incompleteness of our present knowledge and of the manifold of speculative possibilities. . . . What is the significance or the pattern behind all these interrelated symmetries, partial symmetries, and asymmetries?

The yearly Rochester conference had also changed venue for the occasion, and he discussed the weak-interaction theory, impressing listeners with the body language he used to demonstrate the appropriate spins and handednesses. He had just turned forty. It was spring, and the young woman in the blue bikini volunteered that Lake Geneva was cold. "You speak English!" he said. She was Gweneth Howarth, a native of a village in Yorkshire, England. She had left home to see Europe by working as an au pair. That evening he took her to a nightclub.

The violation of parity had reached newspapers and magazines briefly. For readers who looked to science for a general understanding of the nature of the universe, the fall of left-right symmetry may have been the last genuinely meaningful lesson to emerge from high-energy physics, circumscribed though it was in the domain of certain very short-lived particle interactions. By contrast, though the universal theory of weak interactions commandeered the attention of theorists and experimenters a year later, the replacement of S and T with V and A made no ripple in the cultural consciousness. By then the American public was busy anyway, assimilating the most shocking scientific development of the 1950s, the piece of news establishing once again in the public mind the truism that science is power.

The beachball-sized aluminum sphere called Sputnik began orbiting the earth on October 4, 1957. Its unexpected presence overhead and the insouciant beep-beep-beep played again and again on American radio and television broadcasts set off a wave of anxiety like nothing since the atomic bomb itself. (Feynman arrived at a picnic that evening in the biologist Max Delbrück's backyard with a small gray radio receiver that looked as if he had built it himself. He called for an extension cord, tuned the receiver quickly, held up a finger to demand silence, and grinned as the beeps played out over the crowd.) "Red Moon over U.S.," said *Time* magazine, immediately announcing "a new era in history" and "a grim new chapter in the cold war." *Newsweek* called it "The Red Conquest"—with "all the mastery that it implies in the affairs of men on earth." Why had the United States established no comparable space program? A worried-looking President Eisenhower said at a news conference, "Well, let's get this straight. I am not a scientist." The director of the American Institute of Physics seized the occasion to say that unless his country's science education caught up with the Soviet Union's, "our way of life is doomed." That message was heard: Sputnik produced a rapid new commitment to the teaching of science. Magazines focused new attention on American physicists. Among the younger generation, *Time* singled out Feynman—

> Curly-haired and handsome, he shuns neckties and coats, is an enormously dedicated adventurer . . . became fascinated with samba rhythms . . . playing bongo drums, breaking codes, picking locks . . .

and Gell-Mann—

> he formulated the "Strangeness Theory," *i.e.* assigned physical meanings to the behavior of newly discovered particles. At CalTech Gell-Mann works closely with Feynman on weak couplings. At the blackboard the two ex-

plode with ideas like sparks flying from a grindstone, alternately slap their foreheads at each other's simplifications, quibble over the niceties.

But the physicist who received most of the public's attention that fall was Edward Teller. He was in tune with the cold war. Sputnik led him to declare—though there was evidence to the contrary—"Scientific and technical leadership is slipping from our hands." A direct Soviet attack on the United States was possible, but he saw an even greater threat. "I do not think this is the most probable way in which they will defeat us," he said. He predicted that the Soviet Union would gain a broad technological dominance over the free world. "They will advance so fast in science and leave us so far behind that their way of doing things will be *the* way, and there will be nothing we can do about it."

With the winter's excitement barely waning—the *Reader's Digest* had now faced into the wind with an article titled "No Time for Hysteria"—a State Department official let Caltech know that the department would appreciate a presentation at the Geneva conference in the name of both Feynman and Gell-Mann, to balance the expected Soviet scientific presence there. Feynman acquiesced, although the mixing of propaganda and science disturbed him.

He declined to let the State Department make his hotel reservation; he found a walk-up room in an establishment called, in English, Hotel City. It reminded him of the flophouses he had known in Albuquerque and on his cross-country trip with Freeman Dyson. He had hoped to bring a woman with whom he had been having a sporadic and tempestuous yearlong love affair—the wife of a research fellow. She had accompanied him on a trip the summer before, when he was working on weak interactions. Now she agreed to meet him afterward in England but refused to come to Geneva. Instead, he met Gweneth Howarth on the beach.

She told him she was making her way around the world. She was twenty-four years old, the daughter of a jeweler in a town called Ripponden. She had worked as a librarian for a salary of three pounds weekly and then as a yarn tester at a cotton mill before deciding life in the backwaters of Yorkshire was too dull. She let Feynman know that she had two current boyfriends, a semiprofessional miler from Zurich, always in training, and a German optician from Saarbrücken. He immediately invited her to come to California and work for him. He needed a maid, he said. He would sponsor her with the immigration authorities and pay her twenty dollars a week. It seemed to her that he was not behaving like a forty-year-old; nor

like other Americans she had met. She said she would consider it, and an
unusual courtship began.

"I've decided to stay here after all," she wrote him that fall. One of the
boyfriends, Johann, had decided to marry her—out of jealousy, she suspected—

> so you see what a good turn you did for me. . . . we talked for hours and
> hours, planning our life together. We shall probably start married life
> living in one room. . . . Were you really expecting me to come? . . .
> You'll just have to get married again, or find a nice solid middle-aged
> housekeeper so people won't gossip.

His love affairs were going badly, meanwhile. That same week a letter
arrived from the other woman, making it clear that their relationship was
over. She demanded money—five hundred dollars—"I will be frank, the
chances of your getting it all back within a year are nil." She had asked
for money before, saying that she needed it for an abortion, but now she
said that that had been a ruse. His money had actually gone for furniture
and house painting.

> You were too much of the "playboy." But I was both embarrassed &
> intrigued by the effects that your girl friends had on you when they called
> you in my presence. Sometimes you left the phone, shaking & foaming
> at the mouth. . . . I recognized a baseness in you and was frightened that
> you took my love and affection for you cheaply, and so I wanted to com-
> pensate against that horrible feeling.

She knew too much about the women he had been seeing since his divorce.
She named four of them and described an anonymous note that had come
addressed to "Occupant":

> Dirty Dick, Filthy Fucking Feynman dates you. He will never marry you.
> Tell him he has made you pregnant. You'll make a quick $300–$500.

She had been devastated by nasty physicist-gossip she had overheard about
Feynman and his women, Feynman and "the pox." He should get married,
she said.

The baseness you talk of is due to the fact you aren't married. You try to sublimate your desires by attending Burlesque Shows, Night Clubs, etc. These are fun for the healthy, but only an escape for the dissatisfied. I know this, because last year you were content in Rio, & as a result produced Beta Decay. . . .

Find yourself a real companion, someone you can really love & respect. Then capture love whilst it is fresh & spontaneous. . . .

At some point she had walked off with the gold medal he had received with the Einstein Award. She still had it, she reminded him.

Feynman implored Gweneth Howarth to reconsider. By November, as it happened, she and Johann were no longer on speaking terms and she had begun the immigration paperwork through the United States Consulate in Zurich. He consulted a lawyer, who warned that there were dangers in transporting women for immoral purposes and advised him to find a third-party employer; a Caltech friend, Matthew Sands, agreed to lend his name on the required documents. Feynman calculated fares (more than a year's salary for a Yorkshire librarian, she noticed): $394.10 to Los Angeles; or $290.10 to New York and then $79.04 including tax for a bus from New York to Los Angeles.

She was excited but unsure. "You'll write & tell me if you decide to get married again, or if there is any other reason why I should not come?" She wanted him to realize she had other possibilities—Armando, whom she met skiing, or a fellow who had been watching her at language class ("he walked part of the way home with me . . . I'd like it to be a platonic friendship, but I don't suppose he will want it like that . . .") and yet there were always hints of the domestic future Feynman so craved now—she was caring for "a beautiful baby now, I wish I could have one exactly like him." A new friend, Engelbert, was buying skis for her; meanwhile she could now cook pheasant, chicken, goose, and hare with the appropriate sauces ("I'm improving, am I not?"). Feynman kept hearing from the other woman, too. She was telling her husband everything; they had left California for the East Coast. She wanted more money. She felt used. He let her know how angry he was. She told him, "altho' you are clever at your own special work, you are very dim at human relationships." She assured him that his Einstein Award medal was "safe"; also his copy of the *Rubáiyát* of Omar Khayyám, with drawings that had been carefully colored, so long ago, by Arline.

He begged her to come see him again. "I only mentioned my inner feelings for revenge, etc. to explain why it would be hard to guarantee you something that you asked," he wrote. He still wanted to marry her.

> I know where the right is—but emotions, like anger and hate and vengeance etc. are like a bunch of snakes in a barrel—with reason and good heart as a lid. . . . it is frightening and uncertain. Worth a good try tho.

She refused, despite the warm memories that now came back to her: building a sandcastle at the beach, surrounded by a mob of small boys; camping under the stars at Joshua Tree National Monument, where Feynman had tinkered delightedly with his gleaming green Coleman stove. On a wet Sunday night he had shown her a battered suitcase with all of Arline's letters and photographs. Once in a flash of anger he had called her a prostitute—a cruel rhetorical weapon he had used before. "And," she wrote, "I did enjoy my boss & my work."

Her husband's memories were not so warm. At a party he listened to someone telling a story about Feynman and blurted out that he knew a better one—but stopped. A few days later he wrote Feynman a formal letter demanding compensation. "You have taken callous & unscrupulous advantage of your position & salary to seduce an impressionable girl away from her husband," he wrote. Could Feynman not remember the harder times of his own first marriage? "You alienated my wife's affections. You flattered her with your attentions and your gifts. You made clandestine plans for exciting vacations. . . . I think you should pay for indulging your selfish pleasure." He demanded $1,250. Feynman refused.

Gweneth Howarth was reporting that Engelbert had brought cognac and chocolate to celebrate her twenty-fifth birthday; she decided to improve her shorthand and typing ("You do need someone to look after you, don't you?"). Feynman sent the consulate in Zurich an affidavit vouching for her ("she is an intelligent girl with a fine personality and is an excellent cook and domestic servant") and guaranteeing to undertake her financial support if necessary. Gweneth thanked him, mentioning that she had now met an Arab boy, beautifully polite, but he had started to make love to her. She had to avoid Engelbert because she could not hide a love bite on her neck. She was making her way through the immigration paperwork: pages of questions designed to ensure that she was not a Communist and then—infuriating her—questions about whether she was a woman of good character where sex was concerned. From what moral high ground—and with what bureaucratic logic—did the American authorities require her to swear that she was neither a prostitute nor an adulterer?

Feynman, meanwhile, tried to placate his former lover's husband: ". . . forgive her and make her happy. . . . your love will be deeper for the forgiveness and greater because you each know how you have suffered."

"Good thought," the husband retorted, "but why don't you apply it to yourself since you have enjoyed her for so long. . . . Don't give me the story of your parents' teachings, society etc. for I don't go for that." He engaged attorneys, who sent threatening letters on his behalf. But Feynman's attorneys advised him not to settle, guessing that the matter would fade away on its own. The last word belonged to his lover.

> I hope you are happy with your maid. Now you will always have your sex laid on. I think I begin to understand what you mean by a "good rela-tionship." . . . But I can't understand why you are so afraid of marriage? Is it too dull for you? I ~~thought~~ think sex without love ~~wasn't~~ isn't very satisfying, that the satisfaction only came by both parties desiring the happiness of the other, given in complete faith, truth & love without reserve. Anything short of that, I thought, was lust or fucking like ani-mals.—Perhaps that is why you have such a large turn-over with your women.

A half-year later she finally returned his medal.

He surprised Gweneth with his excitement at the news that her visa had finally cleared the consulate. "Well, at last!" he wrote. "I was overjoyed to hear that you are coming at last."

> I need you more than ever. . . . I'm looking forward to being much happier. . . . I have to take care of you too, you know. As soon as you arrive here you are a responsibility of mine to see you are happy & not scared.

He had pared back the domestic side of daily life in minimalist fashion, striving for the least drain on his consciousness. When Gweneth Howarth finally arrived in the summer of 1959 she found a man with five identical pairs of shoes, a set of dark blue serge suits, and white shirts that he wore open at the neck. (She surreptitiously introduced colored shirts in deliberate stages, beginning with the palest of pastels.) He owned neither a radio nor a television. He carried pens in a standard slip-in shirt-pocket protector. He taught himself to keep keys, tickets, and change always in the same pocket so that he would never have to give them an instant's thought.

. At first he kept her presence secret from all but a few close colleagues. She took charge of the household as promised. He reveled in his pretty English domestic servant. He taught her to drive and experimented with letting her drive him about chauffeur-style, while he sat in the rear seat. She worried that he thought she was fluffy-minded; in fact he discovered that she was cool and independent. She made a point of finding men to date—a Beverly Hills stockbroker replaced the German optician—but Feynman's friends gradually realized that their arrangement was turning romantic. They would appear at parties together and then make a show of departing separately, as though they had different places to go. Sometime in the next spring he realized how contented he felt, but he was not sure how to make the next decision. He marked a date on the calendar several weeks ahead and told himself that if his feelings had not changed by then, he would ask Gweneth to marry him. As the day approached, he could hardly wait. The evening before, without telling her why, he kept her awake until midnight. Then he proposed. They were married on September 24, 1960, at Pasadena's grand Huntington Hotel. He hid his car so that no one could tie tin cans to the fenders, and moments after the reception he ran out of gasoline on the Pasadena Freeway. He told Gweneth cheerfully: So this is how we're starting life. Murray Gell-Mann, who had married an Englishwoman he met at the Institute for Advanced Study several years before, thought Feynman was playing catch-up—now he, too, had acquired an English wife and a small brown dog.

The Feynmans and the Gell-Manns bought houses not far from each other in Altadena, north of the campus, nestled in the high hills that cup the smog drifting up from Los Angeles. Richard spent long hours teaching the dog, Kiwi, increasingly circuitous tricks; Feynman's mother, who had moved out to Pasadena to be near her son, made droll remarks about what a child would be up against. Gweneth began a garden with citrus scents and exotic colors that could never have survived a Yorkshire winter. In 1962 a son, Carl, was born; six years later they adopted a daughter, Michelle. It was instantly clear to Richard's friends how much he had wanted children. At first Murray and his wife, Margaret, visited from time to time, and the friendship was never warmer. An image lodged in Gell-Mann's memory of his friend pitching wads of newspaper into the fireplace for kindling, one after another—and making an ebullient game of it, as he made a game of every mundane gesture. The dog bounded here and there at his command, and he called out happily to Gweneth, and Murray felt magic in his presence.

From QED to Genetics

"Hello, my sweetheart,

"Murray and I kept each other awake arguing until we could stand it no longer. We woke up over Greenland . . ."

They were off to Brussels together for a conference, partly nostalgic, on "the present state of quantum electrodynamics." Dirac was there, and Feynman spoke once again with his old hero—Dirac still wholly unreconciled to the renormalization program for evading the infinities that had plagued his old theory. Renormalization seemed an ugly gimmick, an arbitrary and unphysical device for merely discarding inconvenient quantities in one's equations. To most physicists Dirac's qualms sounded like the intolerance of the old in the face of new ideas—in this case ideas that succeeded where Dirac's own theory had broken down. He reminded them of Einstein, with his famous crotchety unwillingness to accept quantum mechanics, and like Einstein he could hardly be dismissed. Honest physicists at least understood his qualms, even if they attributed them, ultimately, to a generational hardening of the intuitions. Age was no friend of the physicist. Wisdom counted for nothing. Feynman was acutely and painfully aware of the truth expressed in a ditty sometimes attributed to Dirac himself; it appeared from time to time, over the years, on Caltech office doors:

> Age is, of course, a fever chill
> That every physicist must fear.
> He's better dead than living still
> When once he's past his thirtieth year.

Feynman also sympathized with Dirac's qualms about renormalization, more so than any of his coinventors of the modern methods. Quantum electrodynamics had become a singular triumph of theoretical physics. The computations that had taken Feynman and Schwinger hours or weeks to accomplish in their first and second approximations could now be extended to many deeper levels of accuracy, using electronic computers and hundreds of Feynman diagrams to organize the work. Some theorists and their graduate students spent years on these calculations. They added and subtracted hundreds of terms, deeper and deeper into infinite series. It struck some of them as bizarrely unsatisfying work: some of the terms were enormous, positive or negative, compared to the final result. Yet presumably they would cancel out in the end, leaving a small, finite number. The mathematical status of such computation remained uneasy. It was not mathematically certain that the calculations would converge. Yet for practical calculations in quantum electrodynamics they always seemed to, and when the increasingly precise results were compared with the results of increasingly sensitive experiments, they matched. To convey a sense of how "delicately" experiment and theory agreed, Feynman would say it was like measuring the distance from New York to Los Angeles to within the thickness of a single hair. Yet the unphysical nature of the computing process troubled him, the corrections upon corrections with no sense of whether the next correction must be large or small. "We have been computing terms like a blind man exploring a new room," he said in his keynote talk in Brussels.

Other theorists, meanwhile, had begun to use the very concept of "renormalizability" as a way of distinguishing between possible theories for the esoteric particles to which quantum electrodynamics did not apply. Dyson had first recognized that it might be fruitful to think of renormalizability this way, as a criterion for judgment. A renormalizable theory was one by which, practically speaking, calculations could be made. "Note the cunning of reason at work," said the physicist and historian Silvan S. Schweber. "The divergences that had previously been considered a disastrous liability now became a valuable asset." Gell-Mann and younger theorists applied the notion with real success. "We very much need a guiding principle like renormalizability to help us pick the quantum field theory of the real world out of the infinite variety of conceivable quantum field theories," said Steven Weinberg years later—recognizing, however, that he was begging the question of *why?* Why should the correct theories be the computable ones? Why should nature make matters easy for human physicists? Feynman himself remained nearly as uncomfortable as Dirac. He continued to say that renormalization was "dippy" and "a shell game" and "hocus-pocus."

By the 1960s he seemed to be withdrawing from the most esoteric frontiers of high-energy physics. Quantum electrodynamics had achieved the quiet stature of a solved problem. As a practical theory it had entered applied, solid-state fields like electrical engineering, where, for example, quantum mechanics gave rise to the maser, a device for creating intense beams of coherent radiation, and its successor, the laser. Feynman drifted into the theory of masers for a while, using his path integral methods to lay some of the foundation. He had also worked persistently on another solid state problem, the problem of the so-called polaron, an electron moving through a crystal lattice. The electron distorts the lattice and interacts with its own cloud of distortion, creating, as Feynman realized, a kind of case study for examining the interaction of a particle with its field. Again his diagrams and path integrals found fertile ground. Yet this was minor work, not the special outpouring of someone already regarded as a legend (though each fall, it seemed, younger men won the Nobel Prize).

He could not find the right problem to work on. His Caltech salary passed the twenty-thousand-dollar mark—he was the highest paid member of the faculty. He started telling people jovially that that was a lot of money to be paid for theoretical physics; it was time to do some *real work*. He had a sabbatical year coming. He did not want to travel. His friend Max Delbrück, himself a physicist turned geneticist, was always trying to lure physicists into his group at Caltech, saying that the interesting questions now lay in molecular biology. Feynman told himself that he would go into a different field instead of a different country.

In biology the theorists and the laboratory workers were still largely one and the same. Feynman began in the summer of 1960 by learning how to grow strains of bacteria on plates, how to suck drops of solution into pipettes, how to count bacteriophages—viruses that infect bacteria—and how to detect mutations. He planned experiments at first to teach himself the techniques. Much of Delbrück's laboratory devoted itself to the genetics of such microcreatures: tiny, efficient DNA-replicating machines. The most popular virus when Feynman arrived in the upper basement of Church Hall was a bacteriophage called T4, which grew on the common strain of *E. coli* bacteria.

Less than a decade had passed since James Watson and Francis Crick had elucidated the structure of DNA, the molecule that carried the genetic code. *Code* was one word for this storing of information; geneticists also thought in terms of maps and blueprints, printed text and recording tape— the mechanics were far from clear. Mutations were known to be changes in the DNA sequence, but no one understood how a developing organism actually "read" the altered map, text, or tape. Was there a biological copying, splicing, folding? Feynman began to feel at home in the basement laboratory. He took comfort from the knowledge that everything around was made of matter. He felt well acquainted with the essence of evaluating experiments—as he said, "understanding when a thing is really known and when it is not really known." He could see at once how the centrifuge worked and how ultraviolet absorption would show how much DNA remained in a test tube. Biology was messier—things grew and wiggled, and he found it difficult to repeat experiments as exactly as he wished.

He focused on a particular mutation of the T4 virus called rII. This mutant had the useful quality of growing abundantly on one strain of the E. coli bacteria, strain B, while not growing at all on strain K. So a researcher could infect strain K bacteria with the mutants and watch for signs of T4. If any appeared, it must mean that something had happened to the rII mutation—presumably, it had reverted back to its original form. Such "backmutation" was relatively rare, but when it happened, giving the virus the ability to grow again in the K bacteria, it could be detected with extreme sensitivity, rates as low as one in a billion. Feynman compared finding a T4 backmutation to finding one man in China with elephant ears, purple spots, and no left leg. He collected them, isolated them, and injected them back into bacteria of strain B to see how they would grow.

Odd-looking plaques appeared. Among the normal, backmutated T4, he began to see phages that did not grow as they should have. He called them "idiot r's." He could only guess what might be happening at the level of the DNA itself to create the idiot r's. He saw two possibilities: the site of the rII mutation in the DNA strand might have undergone a second, further mutation. Or a second mutation might have occurred at a different site, somehow acting to partially cancel the effect of the first mutation.

Tools for directly examining the genetic sequence, letter by letter, base pair by base pair, did not exist. But by painstakingly crossing the idiot r's with the original virus, Feynman was able to show that his second guess was correct: two mutations, situated close to each other on the gene, were interacting. Furthermore, he showed that the second mutation had the same character as the first; it was another rII mutation. He had discovered a new phenomenon, mutations that suppressed each other within the same gene. Friends of his in the laboratory called these "Feyntrons" and tried to persuade him to write up his work for publication. Elsewhere, discovered independently, the phenomenon came to be called intragenic suppression. Feynman could not explain it. The Caltech biologists had no clear model for understanding how the genetic code was read, how the information encoded in DNA actually transformed itself into working proteins and more complex organisms. And Feynman's time as a geneticist had come to an end. He desperately wanted to return to physics. When he was not grinding microsomes, he had been working more and more intently on a quantum theory of gravity.

Without realizing it, Feynman had come to the brink of the next great breakthrough in modern genetics. The specialists had an advantage after all: a year later, Francis Crick's team at Cambridge, England, used the discovery of intragenic suppression as the touchstone for an explanation of how the genetic code was read. They guessed, correctly, that the mutations actually added or deleted a unit of DNA, thus shifting the message back or forward. One mutation threw the message temporarily out of phase; the next mutation put it back in phase. This interpretation suggested—or perhaps Crick already had it in mind—one of the simplest, yet strangest, mechanical models for genetic decoding: that the message of the gene is read in linear fashion, one base pair after another, from beginning to end. By 1966 Crick was declaring, "The story of the genetic code is now essentially complete."

Ghosts and Worms

The problem of gravity had the finest pedigree—it came in a direct line of descent from Einstein's greatest work—yet it lay outside the mainstream of high-energy theoretical physics in the early 1960s. As the general theory of relativity neared its fiftieth anniversary, some relativists and mathematical physicists continued to struggle with the natural problem of trying to create a quantum theory of gravitation—to quantize the gravitational field, as the fields associated with other forces had been quantized. It was difficult, involuted work. A quantum field theory of Einsteinian gravitation meant, as Gell-Mann said, a "quantum mechanical smearing of space-time" itself. No experimental evidence demanded that gravity must be quantized, but physicists did not wish to imagine a world in which some fields obeyed the laws of quantum mechanics and others did not.

The difficulty, from an experimentalist's perspective, was that gravity was so weak compared to the other forces. A bare handful of electrons can create a palpable electromagnetic force, while it takes a mass as great as the earth to create the gravity that draws a leaf from a tree. The orders of magnitude separating these forces strain the imagination and cause immense mathematical difficulties for theorists trying to reconcile them. The difference is 10^{42}, a number that defied even Feynman's ability to find illustrative analogies. "The gravitational force is weak," he said at one conference, introducing his work on quantizing gravity. "In fact, it's *damned* weak." At that instant a loudspeaker demonically broke loose from the ceiling and crashed to the floor. Feynman barely hesitated: "Weak—but not negligible."

He had begun with Einstein's theory and simply started calculating, as he had done in electrodynamics. He pushed his way into different corners of the problem in original fashion. The late 1950s were a time when relativity specialists were confused about the nature of gravitational radiation, and the high levels of mathematical rigor they demanded were blocking them from the right approximations. To Feynman it seemed straightforward that gravitational waves were real. Once again he began with a palpable physical intuition and charged forward. He found answers—decisive, he believed—to questions that relativists argued about: Do gravity waves carry energy? (Yes, he showed.) Can gravity waves be detected by small-scale measurements inside the wavelength? (No, he argued. "Only beyond the wave length can a clear proof of waves be found," he wrote Victor Weisskopf when he heard that his old friend was interested in his gravity work. "I have not seen any plans for any such experiments, except by crackpots.") For the sake of argument, at least, he refused to abandon altogether the possibility that gravity could not be quantized after all. "Maybe gravity is a way that quantum mechanics fails at large distances. Isn't it interesting to live in our time and have such wonderful puzzles to work on?" He wrote down Feynman diagrams and computed integrals, and he could see that he was producing answers that could not be right. The probabilities did not add up to one. Yet he realized—with a combination of physical and diagrammatic intuition—that he could make up the deficits all at once if he resorted to a gimmick. He had to add "ghosts," fictitious particles that would circle around the Feynman diagrams, appearing just long enough to form loops and then vanishing once more into mathematical oblivion. It was a curious idea, but it worked, and he reported it in Warsaw, Poland, at a conference on gravitation in July 1962.

The subject was on the eve of a rebirth, when discoveries from astro-physicists and theories from relativists would come together in a shower of black holes, white dwarfs, quasars, and other cosmological treasures. Feyn-man himself continued his gravitational work for years. He applied the gauge-symmetry machinery known as Yang-Mills. He made an influential contribution without ever reaching a complete enough theory to publish whole. For the moment, he found no more joy in a gathering of relativists than in the conclaves on high-energy physics he was temporarily fleeing. One of the speakers began seriously: "Since 1916 we have had a slow, rather painful accumulation of minute technical improvements. . . . I think that the attempt to continue obtaining such minute improvements constitutes a legitimate and fascinating part of mathematical physics. If something really exciting turns up, fine. . . ." The American physicists mingled uneas-ily with their Russian counterparts. They teased each other about searching their rooms for microphones; Feynman actually took apart his telephone at the Grand Hotel and decided that if it contained no bugs the Poles were wasting wire. He was overheard during a break baiting one of the Russians: "What have you ever done in physics, Ivanenko?"

"I've written a book with Sokolov."

"How do I know what you contributed to it? Ivanenko, what is the integral of e to the minus x squared from minus to plus infinity?" Silence. "Ivanenko, what is one and one?" Feynman was dismayed by the work offered up. His own presentation drew little immediate notice, though his "ghosts," ex-tended by other theorists, later became crucial to modern theory. "I am learning nothing," he wrote home in frustration, and he gave Gweneth a scathing taxonomy of pretentious science:

The "work" is always: (1) completely un-understandable, (2) vague and indefinite, (3) something correct that is obvious and self-evident, worked out by a long and difficult analysis, and presented as an important dis-covery, or (4) a claim based on the stupidity of the author that some obvious and correct fact, accepted and checked for years is, in fact, false (these are the worst: no argument will convince the idiot), (5) an attempt to do something, probably impossible, but certainly of no utility, which, it is finally revealed at the end, fails or (6) just plain wrong. There is a great deal of "activity in the field" these days, but this "activity" is mainly in showing that the previous "activity" of somebody else resulted in an error or in nothing useful or in something promising.

He never had liked crowds in science. "It is like a lot of worms trying to get out of a bottle by crawling all over each other."

Dissatisfied though Feynman remained, his Warsaw talk marked the beginning of a turn toward his path integrals as a fundamental approach to the deepest of cosmological issues. Neither he nor other theorists had relied on this viewpoint in the high-energy physics of the late 1950s. Much later, however, some physicists applied path integrals to the very structure of space-time. They sought to unify its conceivable topologies by, in a sense, summing over all possible universes. Gell-Mann himself speculated that Feynman's path integrals might prove to be more than a method, more than an equivalent alternative formulation: "the real foundation of quantum mechanics and thus of physical theory."

Room at the Bottom

So little of modern physics seemed dedicated to the world of human scales. High-energy theorists had skipped far down a ladder of sizes, past the merely microscopic into a realm of the unimaginably small and short-lived. "Miniaturization" was a catchword of the day, but tininess meant something more modest to engineers and manufacturers than to particle physicists. The transistor, invented just over a decade before at the Bell Telephone Laboratories, was becoming a commodity. Transistors meant radios, battery-powered, with brittle plastic casings, small enough to fit in one's hand. Researchers were beginning to consider ways of further reducing suitcase-sized devices like tape recorders. Electronic computers that had filled large rooms could now be squeezed into cabinets barely larger than an automobile. It occurred to Feynman that engineers had barely begun to imagine the possibilities. "There is a device on the market, they tell me," he said at the end of 1959, when the American Physical Society held its annual meeting at Caltech, "by which you can write the Lord's Prayer on the head of a pin. But that's nothing. . . ." On toward the atom, he urged them. "It is a staggeringly small world that is below."

That same pinhead could hold the twenty-four volumes of the *Encyclopaedia Britannica*, pictures and all, if the encyclopedia were reduced 25,000 times in each direction. A modest reduction, considering that the barely visible dots making up a halftone photoengraving would still contain a thousand or so atoms. For writing and reading this tiny *Britannica*, he proposed engineering techniques within the limits of contemporary technology: reversing the lenses of an electron microscope, for example, and focusing a beam of ions to a small spot. At this scale, the world's entire store of book knowledge could be carried about in a small pamphlet. But direct reduction would be crude, he continued. Telephones and computers had given rise to a new way of thinking about information, and in terms of raw information—allowing six or seven "bits" per letter and a generous one hundred atoms per bit—all the world's books could be written in a cube no larger than a speck of dust. His audience, unaccustomed to lectures of this kind at American Physical Society meetings, was enthralled. "Don't tell me about microfilm!" Feynman declared.

He had several reasons for thinking about the mechanics of the atomic world. Although he did not say so, he had been pondering the second law of thermodynamics and the relationship between entropy and information; at atomic scales came the threshold where his calculations and thought experiments took place. The new genetics also brought such issues to the surface. He talked about DNA (fifty atoms per bit of information) and about the capacity of living organisms to build tiny machinery, not just for information storage but for manipulation and manufacturing. He talked about computers: given millions of times more power, they would not just calculate faster but would reveal qualitatively different abilities, such as the ability to make judgments. "There is nothing I can see in the physical laws that says the computer elements cannot be made enormously smaller than they are now," he said. He talked about problems of lubrication, and he talked about the realm where quantum-mechanical laws would take over. He envisioned machines that would make smaller machines, each of which would make machines that were smaller still. "It doesn't cost anything for materials, you see. So I want to build a billion tiny factories, models of each other, which are manufacturing simultaneously, drilling holes, stamping parts, and so on." He concluded by offering a pair of one-thousand-dollar prizes: one for the first microscope-readable book page shrunk 25,000 times in each direction, and one for the first operating electric motor no larger than a 1/64th-inch cube.

Caltech's magazine *Engineering and Science* printed Feynman's talk, and it was widely reprinted elsewhere. (*Popular Science Monthly* retitled it "How to Build an Automobile Smaller than This Dot.") Twenty years later there was a name for the field Feynman had been trying to invent: nanotechnology. Nanotechnologists, partly inspired and partly crackpot, made tiny silicon gears with carefully etched teeth and displayed them proudly in their microscopes; or imagined tiny self-replicating robot doctors that would swim through one's arteries. They thought of Feynman as their spiritual father, although he himself never returned to the subject. In the crude mechanical sense, tiny machines seemed a feature of a future just as distant as in 1959. The mechanical laws of physics meant that friction, viscosity, and electrical forces did not scale down as neatly as Feynman's imagined billion tiny factories. Wheels, gears, and levers tended to glue themselves together. Tiny machines had come into being, storing and manipulating information even more efficiently than he had predicted. But they were electronic, not mechanical, using quantum mechanics, not fighting it. Not until 1985 did Feynman have to pay the thousand dollars for tiny writing: a Stanford University graduate student, Thomas H. Newman, spent a month shrinking the first page of A *Tale of Two Cities* onto silicon by almost exactly the technique Feynman had outlined.

The tiny motor did not take so long. Feynman had underestimated existing technology. A local engineer, William McLellan, read the *Engineering and Science* article in February. By June, when he had not heard any more, he decided he had better make the motor himself. It took two months of working in his spare time, using a watchmaker's lathe and a microdrill press, drilling invisible holes and wrapping 1/2000th-inch copper wire. Tweezers were too crude. McLellan used a sharpened toothpick. The result was a one-millionth-horsepower motor.

One day in November he visited Feynman, who was working alone in a Caltech laboratory. McLellan brought his equipment in a large wooden box. He saw Feynman's eyes glaze; too many cranks had turned up, typically bringing toy automobile engines that they could hold in the palm of a hand. But McLellan opened his box and pulled out a microscope.

"Uh-oh," Feynman said. He had neglected to make any arrangements for funding the prize. He sent McLellan a personal check.

All His Knowledge

He could not let go of the simple questions. He had spent much of a lifetime assembling a picture of how the world worked, how atoms and forces conjoined to create ice crystals and rainbows. In conjuring a world of miniature machines, he continued to work out possibilities at the level of long-lived molecules, not ephemeral strange particles. He had made himself a member of the community of theoretical physics, and he accepted their goals and their rhetoric: he had told the American Physical Society apologetically that miniaturization was not "fundamental physics (in the sense of, 'What are the strange particles?')." Indeed, his community now assigned a kind of intellectual primacy to phenomena that could be observed only in the searing less-than-an-instant of a particle collision. But a part of him still preferred to give *fundamental* a different definition. "What we are talking about is real and at hand: Nature," he wrote to a correspondent in India, who had, he thought, spent too much time reading about esoteric phenomena.

Learn by trying to understand simple things in terms of other ideas—always honestly and directly. What keeps the clouds up, why can't I see stars in the daytime, why do colors appear on oily water, what makes the lines on the surface of water being poured from a pitcher, why does a hanging lamp swing back and forth—and all the innumerable little things you see all around you. Then when you have learned what an explanation really is, you can then go on to more subtle questions.

The first plank in every Caltech undergraduate education was a two-year required course in basic physics. By the 1960s the institute administration recognized a problem. The course had grown stale. Too much ancient pedagogy lingered in it. Bright young freshmen arrived from their high schools around the country, ready to tackle the mysteries of relativity and strange particles, and were plunged into the study of—as Feynman put it—"pith balls and inclined planes." There was no main lecturer; the course met in sections taught by graduate students. The administration decided in 1961 to revise the course from the bottom up and asked Feynman to take it on for one year. He would have to lecture twice a week.

Caltech was not alone; nor was physics. The pace of change in modern science had accelerated as most college syllabuses had hardened. It was no longer possible, as it had been a generation before, to bring undergraduates up to the live frontier of a field like physics or biology. Yet if quantum mechanics or molecular genetics could not be integrated into undergraduate education, science risked becoming a historical subject. Many first-year physics courses did begin with history: physics in ancient Greece; the pyramids of Egypt and the calendars of Sumeria; medieval physics through nineteenth-century physics. Virtually all began with some form of mechanics. A typical program went:

1. Historical Development of Physical Science
2. Present Status of Physical Science
3. Kinematics: The Study of Motion
4. The Laws of Dynamics
5. Application of the Laws of Motion: Momentum and Energy
6. Elasticity and Simple Harmonic Motion
7. Dynamics of Rigid Bodies
8. Statics of Rigid Bodies

and so on, until in its final weeks the course would reach

26. Atoms and Molecules

in time to touch upon Nuclear Physics and Astrophysics. Caltech was still using a generation-old text by its own luminary, Robert Millikan, that remained soundly mired in the physics of the eighteenth and nineteenth centuries.

Feynman began with atoms, because that was where his own understanding of the world began—not the world of quantum mechanics but the quotidian world of floating clouds and colors shimmering in oily water. Moments after nearly two hundred freshmen entered the hall for his first lecture in the fall of 1961, they heard these words from the grinning physicist striding back and forth upon the stage:

So, what *is* our over-all picture of the world?

If, in some cataclysm, all of scientific knowledge were to be destroyed, and only one sentence passed on to the next generation of creatures, what statement would contain the most information in the fewest words? I believe it is the *atomic hypothesis* (or the atomic *fact*, or whatever you wish to call it) that *all things are made of atoms—little particles that move around in perpetual motion, attracting each other when they are a little distance*

apart, but repelling upon being squeezed into one another. In that one
sentence, you will see, there is an *enormous* amount of information about
the world, if just a little imagination and thinking are applied.

Imagine a drop of water, he said. He took them on a tour inward through
the length scales, magnifying the drop until it was forty feet across, then
fifteen miles across, then 250 times larger still, until the teeming molecules
came into view, each with a pair of hydrogen atoms stuck like round arms
upon a larger oxygen atom. He discussed the contrary forces holding the
molecules together and forcing them apart. He described heat as atoms in
motion . . . pressure . . . expansion . . . steam. He described ice, with its
molecules held in a rigid crystalline array. He described the surface of water
in air, absorbing oxygen and nitrogen and giving off vapor, and he immediately raised issues of equilibrium and disequilibrium. Instead of Ar-
istotle and Galileo, instead of levers and projectiles, he was building a
tangible sense of how atoms create the substances around us and why
substances behave as they do. Solution and precipitation, fire and odor—
he kept moving, displaying the atomic hypothesis not as a reductive end
point but as a road toward complexity.

If water—which is nothing but these little blobs, mile upon mile of the
same thing over the earth—can form waves and foam, and make rushing
noises and strange patterns as it runs over cement; if all of this, all the life
of a stream of water, can be nothing but a pile of atoms, *how much more
is possible?* . . . Is it possible that the "thing" walking back and forth in
front of you, talking to you, is a great glob of these atoms in a very complex
arrangement . . . ? When we say we are a pile of atoms, we do not mean
we are *merely* a pile of atoms, because a pile of atoms which is not repeated
from one to the other might well have the possibilities which you see before
you in the mirror.

He found that he was working harder than at any time since the atomic
bomb project. Teaching was only one of his goals. He realized also that
he wished to organize his whole embracing knowledge of physics, to turn
it end over end until he could find all the interconnections that were usually,
he believed, left as loose ends. He felt as though he were making a map.
In fact, for a while he considered actually trying to draw one, a diagram—
a "Guide to the Perplexed," as he put it.

A team of Caltech physics professors and graduate students scrambled to keep up, week after week, designing problem sets and supplementary material, as his guide to the perplexed took shape. They met with him at lunch after each lecture to piece together what Feynman had spun from as little as a single sheet of cryptic notes. Despite the homespun lyricism of his voice, the stress on ideas rather than technique, he was moving quickly, and his fellow physicists had to work to keep up with some of his leaps.

As every physics course recapitulated the subject's history, so did Feynman's, but instead of surveying the Sumerians or the Greeks he chose—in his second lecture—to sum up "Physics before 1920." Less than a half-hour later he was on to a quick tour of quantum physics and then the nuclei and the strange particles according to Gell-Mann and Nishijima. This was what many students wanted to hear. Yet he did not want to leave them with the easy sense that here, at the microlevels, lay the most fundamental laws or the deepest unanswered questions. He described another problem, crossing the artificial boundaries that divide scientific disciplines, "not the problem of finding new fundamental particles, but something left over from a long time ago."

It is the analysis of *circulating or turbulent fluids*. If we watch the evolution of a star, there comes a point where we can deduce that it is going to start convection, and thereafter we can no longer deduce what should happen. . . . We cannot analyze the weather. We do not know the patterns of motions that there should be inside the earth.

No one knew how to derive this chaos from the first principles of atomic forces or fluid flow. Simple fluid problems were for textbooks, he told the freshmen.

What we really cannot do is deal with actual, wet water running through a pipe. That is the central problem which we ought to solve some day.

Feynman designed his lectures as self-contained dramas. He never wanted to end by saying, "Well, the hour is up, we will continue this discussion next time . . ." He timed his diagrams and equations to fill the sliding two-tier blackboard so definitively that an image of the final chalk tableau seemed to have been in his head from the start. He chose grand themes with tentacles that spread into every corner of science: Conservation of Energy; Time and Distance; Probability . . . Before a month was out he introduced the deep and timely issue of symmetry in physical laws. His approach to the conservation of energy was revealing. This principle was never far from the consciousness of a working theoretical physicist, yet most textbooks let it arise in passing, toward the end of chapters on mechanical energy or thermodynamics. First they would note that mechanical energy is *not* conserved, since friction inevitably drains it away. Not until the Einsteinian equivalence of matter and energy does the principle fully come into its own.

Feynman took the conservation of energy as a starting point for discussing conservation laws in general (as a result, his syllabus managed to introduce the conservation of charge, baryons, and leptons weeks before reaching the subject of speed, distance, and acceleration). He put forward an ingenious analogy. Imagine, he said, a child with twenty-eight blocks. At the end of every day, his mother counts them. She discovers a fundamental law, the conservation of blocks: there are always twenty-eight.

One day she sees only twenty-seven, but careful investigation reveals one under the rug. Another day she finds twenty-six—but a window is open, and two are outside. Then she finds twenty-five—but there is a box in the room, and upon weighing the box and weighing individual blocks she surmises that three blocks are inside. The saga continues. Blocks vanish beneath the dirty water of a bathtub, and further calculations are needed to infer the number from the rising water level. "In the gradual increase in the complexity of her world," Feynman said, "she finds a whole series of terms representing ways of calculating how many blocks are in places where she is not allowed to look." One difference, he warned: in the case of energy, there are no blocks—just a set of abstract and increasingly intricate formulas which must always, in the end, return the physicist to his starting point.

With the vivid analogies and large themes immediately came computation. In the same one-hour lecture on the conservation of energy, Feynman had his students calculating potential and kinetic energy in a gravitational field. A week later, when he introduced the uncertainty principle of quantum mechanics, he not only conveyed the philosophical drama of this "inherent fuzziness" in the description of nature but also leapt through the calculation of the probability density of an undisturbed hydrogen atom. He still had not reached the basics of speed, distance, and acceleration.

No wonder his colleagues found their nerves jangling as they tried to write problem sets. Before a half-year was gone, he was teaching an un-

compromising version of the geometry of relativistic space-time, complete
with particle diagrams, geometrical transformations, and four-vector algebra. For college freshmen this was difficult. Along with the mathematics
Feynman tried to convey a feeling for how he visualized such problems,
placing his "brain" into his diagrams like Alice plunging through the
Looking-Glass. He tried to make his students imagine the apparent width
and depth of an object:

> *They* depend upon *how* we look at it; when we move to a new position,
> our brain immediately recalculates the width and the depth. But our brain
> does not immediately recalculate coordinates and time when we move at
> high speed, because we have had no effective experience of going nearly
> as fast as light to appreciate the fact that time and space are also of the
> same nature.

The students were sometimes terrified. Yet Feynman also returned to the
standard fare of an introductory physics course. When he covered centers
of mass and spinning gyroscopes, experienced physicists realized that he
was giving the students not just the mathematical methods but also original,
physical understanding. Why does a spinning top stand upright on your
fingertip and then, as gravity pulls its axis downward, slowly circle about?
Even physicists felt they were learning the *why* for the first time when they
heard Feynman explain that the gyroscope began by "falling" an invisibly
small distance . . . (He did not want to leave the students thinking a gyroscope was a miracle: "It *is* a wonderful thing, but it is not a miracle.")
    No realm of science was out of bounds. After consulting with experts
in other fields, he gave two lectures on the physiology of the eye and the
physiochemistry of color vision, making a profound connection between
psychology and physics. He described the view of time and fields that arose
from advanced and retarded potentials, his graduate work with Wheeler.
He delivered a special lecture on the principle of least action, beginning
with his high-school memories of his teacher Mr. Bader—how does a ball
know what path to follow?—and ending with least action in quantum
mechanics. He devoted an entire lecture to one of the simplest of mechanical gadgets, the ratchet and pawl, the sawtoothed device that keeps a
watch spring from unwinding—but it was a lesson in reversibility and
irreversibility, in disorder and entropy. Before he was done he had linked
the macroscopic behavior of the ratchet and pawl to the events occurring

at the level of its constituent atoms. The history of one ratchet was also the thermodynamic history of the universe, he showed:

> The ratchet and pawl works in only one direction because it has some ultimate contact with the rest of the universe. . . . Because we cool off the earth and get heat from the sun, the ratchets and pawls that we make can turn one way. . . . It cannot be completely understood until the mystery of the beginnings of the history of the universe are reduced still further from speculation to scientific understanding.

The course was a magisterial achievement: word was spreading through the scientific community even before it ended. But it was not for freshmen. As the months went on, the examination results left Feynman shocked and discouraged. Still, when the year ended, the administration pleaded with him to keep on for a second year, teaching the same students, now sophomores. He did, finally trying to teach a thorough subcourse in quantum mechanics, again reversing the conventional order. Another Caltech physicist, David Goodstein, said long afterward, "I've spoken to some of those students in recent times, and in the gentle glow of dim memory, each has told me that having two years of physics from Feynman himself was the experience of a lifetime." The reality was different:

> As the course wore on, attendance by the kids at the lectures started dropping alarmingly, but at the same time, more and more faculty and graduate students started attending, so the room stayed full, and Feynman may never have known he was losing his intended audience.

This was the world according to Feynman. No scientist since Newton had so ambitiously and so unconventionally set down the full measure of his knowledge of the world—his own knowledge and his community's. With intensive editing by other physicists, chiefly Robert B. Leighton and Matthew Sands, the lectures became the famous "red books"—the three-volume *Feynman Lectures on Physics*. Colleges and universities worldwide tried to adopt them as textbooks and then, inevitably, gave them up for more manageable and less radical alternatives. Unlike true textbooks, however, Feynman's volumes continued to sell steadily a generation later.

Adorning each volume was a picture of Feynman in shirtsleeves, gleefully pounding a bongo drum. He came to regret that. "It is odd," he said after hearing himself introduced yet again as a bongo player, "but on the infre-

quent occasions when I have been called upon in a formal place to play
the bongo drums, the introducer never seems to find it necessary to mention
that I also do theoretical physics. I believe that is probably because we
respect the arts more than the sciences." And when yet another request
came in for a copy of the photograph—from a Swedish encyclopedia publisher who wished to "give a human approach to a presentation of the
difficult matter that theoretical physics represents"—he exploded. "Dear
Sir," he scrawled,

> The fact that I beat a drum has nothing to do with the fact that I do
> theoretical physics. Theoretical physics is a human endeavor, one of the
> higher developments of human beings—and this perpetual desire to prove
> that people who do it are human by showing that they do other things
> that a few other humans do (like playing bongo drums) is insulting to me.
>
> I am human enough to tell you to go to hell.

The Explorers and the Tourists

"When you have learned what an explanation really is," Feynman had
said, "you can then go on to more subtle questions."

Creeping philosophy. What is an explanation? Science and scientists had
commandeered the practice of explanation, but the theory they left mainly
to philosophers. The *why* seemed to fall in their domain. "With this question philosophy began and with this question it will end," Martin Heidegger
had recently said, "provided that it ends in greatness and not in an impotent
decline." Feynman, who believed that the impotent decline was well under
way in the academies that supported philosophers, realized that he had had
to develop a view of what constituted explanation, what legitimized explanation, and which phenomena did and did not require explanation.

His understanding of explanation did not depart far from the modern
philosophical mainstream, though its jargon of *explanans* and *explanandum*
was an alien language to him. Like most philosophers, he found explanations most satisfactory when they called upon a generalizing, underlying
"law." A thing is the way it is because other things of its kind are all that
way. Why does Mars travel around the sun in an ellipse? Feynman explained—and ventured deep into philosophical territory—in an invited

lecture series at Cornell University in 1964. He began by speaking, nominally, about the law of gravitation. In reality his subject was explanation
itself.

All satellites travel in elliptical orbits. Why? Because objects tend to
travel in a straight line when left alone (the law of inertia) and the combination of that unchanging motion and a force exerted toward a center of
gravity—by the law of gravitation—creates an ellipse. What validates the
law of gravitation? Feynman expressed the scientist's modern view, a blend
of the pragmatic and the aesthetic. He cautioned that even so beautiful a
law was provisional: Newton's law of gravitation gave way to Einstein's,
and a necessary quantum modification eluded physicists even now.

> That is the same with all our other laws—they are not exact. There is
> always an edge of mystery, always a place where we have some fiddling
> around to do yet. This may or may not be a property of Nature, but it
> certainly is common to all the laws as we know them today.

Yet in its unfinished form the law of gravitation explained so much. To a
practicing scientist, that validated it. The same small parcel of mathematics
explained Tycho Brahe's nightly observations of the planets in the sixteenth
century and Galileo's measurements of balls rolling down inclined planes,
timed against the beat of his own pulse. The planets are falling, Newton
reasoned; the moon feels the same force as an earthly projectile, the force
weakening with the square of the distance. A law is not a *cause*—philosophers still wrestled with this distinction—yet it is more than merely a
description. It precedes the thing explained, not in time but in generality
or in profundity. The same law explained the earth's symmetrically bulging
tides, rising both toward and away from the moon, and the newly measured
orbits of the moons of Jupiter. It made new predictions that scientists could
confirm or disprove with experiments on balls hanging delicately in a
laboratory or observations of majestically rotating galaxies a hundred million
million times larger. "Exactly the same law," Feynman said, and added—
having struggled to find the right wording—

> Nature uses only the longest threads to weave her pattern, so each small
> piece of the fabric reveals the organization of the entire tapestry.

Meanwhile, *why* does an object in motion tend to travel forever in a straight
line? That, Feynman said, nobody knows. At some deep stage, the explanations must end.

"Science repudiates philosophy," Alfred North Whitehead had said. "In other words, it has never cared to justify its truth or explain its meaning." Feynman's colleagues liked to think of their gruffly plain-spoken pragmatist hero as the perfect antiphilosopher, *doing* rather than justifying. His own rhetoric encouraged them. He lacked patience for the now-popular *What is reality?* brand of speculation arising from quantum-mechanical paradoxes. Yet he could not repudiate philosophy; he had to find ways to justify the truth that he and his colleagues sought. The modern physics had banished any possibility of discovering a system of laws unambiguously tying effects to causes; or a system of laws deduced and conjoined with perfect logical consistency; or a system of laws rooted in the objects that people can see and feel. For philosophers, these had all been marks of a sound explanatory law. Now, however, a particle might or might not decay, an electron might or might not pass through a slit in a screen. A minimum principle like the principle of least action might be derived from laws of forces and motion, or those laws might depend on the principle: who could say with logical certainty? And the basic stuff of science had grown inexorably more abstract. As the physicist David Park put it: "None of the entities that appear in fundamental physical theory today are accessible to the senses. Even more . . . there are phenomena that apparently are not *in any way* amenable to explanation in terms of things, even invisible things, that move in the space and time defined by the laboratory." With all these traditional virtues removed—or worse, partly removed while still partly necessary—it fell to science to build a new understanding of the nature of explanation. Or so Feynman argued: the philosophers themselves, he said, were always a tempo behind, like tourists moving in after the explorers have left.

Scientists had their own forms of blindness. It was often said in the quantum-mechanical era—Feynman had said it himself—that the only true test of a theory was its ability to produce good numbers, numbers agreeing with experiment. The American pragmatism of the early twentieth century had brought forth views like Slater's at MIT: "Questions about a theory which do not affect its ability to predict experimental results correctly seem to me quibbles about words." Yet Feynman now felt a hollowness in the purely operational view of what a theory means to a scientist. He recognized that theories came laden with mental baggage, with what he called a philosophy, in fact. He had trouble defining this: "an understanding of the law"; "a way that a person holds the laws in his mind . . ." The philosophy could not be discarded as readily as a pragmatic scientist might suggest.

Consider a Mayan astronomer, he suggested. (In Mexico he had grown interested in the deciphering of the great ancient codices, hieroglyphic manuscripts that employed long tables of bars and dots to set down an intricate knowledge of the movements of sun, moon, and planets. Codes, mathematics, and astronomy—eventually he delivered a lecture at Caltech on deciphering Mayan hieroglyphics. Afterward, Murray Gell-Mann "countered," Feynman said, with a series of six lectures on the languages of the world.) The Maya had a theory of astronomy that enabled them to explain their observations and to make predictions long into the future. It was a *theory* in the utilitarian modern spirit: a set of rules, quite mechanical, which when followed produced accurate results. Yet it seemed to lack a kind of understanding. "They counted a certain number and subtracted some numbers, and so on," he said. "There was no discussion of what the moon was. There was no discussion even of the idea that it went around."

Now a "young man" approaches the astronomer with a new idea. What if there are balls of rock out there, far away, moving under the influence of forces just like the forces that pull rocks to the ground? Perhaps it would make possible a different way of calculating the motions of the heavenly bodies. (Feynman certainly had memories of a young man confronting his elders with new, half-formed physical intuitions.)

"Yes," says the astronomer, "and how accurately can you predict eclipses?" He says, "I haven't developed the thing very far yet." Then says the astronomer, "Well, we can calculate eclipses more accurately than you can with your model, so you must not pay any attention to your idea because obviously the mathematical scheme is better."

The notion that alternative theories could account plausibly for the same observations had slipped into a central position in the working philosophy of scientists. Philosophers called it *empirical equivalence*, when they began to catch up. The recent history of quantum mechanics had pivoted on the empirical equivalence of Heisenberg's and Schrödinger's versions. The empirical equivalence of very different-seeming theories could be demonstrated mathematically, as Dyson had shown for Feynman's and Schwinger's quantum electrodynamics. Scientists knew, usually without thinking

about it, that empirically equivalent theories could have different consequences, mathematics and logic notwithstanding.

For Feynman, especially, the tension between alternative theories served
as a creative force, an engine for generating new knowledge. Perhaps more
than any living physicist, he had made a specialty of learning what models
could be derived from which principles, and what models from each other.
To Dyson's astonishment, he had stood at a blackboard one day in 1948
and interrupted their heady discussions of quantum electrodynamics to
show him something different. Sketching quickly, he derived the nineteenth-century Maxwell field equations—the classical understanding of
electricity and magnetism—backward from the new quantum mechanics.
Einstein had started with the Maxwell equations and then shifted the perspective of the observer to arrive at his theory of relativity; Feynman went
the other way in a fit of ahistorical perversity. He began with a void, no
fields or waves, no concept of relativity, not even a notion of light itself,
just a single particle obeying quantum mechanics' odd rules. Before Dyson's
eyes he traveled back mathematically from the new physics, with its riddles
of uncertainty and immeasurability, to the comforting exactitude of the
previous century. He showed that Maxwell's field equations were not a
foundation but a consequence of the new quantum mechanics. Startled
and impressed, Dyson urged him to publish. Feynman just laughed and
said, "Oh, no, it's not serious." As Dyson understood it later, Feynman
had been trying to create a new theory "outside the framework of conventional physics."

His motivation was to discover a new theory, not to reinvent the old
one. . . . His purpose was to explore as widely as possible the universe of
particle dynamics. He wanted to make as few assumptions as he could.

A theorist who can juggle different theories in his mind has a creative
advantage, Feynman argued, when it comes time to change the theories.
The path-integral formulation of quantum mechanics might be empirically
equivalent to other formulations and yet—given less-than-omniscient
human physicists—find more natural-seeming application to realms of science not yet explored. Different theories tended to give a physicist "different
ideas for guessing," Feynman said. And the century's history had shown
that when even so elegant and pure a theory as Newton's had to be replaced,
slight modifications could not suffice.

To get something that would produce a slightly different result it had to be completely different. In stating a new law you cannot make imperfections on a perfect thing; you have to have another perfect thing.

He understood explanations as a surgeon understands knives. He had a set of practical tests, heuristics, that he applied when reaching a judgment about a new idea in physics: for example, did it explain something unrelated to the original problem. He would challenge a young theorist: *What can you explain that you didn't set out to explain?* He knew that *why?* is a question without an end and that our knowledge of things is inextricable from the language we use. The words and analogies from which we build our explanations are culpably linked with the things explained. *Explanans* and *explanandum* are inextricable after all. An interviewer for the British Broadcasting Corporation, Christopher Sykes, once asked him to explain magnets: "If you get hold of two magnets and you push them you can feel this pushing between them. . . . Now what is it, the feeling between those two magnets?"

"What do you mean, what's the feeling?" Feynman growled. His hair, swept back in dramatic gray waves, had receded high atop his head, leaving a statue's high brow above a pair of heavy eyebrows that curled more impishly than ever. His pale blue shirt was open at the collar. A pen and eyeglass case rested in his front pocket, as always. Off camera, a defensive note entered the interviewer's voice.

"Well, there's something there, isn't there? The sensation is that there's something *there* when you push these two magnets together."

"Listen to my question," Feynman said. "What is the meaning when you say there's a *feeling*? Of course you feel it. Now what is it you want to know?"

"What I want to know is what's going on between these two bits of metal."

"The magnets repel each other."

"But what does that *mean*? Or *why* are they doing that? Or *how* are they doing that?" Feynman shifted in his easy chair, and the interviewer added, "I must say I think that's a perfectly reasonable question to ask."

"Of course it's a reasonable—— it's an excellent question, okay?" Reluctantly, Feynman now stepped into metaphysics. Particle theorists were toying with a "bootstrap" model, in which no particle lies at a deepest level, but all are interdependent composites. The name *bootstrap* paid homage to the paradoxical circularity of having to build each fundamental particle from all the others. Feynman, as he now made clear, believed in a kind of bootstrap model of explanation itself.

You see, when you ask why something happens, how does a person answer why something happens?

For example, Aunt Minnie is in the hospital. Why? Because she went out on the ice and slipped and broke her hip. That satisfies people. But it wouldn't satisfy someone who came from another planet and knew nothing about things. . . . When you explain a *why*, you have to be in some framework that you've allowed something to be true. Otherwise you're *perpetually* asking why. . . . You go deeper and deeper in various directions.

Why did she slip on the ice? Well, ice is slippery. Everybody knows that—no problem. But you ask *why* is ice slippery. . . . And then you're involved with something, because there aren't many things as slippery as ice. . . . A solid that's so slippery?

Because it is in the case of ice that when you stand on it, they say, momentarily the pressure melts the ice a little bit so that you've got an instantaneous water surface on which you're slipping. Why on ice and not on other things? Because water expands when it freezes. So the pressure tries to undo the expansion and melts it. . . .

I'm not answering your question, but I'm telling you how difficult a *why* question is. You have to know what it is that you're permitted to understand . . . and what it is you're not.

You'll notice in this example that the more I ask why, it gets interesting after a while. That's my idea, that the deeper a thing is, the more interesting. . . .

Now when you ask why two magnets repel, there are many different levels. It depends whether you're a student of physics or an ordinary person who doesn't know anything.

If you don't know anything at all, about all I can say is that there's a magnetic force that makes them repel. And that you're feeling that force. Well, you say that's very strange because I don't feel a force like that in other circumstances. . . . You're not at all disturbed by the fact that when you put your hand on the chair it pushes you back. But we found out by looking at it that that's the same force. . . . It turns out that the magnetic and electric force with which I wish to explain these things is the deeper thing that we would start with to explain many other things. . . .

If I said that magnets attract as if they were connected with rubber bands, I would be cheating you, because they're not connected with rubber bands. . . . If you were curious enough you'd ask me why rubber bands tend to pull back together again, and I would end up explaining *that* in terms of electrical forces—which are the very things I was using the rubber bands to explain, so I have cheated very badly, you see.

So I am not going to be able to give you an answer to why magnets attract. Except to tell you that they do . . . I really can't do a good job—any job—of explaining the electromagnetic force in terms of something you're more familiar with, because I don't *understand* it in terms of anything else that you're more familiar with.

He sat back and grinned.

To the professionals Feynman's musings were not philosophy but a charmingly naïve folk wisdom. He was both after and ahead of his time. Academic epistemology was still wrestling with *un*knowability. What choice did they have, in light of scientific relativity and uncertainty, the abandonment of strict causality and the pervasiveness of ever-qualified probabilities? No more certainties, no more absolutes. The Harvard philosopher W. V. Quine mused, "I think that for scientific or philosophical purposes the best we can do is give up the notion of knowledge as a bad job. . . ." Not knowing had its ironies as well as its pleasures. For philosophers this was "the post-scholastic era," as a later physicist, John Ziman, put it, "when it seemed essential to (dis)prove the peculiar (un)reality of scientific knowledge (theories/facts/data/hypotheses) by analysing (deconstructing) the arguments on which it was (supposedly) based." Scientists themselves, in the knowledge business, had no use for this mode of discourse. Judged by results, their understanding of nature seemed richer and more efficacious than ever, the quantum paradoxes notwithstanding. They had rescued knowledge from uncertainty after all. "The scientist has a lot of experience with ignorance and doubt and uncertainty," Feynman said. ". . . we take it for granted that it is perfectly consistent to be unsure—that it is possible to live and *not* know. But I don't know whether everyone realizes that this is true."

Feynman's gift to his coworkers was a credo, accreted over time and disbursed both formally and informally, in lectures and books like the 1965 *Character of Physical Law* and in a stance, an attitude, that seemed too natural to constitute a philosophy.

He believed in the primacy of doubt, not as a blemish upon our ability to know but as the essence of knowing. The alternative to uncertainty is authority, against which science had fought for centuries. "Great value of a satisfactory philosophy of ignorance," he jotted on a sheet of notepaper one day. ". . . teach how doubt is not to be feared but welcomed."

He believed that science and religion are natural adversaries. Einstein said, "Science without religion is lame; religion without science is blind." Feynman found this style of accommodation to be intolerable. He repudiated the conventional God: "the kind of a personal God, characteristic of Western religions, to whom you pray and who has something to do with creating the universe and guiding you in morals." Some theologians had retreated from the conception of God as a kind of superperson—Father and King—willful, white-haired, and male. Any God who might take an interest in human affairs was too anthropomorphic for Feynman—implausible in the less and less human-centered universe discovered by science. Many scientists agreed, but his views were so rarely expressed that in 1959 a local television station, KNXT, felt obliged to suppress an interview in which he declared:

> It doesn't seem to me that this fantastically marvelous universe, this tremendous range of time and space and different kinds of animals, and all the different planets, and all these atoms with all their motions, and so on, all this complicated thing can merely be a stage so that God can watch human beings struggle for good and evil—which is the view that religion has. The stage is too big for the drama.

Religion meant superstition: reincarnation, miracles, virgin birth. It replaced ignorance and doubt with certainty and faith; Feynman was happy to embrace ignorance and doubt.

No scientist liked the God of Sunday school stories or the "God of the gaps"—the last-resort explanation for the unexplainable, called on through the ages to fill holes in current knowledge. Those who did turn to faith as a supplement to science preferred grander and less literal gods: "the ground of all that is," as John Polkinghorne, a high-energy physicist turned Anglican priest, said: "Those who are seeking understanding through and through— a natural instinct for the scientist—are seeking God, whether they name him or not." Their God did not fill gaps in the sense of particular lacunae for evolutionary theory or astrophysics—how did the universe begin?—but hovered over whole domains of knowledge: ethics, aesthetics, metaphysics. Feynman conceded the existence of genuine knowledge outside the range of science. He admitted that there were questions science could not answer, but grudgingly: he saw a danger in tying moral guidance to unpalatable myths, as religion did, and he resented the common view that science, with its merciless unraveling and explaining, was an enemy of the emotional appreciation of beauty. "Poets say science takes away from the beauty of the stars—mere globs of gas atoms," he wrote in a famous footnote.

> I too can see the stars on a desert night, and feel them. But do I see less or more? The vastness of the heavens stretches my imagination—stuck on this carousel my little eye can catch one-million-year-old light. A vast pattern—of which I am a part. . . . What is the pattern, or the meaning, or the *why*? It does not do harm to the mystery to know a little about it. For far more marvelous is the truth than any artists of the past imagined it. Why do the poets of the present not speak of it? What men are poets who can speak of Jupiter if he were a man, but if he is an immense spinning sphere of methane and ammonia must be silent?

He believed, too, in an independence of moral belief from any particular theory of the machinery of the universe. An ethical system that depended on faith in a watchful or vengeful God was unnecessarily fragile, prone to collapse when doubt began to undermine faith.

He believed that it was not certainty but freedom from certainty that empowered people to make judgments about right and wrong: knowing that they could never be more than provisionally right, but able to act nonetheless. Only by understanding uncertainty could people learn how to evaluate the many kinds of false knowledge that bombard them: claims of mind reading and spoon bending, belief in flying saucers bearing alien visitors. Science can never disprove such claims, any more than it can disprove God. It can only devise experiments and explore alternative explanations until it gains a commonsense sureness. "I have argued flying saucers with lots of people," Feynman once said. "I was interested in this: they keep arguing that it is possible. And that's true. It is possible. They do not appreciate that the problem is not to demonstrate whether it's possible or not but whether it's going on or not."

How could one evaluate miracle cures or astrological forecasts or telekinetic victories at the roulette wheel? By subjecting them to the scientific method. Look for people who recovered from leukemia *without* having prayed. Place sheets of glass between the psychic and the roulette table. "If it's not a miracle," he said, "the scientific method will destroy it." It was essential to understand coincidence and probability. It was noteworthy that flying-saucer lore involved a considerably greater variety of saucer than of creature: "orange balls of light, blue spheres which bounce on the floor, gray fogs which disappear, gossamer-like streams which evaporate into the air, thin, round flat things out of which objects come with funny shapes that are something like a human being." It was fantastically improbable, he noted, that alien visitors should come in near-human form and just at the moment in history when people discovered the possibility of space travel.

He subjected other forms of science and near-science to the same scrutiny: tests by psychologists, statistical sampling of public opinion. He had developed pointed ways of illustrating the slippage that occurred when experimenters allowed themselves to be less than rigorously skeptical or failed to appreciate the power of coincidence. He described a common experience: an experimenter notices a peculiar result after many trials— rats in a maze, for example, turn alternately right, left, right, and left. The experimenter calculates the odds against something so extraordinary and decides it cannot have been an accident. Feynman would say: "I had the most remarkable experience. . . . While coming in here I saw license plate ANZ 912. Calculate for me, please, the odds that of all the license plates . . ." And he would tell a story from his days in the fraternity at MIT, with a surprise ending.

I was upstairs typewriting a theme on something about philosophy. And I was completely engrossed, not thinking of anything but the theme, when all of a sudden in a most mysterious fashion there swept through my mind the idea: my grandmother has died. Now of course I exaggerate slightly, as you should in all such stories. I just sort of half got the idea for a minute. . . . Immediately after that the telephone rang downstairs. I remember this distinctly for the reason you will now hear. . . . It was for somebody else. My grandmother was perfectly healthy and there's nothing to it. Now what we have to do is to accumulate a large number of these to fight the few cases when it could happen.

Feynman, who had once astonished the Princeton admissions committee with his low scores in every subject but physics and mathematics, did believe in the primacy of science among all the spheres of knowledge. He would not concede that poetry or painting or religion could reach a different kind of truth. The very idea of different, equally valid versions of truth struck him as a modern form of cant, another misunderstanding of uncertainty.

That any particular knowledge—quantum mechanics, for example—must be provisional and imperfect does not mean that competing theories cannot be judged better or worse. He was not what philosophers called a realist— by one definition, someone who, in asserting the existence of, say, electrons, adds "a desk-thumping, foot-stamping shout of 'Really!' " Real though electrons seemed, Feynman and some other physicists recognized that they are part of a never-perfect, always-changing scaffolding. Do electrons *really* travel backward in time? Are those nanosecond resonances *really* particles? Do particles really spin? Do they really have strangeness and charm? Many scientists believed in a straightforward reality. Others, including Feynman, felt that in the late twentieth century it was not necessary or possible to answer a final *yes*. It was preferable to hold one's models delicately in the mind, weighing alternative viewpoints and letting assumptions slide here and there. But to physicists the scaffolding was not *all*. It did imply a truth within, toward which humans might perpetually strive, however imperfectly. Feynman did not believe, as many philosophers did, that the now-famous "conceptual revolutions" or "paradigm shifts" to which science seemed so prone—Einstein's relativity replacing Newton's dynamics— amounted to the replacing of one socially bound fashion by another, like hemlines rising and falling year to year. Like most members of his community, he could not abide in his business what one philosopher, Arthur Fine, called "the great lesson of twentieth-century analytic and continental philosophy, namely, that there *are* no general methodological or philosophical resources for deciding such things." Scientists do have methods. Their theories are provisional but not arbitrary, not mere social constructions. By means of the peculiar stratagem of refusing to acknowledge that any truth may be as valid as any other, they succeed in preventing any truth from becoming as valid as any other. Their approach to knowledge differs from all others—religion, art, literary criticism—in that the goal is never a potpourri of equally attractive realities. Their goal, though it always recedes before them however they approach it, is consensus.

The Swedish Prize

When Einstein won the 1921 Nobel Prize, it did not create a stir. Although Einstein could command front-page coverage in the *New York Times* merely by delivering a public lecture, the detail of the prize impressed the editors only to the extent of a one-sentence notice inside the newspaper, lumping him with the next year's winner, a more obscure professor whose name they misspelled:

The Nobel Committee has awarded the physics prize for 1921 to Professor Dr. Albert Einstein of Germany, identified with the theory of relativity, and that for 1922 to Professor Neils Bohr, Copenhagen.

Gradually the awards gained in stature. Longevity contributed: there were other prizes, but the foresighted Alfred Nobel, inventor of dynamite, had established his early. The particular contributions of scientists grew more difficult to describe to a lay public, and the awarding of such a distinguished international honor provided a useful benchmark. A physicist's obituary in the late twentieth century would almost have to begin with the phrase "won the Nobel Prize for . . ." or the phrase "worked on the atomic bomb," or both. The prize committee arrived at its judgments with care: it made errors, sometimes serious ones, but it generally reflected a conservative consensus of leading scientists in many countries. Scientists began to covet the prize with an intensity that they suppressed as well as they could. Their interest could be felt nonetheless in the ways scientists did and did not discuss the prize. Any potential prizewinner exhibited an extreme reluctance to mention its name. The distinguished group of those who had almost won revealed a forlorn tendency to rehearse for the rest of their lives the slight contingencies that had stood between them and the prize—the indecision that made them delay a paper for a crucial few months, or the timidity that kept them from joining a team embarked on an all-too-promising experiment. Even winners showed how much they cared through small mannerisms, such as the euphemism winkingly employed by Gell-Mann, among others: "the Swedish prize." The winners formed an elite group—but *elite* was too weak a word. A sociologist assessing the prize's stature found herself having to multiply superlatives: "As the *ne plus ultra* of honors in science, the Nobel Prize elevates its recipients not merely to the scientific elite but to the uppermost rank of the scientific ultra-elite, the thin layer of those at the top of the stratification hierarchy of elites who exhibit especially great influence, authority, or power and who generally have the highest prestige within what is a prestigious collectivity to begin with." Physicists always knew who among their colleagues had won and who had not.

Few scientists after Einstein, if any, remained larger than the prize—capable of adding as much to its stature as it added to theirs. In 1965 several active physicists at least seemed to be sure future winners, as much because of their dominance in the community as because of their particular accomplishments. Feynman, Schwinger, Gell-Mann, and Bethe were chief among them. The Nobel committee traditionally found it easier to identify worthy candidates than to pinpoint their most worthy particular achievements. Most notoriously, Einstein had won specifically for his work on the photoelectric effect, not for relativity. When Bethe finally did win, in 1967, the prize singled out his parsing of the thermonuclear reactions in stars—important work, but an arbitrary choice from an unusually broad and influential career spanning decades. Feynman could plausibly have won for his liquid-helium work, had that been his only achievement. Each fall, as the announcement neared, Feynman had been alive to the possibility. He and Gell-Mann might have won for their theory of weak interactions, yet Gell-Mann had already moved on to a more sweeping model of high-energy particle physics. The committee found it easier to reward particular experiments or discoveries, and experimenters tended to win their prizes far more promptly than theorists. Broad theoretical conceptions like relativity were the most difficult of all. Even so, it was odd that the Nobel committee had not yet recognized the theoretical watershed reached almost twenty years before with quantum electrodynamics and renormalization. The experimenters Willis Lamb and Polykarp Kusch had long since been recognized, in 1955, for their contributions to quantum electrodynamics.

No more than three people may share a Nobel Prize. That rule may have added to the complications in the case of quantum electrodynamics. Feynman and Schwinger were two. Tomonaga had matched or anticipated the essence of Schwinger's theory, even if his version had not been quite as panoramic. Dyson was a problem. His contribution had been the most mathematical, and the Nobel Prize abhorred mathematics. Some physicists felt vehemently that Dyson had done no more than analyze and publicize work created by others. Dyson, having settled at the Institute for Advanced Study, drifted away from the theoretical physics community. He had no taste for the involutions of particle physics. He indulged his lifelong passion for space travel by participating in various visionary projects. He grew fascinated with the global politics of nuclear weapons and with the origin of life. The Nobel recommendations of influential American physicists—his old antagonist Oppenheimer among them—may have omitted Dyson, although to a knowledgeable minority it seemed that no one, during the tumultuous birth of modern quantum electrodynamics, had understood the problem more broadly or influenced the community more deeply.

Thus, when the Western Union "telefax" arrived at 9 A.M. on October 21, 1965, it named Feynman, Schwinger, and Tomonaga for their "fundamental work in quantum electrodynamics with deep ploughing consequences for the physics of elementary particles." By then Feynman had been awake for more than five hours. The first call had come at 4 A.M. from a correspondent of the American Broadcasting Corporation shortly after the announcement in Stockholm. He rolled over and told Gweneth. At first she thought he was joking. The telephone kept ringing until finally they left it off the hook. They could not get back to sleep. Feynman knew his life would not be the same. Photographers from the Associated Press and the local newspaper were at his house before sunrise. He posed outdoors in the dark with Carl, his sleepy three-year-old, and gamely held a telephone receiver to his ear as the flashbulbs popped.

Since the press now had to give an account of quantum electrodynamics for the first time, Feynman rapidly learned to field a sequence of variations on what seemed to him a single question: "Will you please tell us what you won the prize for—— but don't tell us! Because we'll not understand it." The actual questions were impossible to answer: "What applications does this paper have in the computer industry?" "I'm going to ask you also to comment on the statement that your work was to convert experimental data on strange particles into hard mathematical fact." And then the one question he *could* answer: "What time did you hear about the award?" In a private moment a reporter for *Time* made a suggestion he loved: that he simply say, "Listen, buddy, if I could tell you in a minute what I did, it wouldn't be worth the Nobel Prize." He realized that he could work up a stock phrase about the interaction of matter and radiation but felt it would be a fraud. He did make a serious remark—and repeated it all day—that reflected his inner feeling about renormalization. The problem had been to eliminate infinities in calculations, he said, and "We have designed a method for sweeping them under the rug."

Julian Schwinger called, and they shared a happy moment. Schwinger, still at Harvard, was pursuing an ever more solitary road in his theoretical physics but, unlike Feynman, had brought forth a long and distinguished string of graduate students working on the frontier problems of high-energy physics. A decade earlier, when Feynman won the Einstein Award, he wrote his mother: "I thought you would be happy that I beat Schwinger out at last, but it turns out he got the thing 3 yrs ago. Of course, he only got ½ a medal, so I guess you'll be happy. You always compare me with Schwinger." Now their rivalry was over, if not forgotten. Feynman called Tomonaga in Japan and then reported to a student journalist a capsule caricature of the Nobel Prize–day telephone conversation:

[FEYNMAN:] Congratulations.
    [TOMONAGA:] Same to you.
How does it feel to be a Nobel Prize winner?
    I guess you know.
Can you explain to me in layman's terms exactly what it was you did to win the prize?
    I am very sleepy.

By afternoon students had raised across the dome of Throop Hall an enormous cloth banner reading, "Win big, RF."

Hundreds of letters and telegrams came in over the next weeks. He heard from childhood friends who had not seen him in almost forty years. There were cables from shipboard and muffled telephone calls from Mexico. He told reporters that he planned to spend his third of the $55,000 prize money to pay his taxes on his other income (actually he used it to buy a beach house in Mexico). He felt himself under stress. He had always felt that honors were suspect. He liked to ridicule pomp and talk about his father, the uniform salesman who taught him to see past the uniforms. Now he would be traveling to Sweden to appear before the king. The mere thought of buying a tuxedo made him nervous. He did not want to bow before a foreign potentate. For several weeks he grew obsessed with an odd fantasy that one was forbidden to turn one's back on the king and therefore had to back up a flight of steps after receiving the award. He practiced jumping backward up steps, both feet at once, because he decided that he would invent a method that no one had used before. He planned to examine the actual steps in advance and rehearse. One friend sent him a rear-view mirror from an automobile as a joke; Feynman took it as evidence that other people knew about this rule. When Sweden's ambassador paid him a courtesy call, Feynman took the opportunity to confess his worry. The ambassador assured him that he could face any direction he chose; no one climbed stairs backward.

In the event, he put on white tie and tails, slicked his hair down, and grinned as he accepted the award from a bespectacled King Gustav VI

Adolf. The prizewinners sped through a week of banquets, dances, formal toasts, and impromptu speeches in Sweden's ornate and palatial civic buildings. They traveled from Stockholm to Uppsala and back, partied with students in a beer cellar, and made conversation with ambassadors and princesses. They collected their medals, certificates, and bank checks. They delivered their Nobel Prize lectures. Feynman realized that he had never read anyone's Nobel lecture. Scientists', especially, seemed automatically obscure. Friends told him about William Faulkner's famous speech in 1950 ("I believe that man will not merely endure: he will prevail"); he did not think he could produce anything so grand, but he wanted to say something memorable, and he did not want to give the précis of quantum electrodynamics that might also be coming from his fellow winners.

He believed that historians, journalists, and scientists themselves all participated in a tradition of writing about science that obscured the working reality, the sense of science as a process rather than a body of formal results. Real science was confusion and doubt, ambition and desire, a march through fog. With hindsight, the polished histories tended to impose a post facto logic on the sequence of reasoning and discovery. The appearance of an idea in the scientific literature and the actual communication of the same idea through the community could be sharply different, Feynman knew. He decided to give a personal, anecdotal, and—he claimed—unpolished version of his route to the space-time view of quantum electrodynamics. "We have a habit in writing articles published in scientific journals to make the work as finished as possible," he began, "to cover up all the tracks, to not worry about the blind alleys or to describe how you had the wrong idea first."

He described the historic difficulty of infinities in the self-interaction of the electron. He confessed his secret desire as a graduate student to eliminate the field altogether—to produce a theory of direct action between charges. He recounted his collaboration with Wheeler: "as I was stupid, so was Professor Wheeler that much more clever." He tried to give his listeners a feeling for what had seemed a new philosophical stance—the willingness of a physicist in the post-Einstein era to accept paradoxes without stopping to say, "Oh, no, how could that be?"—and offered his memory of the way his physical viewpoint had evolved. He repeated his view of renormalization: "I think that the renormalization theory is simply a way to sweep the difficulties of the divergences of electrodynamics under the rug. I am, of course, not sure of that."

He pointed out a remarkable irony of the story. So many of the ideas he nursed on his way to his Nobel Prize–winning work had themselves proved faulty: his first notion that a charge should not act on itself; the whole Wheeler-Feynman half-advanced, half-retarded electrodynamics. Even his path integrals and his view of electrons moving backward in time were only aids to guessing, not essential parts of the theory, he said.

> The method used here, of reasoning in physical terms, therefore, appears to be extremely inefficient. On looking back over the work, I can only feel a kind of regret for the enormous amount of physical reasoning and mathematical re-expression. . . .

But he also believed that the inefficiency, the guessing of equations, the juggling of alternative physical viewpoints were, even now, the key to discovering new laws. He concluded with advice to students:

> The chance is high that the truth lies in the fashionable direction. But, on the off-chance that it is in another direction—a direction obvious from an unfashionable view of field theory—who will find it? Only someone who has sacrificed himself by teaching himself quantum electrodynamics from a peculiar and unfashionable point of view; one that he may have to invent for himself.

He left Stockholm for Geneva, where he repeated the talk before a jubilant, reverent audience at Europe's great new accelerator center, CERN, the European Center for Nuclear Research. He said, standing before them in his new dress suit, that the new laureates had been talking about whether they would ever be able to return to normal. Jacques Monod, who shared the prize for medicine, had declared it was a biological fact that an organism is changed by experience. "I discovered a great difficulty," Feynman said, grinning malevolently. "I always took off my coat in giving a lecture, and I just don't feel like taking it off." As he continued, "I've changed! I've changed!" the audience erupted in laughter and catcalls. He took off the coat.

Once more, he said he would speak as an old man to the young scientists and urge them to break away from the pack. At CERN, as at all the laboratories of high-energy physics, the pack was growing rapidly. Every experiment required enormous teams. Author lists for articles in the *Physical Review* were beginning to take up a comically large portion of the page.

"It will not do you any harm whatever to think in an original fashion," Feynman said. He offered a probabilistic argument.

> The odds that your theory will be in fact right, and that the general thing that everybody's working on will be wrong, is low. But the odds that you, Little Boy Schmidt, will be the guy who figures a thing out, is *not* smaller. . . . It's very important that we do not all follow the same fashion. Because although it is ninety percent sure that the answer lies over there, where Gell-Mann is working, what happens if it doesn't?

"If you give more money to theoretical physics," he added, "it doesn't do any good if it just increases the number of guys following the comet head. So it's necessary to increase the amount of variety . . . and the only way to do it is to implore you few guys to take a risk with your lives that you will never be heard of again, and go off in the wild blue yonder and see if you can figure it out."

Most scientists knew the not-so-amusing metalaw that the receipt of the Nobel Prize marks the end of one's productive career. For many recipients, of course, the end came long before. For others the fame and distinction tend to accelerate the waning of a scientist's ability to give his creative work the time-intensive, fanatical concentration it often requires. Some prize-winners fight back. Francis Crick designed a blunt form letter:

> Dr. Crick thanks you for your letter but regrets that he is unable to accept your kind invitation to:

send an autograph	help you in your project
provide a photograph	read your manuscript
cure your disease	deliver a lecture
be interviewed	attend a conference
talk on the radio	act as chairman
appear on TV	become an editor
speak after dinner	write a book
give a testimonial	accept an honorary degree

Requests in most of these categories now filled Feynman's mail (except that his correspondents tended more toward *hear my theory of the universe* than *cure my disease*). Mature scientists did become laboratory heads, department chairmen, foundation officials, institute directors. Victor Weisskopf, one of those whom the prize had just barely eluded, was now director of CERN, and he thought Feynman, too, would be driven willy-nilly into administration. He goaded Feynman into accepting a wager, signed before witnesses: "Mr. FEYNMAN will pay the sum of TEN DOLLARS to Mr. WEISSKOPF if at any time during the next TEN YEARS (i.e. before the THIRTY FIRST DAY of DECEMBER of the YEAR ONE THOUSAND NINE HUNDRED AND SEVENTY FIVE), the said Mr. FEYNMAN has held a 'responsible position.'" They had no disagreement about what that would mean:

> For the purpose of the aforementioned WAGER, the term "responsible position" shall be taken to signify a position which, by reason of its nature, compels the holder to issue instructions to other persons to carry out certain acts, notwithstanding the fact that the holder has no understanding whatsoever of that which he is instructing the aforesaid persons to accomplish.

Feynman collected the ten dollars in 1976.

He already tried to avoid encumbrances as though every invitation, honor, professional membership, or knock at his door were another vine wrapping itself around his creative center. By the time he won the Nobel Prize he had been trying for five years to resign from the National Academy of Sciences. This simple task was taking on a life of its own. He began by scribbling a note with his dues bill: he paid the forty dollars, but he resigned. Almost a year later he received a personal letter from the academy's president, the biologist Detlev W. Bronk (whose original paper on the single nerve impulse he had read as a Princeton student). He felt obliged to write a polite explanation:

> My desire to resign is merely a personal one; it is not meant as a protest of any kind. . . . My peculiarity is this: I find it psychologically very distasteful to judge people's "merit." So I cannot participate in the main activity of selecting people for membership. To be a member of a group, of which an important activity is to choose others deemed worthy of membership in that self-esteemed group, bothers me. . . .
>
> Maybe I don't explain it very well, but suffice to say that I am not happy as a member of a self-perpetuating honorary society.

It was 1961. Bronk let Feynman's letter sit for months. Then he answered with calculated obtuseness:

Thank you for your willingness to continue as a member of the academy. . . . I have done my best to reduce the emphasis on the "honor" of election. . . . I am grateful that you will continue a member at least during my last year as president.

Eight years later, Feynman was still trying. He re-resigned. A reply came from the president-elect, Philip Handler, who mused talmudically, "I suppose that we truly have no alternative, in the sense that surely the Academy must adhere to your wishes," and deftly slid Feynman's resignation into the subjunctive mood:

I would consider your resignation a most sorrowful event indeed. . . . I write to hope that you will reconsider. . . . I am reluctant to endorse such an action. . . . Before processing your request, a procedure for which I trust that the Office of the Home Secretary is in some manner prepared, I very much hope that you will let us hear from you further. . . .

Feynman wrote again, as plainly as he could. Handler replied:

I have your somewhat cryptic note. . . . We are seeking to increase the meaningful roles of the Academy. . . . Wouldn't you rather join us in that effort?

Finally, by 1970, Feynman's resignation began to seem real even to the academy, though he continued to hear from scientists who wondered whether he would confirm the rumor and explain why.

He turned down honorary degrees offered by the University of Chicago and by Columbia University and thus finally kept the promise he had made to himself on the day he received his doctorate from Princeton. He turned down hundreds of other propositions with a curtness that impressed even his protective secretary. To a book publisher who had invited him to "introduce a draft of fresh air into a rather stuffy area," he wrote: "No sir. The area is stuffy from too much hot air already." He refused to sign petitions and newspaper advertisements; the Vietnam War was now drawing the opposition of many scientists, but he would not join them publicly. Feynman, Nobel laureate, found that even canceling a magazine subscription took an entire correspondence. "Dear Professor Feynman," began a long letter from the editor of *Physics Today*, the magazine whose second issue had carried his article about the Pocono conference in 1948:

The comment you sent back with our questionnaire on our May issue ("I never read your magazine. I don't know why it is published. Please take me off your mailing list. I don't want it.") poses some interesting questions for us. . . .

Four hundred words later, the editor had not given up:

I apologize for asking any more of your time, but all of us at *Physics Today* will appreciate it very much if we can have amplification of your earlier comments.

So Feynman amplified:

Dear Sir,

I'm not "physicists," I'm just me. I don't read your magazine so I don't know what's in it. Maybe it's good, I don't know. Just don't send it to me. Please remove my name from the mailing list as requested. What other physicists need or don't need, want or don't want, has nothing to do with it. . . . It was not my intention to shake your confidence in your magazine—nor to suggest that you stop publication—only that you stop sending it here. Can you do that please?

He was hardening his shell. He knew he could seem cold. His secretary, Helen Tuck, protected him, sometimes sending away visitors while Feynman hid behind her door. Or he would just shout at a hopeful student to go away—he was working. He almost never participated in the business of his department at Caltech: tenure decisions, grant proposals, or any of the other administrative chores that constitute overhead on most scientists' time. Caltech's divisions, like the science departments at every American university, were largely financed through a highly structured process of applications to the Department of Energy, the Department of Defense, and other government agencies. There were group applications and individual applications, supporting salaries, students, equipment, and overhead. At Caltech a senior professor who could arrange to have the air force, for example, pay a portion of his salary was rewarded with a discretionary kitty with which he could travel, buy a computer, or support a graduate student. Alone at Caltech, and virtually alone in physics, Feynman was humored in his refusal to participate in this process. To some colleagues he seemed selfish. It occurred to the historian of science Gerald Holton, however, that

Feynman had put on a kind of hair shirt. "It must have been very difficult to live that way," Holton said. "It does not come easy to make that conscious decision to remain unadulterated. Culture by definition is very seductive. He was a Robinson Crusoe in the big city, and that isn't easy to do." I. I. Rabi once said that physicists are the Peter Pans of the human race. Feynman clutched at irresponsibility and childishness. He kept a quotation from Einstein in his files about the "holy curiosity of inquiry": "this delicate little plant, aside from stimulation, stands mainly in need of freedom; without this it goes to wrack and ruin without fail." He protected his freedom as though it were a dying candle in a hard wind. He was willing to risk hurting his friends. Hans Bethe turned sixty the year after Feynman won his Nobel Prize, and Feynman refused to send a contribution to the customary volume of articles in his honor.

He was frightened. In the years after the prize he felt uncreative. His Caltech colleague David Goodstein traveled with him to the University of Chicago when he went to address the undergraduates there early in 1967. Goodstein thought he seemed depressed and worried. When Goodstein came down to breakfast at the faculty club, he found Feynman already there, talking with someone who Goodstein gradually realized was the codiscoverer of DNA, James Watson. Watson gave Feynman a manuscript tentatively titled *Honest Jim*. It was a tame memoir by later standards, but when it was published—under a different title, *The Double Helix*—it caused an enormous popular stir. With a candor that shocked many of Watson's colleagues, it portrayed the ambition, the competitiveness, the blunders, the miscommunications, and the raw excitement of real scientists. Feynman read it in his room at the Chicago faculty club, skipping the cocktail party in his honor, and found himself moved. Later he wrote Watson:

> Don't let anybody criticize that book who hasn't read it through to the end. Its apparent minor faults and petty gossipy incidents fall into place as deeply meaningful. . . . The people who say "that is not how science is done" are wrong. . . . When you describe what went on in *your* head as the truth haltingly staggers upon you and passes on, finally fully recognized, you are describing how science *is* done. I know, for I have had the same beautiful and frightening experience.

Late that night in Chicago he startled Goodstein by pressing the book into his hands and telling him he had to read it. Goodstein said he would look forward to it. No, Feynman said. You have to read it *now*. So Goodstein did, turning pages until dawn as Feynman paced nearby or sat and doodled on a sheet of paper. At one point Goodstein remarked, "You know, it's amazing that Watson made this great discovery even though he was so out of touch with what everyone in his field was doing."

Feynman held up the paper he had been writing on. Amid scribbling and embellishments he had inscribed one word: *DISREGARD*.

"That's what I've forgotten," he said.

Quarks and Partons

In 1983, looking back on the evolution of particle physics since the now-historic Shelter Island conference, Murray Gell-Mann said, uncontroversially, that he and his colleagues had developed a theory that "works." He summed it up in one intricately crafted sentence (rather more refined than "All things are made of atoms . . ."):

> It is of course a Yang-Mills theory, based on color $SU(3)$ and electroweak $SU(2) \times U(1)$, with three families of spin ½ leptons and quarks, their antiparticles, and some spinless Higgs bosons in doublets and antidoublets of the weak isotopic spin to break the electroweak group down to U_1 of electromagnetism.

His listeners recognized vintage Gell-Mann, from the "of course" onward. For aficionados there was a poetry in the jargon, much of which Gell-Mann had invented personally. He loved language more than ever. As always, during the next hour he punctuated his physics with a stream of abstruse and punning nomenclatural asides: "By the way, some people have called the higglet by another name [holds up a box of Axion laundry presoak], in which case it's extremely easy to discover in any supermarket"; ". . . many physicists—Dimopoulos, Nanopoulos, and Iliopoulos, and for the benefit of my French friends I add Rastopopoulos"; ". . . O'Raifertaigh. (His name, by the way, is written in a simplified manner; the 'f' should really be 'thbh')"; and so on.

Some people found his style irritating—among them, those whose names he tried to correct—but that was a minor detail. Gell-Mann, more than any other physicist of the sixties and seventies, defined the mainstream of the physics that Feynman had reminded himself to *disregard*. In so many ways these two scientific icons had come to seem like polar opposites—the Adolphe Menjou and Walter Matthau of theoretical physics. Gell-Mann loved to know things' names and to pronounce them correctly—so correctly that Feynman would misunderstand, or pretend to misunderstand, when Gell-Mann uttered so simple a name as *Montreal.* Gell-Mann's conversational partners often suspected that the obscure pronunciations and cultural allusions were designed to place them at a disadvantage. Feynman pronounced *potpourri* "pot-por-eye" and *interesting* as if it had four syllables, and he despised nomenclature of all kinds. Gell-Mann was an enthusiastic and accomplished bird-watcher; the moral of one of Feynman's classic stories about his father was that the name of a bird did not matter, and the point was hardly lost on Gell-Mann.

Physicists kept finding new ways to describe the contrast between them. Murray makes sure you know what an extraordinary person he is, they would say, while Dick is not a person at all but a more advanced life form pretending to be human to spare your feelings. Murray was interested in almost everything—but not the branches of science outside high-energy physics; he was openly contemptuous of those. Dick considered all science to be his territory—his responsibility—but remained brashly ignorant of everything else. Some well-known physicists resented Feynman for his cherished irresponsibility—it was, after all, irresponsibility to his academic colleagues. A larger number disliked Gell-Mann for his arrogance and his sharp tongue.

There was always more. Dick wore shirtsleeves, Murray wore tweed. Murray ate at the Atheneum, the faculty club, while Dick ate at "the Greasy," the cafeteria. (This was only half true. Either man could be found at either place on occasion, although Feynman, when the Atheneum still required ties and jackets, would show up in shirtsleeves and demand the most garish and ill-fitting of the spare items kept on hand for emergencies.) Feynman talked with his hands—with his whole body, in fact—whereas Gell-Mann, as the physicist and science writer Michael Riordan observed, "sits calmly behind his desk in a plush blue swivel chair, hands folded, never once lifting them to make a gesture. . . . Information is exchanged by words and numbers, not by hands or pictures." Riordan added:

> Their personal styles spill over into their theoretical work, too. Gell-Mann insists on mathematical rigor in all his work, often at the expense of comprehensibility. . . . Where Gell-Mann disdains vague, heuristic models that might only point the way toward a true solution, Feynman

revels in them. He believes that a certain amount of imprecision and ambiguity is *essential* to communication.

Yet they were not so different in their approach to physics. Those who knew them best as physicists felt that Gell-Mann was no more likely than Feynman to hide behind formalism or to use mathematics as a stand-in for physical understanding. Those who considered him pretentious about language and cultural trivia felt nonetheless that when it came to physics he was as honest and direct as Feynman. Over a long career Gell-Mann made his vision not only comprehensible but irresistible. Both men were relentless on the trail of a new idea, able to concentrate absolutely, willing to try anything.

Both men, it seemed to a few perceptive colleagues, presented a mask to the world. "Murray's mask was a man of great culture," Sidney Coleman said. "Dick's mask was Mr. Natural—just a little boy from the country that could see through things the city slickers can't." Both men filled their masks until reality and artifice became impossible to pry apart.

Gell-Mann, as naturalist, collector, and categorizer, was well primed to interpret the exploding particle universe of the 1960s. New technology in the accelerators—liquid hydrogen bubble chambers and computers for automating the analysis of collision tracks—seemed to have spilled open a bulky canvas bag from which nearly a hundred distinct particles had now tumbled forth. Gell-Mann and, independently, an Israeli theorist, Yuval Ne'eman, found a way in 1961 to organize the various symmetries of spins and strangeness into a single scheme. It was a group, in the mathematicians' sense of the word, known as $SU(3)$, though Gell-Mann quickly and puckishly dubbed it the Eightfold Way. It was like an intricate translucent object which, when held to the light, would reveal families of eight or ten or possibly twenty-seven particles—and they would be different, though overlapping, families, depending on which way one chose to view it. The Eightfold Way was a new periodic table—the previous century's triumph in classifying and thus exposing the hidden regularities in a similar number of disparate "elements." But it was also a more dynamic object. The operations of group theory were like special shuffles of a deck of cards or the twists of a Rubik's cube.

Much of $SU(3)$'s power came from the way it embodied a concept increasingly central to the high-energy theorist's way of working: the concept of inexact symmetry, almost symmetry, near symmetry, or—the term that won out—*broken symmetry*. The particle world was full of near misses in its symmetries, a dangerous problem, since it seemed to permit an ad hoc escape route whenever an expected relationship failed to match. *Broken symmetry* implied a process, a change in status. A symmetry in water is broken when it freezes, for now the system does not look the same from every direction. A magnet embodies symmetry breaking, since it has made a kind of choice of orientation. Many of the broken symmetries of particle physics came to seem like choices the universe made when it condensed from a hot chaos into cooler matter, spiked as it is with so many hard-edged, asymmetrical contingencies.

Once again Gell-Mann trusted his scheme enough to predict, as a consequence of broken symmetry, a specific hitherto-unseen particle. This, the omega minus, duly turned up in 1964—a thirty-three-experimenter team had to canvass more than one million feet of photographs—and Gell-Mann's Nobel Prize followed five years later.

His next, most famous invention came in an effort to add explanatory understanding to the descriptive success of the Eightfold Way. $SU(3)$ should have had, along with its various eight-member and ten-member and other families, a most-basic three-member family. This seemed a strange omission. Yet the rules of the group would have required this threesome to carry fractional electric charges: $\frac{2}{3}$ and $-\frac{1}{3}$. Since no particle had ever turned up with anything but unit charge, this seemed implausible even by modern standards. Nevertheless, in 1963 Gell-Mann and, independently, a younger Caltech theorist, George Zweig, proposed it anyway. Zweig called his particles *aces*. Gell-Mann won the linguistic battle once again: his choice, a croaking nonsense word, was *quark*. (After the fact, he was able to tack on a literary antecedent when he found the phrase "Three quarks for Muster Mark" in *Finnegans Wake*, but the physicist's quark was pronounced from the beginning to rhyme with "cork.")

It took years for Gell-Mann and other theorists to generate all the contrivances needed to make quarks work. One contrivance was a new property called *color*—purely artificial, with no connection to everyday color. Another was *flavor*. Gell-Mann decided that the flavors of quarks would be called *up*, *down*, and *strange*. There had to be antiquarks and anticolors. A new mediating particle called the *gluon* would have to carry color from one quark to another. All this encouraged skepticism among physicists. Julian Schwinger wrote that he supposed such particles would be detected by "their palpitant piping, chirrup, croak, and quark." Zweig, far more vulnerable than Gell-Mann, felt that his career was damaged. The quark theorists had to wrestle with the fact that their particles never appeared anywhere, though people did begin a dedicated search in particle accelerators and supposed cosmic-ray deposits in undersea mud.

There was a reality problem, distinctly more intense than the problem posed by more familiar entities such as electrons. Zweig had a concrete, dynamical view of quarks—too mechanistic for a community that had learned as far back as Heisenberg to pay attention only to *observables*. Gell-Mann's comment to Zweig was, "The concrete quark model—that's for blockheads." Gell-Mann was wary of the philosophical as well as the sociological problem created by any assertion one way or the other about quarks being real. For him quarks were at first a way of making a simple toy field theory: he would investigate the theory's properties, abstract the appropriate general principles, and then throw away the theory. "It is fun to speculate about the way quarks would behave if they were physical particles (instead of purely mathematical entities as they would be in the limit of infinite mass)," he wrote. As *if* they were physical particles; then again, *as if* they were conveniences of mathematics. He encouraged "a search for stable quarks"—but added with one more twist that it "would help reassure us of the nonexistence of real quarks." His initial caveats were quoted by commentators again and again in the years that followed. One physicist's typically uncharitable interpretation: "I always considered that to be a coded message. It seemed to say, 'If quarks are not found, remember I never said they would be; if they are found, remember I thought of them first.' " For Gell-Mann this became a permanent source of bitterness.

Feynman, meanwhile, had disregarded so much of the decade's high-energy physics that he had to make a long-term project of catching up. He tried to pay more attention to experimental data than to the methods and language of theorists. He tried, as always, to read papers only until he understood the issue and then to work out the problem for himself. "I've always taken an attitude that I have only to explain the regularities of nature—I don't have to explain the methods of my friends," he told a historian during these years. He did manage to avoid some passing fashions. Still, he was turning back to a community after having drifted outside, and he had to learn its shared methods after all. It was no longer possible to approach these increasingly formidable, specialized problems as an outsider. He had stopped teaching high-energy physics; in the late sixties he began again. At first his syllabus contained no quarks.

By the late sixties and early seventies a new accelerator embedded in the rolling hills near Stanford University in northern California had taken the dominant role in the strong-interaction experiments that were so central to the search for quarks. The Stanford Linear Accelerator Center (SLAC) made a straight two-mile cut in the grassy landscape. Aboveground, cows grazed and young physicists in jeans and shirts—nearly a hundred of them—sat at picnic tables or walked in and out of the center's many buildings. Below, inside a knife-straight evacuated copper tube, a beam of electrons streamed toward targets of protons. The electrons achieved energies far greater than theorists had ever had to manage. They struck their targets inside an end station like a giant airplane hangar and then, with luck, entered a detector inside a concrete blockhouse, lined with lead bricks, riding on railroad tracks and angled upward toward the ceiling. Sometimes high-speed motion-picture cameras recorded the results, and elsewhere in the laboratory teams of human scanners guided an automatic digitizer that could read the particle tracks from—for a given monthlong experiment— hundreds of millions of filmed images. A single bubble chamber at the end of the particle beam, in its five-and-a-half-year useful lifetime, saw the discovery of seventeen new particles.

It was a tool for exploring the strong force—so called because, at the very short distances in the domain of the nucleus, it must dominate the force of electromagnetic repulsion to bind protons and neutrons (*hadron* was now the general term for particles that felt the strong force). Feynman had been thinking about how to understand the working of the strong force in collisions of hadrons with other hadrons. These were complex: at the high energies now available for studying short distances, hadron-hadron collisions produced gloriously messy sprays of detritus. The hadrons themselves were neither simple nor pointlike. They had size, and they seemed to have internal constituents—a whole swarming zoo of them. As Feynman said, the hadron-hadron work was like trying to figure out a pocket watch by smashing two of them together and watching the pieces fly out. He began visiting SLAC regularly in the summer of 1968, however, and saw how much simpler was the interaction offered by electron-proton collisions, the electron tearing through the proton like a bullet.

He stayed with his sister; she had moved to the Stanford area to work for a research laboratory, and her house was just across Sand Hill Road from the accelerator center. The physicists who would gather on the outdoor patio to listen to his stories that summer would see him slamming his open hands together in a boisterous illustration of a new idea he had. He was talking about "pancakes"—flat particle pancakes with hard objects embedded in them.

The Caltech connection was important to experimenters at SLAC, and by the late sixties the connection meant Gell-Mann far more than Feynman. Gell-Mann had created the scientific subculture of current algebra, the mathematical framework surrounding his quarks, and SLAC theorists thought of themselves as trying to generalize these tools to smaller distances, higher energies. At accelerators like SLAC, most of the thinking focused on the simplest reactions—two particles in, two particles out—although most of the actual collisions produced enormous flashes of many more particles. Experimenters wanted the most precise possible data, and precision was impossible in these bursts of detritus. Feynman chose a different point of view. He introduced a formalism in which one could look at the distributions of twenty or fifty or more particles. One did not have to be able to measure the momentum of each particle; in effect one could sum over all the possibilities. A Stanford theorist, James D. Bjorken, had been thinking along similar lines. An electron hits a proton; an electron comes out, along with a burst of immeasurable fragments. The emerging electron was a common factor. Bjorken decided to set aside the miscellaneous spray and simply plot the distribution of the energies and angles of the emerging electrons, averaged over many collisions.

He isolated a remarkable regularity in the data, a phenomenon he called "scaling"—the data looked the same at different energy scales. He did not know just how to interpret this. He had a variety of guesses, most framed in the language of current algebra. When Feynman arrived, Bjorken happened to be away; Feynman saw the graphed data without hearing a clear explanation of its origin. He suddenly recognized it, however, and he calculated long into the evening. It could be viewed as a graph of his pancake theory, the theory he had been toying with all summer on his own.

He had decided to cut through the incalculable swarming muddle of proton pieces by positing a mysterious new constituent that he called a *parton*, a name based inelegantly on the word *part*. (Finally he had an entry of his own in the *Oxford English Dictionary*.) Feynman made almost no assumptions about his partons except two: they were pointlike, and they did not interact meaningfully with one another but floated freely about inside the proton. They were an abstraction—just the kind of unobservable entity that physicists hoped not to have to fall back on—yet they were tantalizingly visual in spirit. They were pegs on which to hang a field theory of the old, manageable sort, with wave functions and calculable probability amplitudes. By analogy, quantum electrodynamics had its partons, too: the bare electrons and photons.

Feynman showed that collisions with these hard nuggets inside the proton would produce the scaling relations in a natural manner, unlike collisions with the puffy whole proton. He chose not to decide what quantum numbers they did or did not carry, and he most emphatically decided not to worry one way or the other about whether his partons were the fractionally charged quarks of Gell-Mann and Zweig.

By the time Bjorken returned, he found the theory group awash in partons. Feynman buttonholed him. He had idolized Feynman ever since taking an old-fashioned, historically organized quantum electrodynamics course at Stanford. "When Feynman diagrams arrived," he said, "it was the sun breaking through the clouds, complete with rainbow and pot of gold. Brilliant! Physical and profound!" Now here was Feynman in the flesh, explaining Bjorken's own theory to him with a new language and a new visual image. As he could instantly see, Feynman's essential insight was to place himself once again *in* the electron, to see what the electron would see at light speed. He would see the protons flashing toward him—and they were therefore flattened relativistically into pancakes. Relativity also slowed their internal clocks, in effect, and, from the electron's point of view, froze the partons into immobility. His scheme reduced the messy interaction of an electron with a fog of different particles to a much simpler interaction of an electron with a single pointlike parton emerging from the fog. Bjorken's scaling pattern flowed directly from the physics of this picture. The experimenters grasped it instantly.

The parton model was oversimplified. It explained nothing that Bjorken could not explain, although Bjorken's explanation *seemed* less fundamental. Partons required considerable hand-waving. Yet physicists clutched at them like a lifeboat. Three years passed before Feynman published a formal paper and many more before his partons finally and definitively blended with quarks in the understanding of physicists.

Zweig's aces, Gell-Mann's quarks, and Feynman's partons became three paths to the same destination. These constituents of matter served as the quanta of a new field, finally making possible a field theory of the strong force. Quarks had not been seen or detected in the direct fashion of more venerable particles. They became real nonetheless. Feynman took on a project in 1970 with two students, assembling a vast catalog of particle data in an effort to make a judgment about whether a simple quark model could underlie it all. He chose an unconventional model once again, using data that let him think in terms of the electromagnetic field theory of the last generation, instead of the hadron-collision data that interested most theorists. For whatever reason, he was persuaded—converted into a quarkerian, as he said—although he continued to stress the tentativeness of any one model. "A quark picture may ultimately pervade the entire field of hadron physics," this paper concluded. "About the paradoxes of the quark model we have nothing to add, except perhaps to make these paradoxes more poignant by exhibiting the mysteriously good fit of a peculiar model." Younger theorists learned how to explain confinement—the quark's inability to appear as free particles—in terms of a force that grew rapidly with distance, in strange contrast to forces such as gravity and electromagnetism. Quarks became *real* not only because ingenious experiments gave an indirect look at them, but because it became harder and harder for theorists to construct a coherent model in which they did not figure. They became so real that Gell-Mann, their inventor, had to endure the after-the-fact criticism that he had not fully believed in them. He never understood why Feynman had created his own alternative quark and maintained a distinction that faded in the end. He missed no opportunity to call Feynman's particles "put-ons." Like Schwinger years before, he disliked the fanfare over a picture that he thought was oversimplified—anyone could use it.

Quarks were *real*, at least to physicists of the last years of this century. Partons were not, in the end. What *is* real? Feynman tried to keep this question from disappearing into the background. In a book assembled from his lectures, *Photon-Hadron Interactions*, he concluded:

We have built a very tall house of cards making so many weakly based conjectures one upon the other. . . . Even if our house of cards survives and proves to be right we have not thereby proved the existence of partons. . . . On the other hand, the partons would have been a useful psychological guide . . . and if they continued to serve this way to produce other valid expectations they would of course begin to become "real," possibly as real as any other theoretical structure invented to describe nature.

Once again Feynman had placed himself at the center of modern theoretical physics. His language, his framework, dominated high-energy physicists' discourse for several years. He wanted to move on again, or so he told himself. "I'm a little bit frustrated," he said to a historian soon after he published his first parton paper.

I'm tired of thinking of the same thing. I need to think of something else. Because I got stuck—— see, if it would keep going it would be all right, but it's hard to get any new results. . . . This parton thing has been so successful that I have become fashionable. I have to find an unfashionable thing to do.

Feynman routinely refused to recommend colleagues for the Nobel Prize, but he broke his rule in 1977—after Gell-Mann had already won the prize once—and quietly nominated Gell-Mann and Zweig for their invention of quarks.

Teaching the Young

RICHARD. [*Humming softly to himself*] Jee-jee-jee-ju-ju. Jee-jee-jee-ju-ju. [*He is working. Dishes are being cleared from the breakfast table. A tape recorder makes a faint whirr as it eavesdrops: a friend has taken to leaving it running in hopes of capturing stories about Feynman's past.*] Jee-jee-jee-ju-ju. [*Stops abruptly.*] There's some fool has made a mistake here. Some damn fool made a mistake here.

MICHELLE. Prob'ly you.

RICHARD. Me? What do you mean, me? [*Pause.*] Some idiot has made an error. [*Sings*] I have an idiot here who made an error.

MICHELLE. Yeah—you!

RICHARD. Michelle, dear, be careful what you say. After all your father is a nice fellow and he doesn't want that kind of trouble. [*Pause.*] He's made a mis-too-ko. You know, mistookos happen. You know. You don't want your daddy to be a bad boy. [*Drums a sharp tattoo with his fingers.*] That is of course wrong! As any fool can see.

It took years for Feynman's children to realize that their father was not like other fathers. He seemed normally distracted, lounging in his dog-chewed recliner or lying on the floor, writing on notepads, humming to himself in flights of concentration that were hard to break through. He doted on them and told them fantastically imaginative stories. In one ongoing saga they became tiny inhabitants of a gigantic household world; Feynman would describe the forest of brown leafless trees rising around them, for example, until suddenly they would guess that those were the fibers of the carpet. Or he would hold them on his lap and say, "What do you know about? You know about concrete and you know about rubber and you know about glass . . ." He taught them what he considered the basics of economics: that when prices go up, people buy less; that manufacturers set prices to maximize profits; that economists know very little. There were times when they thought he had been placed on earth mainly to embarrass them in public—pretending to beat them about the head with a newspaper or talking to waiters in his mock Italian. He was always what Michelle thought of as borderline boisterous, singing and whistling to himself. He would make up rhymes under his breath as he walked around the house—"I'm going to pick up my shoe, that's what I'm going to do"—and when challenged he would be unable to repeat what he had just said. Belatedly it dawned on them that not all their friends could look up their fathers in the encyclopedia. His own mother was still alive, and he seemed to revert to a child in her presence. Lucille would say, "Richard, I'm cold—would you please put on a sweater?" When *Omni* magazine called him the world's smartest man, she remarked, "If that's the world's smartest man, God help us."

Carl showed an early gift for science, to Feynman's immense delight. When he was twelve, Feynman showed him an odd-looking photograph he had brought home from a Canadian laboratory and Carl guessed—correctly—that it was "probably a diffraction pattern from a laser from a regular pattern of square holes," and Feynman could not help boasting to a friend, "I could have killed him—I was afraid to ask him for the focal length of the lens used!" He tried not to prod too clumsily, and he told himself that he would be happy with any careers his children chose ("trumpet playing—social worker—zygophalatelist—or whatever," he wrote Carl), as long as they were happy and good at what they did. When Carl reached college, however—MIT—he found the one career ambition guaranteed to break his father's equilibrium. "Well," Feynman wrote, "after much effort at understanding I have gradually begun to accept your decision to become a philosopher." But he hadn't. He felt as betrayed and put upon as a business executive whose child wants to be a poet.

I find myself asking, "How can you be a good philosopher?" I see now that, like the poet son who never thinks of money (because he expects his old man to pay) you have chosen philosophy, over clear thought (and so your old man goes on with his clear thoughts) so that you can fly above common sense to far higher and more beautiful aspects of the intellect.

"Well," he added sarcastically, "it must be wonderful to be able to do that."

Educating his children made him think again about the elements of teaching and about the lessons his own father had taught. By the time Carl was four, Feynman was actively lobbying against a first-grade science book proposed for California schools. It began with pictures of a mechanical wind-up dog, a real dog, and a motorcycle, and for each the same question: "What makes it move?" The proposed answer—"Energy makes it move"—enraged him.

That was tautology, he argued—empty definition. Feynman, having made a career of understanding the deep abstractions of *energy*, said it would be better to begin a science course by taking apart a toy dog, revealing the cleverness of the gears and ratchets. To tell a first-grader that "energy makes it move" would be no more helpful, he said, than saying "God makes it move" or "moveability makes it move." He proposed a simple text for whether one is teaching ideas or mere definitions:

> You say, "Without using the new word which you have just learned, try to rephrase what you have just learned in your own language. Without using the word *energy*, tell me what you know now about the dog's motion."

Other standard explanations were just as hollow: *gravity makes it fall*, or *friction makes it wear out*. Having tried to impart fundamental knowledge to Caltech freshmen, he also believed it was possible to teach real knowledge to first-graders. "Shoe leather wears out because it rubs against the sidewalk and the little notches and bumps on the sidewalk grab pieces and pull them off." That is knowledge. "To simply say, 'It is because of friction,' is sad, because it's not science."

Feynman taught thirty-four formal courses during his Caltech career, roughly one a year. Most were graduate seminars called Advanced Quantum Mechanics or Topics in Theoretical Physics. That often meant his current research interest: graduate students sometimes heard, without realizing it, the first and last report of substantial work that another physicist would have published. For almost two decades he also taught a course, listed in no catalog, known as Physics X: one afternoon a week, undergraduates would gather to pose any scientific question they wished, and Feynman would improvise. His effect on these students was immense; they often left the Lauritsen Laboratory basement feeling that they had had a private pipeline to an oracle with an earthy kind of omniscience. He believed—in the face of the increasing esotericism of his own subject—that true understanding implied a kind of clarity. A physicist once asked him to explain in simple terms a standard item of the dogma, why spin-one-half particles obey Fermi-Dirac statistics. Feynman promised to prepare a freshman lecture on it. For once, he failed. "I couldn't reduce it to the freshman level," he said a few days later, and added, "That means we really don't understand it."

It was his own children, however, who crystallized many of his attitudes toward teaching. In 1964 he had made the rare decision to serve on a public commission, responsible for choosing mathematics textbooks for California's grade schools. Traditionally this commissionership was a sinecure that brought various small perquisites under the table from textbook publishers. Few commissioners—as Feynman discovered—read many textbooks, but he determined to read them all, and had scores of them delivered to his house. This was the era of the so-called new mathematics in children's education: the much-debated effort to modernize the teaching of mathematics by introducing such high-level concepts as set theory and non-decimal number systems. New math swept the nation's schools startlingly fast, in the face of parental nervousness that was captured in a *New Yorker* cartoon: "You see, Daddy," a little girl explains, "this set equals all the dollars you earned; your expenses are a sub-set within it. A sub-set of *that* is your deductions."

Feynman did not take the side of the modernizers. Instead, he poked a blade into the new-math bubble. He argued to his fellow commissioners that sets, as presented in the reformers' textbooks, were an example of the most insidious pedantry: new definitions for the sake of definition, a perfect case of introducing words without introducing ideas. A proposed primer instructed first-graders: "Find out if the set of the lollipops is equal in number to the set of the girls." Feynman described this as a disease. It removed clarity without adding any precision to the normal sentence: "Find out if there are just enough lollipops for the girls." Specialized language should wait until it is needed, he said, and the peculiar language of set theory never *is* needed. He found that the new textbooks did not reach the areas in which set theory does begin to contribute content beyond the definitions: the understanding of different degrees of infinity, for example.

> It is an example of the use of words, new definitions of new words, but in this particular case a most extreme example because *no facts whatever* are given. . . . It will perhaps surprise most people who have studied this textbook to discover that the symbol $\cup$ or $\cap$ representing union and intersection of sets . . . all the elaborate notation for sets that is given in

these books, almost never appear in any writings in theoretical physics, in engineering, business, arithmetic, computer design, or other places where mathematics is being used.

Feynman could not make his real point without drifting into philosophy. It was crucial, he argued, to distinguish *clear* language from *precise* language. The textbooks placed a new emphasis on precise language: distinguishing "number" from "numeral," for example, and separating the symbol from the real object in the modern critical fashion—pilpul for schoolchildren, it seemed to Feynman. He objected to a book that tried to teach a distinction between a ball and a picture of a ball—the book insisting on such language as "color the picture of the ball red."

"I doubt that any child would make an error in this particular direction," Feynman said dryly.

As a matter of fact, it is impossible to be precise . . . whereas before there was no difficulty. The picture of a ball includes a circle and includes a background. Should we color the entire square area in which the ball image appears all red? . . . Precision has only been pedantically increased in one particular corner when there was originally no doubt and no difficulty in the idea.

In the real world, he pointed out once again, absolute precision is an ideal that can never be reached. Nice distinctions should be reserved for the times when doubt arises.

Feynman had his own ideas for reforming the teaching of mathematics to children. He proposed that first-graders learn to add and subtract more or less the way he worked out complicated integrals—free to select any method that seems suitable for the problem at hand. A modern-sounding notion was, *The answer isn't what matters, so long as you use the right method.* To Feynman no educational philosophy could have been more wrong. The answer is all that does matter, he said. He listed some of the techniques available to a child making the transition from being able to count to being able to add. A child can combine two groups into one and simply count the combined group: to add 5 ducks and 3 ducks, one counts 8 ducks. The child can use fingers or count mentally: 6, 7, 8. One can memorize the standard combinations. Larger numbers can be handled by making piles—one groups pennies into fives, for example—and counting the piles. One can mark numbers on a line and count off the spaces—a method that becomes useful, Feynman noted, in understanding measurement and fractions. One can write larger numbers in columns and carry sums larger than 10.

To Feynman the standard texts seemed too rigid. The problem 29 + 3 was considered a third-grade problem, because it required the advanced technique of carrying; yet Feynman pointed out that a first-grader could handle it by thinking 30, 31, 32. Why should children not be given simple algebra problems (2 times what plus 3 is 7?) and encouraged to solve them by trial and error? That is how real scientists work.

> We must remove the rigidity of thought. . . . We must leave freedom for the mind to wander about in trying to solve the problems. . . . The successful user of mathematics is practically an inventor of new ways of obtaining answers in given situations. Even if the ways are well known, it is usually much easier for him to invent his own way—a new way or an old way—than it is to try to find it by looking it up.

Better to have a jumbled bag of tricks than any one orthodox method. That was how he taught his own children at homework time. Michelle learned that he had a thousand shortcuts; also that they tended to get her into trouble with her arithmetic teachers.

Do You Think You Can Last On Forever?

Although he had never liked athletic activity, he tried to stay fit. After he broke a kneecap falling over a Chicago curb, he took up jogging. He ran almost daily up and down the steep paths above his house in the Altadena hills. He owned a wet suit and swam often at the beachfront house in Mexico that he had bought with his Nobel Prize money. (It had been a shambles when he and Gweneth first saw it. He told her that they did not want it. She looked at the glass wall facing the warm currents sweeping up from the Tropic of Cancer and replied, "Oh yes, we do.")

Traveling in the Swiss Alps in the summer of 1977, he frightened Gweneth by suddenly running to the bathroom of their cabin and vomiting—something he never did as an adult. Later that day he passed out in the téléphérique. Twice that year his physician diagnosed "fever of undetermined origin." It was not until October 1978 that cancer was discovered:

a tumor that had grown to the size of a melon, weighing six pounds, in the back of his abdomen. A bulge was visible at his waistline when he stood straight. He had ignored the symptoms for too long. He had had other worries: just months before, Gweneth had herself undergone surgery for cancer. Feynman's tumor pushed his intestines aside and destroyed his left kidney, his left adrenal gland, and his spleen.

It was a rare cancer of the soft fat and connective tissue, a myxoid liposarcoma. After difficult surgery, he left the hospital looking gaunt and began a search of the medical literature. There he found no shortage of probabilistic estimates. The likelihood of a recurrent tumor was high, though his had appeared well encapsulated. He read a series of individual case studies, none with a tumor as large as his. "Five-year survival rates," one journal said in summary, "have been reported from 0% to 11%, with one report of 41%." Almost no one survived ten years.

He returned to work. "You are old, Father Feynman," wrote a young friend in a mocking bit of verse,

> "And your hair has turned visibly gray;
> And yet you keep tossing ideas around—
>   At your age, a disgraceful display!"

> "In my youth," said the Master, as he shook his long locks,
>   "I took a great fancy to sketching;
> I drew many diagrams, which most thought profound
>   While others thought just merely fetching."

> "Yes, I know," said the youth, interrupting the sage,
>   "That you once were so awfully clever;
> But now is the time for quark sausage with chrome.
>   Do you think you can last on forever?"

Younger physicists, including Gell-Mann, had already stepped aside from the research frontier, but Feynman turned to problems in quantum chromodynamics—the latest synthesis of field theories, so named because of the central role of quark color. With a postdoctoral student, Richard Field, he studied the very-high-energy details of quark jets. Other theorists had realized that the reason quarks never emerged freely was that they were confined by a force unlike those with which physics was familiar. Most forces diminished with distance—gravity and magnetism, for example. It seemed obvious that this must be so, but the opposite was true for quarks. When they were close together, the force between them was negligible;

when they were drawn apart, the force grew extremely strong. Jets, as Feynman and Field understood them, were a by-product. In a high-energy collision, before a quark could be broken free of these bonds, the force would become so great that it would create new particles, pulling them into existence out of the vacuum in a burst traveling in the same direction— a jet.

At first Field met with Feynman one afternoon a week. Feynman did not realize that Field was spending almost every waking hour preparing for their meeting. Their work took the form of predictions in a language well suited to experimenters. It was not abstruse theory but a realistic guide to what experimenters should see. Feynman insisted that they calculate only experiments that had not yet been performed; otherwise, he said, they would not be able to trust themselves. Gradually they found that they were able to stay a few months ahead of the experiments and provide a useful framework. As the accelerators reached higher energies, jets of the kind Feynman and Field had described came into existence.

Theorists meanwhile continued to struggle with their understanding of quark confinement: whether quarks must always be confined under every circumstance and whether confinement could be derived naturally from the theory. Victor Weisskopf urged Feynman to work on this, too, by saying that all he could see in the literature was formal mathematics. "I don't get any physics out of it. Why don't you attack the problem? You are just the right guy for it and you would find the essential physical reasons why QCD confines the quarks." Feynman made an original effort in 1981 to solve this problem analytically in a toy model of two dimensions. Quantum chromodynamics, as he noted, had become a theory of such internal complexity that usually even the fastest supercomputers could not generate specific predictions to compare with experiments. "QCD field theory with six flavors of quarks with three colors, each represented by a Dirac spinor of four components, and with eight four-vector gluons, is a quantum theory of amplitudes for configurations each of which is 104 numbers at each point in space and time," he wrote. "To visualize all this qualitatively is too difficult." So he tried removing a dimension. This turned out to be a blind alley, although the freshness of his approach kept the work on some theorists' reading lists long after they had passed by its conclusions.

In September 1981 a tumor recurred, this time entwined about Feynman's intestines. The doctors tried a combination of doxorubicin, radiation treatment, and heat therapy. Then he underwent his second major surgery. The radiation had left his tissues spongy. The surgery lasted fourteen and a half hours and involved what the physicians described euphemistically as a "vascular incident"—his aorta split. An emergency request for blood went out at Caltech and the Jet Propulsion Laboratory, and donors lined up. Feynman needed seventy-eight pints. When Caltech's president, Marvin Goldberger, entered his hospital room afterward, Feynman said, "I'd rather be where I am than where you are" and added that he still was not going to do anything Goldberger asked. In visible pain, he entertained his hospital visitors with new stories. Before the operation, the surgeon, Donald Morton of the UCLA Medical Center, had appeared with a halo of residents and nurses. Feynman asked what his chances were. "It's impossible to talk about the probability of a single event," he recounted the surgeon as saying, and he replied, "From one professor to another, it is possible if it's a future event."

Caltech's influence in physics had waned. It drew the same extraordinary collection of bright, naïve, gangly undergraduates, all assuming that they would be taking graduate courses by their junior years. The best graduate students, however, went elsewhere. The physics colloquium remained an institution—Feynman usually sitting like a magnet in the front row, capable of dominating every session, visitors knew, entertainingly or ruthlessly. He could reduce an unwary speaker to tears. He shocked colleagues by tearing the flesh off an elderly Werner Heisenberg, made the young relativist Kip Thorne physically ill—the stories reminded older physicists of Pauli ("*ganz falsch*"). Douglas Hofstadter, a pioneer in artificial intelligence, gave an unusual talk on the slippery uses of analogy. He began by asking the audience to name the First Lady of England, looking for such answers as Margaret Thatcher, Queen Elizabeth, or Denis Thatcher. "My wife," came the cry from the front row. Why? "Because she's English and she's great." Through the rest of his talk, it seemed to Hofstadter that Feynman continued heckling in the manner of the village idiot. He was no less an institution than ever, but the center of gravity of elementary particle physics had drifted eastward again, toward Harvard and Princeton and other universities. A combined theory of electromagnetism and weak interactions had led to the gauge theories that brought together the strong interactions under the same quantum-chromodynamical umbrella. This resurgence of quantum theory also brought a new appreciation of Feynman's path integrals, because path integrals proved essential in quantizing the gauge theories. Feynman's discovery now seemed not just a useful tool but an organizing principle at nature's deepest levels. Yet he did not pursue the new implications of path integrals himself. At the forefront were such theorists as Steven Weinberg,

Abdus Salam, Sheldon Glashow, and younger colleagues who had seen neither Feynman nor Gell-Mann as the magnets they had once been. Caltech physicists, concerned about the loss of their department's preeminence, sometimes blamed Feynman for not involving himself enough in hiring and Gell-Mann for involving himself too much.

Ever since his return to high-energy physics with his parton model, Feynman had been struggling against the pull of gray-eminence, elderstatesman status. In 1974 he replied unnecessarily to a standard departmental inquiry by writing a one-sentence memorandum: "I have not accomplished anything this year in the way of research!" Two years later, when a friend, Sidney Coleman, put him on the participant list for a quantum field theory conference sponsored by Werner Erhard's est Foundation, Feynman summed up his ambivalence about his insider and outsider status by replying in Groucho Marx fashion:

> What the hell is Feynman invited for? He is not up to the other guys and is doing nothing as far as I know. If you clean up the invitation list, to just the hard-core workers, I might begin to think about attending.

Coleman duly removed him from the list, and Feynman attended.

He was untroubled by the association with est's vaguely humbug sixties-inspired self-improvement seminars, suffused though they were with the pseudoscientific jargon that he ordinarily despised—"another piece of evidence," as Coleman had said, "that we are living in the Golden Age of Silliness." Erhard's organization and other postsixties institutions were attracted to quantum theory for what appeared—misleadingly—to be a mystical view of reality, reminiscent, they thought, of Eastern religions and anyway more intriguing than the old-fashioned view that things are more or less what they seem. Such organizations, struggling to emerge from the sixties as ongoing business enterprises, were attracted to quantum physicists for the respectability they could lend. Meanwhile, Feynman was drawn to Erhard and other "flaky people"—as Gweneth referred to some of his new friends—partly because curiosity and nonconformity had long been his own trademarks. The youth movements of the sixties had caught up with him. They had brought his own style into vogue—his tieless, pomp-free outlook, the persona that he and Carl privately spoke of as "aggressive dopiness." He grew his graying hair in a long mane. As much as he reviled organized psychology for what he considered its slippery use of the forms and methods of experimental science, he loved the introspective, self-

examining kind of psychology. He let not only Werner Erhard but also
John Lilly, an aficionado of dolphins and sensory-deprivation tanks, be-
friend him. He tried to ignore what he called Lilly's "mystic hokey-poke"
but nonetheless submerged himself in his tanks in the hope of having
hallucinations, just as he had tried so hard to observe his own dream states
forty years before. Death was not far from his thoughts. He recovered the
earliest childhood memories he could dredge from his mind. He tried
marijuana and (he was more embarrassed about this) LSD. He listened
patiently as Baba Ram Das, the former Richard Alpert of Harvard, author
of the cult book *Be Here Now*, instructed him on how to attain out-of-body
experiences. He practiced these—OBE's, in the current jargon—not willing to believe any of the mystical paraphernalia but happy and interested
to imagine his ego floating here or there, outside himself, outside the room,
outside the sixty-five-year-old body that was failing him so grievously.

Physicists did not make natural hippies. They had played too great a role
in creating the technology-worshiping, nuclear-shadowed culture against
which the counterculture set itself. When Feynman spoke now about his
experience in the Manhattan Project, he stressed more than ever his crack-
ing of safes and baiting of censors. He was more a rebel than an ambitious
and effective group leader. Other people, "people in higher echelons,"
made the decisions, he said, prefacing a 1975 talk at Santa Barbara. "I
worried about no big decisions. I was always flittering about underneath."
He was hardly an enemy of technology; nor, despite his distaste for the
bureaucracy of science, was he an enemy of what was now called the
military-industrial complex. He had always refused to attach his name to
Caltech's grant proposals to the federal funding agencies that kept all university physics departments solvent. Still, he would emerge from Lilly's
sensory-deprivation tank, rinse off the Epsom salts in the shower, dress,
and drive over to Hughes Aircraft Company, a military contractor, to deliver
lectures on physics. He was not guarding his time as he had in the past.
Sporadically, he worked for Hughes and several other companies as a consultant; he advised Hughes on a neural-net project sponsored by the Department of Defense and consulted with 3M Company engineers on
nonlinear optical materials. For less than four hours of conversation he
earned fifteen hundred dollars. These were scattered jobs, chosen with no
special thought. Many of his colleagues arranged their consulting far more
carefully and earned far more money. Feynman's clients often seemed more
grateful for the thrill of meeting him than for any particular technical
contribution he made. He knew he was no businessman. He was Caltech's

highest paid professor, along with Gell-Mann; but Caltech kept all the royalties from *The Feynman Lectures on Physics*. When his old friend Philip Morrison sent him an advertisement for "seventeen towering lectures by two physics giants," available from Time-Life Films, he wondered whether Morrison received any royalties. "I don't," Feynman said. "Are we physics giants business dwarfs?"

His favorite extracurricular patron in the early 1980s was the Esalen Institute at Big Sur on the California coast, a hub for many varieties of self-actualization, self-enrichment, and self-fulfillment: Rolfing, Gestalt therapy, yoga, meditation. Under the giant trees on cliffs overlooking the Pacific were the original hot tubs, fed by natural sulfur springs. For its many patrons Esalen offered an expensive form of relaxation—a "lube job for the mind," as Tom Wolfe once put it. Feynman described it as a hotbed of antiscience: "mysticism, expanded consciousness, new types of awareness, ESP, and so forth." He became a regular visitor. He soaked in the hot tubs, stared gleefully at the nude young women sunbathing, and learned to give massages. He gave some of his standard lectures, adjusted to fit the mental state of the audience. Barefoot, with his thin legs emerging from khaki shorts, he began his "Tiny Machines" talk:

It has to do with the question of how small can you make machinery. Okay? That's the subject. Because I've heard people around, in the baths, saying, "Tiny machines? What's he talking about?" and I say to them, "You know, very small *machines*" [pinching an invisible tiny machine between thumb and forefinger] and it doesn't work. [Pause.]

I am talking about very—tiny—machines. Okay?

And on he would go, to occasional cries of "All right!" from the audience. In the question period, the conversation would invariably turn to antigravity devices, antimatter, and faster-than-light travel—if not in the world of physicists then in the spiritual world. Feynman always answered soberly, explaining that faster-than-light travel was impossible, antimatter was routine, and antigravity devices were unlikely—except, as he said, "that pillow and the floor under your behind will support you effectively for a long time." For several years he conducted a workshop in "idiosyncratic thinking." Esalen's catalog copy promised a route to "peace of mind and enjoyment of life's contradictions" and added: "You are invited to bring rhythm instruments."

Late in spring 1984, on his way to pick up one of the first available IBM personal computers in Pasadena, he leapt excitedly out of his car, tripped on the sidewalk, and struck his head on the side of the building. A passerby told him he had a gory enough gash to go to the hospital for stitches. For a few days he felt fuzzy, but he told himself nothing was wrong.

More days went by. It seemed to Gweneth that he was behaving strangely. He awoke in the night and wandered through Michelle's room. He spent forty-five minutes one day looking for his car, which was parked outside the house. At the house of a model he was drawing, he suddenly undressed and tried to go to sleep; she anxiously told him that he was not at his own home. Finally, after beginning a classroom lecture, he suddenly realized he was speaking disjointed nonsense. He stopped, apologized, and left the room.

A scan of his brain revealed a massive subdural hematoma, slow bleeding inside the skull that was putting strong pressure on the brain tissue. The doctors sent him directly into surgery, where the standard procedure was performed at once: two holes drilled through the cranium to drain the liquid. By the early hours of the next morning Gweneth was relieved to find him sitting up and speaking normally. He had no memory of the lost three weeks. Afterward the specialist who had performed the scan repeated it to rule out a recurrence. He could not resist scrutinizing this remarkably detailed image of Feynman's brain, the convoluted gray tissue, the wrapped bundles of nerve fiber ("But you can't see what I am thinking," Feynman told him), looking for a sign of something different from all the other sixty-five-year-old brains he had scanned. Were the blood vessels larger? The doctor was not sure.

Surely You're Joking!

Feynman had begun to have autobiographical thoughts around the time of the Nobel Prize. Historians came by to record his recollections, and they treated his notes as artifacts too important to be piled in boxes or strewn about on the shelves in the home office he had made in his basement. Sitting there was *Arithmetic for the Practical Man*, a relic of his childhood. He still had the adolescent notebook he had sent back and forth to T. A. Welton in the course of reinventing early quantum mechanics. Interviewers set up tape recorders to capture every word of the same stories he had entertained his friends with for decades.

An MIT historian, Charles Weiner, persuaded him to cooperate in what became the most thorough and serious of his interviews. For a while Feynman considered collaborating with Weiner on a biography. They sat in Feynman's screened back patio while Carl played in a tree house nearby. He not only told his stories but also demonstrated them: "Okay, start your watch," he told Weiner; then, after they had conversed for eight minutes and forty-two seconds, he interrupted himself and said, "Eight minutes forty-two seconds." After many hours the conversation sometimes grew intimate. He rummaged through one box and pulled out a photograph of Arline, reclining almost nude, wearing only translucent lingerie. He almost wept. They shut off the tape recorder and remained silent for a time. Feynman kept most of those memories to himself even now.

He began dating his scientific notes as he worked, something he had never done before. Weiner once remarked casually that his new parton notes represented "a record of the day-to-day work," and Feynman reacted sharply.

"I actually did the work on the paper," he said.

"Well," Weiner said, "the work was done in your head, but the record of it is still here."

"No, it's not a *record*, not really. It's *working*. You have to work on paper, and this is the paper. Okay?" It was true that he wrote in astonishing volume as he worked—long trains of thought, almost suitable to serve immediately as lecture notes.

He told Weiner that he had never read a scientific biography he had liked. He thought he would be portrayed either as a bloodless intellectual or a bongo-playing clown. He vacillated and finally let the idea drop. Still, he sat for interviews with historians interested in Far Rockaway and Los Alamos and filled out questionnaires for psychologists interested in creativity. ("Is your scientific problem-solving accompanied by any of the following?" He checked *visual images, kinesthetic feelings*, and *emotional feelings* and added "(1) acoustic images, (2) talk to self." Under "major illnesses" he reported: "Too much to list. . . . Only adverse effects are laziness during recovery period.")

For several years he had played drums regularly with a young friend, Ralph Leighton, the son of another Caltech physicist. Leighton had begun taping their sessions, and then he began taping the stories Feynman would tell. He urged him on, calling him Chief and begging to hear the same stories again and again. Feynman told them: how he became known in Far Rockaway as the boy who fixed radios by thinking; how he asked a Princeton librarian for the map of the cat; how his father taught him to see through the tricks of circus mind readers; how he outwitted painters, mathematicians, philosophers, and psychiatrists. Or he would just ramble while Leighton listened. "Today I went over to the Huntington Medical Library," he said one day—his remaining kidney was presenting problems. "But it's all interesting, how the kidney works, and everything else. You want me to tell you some interesting things? The damn kidney is the craziest thing in the world!"

Gradually a manuscript began to take shape. Leighton transcribed the tapes and presented them to Feynman for editing. Feynman had strong views about the structure of each story; Leighton realized that Feynman had developed a routine of improvisational performance in which he knew the order and pacing of every laugh. They consciously worked on the key themes. Feynman talked about Arline's having embarrassed him with a box of "Richard darling, I love you! Putzie" pencils:

RICHARD. And the next morning, all right? Next morning, in the mail, there's this letter, all right, this postcard, which starts out, "What's the idea of trying to cut the name off the pencils?"
RALPH. [*Laughs*] Oh, boy! [*Laughs.*]
RICHARD. "What do you care what other people think?"
RALPH. Oh, this is—— Yeah, this is a good theme.
RICHARD. Hmmm?
RALPH. This is a good theme, because there's a theme in here. You know, what other people think . . .

They knew they had a remarkable central figure, a scientist who prided himself not on his achievements in science—these remained deep in the background—but on his ability to see through fraud and pretense and to master everyday life. He underscored these qualities with an exaggerated humility; he took the tone of a boy calling the grownups Mr. and Mrs. and asking politely dangerous questions. He was Holden Caulfield, a plain old straight shooter trying to figure out why so many other people are phonies.

"Pompous fools—guys who are fools and are covering it all over and impressing people as to how wonderful they are with all this hocus pocus—THAT, I CANNOT STAND!" Feynman said. "An ordinary fool isn't a faker; an honest fool is all right. But a dishonest fool is terrible!"

His favorite sort of triumph in the world of these stories came in the realm of everyday cleverness—as when he arrived at a North Carolina airport, late for a meeting of relativists, and worked out how to get help from a taxi dispatcher:

> "Listen," I said to the dispatcher. "The main meeting began yesterday, so there were a whole lot of guys going to the meeting who must have come through here yesterday. Let me describe them to you: They would have their heads kind of in the air, and they would be talking to each other, not paying attention to where they were going, saying things to each other like 'G-mu-nu. G-mu-nu.'"
>
> His face lit up. "Ah, yes," he said. "You mean Chapel Hill!"

Feynman chose as a title the odd phrase uttered by Mrs. Eisenhart at his first Princeton tea when he asked for both cream and lemon: "Surely you're *joking*, Mr. Feynman!" Those words had stayed in his mind for forty years, a reminder of how people used manners and culture to make him feel small, and now he was taking revenge. W. W. Norton and Company bought the manuscript for an advance payment of fifteen hundred dollars, a tiny sum for a trade book. Its staff did not like Feynman's title at all. They proposed *I Have to Understand the World* or *I Got an Idea* ("a nice Brooklyn ring and a little double meaning," the editor said). But Feynman would not budge. Norton released *Surely You're Joking, Mr. Feynman!* in a small first printing early in 1985. It sold out quickly, and within weeks the publisher had a surprising best-seller.

One unhappy reader was Murray Gell-Mann. His attention focused on Feynman's description of the joy of discovering the "new law" of weak interactions in 1957: "It was the first time, and the only time, in my career that I knew a law of nature that nobody else knew." Gell-Mann's rage could be heard through the halls of Lauritsen Laboratory, and he told other physicists that he was going to sue. For late editions of the paperback Feynman added a parenthetical disclaimer: "Of course it wasn't true, but finding out later that at least Murray Gell-Mann—and also Sudarshan and Marshak—had worked out the same theory didn't spoil my fun."

Surely You're Joking gave offense in another way. Feynman spoke of women as he always had—"a nifty blonde, perfectly proportioned"; "a cornfed, rather fattish-looking woman." They appeared as objects of flirtation, nude models for his drawings, or "bar girls" to be tricked into sleeping with him. He knew that his diction was not wholly innocent. Sexual politics had caught up with him before, at the 1972 meeting of the American

Physical Society in San Francisco, where he accepted the Oersted Medal for contributions to the teaching of physics. His personal relationships were not the issue, although in the male world of Caltech a part of his glamorous reputation with envious students came from his apparent sway over women. He continued to flirt with young women at parties and encouraged Don Juan–style rumors. He frequented one of the first California topless bars, Gianonni's—he filled its scalloped paper placemats with chains of equations—and amused the local press by testifying in court on its behalf in 1968. There was genuine machismo in the hero-worship of the male graduate students.

He had received a letter the previous fall suggesting that some of his language tended to "reinforce many 'sexist' or 'male-chauvinist' ideas." For example, he told an anecdote about a scientist who was "out with his girl friend the night after he realized that nuclear reactions must be going on in the stars."

She said "Look at how pretty the stars shine!" He said "Yes, and right now I am the only man in the world who knows *why* they shine."

The letter writer, E. V. Rothstein, cited another anecdote about a "lady driver" and asked him, please, not to contribute to discrimination against women in science. In replying, Feynman decided not to emphasize his sensitivity:

Dear Rothstein:
    Don't bug me, man!
    R. P. Feynman.

The result was a demonstration organized by a Berkeley group at the APS meeting, with women carrying signs and distributing leaflets titled "PR ♀ TEST" and addressed to "Richard P. (for Pig?) Feynman."

Despite the women's movement that emerged in the sixties, science remained forbiddingly male in its rhetoric and its demographics. Barely 2 percent of American graduate degrees in physics went to women. Caltech did not hire its first female faculty member until 1969, and she did not receive tenure until she forced the issue in court in 1976. (Feynman, to the surprise and displeasure of some of his humanities colleagues, had taken her side; he had spent many pleasant hours in her office reading aloud such poems as Theodore Roethke's "I Knew a Woman": "I measure time by how a body sways. . . .") Like most men in physics, Feynman had known a few women as professional colleagues and believed that he had treated them, individually, as equals. They tended to agree. What more, he wondered, could anyone ask?

The Berkeley protesters had discovered his lady-driver anecdotes but had overlooked other examples of a style of speaking in which, habitually, the scientist is male and nature—her secrets waiting to be penetrated—is female. In his Nobel lecture Feynman had recalled falling in love with his theory: "And, like falling in love with a woman, it is only possible if you do not know much about her, so you cannot see her faults." And he had concluded:

So what happened to the old theory that I fell in love with as a youth? Well, I would say it's become an old lady, that has very little attractive left in her and the young today will not have their hearts pound when they look at her anymore. But, we can say the best we can for any old woman, that she has been a good mother and she has given birth to some very good children.

In 1965 a large audience of men and women could listen to these words without taking offense or hearing a politically charged subtext. In 1972 Feynman was able to defuse the protest easily when he took the podium, by declaring: "There is in the world of physics today a tremendous prejudice against women. This is a ridiculous thing and should stop, as there is no sense to it whatsoever. I love the subject of physics and it has always been my desire to try to share the delights of understanding it with any minds that were able to—male or female. . . ." Many of the demonstrators applauded. In 1985 Feynman once again seemed to some feminists a symbol of male dominance in physics. Real life was complex: one tough-minded Caltech professional would close her door and confide to a stranger that Feynman, even in his sixties, was the sexiest man she had ever known; others, wives of colleagues, resented their husbands for loving him so uncritically. Meanwhile, the status of women in the profession of physics had barely changed.

Despite himself, he was stung by the occasional criticism of *Surely You're Joking*. He knew, too, that some of the physicists who had known him longest were disappointed by a self-portrait that made Feynman seem more joker than scientist. His old friends in Hans Bethe's generation were often pained, or shocked, though they did repeat Feynman's stories about them with relish, detail for detail, as though from their own memory, Feynman's voice having transplanted itself into their brains. Others saw through to the essence of what they loved in Feynman. Philip Morrison, writing in *Scientific American*, said: "Generally Mr. Feynman is not joking; it is we, the setters of ritual performance, of hypocritical standards, pretenders to care and understanding, who are joking instead. This is the book of a powerful mind honest beyond everything else, a specialist in spade-naming." Feynman nonetheless upbraided people who called the book his autobiography. He wrote in the margin of a science writer's draft manuscript about modern particle physics: "Not An Autobiography. Not So. Simply A Set Of Anecdotes." And when he came across a sentence describing him, at Los Alamos, as "a curiously tragic joker," he scrawled angrily, "What I really was under such circumstances is far deeper than you are likely to understand."

A Disaster of Technology

In 1958, a hasty four months after Sputnik, Americans entered what was called the space race by sending into orbit the first of a series of Explorer satellites from Cape Canaveral, Florida. Explorer I weighed as much as a fully packed overnight bag. It was hurled skyward on January 31 by a four-stage Jupiter-C rocket—more reliable than the navy's Vanguard rockets, which had been exploding at liftoff. It sent back radio signals much like Sputnik's.

Explorer II, bearing a cosmic-ray detector that pushed its weight up to thirty-two pounds, soared skyward five weeks later and disappeared into the clouds. An army team watched under the guidance of Wernher von Braun, resilient veteran of the Nazi rocket program at Peenemünde. They listened to the fading rumble of the rocket and the rising beep of the radio signal transmitted to their squawk box. All seemed well. A half hour after the launch, they held a confident news briefing.

Across the continent, where the Jet Propulsion Laboratory in Pasadena served as the army's main collaborator in rocket research, a team was struggling with the task of tracking the satellite's course. They used a room-size IBM 704 digital computer. It was temperamental. They entered the primitively sparse data available for tracking the metal can that the army's rocket had hurled forward: the frequency of the radio signal, changing

Doppler-fashion as the velocity in the line of flight changed; the time of disappearance from the observers at Cape Canaveral; observations from other tracking stations. The JPL team had learned that small variations in the computer's input caused enormous variations in its output. Albert Hibbs, the laboratory's young research chief, had complained about this difficulty to his former Caltech thesis adviser: Feynman.

Feynman bet that he could outcompute the computer, if fed the same data at the same rate. So when Explorer II lifted off the pad at 1:28 P.M., he sat in a JPL conference room, surrounded by staff members rapidly sorting the data for the computer. At one point Caltech's president, Lee DuBridge, entered the room and was startled to see Feynman—who snapped, *Go away, I'm busy.* After a half hour Feynman rose to say he was finished: according to his calculations the rocket had plunged into the Atlantic Ocean. He left for a weekend in Las Vegas as the trackers kept trying to coax an unambiguous answer from their computer. Tracking stations at Antigua and Inyokern, California, persuaded themselves that they had picked an orbiting satellite out of the background noise, and "moonwatch" teams in Florida spent the night watching the skies. But Feynman was right. The army finally announced at 5 o'clock the next afternoon that Explorer II had failed to reach orbit.

The space shuttle *Challenger* rose from its launching scaffold into a cloudless sky twenty-eight years later, on January 28, 1986. A half second after liftoff, a puff of dark smoke, invisible to human eyes, spurted from the side of one of the shuttle's two solid-fuel rockets. The launch had been postponed four times. Inside the cabin, as always, the many-gravity acceleration pressed the crew against their seats: the commander, Francis Scobee; the pilot, Michael Smith; the mission specialists, Ellison Onizuka, Judith Resnick, and Ronald McNair; an engineer from the Hughes Aircraft Company, Gregory Jarvis; and a New England schoolteacher, Christa McAuliffe, who had been chosen as "Teacher in Space," the winner of a NASA public-relations program meant to encourage the interest of children and also congressmen. The cargo bay—large enough to have carried the 1950s Jupiter-C rocket—held a pair of satellites, a fluid-dynamics experiment, and radiation-monitoring equipment. Ice had built up overnight, and new delays had been ordered while an ice inspection team made sure it had time to melt. Seven seconds after liftoff the shuttle rolled over in its characteristic fashion, so that it appeared to be hanging from the back of its giant disposable fuel tank, and headed east over the Atlantic, its percussive roar audible over hundreds of square miles. The breeze barely bent its column of smoke. At the one-minute mark—halfway through the brief expected lifetime of the solid-fuel rockets—a flickering light appeared where it did not belong, at a joint in the shell of the right-side rocket. The main engines reached full power, and Scobee radioed, "Roger. Go at throttle up." At seventy-two seconds the two rockets began to pull in different directions. At seventy-three seconds the fuel tank burst open and released liquid hydrogen into the air, where it exploded. The shuttle felt an enormous sudden thrust. A cloud of flame and smoke enveloped it. Fragments emerged seconds later: the left wing, like a triangular sail against the sky; the engines, still firing; and somewhere, intact, a plummeting coffin for seven men and women. The technologies of television, aided by satellites lofted in earlier shuttle missions, let more people witness the event, again and again, than any other disaster in history.

Machinery out of control. The American space agency had made itself seem a symbol of technical prowess, placing teams of men on the moon and then fostering the illusion that space travel was routine—an illusion built into the very name *shuttle*. After the nuclear accident at Three Mile Island, Pennsylvania, and the chemical disaster at Bhopal, India, the space-shuttle explosion seemed a final confirmation that technology had broken free of human reins. Did nothing work any more? The dream of technology that held sway over the America of Feynman's childhood had given way to a sense of technology as not just a villain but an inept villain. Nuclear power plants, once offering the innocent promise of inexhaustible power, had become menacing symbols on the landscape. Automobiles, computers, simple household appliances, or giant industrial machines—all seemed unpredictable, dangerous, untrustworthy. The society of engineers, so hopeful in the America of Feynman's childhood, had given way to a technocracy, bloated and overconfident, collapsing under the weight of its own byzantine devices. That was one message read in the image replayed hundreds of times that day on millions of television screens—the fragmenting smoke cloud, the twin rockets veering apart like Roman candles.

President Ronald Reagan immediately announced his determination to continue the shuttle program and expressed his support for the space agency. Following government custom, he appointed an investigatory commission that would repeatedly be described as independent—the White House officially declared it "an outside group of experts, distinguished Americans who have no ax to grind"—although in actuality it was composed mostly of insiders and figures chosen for their symbolic value: its chairman, William P. Rogers, who had served as attorney general and secretary of state in Republican administrations; Major General Donald J. Kutyna, who had headed shuttle operations for the Department of Defense; several NASA consultants and executives of aerospace contractors; Sally Ride, the first American woman in space; Neil Armstrong, the first man on the moon; Chuck Yeager, a famous former test pilot; and, a last-minute choice, Richard Feynman, a professor who brought to the next day's newspaper accounts the tag "Nobel Prize winner." Armstrong said on the day of his appointment that he did not understand why an independent commission was necessary. Rogers said even more baldly, "We are not going to conduct this investigation in a manner which would be unfairly critical of NASA, because we think—I certainly think—NASA has done an excellent job, and I think the American people do."

The White House named Rogers and selected the rest of the commission from a list provided by the space agency's acting administrator, William R. Graham. As it happened, Graham had attended Caltech thirty years before and had often sat in on Physics X, which he remembered as the best course at Caltech. Later he had attended Feynman's lectures at Hughes Aircraft. But he did not think of Feynman for the shuttle commission until his wife, who had accompanied him to some of the Hughes lectures, suggested the name. When Graham called, Feynman said, "You're ruining my life." Only later did Graham realize what he had meant: *You're using up my very short time*. Feynman was now suffering from a second rare form of cancer: Waldenström's macroglobulinemia, involving the bone marrow. In this cancer, one form of B lymphocyte, a white blood cell, becomes abnormal and produces large amounts of a protein that makes the blood sticky and thick. Clotting becomes a danger, and the blood flows poorly to some parts of the body. Feynman's past kidney damage was a complication. He seemed gray and wan. There was little his doctors could propose. They could not explain the presence of two such unusual cancers. Feynman himself refused to consider the speculation that the cause might lie forty years in the past, at the atomic bomb project.

He immediately arranged a briefing with his friends at the Jet Propulsion Laboratory in Pasadena. The day after his appointment was announced, he sat in a small room in the central engineering building and met with a succession of engineers. The laboratory, with its advanced image-processing facilities, already had the original negatives of the thousands of photographs taken by the range cameras as the shuttle drove skyward.

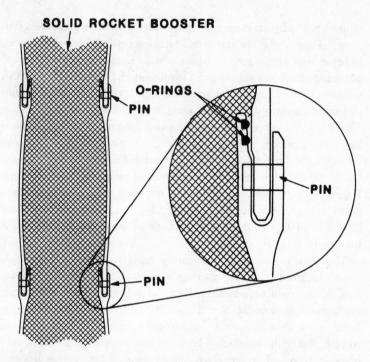

SOLID ROCKET BOOSTER

O-RINGS

PIN

PIN

PIN

The shuttle's solid rocket boosters were made in sections, assembled one atop another at the launch site. The joints holding the sections together had to be sealed to prevent the escape of hot gases from inside the rocket. Pairs of O-rings—a quarter-inch thick— spanned the 37-foot circumference. The pressure of the gas was supposed to wedge them tightly into the joints, creating the seal.

Feynman examined technical drawings and heard from engineers who had worked on the early design studies, on the solid rocket boosters, and on the engines. He learned that the shuttle's engineers, forming a community across the administrative boundaries that separated NASA's various departments and subcontractors, shared a knowledge that every launch was at risk. Recurring cracks had appeared in the turbine blades of the shuttle's engines, at the very edge of engine technology. That first day, February 4, Feynman noted that there were well-known problems with the rubber O-rings that sealed the joints between sections of the tall solid-fuel rockets. These rings represented a remarkable scaling-up of everyday engineering for the high-technology shuttle: they were ordinary rubber rings, thinner than a pencil yet thirty-seven feet long, the circumference of the rocket. They were meant to take the pressure of hot gas and form a seal by squeezing tight into the metal joint. "O-Rings show scorching in Clevis check . . ." Feynman wrote in a shaky, aging hand. "Once a small hole burn thru generates a large hole very fast! few seconds catastrophic failure." He flew to Washington that night.

The commission began in a formal and slow-paced style. Rogers opened the first public meeting with a declaration that NASA officials had been cooperative and that the commission would rely largely on the agency's own investigations. The meeting began with a briefing by NASA's top space-flight official, Jesse Moore. Unexpectedly he found himself interrupted by sharp specific questions from Feynman and several other panel members. They focused on the weather, which had been so cold that ice formed on equipment throughout the launching pad. In response, Moore denied that he had had any warning that cold could pose a problem.

That afternoon, however, another agency official, Judson A. Lovingood, from the Marshall Space Flight Center in Alabama, testified that managers for NASA and for Morton Thiokol, the builder of the solid rockets, had held a telephone conference the night before the launch to discuss, as he said, "a concern by Thiokol on low temperatures." The discussion focused on the O-rings, he said, and Thiokol recommended that the launch proceed. He also mentioned evidence of "blow-by"—soot showing that hot gases had burned through seals that were supposed to contain them. He emphasized, though, that the O-rings were used in pairs and that the secondary O-rings always seemed to hold. "Was that any cause for concern?" asked General Kutyna.

"Oh, yes," Lovingood replied. "That is an anomaly."

Newspaper reports the next day, February 7, focused on the issue of cold weather and noted that NASA had been caught off guard by the aggressive questions. When Moore faced the commission again, Feynman immediately began a new series of questions. The chairman twice asked him to put off the questions until later. But the questioning quickly returned to the seals. Another NASA witness testified that the films showed a puff of dark smoke emerging from the side of the right-hand solid rocket six-tenths of a second after ignition. "This is what we would have called an anomaly?" Feynman asked. The witness, Arnold Aldrich, replied carefully, "It is an anomaly unless we find a film where we have seen one just like it." Pressed by another commissioner, he said:

"Everything that I know about the certification of this seal . . . is that the certification tests run on that joint show that the seal would be somewhat more stiff, but completely adequate for sealing at all temperatures in the ranges. There was never any intention that the system couldn't be launched in freezing conditions."

The chairman commented protectively to Aldrich, "When we ask questions, when we continue to ask questions, we are not really trying to point a finger," and to Moore, "I thought it was a little unfortunate in the paper this morning that they said that—and I don't think you really said that— that you had excluded the possibility that the weather had any effect. . . . If it appears you have excluded that to begin with, particularly because apparently Rockwell did call and gave you a warning which you considered and decided that it was okay to go ahead—suppose that judgment was wrong. Nobody is going to blame anybody. I mean, somebody has to make those decisions."

But Feynman immediately challenged Moore on the view that O-ring blow-by had been acceptable because the secondary rings had held.

"You said we don't expect it on the other O-ring," Feynman said. "On the other hand, you didn't expect it on the first O-ring. . . . If the second O-ring gives just a little bit when the first one is giving, that is a very much more serious circumstance, because now the flow has begun." The air force general, Kutyna, had befriended Feynman when they sat together at the commission's first news conference. ("Co-pilot to pilot," he had said softly, choosing this deferential phrase out of worry that Feynman was nervous beside a general in an imposing uniform, "comb your hair," and Feynman, surprised, growled and asked Kutyna for a comb.) Now Kutyna joined in: "Let me add to your comment. . . . Once it got a path, then it burns like an acetylene torch."

Feynman said, "I have a picture of that seal in cross section here, if anybody wants to see it." No one responded.

For Feynman, for Rogers, for Graham, for the press, and for NASA officials, the weekend of February 8 brought surprises.

Feynman, away from home, thinking of his Los Alamos experience as the prototype for urgent group technical projects, did not want to take Saturday and Sunday off. Through Graham he arranged a series of private briefings on Saturday at NASA's Washington headquarters. He learned more about the engines, the orbiter, and the seals. He found again that the agency's engineers understood a long history of difficulties with the

O-rings; that two- or three-inch segments of the thirty-seven-foot links had repeatedly been burned and eroded; that a critical issue was the speed with which the rubber had to press into the metal gap—in milliseconds; and that the space agency had found a bureaucratic means of simultaneously understanding and ignoring the problem. He was particularly struck by a summary of a meeting between Thiokol and NASA managers the previous August. Its recommendations seemed incompatible:

- The lack of a good secondary seal in the field joint is most critical and ways to reduce joint rotation should be incorporated as soon as possible to reduce criticality. . . .
- Analysis of existing data indicates that it is safe to continue flying. . . .

Elsewhere at NASA headquarters that day, Graham learned that a storm was about to break: the *New York Times* had obtained documents showing urgent warnings within NASA about O-ring problems over a period of at least four years. Graham had taken control of the agency only recently, when the administrator, James Beggs, was indicted on fraud charges unrelated to NASA. He immediately telephoned Rogers.

The article appeared Sunday, quoting warnings even more dire than those the engineers had shown Feynman: that a failure of the seals could cause "loss of vehicle, mission, and crew due to metal erosion, burn-through, and probable case bursting resulting in fire and deflagration," and that

There is little question . . . that flight safety has been and is being com-promised by potential failure of the seals, and it is acknowledged that failure during launch would certainly be catastrophic.

That morning Graham himself took Feynman to the Smithsonian Institution's National Air and Space Museum, where he sat in a cavernous theater and watched an inspirational giant-screen film about the space shuttle. He was surprised at how emotional he felt.

In the afternoon Kutyna called Feynman at his hotel. As shuttle program manager for the military, Kutyna knew the shuttle more intimately than any other commissioner. He also knew how to run a technical commission, because he had headed the air force's own investigation into the explosion of a Titan rocket the year before. And he had his own information sources among the engineers and astronauts—one of whom told him over the weekend that Thiokol had known of a potential loss of resiliency when the rubber O-rings were cold. Kutyna wanted to bring this information into the open without jeopardizing his source. He invited Feynman to his house for Sunday dinner. Afterward they went out to his garage—he collected junk cars as a hobby, and at the moment he was working on an old Opel GT. Its carburetor happened to be sitting on his workbench. He told Feynman, you know, those things leak when it's cold, so do you think cold might have a similar effect on the shuttle O-rings?

Rogers called a closed meeting Monday in reaction to the *New York Times* revelations. He made clear that he considered them a disruption of his proceedings: "I think it goes without saying that the article in the *New York Times* and other articles have created an unpleasant, unfortunate situation. There is no point in dwelling on the past." NASA representatives were asked to respond: "I think that his statement in here where he says that it might be catastrophic I think is overstated," said one, and Rogers remarked, "Well, that may be." Lawrence Mulloy, project manager for the solid rockets. testified that the rubber in the O-rings was required to operate across an enormous temperature range, from minus 30 to 500 degrees Fahrenheit. He did not know of any test results, however, on the actual resiliency of the O-rings at low temperatures.

Mulloy returned the next morning to give the commissioners a briefing— another in the genre that Kutyna thought of as "telling which was the pointy end of the shuttle because they don't know that much about it." He brought more than a dozen charts and diagrams and gave a vivid flavor of the engineering jargon—the tang end up and the clevis end down, the grit blast, the splashdown loads and cavity collapse loads, the Randolph type two zinc chromate asbestos-filled putty laid up in strips—all forbidding to the listening reporters if not to the commissioners themselves. "How are these materials, this putty and the rubber, affected by extremes of temperature? . . ." one commissioner asked.

> Yes, sir, there is a change in the characteristic. As elastomers get colder,
> the resiliency decreases, and the ability to respond——
>   Now, the elastomers are what?
> That is the Viton O-ring.
>   The rubber?

Feynman pressed Mulloy on why resiliency was crucial: a soft metal like lead, squeezed into the gap, would not be able to hold a seal amid the vibration and changing pressure. "If this material weren't resilient for say a second or two," Feynman said, "that would be enough to be a very dangerous situation?"

He was setting Mulloy up. He had been frustrated by the inconclusive and possibly evasive testimony. He had made an official request for test data, through Graham, and had received documents that were irrelevant, showing how the rubber responded over a period of hours instead of milliseconds. Why couldn't the agency answer such a simple question? At dinner Monday night his eyes fell on a glass of ice water, and he had an idea that he first thought might be too easy and gauche. Ice water was a stable 32 degrees, almost exactly the temperature on the pad at the time of the launch. Tuesday morning he rose early and hailed a taxicab. He circled official Washington in search of a hardware store and finally managed to buy a small C-clamp and pliers. As the hearing began, he called for ice water, and an aide returned with cups and a pitcher for the entire commission. As a life-size cross section of the joint was passed along for the commissioners to examine, Kutyna saw Feynman take the clamp and pliers from his pocket and pull a piece of the O-ring rubber from the model. He knew what Feynman meant to do. When Feynman reached for the red button on his microphone, Kutyna held him back—the television cameras were focused elsewhere. Rogers called a short break and, in the men's room, standing next to Neil Armstrong, he was overheard saying, "Feynman is becoming a real pain in the ass." When the hearing resumed, the moment finally arrived.

CHAIRMAN ROGERS: Dr. Feynman has one or two comments he would like to make. Dr. Feynman.

DR. FEYNMAN: This is a comment for Mr. Mulloy. I took this stuff that I got out of your seal and I put it in ice water, and I discovered that when you put some pressure on it for a while and then undo it it doesn't stretch back. It stays the same dimension. In other words, for a few seconds at least and more seconds than that, there is no resilience in this particular material when it is at a temperature of 32 degrees.

I believe that has some significance for our problem.

Before Mulloy could speak, Rogers called the next witness, a budget analyst who had written a memorandum that formed the basis of the *Times* article. The analyst, Richard Cook, had noticed the O-ring problem on a list of "budget threats" month after month, had highlighted it to his superiors, and, when the disaster took place, felt certain that it had been the cause. The chairman, for the first and last time during the shuttle hearings, cross-examined a witness, through the rest of the morning and on into the afternoon, with the cold savagery of a prosecutor:

You didn't, I assume, make any attempt to weigh budgetary considerations and safety considerations, did you?

Not at all.

You weren't qualified for that?

No, sir. . . .

You had no reason to think that people who were weighing those considerations were not qualified to do it? . . . You didn't feel that you were in a position or should you make those decisions about what should be done with the space program?

That's right.

And so that the memo, which has been given a great deal of attention, sort of suggests that you were taking issue with the people who were highly qualified to make those judgments, when in fact you weren't at all? . . . You wrote the memo in the heat of the moment, and I assume you were, like everybody else in the country was, terribly disturbed and upset by the accident, and it was in that spirit or at that time when you wrote the memorandum. You didn't really mean to criticize for public consumption your associates or people around you, did you?

Yet by then it was clear that Cook had described the problems accurately. Feynman's demonstration dominated the television and newspaper reports that evening and the next morning. Mulloy, under further questioning, made the first clear acknowledgment that cold diminished the effectiveness of the seals and that the space agency had known it, although a straightforward test in the manner of Feynman's had never been performed. When such tests were finally performed on behalf of the commission, in April, they showed that failure of the cold seals had been virtually inevitable— not a freakish event, but a consequence of the plain physics of materials, as straightforward as Feynman had made it seem with his demonstration. Freeman Dyson said later, "The public saw with their own eyes how science is done, how a great scientist thinks with his hands, how nature gives a clear answer when a scientist asks her a clear question."

One extraordinary week had passed since Feynman boarded the night flight to Washington. The commission had four months of work remaining, but it had arrived at the physical cause of the disaster.

As the seventies began and the last of the moon landings drew near, NASA had become an agency lacking a clear mission but maintaining a large established bureaucracy and a net of interconnections with the nation's largest aerospace companies: Lockheed, Grumman, Rockwell International, Martin Marietta, Morton Thiokol, and hundreds of smaller companies. All became contractors for the space-shuttle program, formally known as the Space Transportation System, initially intended as a fleet of reusable and economical cargo carriers that would replace the individual one-use rockets of the past.

Within a decade, the shuttle had become a symbol of technology defeated by its own complexity, and the shuttle program had become a symbol of government mismanagement. Every major component had been repeatedly redesigned and rebuilt; every cost estimate offered to Congress had been exceeded many times over. Unpublicized audits had found deception and spending abuses costing many billions of dollars. The shuttle had achieved a kind of Pyrrhic reusability: the cost of refurbishing it after each flight far exceeded the cost of standard rockets. The shuttle could barely reach a low orbit; high orbits were out of the question. The missions flown were a small fraction of those planned, and—despite NASA's public claims to the contrary—the scientific and technological products of the shuttle were negligible. The space agency systematically misled Congress and the public about the costs and benefits. As Feynman stated, the agency, as a matter of bureaucratic self-preservation, found it necessary "to exaggerate: to exaggerate how economical the shuttle would be, to exaggerate how often it could fly, to exaggerate how safe it would be, to exaggerate the big scientific facts that would be discovered." At the time of the *Challenger* disaster the program was breaking down internally: by the end of the year both a shortage of spare parts and an overloaded crew-training program would have brought the flight schedule to a halt.

Yet the report of the presidential commission, issued on June 6, began by declaring that the accident had interrupted "one of the most productive engineering, scientific, and exploratory programs in history." It attributed to the public "a determination . . . to strengthen the Space Shuttle program."

When Feynman talked about his role later, he fell back on his boy-from-the-country image of himself: "It was a great big world of mystery to me, with tremendous forces. . . . I hadda watch out." He claimed no understanding of politics or bureaucracies. These were matters beyond the ken of a technical fellow. Alone among the commissioners, however, Feynman worked to expand the scope of the investigation to include precisely the areas about which he disavowed competence: issues of decision making, communication, and risk assessment within the space agency. Kutyna told him he was the only commissioner free of political entanglements. Despite Rogers's disapproval he insisted on conducting his own lines of inquiry and traveled alone to interview engineers at the Kennedy Space Center in Florida, the Marshall Space Flight Center in Alabama, the Johnson Space Center in Houston, and the headquarters of several contractors. In between, he made repeated visits to a Washington hospital for blood tests and medication for his worsening kidney, and he talked by telephone with his doctor in California, who complained about the difficulty of practicing medicine at long distance. "I am determined to do the job of finding out what happened—let the chips fall!" he wrote Gweneth proudly. He enjoyed the thrill of the game, and he suspected that he was being carefully managed. "But it won't work because (1) I do technical information exchange and understanding much faster than they imagine"—he was, after all, a veteran of Los Alamos and the MIT machine shop—"and (2) I already smell certain rats that I will not forget."

He tried to make use of his naïveté. When Rogers showed him a draft final recommendation, effusive in its praise of the space agency—

> The Commission strongly recommends that NASA continue to receive the support of the Administration and the nation. The agency constitutes a national resource and plays a critical role in space exploration and development. It also provides a symbol of national pride and technological leadership. The Commission applauds NASA's spectacular achievements of the past and anticipates impressive achievements to come. . . .

—he balked, saying he lacked expertise about such policy matters, and he threatened to withdraw his signature from the report.

His protest was ineffective. The language appeared virtually intact, as the commission's "concluding thought" rather than a recommendation. Although the commission learned that the decision to launch had been made over the specific objections of engineers who knew of the critical danger from the O-rings, the final report did not attempt to hold senior space-agency officials responsible for the decision. Evidence emerged show-ing that the history of O-ring problems had been reported in detail to top officials, including the administrator, Beggs, in August 1985, but the com-mission chose not to question those officials. Feynman's own findings, substantially harsher than the commission's, were isolated in an appendix to the final report.

Feynman analyzed the computer system: 250,000 lines of code running on obsolete hardware. He also studied in detail the main engine of the shuttle and found serious defects, including a pattern of cracks in crucial turbine blades, that paralleled the problems with the solid rocket boosters. Overall he estimated that the engines and their parts were operating for less than one-tenth of their expected lifetimes. And he documented a history of ad hoc slippage in the standards used to certify an engine as safe: as cracks were found earlier and earlier in a turbine's lifetime, the certification rules were repeatedly adjusted to allow engines to continue flying.

His most important contribution to the understanding of the disaster came in the area of risk and probability. He showed that the space agency and its contractors—although the essence of their decision making lay in weighing uncertainties—had ignored statistical science altogether and had used a shockingly vague style of risk assessment. The commission's official findings could do no better than quote Feynman's comment during the hearings that the decision making became

a kind of Russian roulette. . . . [The shuttle] flies [with O-ring erosion] and nothing happens. Then it is suggested, therefore, that the risk is no longer so high for the next flights. We can lower our standards a little bit because we got away with it last time. . . . You got away with it, but it shouldn't be done over and over again like that.

Science has tools for such problems. NASA was not using them. A scattering of data points—for the depth of erosion in O-rings, for example—tended to be reduced to simplistic, linear rules of thumb. Yet the physical phe-nomenon, a hot jet of gas carving channels in rubber, was highly nonlinear, as Feynman noted. The way to assess a scattered range of data was through probability distributions, not single numbers. "It has to be understood as a probabilistic and confusing, complicated situation," he said. "It is a question of increasing and decreasing probabilities . . . rather than did it work or didn't it work."

On the crucial question of the effect of temperature on O-ring safety,

NASA had made an obvious statistical blunder. Seven flights had shown evidence of damage. The most damage had occurred on the coldest flight— at a still-mild 53 degrees Fahrenheit—but no general correlation could be seen between temperature and damage. Serious damage had occurred at 75 degrees, for example.

The error was to ignore the flights on which no damage had occurred, on the basis that they were irrelevant. When these were plotted—seventeen flights at temperatures from 66 to 81 degrees—the effect of temperature suddenly stood out plainly. Damage was strongly associated with cold. It was as if, to weigh the proposition that California cities tend to be in the westernmost United States, someone made a map of California—omitting the non-California cities that would make the tendency apparent. A team of statisticians formed by the National Research Council to follow up the commission report analyzed the same data and estimated a "gambling probability" of 14 percent for a catastrophic O-ring failure at a temperature of 31 degrees.

Feynman discovered that some engineers had a relatively realistic view of the probabilities involved—guessing that a disaster might occur on one flight in two hundred, for example. Yet managers had adopted fantastic estimates on the order of one in a hundred thousand. They were fooling themselves, he said. They cobbled together such numbers by multiplying absurd guesses—that the chance of a turbine pipe bursting was one in ten million, for example.

He concluded his personal report by saying, "For a successful technology, reality must take precedence over public relations, for nature cannot be fooled." He joined his fellow commissioners for a ceremony at the White House Rose Garden. Then he returned home, as he now knew, to die.

EPILOGUE

· ÷ ·

*God forbid that we should give out a dream
of our own imagination for a pattern of the
world.*

— Francis Bacon

Nothing is certain. Werner Heisenberg wrote this message in the twentieth century's consciousness. The mathematician Kurt Gödel followed with a famous proof that no logical system can ever be consistent and complete. The possibilities of true knowledge seemed to fade.

Heisenberg formulated his uncertainty principle narrowly: A particle cannot have both a definite place and a definite momentum. Still, philosophers took note. The implications seemed to cover a broader territory than the atom and its interior. Yet Feynman scorned philosophers ("rather than embarrass them, we shall just call them 'cocktail-party philosophers' ") who overinterpreted the laws of physics by saying, for example,

"That all is relative is a consequence of Einstein, and it has profound influences on our ideas." In addition, they say, "It has been demonstrated in physics that phenomena depend upon your frame of reference." We hear that a great deal, but it is difficult to find out what it means. . . . After all, that things depend upon one's point of view is so simple an idea that it certainly cannot have been necessary to go to all the trouble of the physical relativity theory in order to discover it.

Einstein's relativity did not speak to human values. Those were, or were not, relative for reasons unrelated to the physics of objects moving at near-light speed. Borrowing metaphors from the technical sciences could be a dangerous practice. Did the uncertainty principle impose its inevitable fuzziness on any description of nature? Perhaps. But Feynman parted company with many of his colleagues. They looked to quantum uncertainty for an explanation of the many kinds of unpredictability that arise in the everyday, human-scale world: unpredictability in the weather, or indeterminacy in human behavior. Perhaps, some speculated, quantum unpredictability was the microscopic loophole through which free will and human consciousness entered the universe.

Stephen Hawking, typically, wrote: "The uncertainty principle signaled an end to Laplace's dream of a theory of science, a model of the universe that would be completely deterministic. . . . Quantum mechanics therefore introduces an unavoidable element of unpredictability or randomness into science." Feynman's view was different. Even in the 1960s he anticipated the understanding that would emerge in the modern study of chaotic phenomena: that unpredictability was already a feature of the classical world. He believed that a universe without a quantum uncertainty principle would behave—on the scales of planetary storm systems and human brains—just as erratically and freely as our own.

> It is usually thought that this indeterminacy, that we cannot predict the future, is a quantum-mechanical thing, and this is said to explain the behavior of the mind, feelings of free will, etc. But if the world *were* classical—if the laws of mechanics were classical—it is not quite obvious that the mind would not feel more or less the same.

Why? Because tiny errors, tiny gaps in our knowledge, are amplified by the interactions of complex systems until they reach large scales.

> If water falls over a dam, it splashes. If we stand nearby, every now and then a drop will land on our nose. This appears to be completely random. . . . The tiniest irregularities are magnified in falling, so that we get complete randomness. . . .
> Speaking more precisely, given an arbitrary accuracy, no matter how precise, one can find a time long enough that we cannot make predictions valid for that long a time. Now the point is that this length of time is not very large. . . . It turns out that in only a very, very tiny time we lose all

our information. . . . We can no longer predict what is going to happen! It is therefore not fair to say that from the apparent freedom and indeterminacy of the human mind, we should have realized that classical "deterministic" physics could not ever hope to understand it, and to welcome quantum mechanics as a release from a "completely mechanistic" universe.

This discrepancy in beliefs—this subtle disagreement with the more standard viewpoint of physicists like Hawking—was no quibble. It formed a fulcrum on which turned, as the century neared its close, an essential disagreement about the achievements and the future of physics.

Particle physicists were awed by the effectiveness of their theories. They adopted a rhetoric of the "grand unified theory," a concept with its own acronym, GUT. Progress in science had long meant *unification* of phenomena that previously had been treated separately: Maxwell's electrodynamics had begun to unify electricity and light, for example. Steven Weinberg and Abdus Salam had unified the realms of electromagnetic and weak interactions with their (inevitably so-called) electroweak theory; however, this latter unification of such distant realms seemed more a mathematical tour de force than a demonstration that the two realms were two sides of one simple coin. Quantum chromodynamics attempted to embrace the strong interactions as well; however, experimental support seemed remote. Physicists now talked as though they could extend unification to cover everything, as though they could conceive of a time when physics would be able to close shop, its work complete. They could imagine—they could almost *see*—"the ultimate theory of the universe"; "nothing less than a complete description of the universe we live in"; "a complete unified theory of everything." The inflation of rhetoric accompanied a noticeable reversal of the physicists' political stature. The aura that had come with the success of the atomic bomb project was fading. To carry out increasingly high-energy experiments, physicists needed exponentially more-expensive machinery, and the question of financing such projects became politically divisive among scientists.

In the year of Feynman's death, a pair of experimental physicists introduced a text with the simple declaration, "Fifty years of particle physics research has produced an elegant and concise theory of particle interactions at the subnuclear level." Particle-physics outsiders could be less generous. Elegant and concise? Why, then, did so many particle masses and other specific numerical parameters have to be fed into the theory, rather than

read out? Why so many overlapping fields, so many symmetries broken—
it seemed—as necessary to fit the data? Quantum numbers such as color
and charm might be elegant simplifications, or they might be last-minute
rubber bands applied to joints that had threatened to spring loose. And if
theorists explained quark confinement, justifying a kind of particle that
could never stand on its own, they surely could explain anything. Was the
theory rigged—as one critic put it provocatively, "a contrived intellectual
structure, more an assembly of successful explanatory tricks and gadgets . . . than a coherently expressed understanding of experience"? Although each piece of the theory might have been tested against experiment,
the whole theory—the style of theory making—had become resistant to
disproof. It was hard to imagine phenomena that could not be explained
with a new symmetry breaking, a new quantum number, or a few extra
spatial dimensions. Perhaps the spare-parts department of modern physics was so well stocked with ingenious devices that a serviceable engine
could now be devised to handle any data the particle accelerators could
offer.

This was a harsh critique—not Feynman's. Still, in another time, Feynman had spoken of the search for the fundamental laws of nature. No
longer:

> People say to me, "Are you looking for the ultimate laws of physics?" No,
> I'm not. . . . If it turns out there is a simple ultimate law which explains
> everything, so be it—that would be very nice to discover. If it turns out
> it's like an onion with millions of layers . . . then that's the way it is.

He believed that his colleagues were claiming more success at unification
than they had achieved—that disparate theories had been pasted together
tenuously. When Hawking said, "We may now be near the end of the
search for the ultimate laws of nature," many particle physicists agreed. But
Feynman did not. "I've had a lifetime of that," he said on another occasion.
"I've had a lifetime of people who believe that the answer is just around
the corner."

> But again and again it's been a failure. Eddington, who thought that with
> the theory of electrons and quantum mechanics everything was going to
> be simple . . . Einstein, who thought that he had a unified theory just
> around the corner but didn't know anything about nuclei and was unable

of course to guess it. . . . People think they're very close to the answer, but I don't think so. . . .

Whether or not nature has an ultimate, simple, unified, beautiful form is an open question, and I don't want to say either way.

In the 1980s a mathematically powerful and experimentally untestable attempt at unification emerged in the form of string theory, using stringlike entities wrapped through many dimensions as their fundamental objects. The extra dimensions are supposed to fold themselves out of the way in a kind of symmetry breaking given the name *compactification*. String theory relies on Feynman's sum-over-histories method as an essential underlying principle; the theory views particle events as topological surfaces and computes probability amplitudes by summing over all possible surfaces. Feynman kept his distance, sometimes saying that perhaps he was too old to appreciate the new fashion. String theory seemed too far from experiment. He suspected that the string theorists were not trying hard enough to prove themselves wrong. In the meantime he never adopted the rhetoric of GUT's. It made him uncomfortable. He retreated into the stance that he himself merely solved problems as they came along.

When a historian of particle physics pressed him on the question of unification in his Caltech office, he resisted. "Your career spans the period of the construction of the standard model," the interviewer said.

" 'The standard model,' " Feynman repeated dubiously.

"$SU(3) \times SU(2) \times U(1)$. From renormalization to quantum electrodynamics to now?"

"The standard model, standard model," Feynman said. "The standard model—— is that the one that says that we have electrodynamics, we have weak interaction, and we have strong interaction? Okay. Yes."

The interviewer said, "That was quite an achievement, putting them together."

"They're not put together."

"Linked together in a single theoretical package?"

"No."

The interviewer was having trouble getting his question onto the table. "What do you call $SU(3) \times SU(2) \times U(1)$?"

"Three theories," Feynman said. "Strong interactions, weak interactions, and electromagnetic. . . . The theories are linked because they seem to have similar characteristics. . . . Where does it go together? Only if you add some stuff that we don't know. There isn't any theory today that has

$SU(3) \times SU(2) \times U(1)$—whatever the hell it is—that we know is right, that has any experimental check. . . . Now, these guys are all trying to put all this together. They're *trying* to. But they haven't. Okay?"

Particle physicists were his community. They were the elite who revered him, who passed along his legend, who lent him so much of his prestige. He rarely dissented publicly from their standard dogma. For the past two decades, he had worked on their problems: try though he might to *disregard*, in the end he had accepted their agenda.

"So we aren't any closer to unification than we were in Einstein's time?" the historian asked.

Feynman grew angry. "It's a crazy question! . . . We're certainly closer. We know more. And if there's a finite amount to be known, we obviously must be closer to having the knowledge, okay? I don't know how to make this into a sensible question. . . . It's all so stupid. All these interviews are always so damned useless."

He rose from his desk and walked out the door and down the corridor, drumming his knuckles along the wall. The writer heard him shout, just before he disappeared: "It's goddamned useless to talk about these things! It's a complete waste of time! The history of these things is nonsense! You're trying to make something difficult and complicated out of something that's simple and beautiful."

Across the hall Murray Gell-Mann looked out of his office. "I see you've met Dick," he said.

Feynman had always set high standards for *fundamental* work, although he meant something broader by the word than many particle physicists did. Liquid helium and other solid-state problems had seemed to him as fundamental as the smallest-scale particle interactions. He believed that fundamentalness, like beauty or intelligence, was a multidimensional quality. He had tried to understand turbulence and quantum gravity. Throughout his career he had suffered painful periods of malaise, when he could not find a suitable problem. In later years he and his colleagues had seen their crowded field thin: bright young students, looking for fundamental issues on their own terms, often turned to biology, computation, or the new study of chaos and complexity. When his son, Carl, ended his flirtation with philosophy and took up computer science, Feynman, too, looked again at the field he had helped pioneer at Los Alamos. He joined two

Caltech authorities on computation, John Hopfield and Carver Mead, in constructing a course on issues from brain analogues and pattern recognition to error correction and uncomputability. For several summers he worked with the founders of Thinking Machines Corporation, near MIT, creating a radical approach to parallel processing; he served as a high-class technician, applying differential equations to the circuit diagrams, and as an occasional wise man among the young entrepreneurs ("Forget all that 'local minima' stuff—just say there's a bubble caught in the crystal and you have to shake it out"). And he began to produce maverick research at the intersection of computing and physics: on how small computers could be; on entropy and the uncertainty principle in computing; on simulating quantum physics and probabilistic behavior; and on the possibility of building a quantum-mechanical computer, with packets of spin waves roaming ballistically back and forth through the logic gates.

His own community had largely left behind questions with the spirit that first drove him toward physics. An intellectual distance had opened between the subatomic particle universe and the realm of ordinary phenomena—the magic that nature reveals to children. In *The Feynman Lectures* he spoke allegorically of the beauty of a rainbow. Imagine a world in which scientists could not see a rainbow: they might discover it, but could they sense its beauty? The essence of a thing does not always lie in the microscopic details. He supposed that the blind scientists learned that, in some weathers, the intensity of radiation plotted against wavelength at a certain direction in the sky would show a bump, and the bump would shift from one wavelength to another as the angle of the instrument shifted. "Then one day," he said, "the physical review of the blind men might publish a technical article with the title 'The Intensity of Radiation as a Function of Angle under Certain Conditions of the Weather.' " Feynman had no quarrel with *beauty*—our human illusion, our projection of sentiment onto a reality of radiation phenomena.

"We are all reductionists today," said Steven Weinberg—meaning that we seek the deepest explanatory principles in the elementary particles that underlie ordinary matter. He spoke for many particle physicists but not for Feynman. Understanding the principles at the lowest level of the hierarchy—the smallest length-scales—is not the same as understanding nature. So much lies outside the accelerators' domain, even if it is in some sense *reducible* to elementary particles. Chaotic turbulence; the large-scale structures that emerge in complex systems; life itself: Feynman spoke of "the infinite variety and novelty of phenomena that can be generated from such simple principles"—phenomena that are "in the equations; we just haven't found the way to get them out."

> The test of science is its ability to predict. Had you never visited the earth, could you predict the thunderstorms, the volcanoes, the ocean waves, the auroras, and the colorful sunset? . . .
>
> The next great era of awakening of human intellect may well produce a method of understanding the qualitative content of equations. Today we cannot. Today we cannot see that the water-flow equations contain such things as the barber pole structure of turbulence that one sees between rotating cylinders. Today we cannot see whether Schrödinger's equation contains frogs, musical composers, or morality—or whether it does not.

Physicists' models are like maps: never final, never complete until they grow as large and complex as the reality they represent. Einstein compared physics to the conception a person might assemble of the interior mechanism of a closed watch: he might build a plausible model to account for the rhythmic ticking, the sweep of the hands, but he could never be certain. "He may also believe in the existence of the ideal limit of knowledge and that it is approached by the human mind," Einstein said. "He may call this ideal limit the objective truth." It was a simpler time. In Feynman's era, knowledge advanced, but the ideal of objective truth receded deeper into the haze beyond the vision of science. Quantum theory had left an impossible question dangling in the air. One physicist chose to answer it by quoting Feynman, "one of the great philosophers of our time, whose view of the matter I have taken the liberty of quoting in the form of the poetry it surely is":

> We have always had a great deal of difficulty
> understanding the world view
> that quantum mechanics represents.
>
> At least I do,
> because I'm an old enough man
> that I haven't got to the point
> that this stuff is obvious to me.
>
> Okay, I still get nervous with it. . . .

You know how it always is,
every new idea,
it takes a generation or two
until it becomes obvious
that there's no real problem. . . .

I cannot define the real problem,
therefore I suspect there's no real problem,
but I'm not sure
there's no real problem.

In October 1987 another abdominal tumor appeared, and his doctors made one last attempt to stall his cancer surgically. When the *Los Angeles Times* sent him an advance copy of his obituary, he thanked the author but said, "I have decided it is not a very good idea for a man to read it ahead of time: it takes the element of surprise out of it." He knew he was not recovering. He was sixty-nine years old. Pain wracked one of his legs. He was exhausted. He had no appetite. In January he began awakening in the night with sweats and chills. In one corner of his dusty office blackboard he had written a pair of self-conscious mottoes: "What I cannot create I do not understand" and "Know how to solve every problem that has been solved." Nearby was a running list under the heading, "To LEARN" ("Bethe Ansatz Prob., 2D Hall . . ."). Physics changed; he talked about it once with his old Los Alamos friend Stanislaw Ulam, who had been watching a few white clouds roll against the blue New Mexico sky. Feynman seemed to read his mind: "It is really like the shape of clouds," he said. "As one watches them they don't seem to change, but if you look back a minute later, it is all very different." He had not accumulated much: a hand-knitted scarf, hanging on a peg, from some students in Yugoslavia; a photograph of Michelle with her cello; some black-and-white pictures of the aurora borealis; his deep leather recliner; a sketch he had made of Dirac; a van painted with chocolate-brown Feynman diagrams. On February 3 he entered the UCLA Medical Center again.

Doctors in the intensive care unit discovered a ruptured duodenal ulcer. They administered antibiotics. But his remaining kidney had failed. One round of dialysis was performed, with little effect. Feynman refused the further dialysis that might have prolonged his life for weeks or months. He told Michelle calmly, "I'm going to die," in a tone that said: *I* have decided. He was watched and guarded now by the three women who had loved him longest: Gweneth, Joan, and his cousin Frances Lewine, who had lived with him in the house in Far Rockaway. Morphine for pain and an oxygen tube were their last concessions to medicine. The doctors said it would take about five days. He had watched one death before—trying to be scientific, observing the descent into coma and the sporadic breathing, imagining the brain clouding as it was starved of oxygen. He had anticipated his own— toying with the release of consciousness in dark sensory-deprivation tanks, telling a friend that he had now taught people most of the good stuff he knew, and making his peace with bottomless nature:

> You see, one thing is, I can live with doubt and uncertainty and not knowing. I think it's much more interesting to live not knowing than to have answers which might be wrong. I have approximate answers and possible beliefs and different degrees of certainty about different things, but I'm not absolutely sure of anything and there are many things I don't know anything about, such as whether it means anything to ask why we're here. . . .
>
> I don't have to know an answer. I don't feel frightened by not knowing things, by being lost in a mysterious universe without any purpose, which is the way it really is as far as I can tell. It doesn't frighten me.

He drifted toward unconsciousness. His eyes dimmed. Speech became an exertion. Gweneth watched as he drew himself together, prepared a phrase, and released it: "I'd hate to die twice. It's so boring." After that, he tried to communicate by shifting his head or squeezing the hand that clasped his. Shortly before midnight on February 15, 1988, his body gasped for air that the oxygen tube could not provide, and his space in the world closed. An imprint remained: what he knew; how he knew.

ACKNOWLEDGMENTS

NOTES

A FEYNMAN BIBLIOGRAPHY

BIBLIOGRAPHY

INDEX

ILLUSTRATION CREDITS

ACKNOWLEDGMENTS

· ÷ ·

I never met Feynman. I've relied on the published (and semipublished) record; on his own accumulation of personal letters, notes to himself, and other documents, released to me in 1988 by Gweneth Howarth Feynman; on letters shared by other family members and friends; on his office files and other documents stored in the California Institute of Technology Archives in Pasadena; on early material collected at the Niels Bohr Library of the American Institute of Physics in New York. I obtained recently declassified notebooks and papers from the archives of Los Alamos National Laboratory. Other material came from the libraries and manuscript collections of the following institutions: the American Philosophical Society (papers of H. D. Smyth and J. A. Wheeler); the Brooklyn Historical Society; Cornell University (papers of H. A. Bethe); Far Rockaway High School; Harvard University; the Library of Congress (papers of J. R. Oppenheimer); the Massachusetts Institute of Technology; Princeton University; Rockefeller University; and the Stanford Linear Accelerator Center.

The leading physicists who play the largest roles in this book agreed to provide their own recollections in interviews that sometimes extended over many sessions: Hans Bethe, Freeman Dyson, Murray Gell-Mann, Julian Schwinger, Victor Weisskopf, John Archibald Wheeler, and Robert R. Wilson.

Feynman's own voice is everywhere in his published work, of course, and toward the end of his life, wherever he went, tape recorders and video cameras seemed to be running. But several interviews of Feynman by historians and others were especially valuable. The deepest and most comprehensive—a central resource for anyone studying Feynman—is an oral history of many hundreds of pages conducted by Charles Weiner for the American Institute of Physics in 1966 and 1973; I used Feynman's copy of the transcript, with his handwritten corrections and comments. I also consulted the AIP's oral-history interviews with Bethe, Dyson, William A. Fowler, Werner Heisenberg, Philip Morrison, and others. The physicist and historian Silvan S. Schweber kindly shared the tape of his revealing 1980 interview on the development of quantum electrodynamics and on Feynman's style of visualization. Lillian Hoddeson conducted a useful interview of Feynman for her technical history of Los Alamos. Robert Crease gave me the transcript of an interview for his and Charles Mann's *The Second Creation*. Christopher Sykes gave me access to the uncut interview he conducted for what became the 1981 BBC-TV production, *The Pleasure of Finding Things Out*. Sali Ann Kriegsman gave me her transcript of Feynman's recollections of Far Rockaway.

Ralph Leighton, who drew from Feynman the reminiscences that became *Surely You're Joking, Mr. Feynman!* and *What Do You Care What Other People Think?*, generously provided the original tapes of these interviews over nearly a decade. These are the stories that Feynman retold and refined over his lifetime—mostly accurate, but strongly filtered. I have tried not to lean on them too heavily, for reasons that I hope emerge in the text.

Feynman's family members also spoke with me at length: Gweneth, Joan, Carl, and Michelle Feynman and Frances Lewine. Helen J. Tuck, his secretary of many years, shared her invaluable memories and perceptive comments.

Among the many other colleagues, students, friends, and observers of Feynman who helped me by submitting to interviews or providing written recollections—and in some cases copies of letters and diary pages—were Jan Anbjørn, Robert Bacher, Michel Baranger, Barry Barish, Henry H. Barschall, Mary Louise Bell, Rose Bethe, Jerry Bishop, James Bjorken, Peter A. Carruthers, Robert F. Christy, Michael Cohen, Sidney Coleman, Monarch L. Cutler, Predrag Cvitanović, Cecile DeWitt-Morette, Russell J. Donnelly, Sidney Drell, Leonard Eisenbud, Timothy Ferris, Richard D. Field, Michael E. Fisher, Evelyn Frank, Steven Frautschi, Edward Fredkin, Sheldon Glashow, Marvin Goldberger, David Goodstein, Frances R. (Rose McSherry) Graham, William R. Graham, Jules Greenbaum, Bruce Gregory, W. Conyers Herring, Simeon Hutner, Albert Hibbs, Douglas R. Hofstadter, Gerald Holton, John L. Joseph, Daniel Kevles, Sándor J. Kovács, Donald J. Kutyna, Janijoy La Belle, Leo Lavatelli, Ralph Leighton, Charles Lifer, Leite Lopes, Edward Maisel, Anne Tilghman Wilson Marks, Robert E. Marshak, Leonard Mautner, Robert M. May, William H. McLellan, Carver Mead, Nicholas Metropolis, Maurice A. Meyer, Philip Morrison, Masako Ohnuki, Paul Olum, Abraham Pais, David Park, John Polkinghorne, Burton Richter, John S. Rigden, Michael Riordan, Daniel Robbins, Matthew Sands, David Sanger, J. Robert Schrieffer, Theodore Schultz, Al Seckel, Barry Simon, Cyril Stanley Smith, Norris Parker Smith, Novera H. Spector, Millard Susman, Kip S. Thorne, Yung-Su Tsai, John Tukey, Tom van Sant, Dorothy Walker, Robert L. Walker, Steven Weinberg, Charles Weiner, Theodore A. Welton, Arthur S. Wightman, Jane Wilson, Stephen Wolfram, and George Zweig.

Two indispensable histories of twentieth-century physics are Kevles, *The Physicists*, and Pais, *Inward Bound*.

I'm especially grateful to Mitchell Feigenbaum and Silvan S. Schweber for patient guidance and sharp insights on matters of physics. I particularly thank Schweber for letting me read the manuscript-in-progress of his forthcoming history of quantum electrodynamics, *QED: 1946–1950: An American Success Story*. I thank Predrag Cvitanović for permission to quote his fable of Quefithe. Robert Chadwell Williams, a biographer of Klaus Fuchs, sent a helpful mass of archival material relating to the

Manhattan Project. I benefited from discussions with Joseph N. Straus and Hugh Wolff about genius, music, and music theory.

Cheryl Colbert lent me her intelligent and resourceful assistance. Emilio Millan shared a useful file of clippings and other documents that he had collected.

This book owes an enormous obligation to the skills of my editor, Daniel Frank, and my agent, Michael Carlisle.

As always, the indescribable debt is to Cynthia Crossen, who for so long endured, among other things, that strange, persistent presence of an extra soul in our household.

J. G.
Brooklyn, New York
8 July 1992

NOTES

· ÷ ·

ABBREVIATIONS

AIP: Niels Bohr Library, Center for the History of Physics, American Institute of Physics.

BET: H. A. Bethe papers, Cornell University.

CIT: California Institute of Technology Archives.

CPL: *The Character of Physical Law.*

F-H: Interview with Lillian Hoddeson and Gordon Baym, 16 April 1979. LANL.

F-L: Interviews with Ralph Leighton. Tapes courtesy of Leighton.

F-Sch: Interview with Silvan S. Schweber, 13 November 1988. Tape courtesy of Schweber.

F-Sy: Interview with Christopher Sykes, recorded in preparation for *The Pleasure of Finding Things Out,* BBC-TV, 1981. Tape courtesy of Sykes.

F-W: Interviews with Charles Weiner, 4 March 1966, 27–28 June 1966, and 4 February 1973. AIP.

FOI: Feynman's FBI files and documents from other federal agencies, obtained through the Freedom of Information Act.

LANL: Los Alamos National Laboratory Archives.

Lectures: *The Feynman Lectures on Physics.*

LOC: Library of Congress.

MIT: Massachusetts Institute of Technology Libraries.

NL: "The Development of the Space-Time View of Quantum Electrodynamics." Nobel lecture (Feynman 1965a; cf. Feynman 1965b and 1965c). For convenience, page numbers refer to the Weaver 1987 reprint.

OPP: J. R. Oppenheimer papers. LOC.

PERS: Personal papers obtained by the author.

PUL: Princeton University Libraries.

QED: QED: *The Strange Theory of Light and Matter.*
SMY: H. D. Smyth papers, American Philosophical Society.
SYJ: *Surely You're Joking, Mr. Feynman!*
WDY: *What Do You Care What Other People Think?*
WHE: J. A. Wheeler papers, American Philosophical Society.

PROLOGUE

The account of the Pocono meeting is based on interviews with several of the participants (Hans Bethe, Robert Marshak, Abraham Pais, Julian Schwinger, Victor Weisskopf, and John Archibald Wheeler), on Feynman's account in *Physics Today* (Feynman 1948*d*) and his recollections in F-W, on Wheeler's handwritten and mimeographed notes (Wheeler 1948), on correspondence in the J. R. Oppenheimer papers, on historical essays by Silvan S. Schweber (1983 and forthcoming), and on my visit to the site.

3 NOTHING IS CERTAIN: Feynman to Arline Feynman, 9 May 1945, PERS.

3 IT GNAWED AT HIM: Feynman 1975, 132.

3 WOMEN SIDLED AWAY: AIP, 423.

3 HALF GENIUS AND HALF BUFFOON: Freeman Dyson to his parents, 8 March 1948; Dyson, interview, Princeton, N.J.

4 NO TRANSCRIPT: John Archibald Wheeler made and later circulated several dozen pages of handwritten notes, however (Wheeler 1948).

5 PRINCIPLES: "Addresses," notebook, PERS.

6 THE MOST BRILLIANT YOUNG PHYSICIST: "He is by all odds the most brilliant young physicist here, and everyone knows this." Smith and Weiner 1980, 268.

6 THE KEY EQUATION: Hans Bethe, interview, Ithaca, N.Y.

6 TWISTING A CONTROL KNOB: Victor Weisskopf had brought the trains from Russia. "He played the following game. The guy with the switches has to avoid an accident and the other one has to produce an accident. It was the most nervewracking game you can imagine, and Dick was absolutely into it. It didn't matter which role he played." Weisskopf, interview, Cambridge, Mass.

6 WHAT ABOUT THE EXCLUSION PRINCIPLE?: F-W, 471.

7 IS IT UNITARY?: Ibid., 472.

7 THIS WONDERFUL VISION OF THE WORLD: Dyson 1979, 62.

7 THANK GOD: W.H. Auden, "After Reading a Child's Guide to Modern Physics," *Selected Poetry of W. H. Auden* (New York: Vintage, 1971), 214.

7 A POEM FEYNMAN DETESTED: Feynman to Mrs. Robert Weiner, 24 October 1967, CIT. Auden wrote, "This passion of our kind/For the process of finding out/Is a fact one can hardly doubt"—and Feynman resented his adding, "But I would rejoice in it more/If I knew more clearly what/We wanted the knowledge for." Feynmen said: "We want it so we can love Nature more. . . . A modern poet is directly confessing not understanding the emotional value of knowledge of nature."

9 WE PUT OUR FOOT IN A SWAMP: Albert R. Hibbs, interview, Pasadena, Calif.

9 A LITTLE BIZARRE: Snow 1981, 142–43.

10　A SHALLOW WAY TO JUDGE: Morrison 1988, 42.

10　WE GOT THE INDELIBLE IMPRESSION: David Park, personal communication.

10　DICK COULD GET AWAY WITH A LOT: Sidney Coleman, interview, Cambridge, Mass.

10　FEYNMAN TRIED TO STAND ON HIS OWN: Kac 1985, 116.

10　THERE ARE TWO KINDS OF GENIUSES: Ibid., xxv.

11　ANGERED HIS FAMILY: E.g., Gweneth Feynman, interview, Altadena, Calif.; Gell-Mann 1989a, 50.

11　HE'S NO FEYNMAN, BUT: Morrison 1988, 42.

12　A HALF-SERIOUS DEBATE: Coleman, interview.

12　BOOK II, CHAPTER 41, VERSE 6: D. Goodstein 1989, 75.

13　PHILOSOPHERS ARE ALWAYS ON THE OUTSIDE: CPL, 173.

13　IT HAS NOT YET BECOME OBVIOUS: Feynman 1982, 471.

13　DO NOT KEEP SAYING TO YOURSELF: CPL, 129

13　NATURE USES ONLY THE LONGEST THREADS: Ibid., 34; draft, PERS.

15　AN OFFICIAL SECRECY ORDER: U. S. Department of Commerce Rescinding Order, 7 January 1966, CIT.

15　HE DID THE TRAINING IN STAGES: Ralph Leighton, interview, Pasadena.

16　A TWO-HANDED POLYRHYTHM: Theodore Schultz, interview, Yorktown Heights, N.Y.

16　AN HONEST MAN: Schwinger 1989, 48.

FAR ROCKAWAY

Family members and childhood friends provided recollections and copies of correspondence from the 1920s and 1930s: Joan Feynman, Frances Lewine, Jules Greenbaum (Arline Greenbaum's brother), Leonard Mautner, Jerry Bishop, Mary D. Lee, and Novera H. Spector. Far Rockaway High School and the Brooklyn Historical Society had records, school newspapers, Chamber of Commerce publications, and other useful documents from the period. Sali Ann Kriegsman and Charles Weiner kindly shared transcripts of oral-history interviews they had conducted with Lucille Feynman.

18　HE ASSEMBLED A CRYSTAL SET: F-W, 35.

18　WHEN ATMOSPHERIC CONDITIONS WERE RIGHT: SYJ, 5.

18　EINSTEIN WAS SHOWING: Einstein 1909.

18　IT SEEMS THAT THE AETHER: Weyl 1922, 172.

19　"THE SHADOW" AND "UNCLE DON": F-W, 35.

19　A COIL SALVAGED FROM A FORD: SYJ, 4.

19　STANDARD EMERGENCY PROCEDURE: Frances Lewine, interview, Washington, D.C., and Far Rockaway.

19　DANGLING HIS METAL WASTEBASKET: Lucille Feynman to Feynman, 8 August 1945, PERS.

19　HIS SISTER, JOAN: Joan Feynman, interview, Pasadena.

20　RICHARD WALKED TO THE LIBRARY: Feynman, interview conducted by Sali Ann Kriegsman, 27 October 1975.

21　WHEN I WAS A CHILD: Kazin 1951, 8–10.

21 IT SOMETIMES SEEMED THAT THE THINGS NEAR THE SEA: Feynman-Kriegsman.

21 SOMETIMES FELT GAWKY: Evelyn Frank, interview, Marina del Rey, Calif.

22 IF WE STAND ON THE SHORE: *Lectures*, II-2-1.

22 IS THE SAND OTHER THAN THE ROCKS?: Ibid.

22 WHEN FEYNMAN RETURNED: Gweneth Feynman, interview, Altadena; Feynman-Kriegsman.

22 THOSE LITTLE HATS THAT THEY WEAR: Feynman-Kriegsman.

23 THAT WAS THE WAY THE WORLD WAS: Ibid.

24 LUCILLE WAS THE DAUGHTER: Lucille Feynman, interview conducted by Charles Weiner, MIT Oral History Program, 4 February 1981.

25 DON'T GET MARRIED: Ibid.

25 DON'T COUNT YOUR CHICKENS: Ibid.

25 BEFORE THE BABY WAS OUT: F-W, 7–8.

25 HE WAS TWO BEFORE HE TALKED: Lucille Feynman–Weiner.

25 TWENTY-FIVE FEET HIGH: F-Sy.

25 HER MOTHER SUFFERED: Joan Feynman, interview.

26 WITHIN DAYS THE BABY: Ibid.

26 A BIRTH CERTIFICATE AND A HAT: Ibid.

26 SOME EVENINGS THE ADULTS: Lewine, interview.

26 THE HOUSEHOLD HAD TWO OTHER: Joan Feynman, interview conducted by Charles Weiner, MIT Oral History Program, 30 July 1981; Lewine, interview.

27 LOOK UP: Joan Feynman, "Relinquishing the Aurora," letter, *Eos*, 1989, 1649.

27 RITTY WIRED HIS LABORATORY: F-W, 35–37.

27 IT WORKS!: Joan Feynman, interview.

27 IT WAS WORTH IT: F-W, 34.

28 SO THAT HE CAN BETTER FACE THE WORLD: Melville Feynman to Feynman, 10 September 1944, PERS.

28 WHEN A CHILD DOES SOMETHING: Ibid.

28 WHEN MELVILLE TOOK HIS SON: F-W, 14.

28 SEE THAT BIRD?: WDY, 13–14.

29 "THAT," HE SAYS, "NOBODY KNOWS": F-Sy; cf. "Inertia," notes, n.d., CIT: "Is inertia an intrinsic fundamental force which will always defy a more ultimate analysis? Or is inertia a force which has its origin in the workings of other recognized forces like gravitation or electricity?"

30 IT'S A WAY OF DOING PROBLEMS: F-W, 15.

30 JOANIE, IF 2^x: Joan Feynman–Weiner.

30 ALGEBRA 2, TAUGHT BY MISS MOORE: Leonard Mautner, interview, Pacific Palisades, Calif.

30 HIS SCORE ON THE SCHOOL IQ TEST: Feynman 1965*d*, 15.

30 AN INTELLECTUAL DESERT: F-W, 39

30 A SET OF FOUR EQUATIONS: Ibid., 23 and 39.

30 ALL FEYNMAN REMEMBERED: Ibid., 38

31 ENERGY PLAYS AN IMPORTANT PART: "Energy," poem, n.d., AIP.

31 SCIENCE IS MAKING US WONDER: "We Are Forgetful," poem, n.d., AIP.

32 SISSY-LIKE: F-L; edited version in *SYJ*, 67.

32   THE SIGHT OF A BALL: WDY, 24.

32   ANXIETY WOULD STRIKE: Ibid., 21.

32   HIS FIRST CHEMISTRY SET: F-W, 33

32   GOODY-GOOD: Ibid., 21; Feynman 1965d, 11.

32   IN PHYSICS CLUB: The Dolphin, Far Rockaway High School, June 1935, 33.

33   MATH TEAM: SYJ, 10–11; Jerry Bishop, telephone interview; Novera H. Spector, telephone interview.

34   A LOUD SIGH: Feynman 1965d, 12.

34   FEYNMAN PLACED FIRST: The Dolphin, Far Rockaway High School, June 1935, 33.

34   TWO CHILDREN IN HIGH SCHOOL: F-W, 63; Mautner, interview.

36   MR. AUGSBURY ABDICATED: Harold I. Lief to Ralph Leighton, 10 December 1988.

36   MAD GENIUS: The Dolphin, Far Rockaway High School, June 1935.

36   SOME OBSERVATIONS SUPPORTED THE NOTION: Melsen 1952, 22.

37   HOW DO SHARP THINGS STAY SHARP: F-W, 46.

37   ALL THINGS ARE MADE OF ATOMS: Lectures, I-1-2

38   BELIEVE THE EXISTENCE OF ATOMS: Bohr 1922, 315.

38   PURE CHEMISTRY, EVEN TO-DAY: Encyclopaedia Britannica, 13th ed., 1926, 870.

39   MATTER IS UNCHANGEABLE: Boscovitch 1922, 36; Park 1988, 200–201.

40   THE SCIENCE KNOWN AS CHEMICAL PHYSICS: Slater 1975, 193.

40   WE HAVE BEEN FORCED TO RECOGNIZE: Bohr was creating publicity for his philosophical interpretation of quantum mechanics. The press cooperated enthusiastically, although it posed difficulties for headline writers. William L. Laurence of the New York Times wrote optimistically: "The new theory is expected . . . to take its place alongside relativity and quantum mechanics as one of the revolutionary developments of modern scientific thought. . . . Professor Bohr, after a lifetime of contemplation of both the ponderables and the imponderables of the physical and mental world, has come to discover an inherent essential duality. . . . In other words the very process of knowing one aspect of nature makes it impossible for us to know the other aspect." "Jekyll-Hyde Mind Attributed to Man," New York Times, 23 June 1933, 1.

40   FOR THE OCCASION: Joan Feynman–Weiner, 28–29.

40   KNOWLEDGE IS POWER: F-W, 78.

41   NEW YORK IN 1982: Chase 1932, 13.

41   ELECTRICITY POWERED THE HUMAN BRAIN: William A. Laurence, "Brain Phone Lines Counted as 1 Plus 15 Million Zeroes," New York Times, 25 June 1933.

41   IN AN OPENING-DAY STUNT: Dedmon 1953, 334.

41   HERE ARE GATHERED THE EVIDENCES: "Chicago Fair Opened by Farley; Rays of Arcturus Start Lights," New York Times, 28 May 1933.

41   A 151-WORD WALL MOTTO: "Science in 151 Words," New York Times, 4 June 1933.

42   EINSTEIN'S SUPPOSED CLAIM: Cf. Kevles 1987, 175, and Pais 1982, 309. Einstein seems not to have disavowed the remark when given the chance.

42   LIGHTS ALL ASKEW IN THE HEAVENS: New York Times, 9 November 1919, quoted in Pais 1982, 309.

42   A SERIES OF EDITORIALS: Pais 1982, 309.

42   MORE THAN ONE HUNDRED BOOKS: Clark 1971, 247.

42   TRANSMITTED BY UNDERWATER CABLE: Kevles 1987, 175.

42   WE HAVE EINSTEIN'S SPACE: Quoted in Clark 1971, 242.

44   THERE ARE NO PHYSICISTS IN AMERICA: Raymond T. Birge to John van Vleck, 10
     March 1927, quoted in Schweber 1986b, 55–56.

45   I BELIEVE THAT MINNEAPOLIS: Quoted in Kevles 1987, 168.

45   ON THE BEACH SOME DAYS: Lewine, interview; Joan Feynman, interview; Joan
     Feynman–Weiner.

45   SHORTLY HE FOUND HIMSELF LYING: F-W, 117.

45   ONE HORRIBLY RUDE BOY: Ibid., 118

46   ALL LEFT HIM FEELING INEPT: WDY, 20–23.

46   WITH THE COMING OF THE DEPRESSION: Joan Feynman–Weiner.

46   TO THE METROPOLITAN MUSEUM: Ibid., 31–32.

46   THE RADIO HAD PENETRATED: "Modernistic Radios," New York Times, 4 June 1933.

46   HE REWIRED A PLUG: F-W, 105–7.

46   HE FIXES RADIOS BY THINKING!: SYJ, 3.

47   WHAT ARE YOU DOING?: F-W, 107–8; SYJ, 7–8.

47   MERELY TO FIND A MATHEMATICS TEXTBOOK: Feynman 1965d, 10.

47   IF A BOY NAMED MORRIE JACOBS: Feynman to Morris Jacobs, 27 January 1987,
     CIT.

47   HE RECOGNIZED THE PLEASURE: Feynman 1965d, 11.

48   SCHWINGER KNEW HOW TO FIND BOOKS: Schweber, forthcoming.

48   THE PHYSICAL REVIEW: Kevles 1987, 218.

48   THAT YEAR HE CAREFULLY TYPED OUT: Julian Schwinger, interview, Bel Air, Calif.;
     Schwinger 1934. He later said (1983), he had been "parrot[ing] the wisdom of my
     elders, to be later rejected."

49   THEY AMAZED A DINNER PARTY: Marvin Goldberger, interview, Pasadena.

49   HE LONG RESENTED THE LOSS: F-W, 113; WDY, 33.

MIT

Among Feynman's fellow students and fraternity brothers, T. A. Welton, Conyers
Herring, John L. Joseph, Monarch L. Cutler, Leonard Mautner, Maurice A. Meyer,
and Daniel Robbins contributed the most revealing interviews. Welton has set down
his recollections of Feynman in a manuscript titled "Memories" (CIT), and the American Institute of Physics has the notebook in which he and Feynman developed their
view of quantum mechanics. Feynman's MIT transcript and some other academic
records were preserved in his personal papers. The archives of MIT provided some
correspondence and yearbooks. Joan Feynman made available her brother's letters to
her and her parents. Other important sources include: on physics at MIT, the memoirs
of John C. Slater (1975) and Philip Morse (1977), and Schweber's profile of Slater
(1989); on the early development of American quantum physics, Kevles 1987, Schweber,
forthcoming, and Sopka 1980; on the principle of least action, Lectures II-19, Park
1988, Gregory 1988, and QED; on anti-Semitism in science, Silberman 1985, Steinberg
1971, Lipset and Ladd 1971; Dobkowski 1979, and the remarkable correspondence

between Feynman's MIT professors and Harry D. Smyth (MIT and a confidential file at PUL).

52   IN THAT CASE YOU ARE COMPLETELY LOST: Heisenberg 1971, 15–16.
52   THE AMERICAN MIND: Menge 1932, 11.
53   FEYNMAN CHANGED: F-W, 131.
53   BUT THE DEPRESSION HAD FORCED: Kevles 1987, 250–51.
53   NIGHTMARE: Ibid.
53   FEEL THE CRAVING: Menge 1932, 10.
53   DESPITE ANTI-SEMITIC MISGIVINGS: Rabi, for example, recalled Columbia's reluctance in appointing him as its first Jew in 1929: "What happened in the American universities was [that] a department was in some sense like a club, very collegiate, family . . . and certainly the Jews were different, they didn't fit in too well." Quoted in Schweber, forthcoming.
53   HE HAD BEEN ONE OF THE YOUNG AMERICANS: Slater 1975, 131.
53   SLATER KEPT MAKING MINOR DISCOVERIES: Ibid., 130–35.
54   I DO NOT LIKE MYSTIQUES: Slater, oral-history interview, AIP. Quoted in Schweber 1989, 53.
54   HE DOES NOT ORDINARILY ARGUE: Quoted in Schweber, forthcoming.
54   THEY STUDY CAREFULLY THE RESULTS: Ibid.
55   ASSEMBLING A PHYSICS DEPARTMENT: Karl T. Compton, "An Adventure in Education," New York Times, 15 September 1935.
55   BARELY A DOZEN GRADUATE STUDENTS: Morse 1977, 125
56   THE INSTRUCTORS TOLD THE STUDENTS: Slater and Frank 1933, v–vii.
56   WHY DON'T YOU TRY BERNOULLI'S: F-W, 136
56   THE FIRST DAY EVERYONE HAD TO FILL OUT: Welton 1983; F-W, 137.
56   COOPERATION IN THE STRUGGLE: Ibid.
56   MR. FEYNMAN, HOW DID YOU: Ibid. Welton added that Feynman's solutions were "always correct and frequently ingenious" and that "Stratton never entrusted his lecture to me or any other student."
57   A LIFEGUARD, SOME FEET UP THE BEACH: QED, 51–52.
58   OUR FRIEND DIRAC, TOO: Quoted in Schweber, forthcoming.
58   THERE CANNOT BE ANY ATOMS: Descartes 1955, 264.
59   AT THE SAME TIME: Ibid., 299.
60   FEYNMAN WOULD RESORT TO INGENIOUS COMPUTATIONAL TRICKS: F-W, 139
60   FEYNMAN HAD FIRST COME ON THE PRINCIPLE: Lectures, II-19.
61   SEEMED TO FEYNMAN A MIRACLE: Ibid., II-19-2.
61   IT SEEMS TO KNOW: Gregory 1988, 32–33.
61   THIS IS NOT QUITE THE WAY: Park 1988, 250.
61   IT IS NOT IN THE LITTLE DETAILS: Quoted in Jourdain 1913, 11.
61   PARK PHRASED THE QUESTION: Park 1988, 252.
62   LET NONE SAY THAT THE ENGINEER: The Tech, MIT, 1938, 275.
62   BUT AFTER THEY HAVE CONQUERED: Ibid.
62   ONE ENJOYED A WOOING PROCESS: SYJ, 17.
62   THEIR FRATERNITY BROTHERS DROVE FEYNMAN: SYJ, 19; F-W, 200–201.

63 OPPORTUNITIES TO HARASS FRESHMEN: Daniel Robbins, telephone interview.

63 THE SECOND AND THIRD FLOORS: Maurice A. Meyer, telephone interview.

63 SO WORRIED ABOUT THE OTHER SEX: *SYJ*, 18.

63 COURSE NOTES TO BE HANDED DOWN: Michael Oppenheimer, interview, New York.

64 DICK FELT HE GOT A GOOD BARGAIN: *SYJ*, 18.

64 LONG HOURS AT THE RAYMORE-PLAYMORE: Robbins, interview.

64 THE FEYNMANS LET HER PAINT A PARROT: Lewine, interview.

64 SPARED DICK THE NECESSITY: *SYJ*, 18.

64 ARLINE WATCHED UNHAPPILY: Meyer, interview.

64 HIS SECOND PROPOSAL OF MARRIAGE: F-W, 302 and 122.

65 THE IMPORTANCE OF SCIENCE IN AVIATION: WDY, 31.

65 AT ONE OF THE FATEFUL MOMENTS: Feynman to Lucille Feynman, 9 August 1945, PERS; Weisskopf, interview.

66 THE INSTITUTE JUSTIFIED: F-W, 164

66 A PAIN IN THE NECK: Ibid.

66 IN ONE COURSE HE RESORTED: He admitted it thirty years later, embarrassed—"I lost my moral sense for a while"—to a scholar taking oral history for a science archive. F-W, 164.

66 WHY DIDN'T THE ENGLISH PROFESSORS: Ibid, 165.

66 HE READ JOHN STUART MILL'S: F-L; *SYJ*, 30.

66 HE READ THOMAS HUXLEY'S: F-W, 170-73.

66 MEANWHILE IN PHYSICS ITSELF: "Subjects taken in physics at Mass. Institute of Technology," typescript, PUL.

67 WHOM ARLINE WAS READING: F-W, 165-66.

67 HE KNEW ALL ABOUT IMPERFECTION: WDY, 29.

67 PEOPLE LIKE DESCARTES WERE STUPID: F-W, 166.

67 HE TOOK A STRIP OF PAPER: WDY, 29-30.

68 IN THE DISCOVERY OF SECRET THINGS: Gilbert, *De Magnete* (1600).

68 LIKE A PRIME MINISTER: F-W, 167.

68 THE PRAGMATIC SLATER: Schweber 1989, 58.

68 NOT FROM POSITIONS OF PHILOSOPHERS: Harvey, *De Motu Cordis et Sanguinis* (1628).

68 UP IN HIS ROOM: F-W, 169-70.

69 I WONDER WHY I WONDER WHY: Ibid., 170; F-L (*SYJ*, 33).

69 A DISMAYED, DISORIENTED MOMENT: F-L.

70 HE DID DEVELOP A RUDIMENTARY THEORY: *SYJ*, 36.

70 HE SAT THROUGH LECTURES: Ibid., 32.

70 SO MUCH STUFF IN THERE: F-W, 166.

70 SPACE OF ITSELF AND TIME OF ITSELF: Quoted by Feynman in *Lectures*, I-17-8.

72 A SMALL FABLE: Dirac 1971.

72 MY WHOLE EFFORT IS TO DESTROY: Quoted in Park 1988, 318.

73 OF COURSE QUITE ABSURD: Dirac 1971, 41.

74 DURING A LATE EROTIC OUTBURST: Pais 1986, 251-52.

74 THEY FILLED A NOTEBOOK: Feynman and Welton 1936-37.

74 JUST AS SCHRÖDINGER HAD DONE: F-W, 146

76 BOTH BOYS WERE WORRYING: Feynman and Welton 1936–37; F-W, 141.

76 WELTON WOULD SET TO WORK: F-W, 210–11.

77 THE CHUG-CHUG-DING-DING: Welton 1983; Welton, interview; F-W, 142–44.

77 THEY WORKED OUT FASTER METHODS: F-W, 152–53.

77 ALL I'VE DONE IS TAKE: Quoted in "Bright Flashes from a Mind of Marvel," *Washington Post*, 6 January 1990.

78 UTTER CERTAINTY: Heisenberg 1971, 11.

78 MORE THAN THAT OF ALL MANKIND: Ibid., 10.

78 FEYNMAN WANTED TO BE A SHOP MAN: F-W, 154–56; F-L.

79 ENRICO FERMI MADE HIS OWN: Segrè 1980, 204–6; Rhodes 1987, 210–12.

79 UNEXPECTEDLY, THE SLOW NEUTRONS: Enrico Fermi, "Artificial Radioactivity Produced by Neutron Bombardment," in Weaver 1987, 2:74.

79 FEYNMAN AND WELTON, JUNIORS: F-W, 162.

80 THERE WAS JUST ONE ESSENTIAL TEXT: Bethe et al. 1986.

80 THAT CLOUDS SCATTERED SUNLIGHT: F-W, 176.

81 IT CAME JUST ONE STEP PAST: *Lectures*, I-32-8.

81 ONE FOGGY DAY: F-W, 176.

82 FEYNMAN'S FIRST PUBLISHED WORK: Vallarta and Feynman 1939.

82 A PROVOCATIVE AND CLEVER IDEA: "Suppose we consider a particle sent into an element of volume dV of scattering matter in a direction given by the vector R. Let the probability of emerging in the direction R' be given by a scattering function $f(R,R')$ per unit solid angle. Conversely a particle entering in the direction R' will have a probability $f(R',R)$ of emerging in the direction R. Let us assume that the scatterer (magnetic field of the star) has the reciprocal property so that $f(R,R')=f(R',R)$. In our case the property is satisfied provided the particle's sign is reversed at the same time as its direction of motion. That is, the probability of electrons going by any route is equal to the probability of positrons going by the reverse route. . . ." Ibid.

82 SUCH AN EFFECT IS NOT TO BE EXPECTED: Heisenberg 1946, 180.

82 YOU'RE THE LAST WORD: F-W, 178.

82 HE CAUGHT ONE CLASSMATE: Monarch L. Cutler, telephone interview and personal communication; F-W, 179; Cutler, "Reflection of Light from Multi-Layer Films," senior thesis, MIT, 1939. The professors were Hawley C. Cartwright and Arthur F. Turner.

83 THE PUTNAM COMPETITION: Joseph Gallian, Andrew Gleason, telephone interview.

83 ONE OF FEYNMAN'S FRATERNITY BROTHERS: Robbins, interview.

83 FEYNMAN LEARNED LATER: F-W, 191.

83 HIS FIRST THOUGHT HAD BEEN TO REMAIN: Ibid., 193–94.

83 PRACTICALLY PERFECT: John C. Slater to Dean of Graduate School, Princeton, 12 January 1939, PUL.

83 THE BEST UNDERGRADUATE STUDENT: Philip Morse to H. D. Smyth, 12 January 1939, PUL.

83 DIAMOND IN THE ROUGH: Wheeler 1989.

84 HAD NEVER BEFORE ADMITTED: Ibid.

84 THE PHYSICS SCORE WAS PERFECT: Individual Report of the Graduate Record Examination: Feynman, Richard P., 1939, PERS. Besides achieving a perfect physics

result, he scored high in the 99th percentile in mathematics; on the other hand, 69 percent of those taking the test outscored him in verbal skills, 85 percent in literature, and 93 percent in fine arts.

Feynman also applied to the University of California at Berkeley; the department there made it clear that he would be accepted but approved him only as the eighth alternate for a $650-a-year fellowship. Robert Sproul to Feynman, 30 March 1939, and Raymond T. Birge to Feynman, 1 June 1939, PERS.

84 IS FEYNMAN JEWISH?: H. D. Smyth to Philip Morse, 17 January 1939, MIT.

84 FEYNMAN OF COURSE IS JEWISH: Slater to Smyth, 7 March 1939, PUL.

84 PHYSIOGNOMY AND MANNER, HOWEVER: Morse to Smyth, 18 January 1939, MIT. Princeton was persuaded. Smyth later heard about Feynman's success in the Putnam competition and wrote: "My colleagues keep insisting that Feynman is not coming here next year because he took an examination and won a prize fellowship at Harvard. My position is that as long as I have his acceptance and no further word from him he is coming here even if he has been offered the *presidency* of Harvard." Smyth to Morse, 8 June 1939, MIT.

85 WE KNOW PERFECTLY WELL: Quoted in Silberman 1985, 90.

85 THEY TOOK OBVIOUS PRIDE: Francis Russell, "The Coming of the Jews," quoted in Steinberg 1971, 71.

85 BECAUSE, BROTHER, HE IS BURNING: Thomas Wolfe, *You Can't Go Home Again* (New York: Dell, 1960), 462. Quoted in Kevles 1987, 279.

85 IT WAS ALSO UNDERSTOOD: Sopka 1980, 4:105.

85 NEW YORK JEWS FLOCKED OUT HERE: Davis 1968, 83.

85 A FRUSTRATED OPPENHEIMER: J. R. Oppenheimer to Raymond T. Birge, 4 November 1943, 26 May 1944, and 5 October 1944, in Smith and Weiner 1980, 268, 275, and 284.

85 IF FEYNMAN EVER SUSPECTED: Silberman 1985, 91–92; F-W, 198.

86 HALF A LINE: F-W, 182.

87 INSTEAD OF SPINNING: Ibid., 180.

87 A SCIENCE OF MATERIALS: C. Smith 1981, 121–22.

87 MATTER IS A HOLOGRAPH OF ITSELF: Ibid., 122

88 AS FEYNMAN CONCEIVED THE STRUCTURE: Feynman 1939a and b.

90 IT IS TO BE EMPHASIZED: Feynman 1939a, 3; Conyers Herring, telephone interview.

90 HE COMPLAINED THAT FEYNMAN WROTE: Robbins interview.

90 SO HE WAS SURPRISED TO HEAR: F-W, 186. Slater, in his textbooks, preferred "Feynman's theorem" as late as 1963, though he had found that a German, H. Hellmann, had made the same discovery two years earlier. Slater 1963, 12–13; H. Hellmann, *Einführung in die Quantenchemie* (Leipzig: Deuticke, 1937).

91 THAT'S ALL I REMEMBER OF IT: F-W, 196.

91 IT SEEMED TO SOME THAT SLATER: Silvan S. Schweber, interview, Cambridge, Mass.

91 MY SON RICHARD IS FINISHING: Morse 1977, 125–26.

91 MORSE TRIED NOT TO LAUGH: Ibid. Although Morse did not say so, part of Melville's concern was whether anti-Semitism would block a career in physics; he expressed this in a similar conversation with John Wheeler a few years later (Wheeler 1989).

PRINCETON

Wheeler and many of his later students gave me some understanding of the relationship between Wheeler and Feynman. Wheeler 1979a and Klauder 1972 are sources of recollections. Wheeler shared the draft of his talk for a 1989 memorial session (Wheeler 1989). H. H. Barschall, Leonard Eisenbud, Simeon Hutner, Paul Olum, Leo Lavatelli, and Edward Maisel provided recollections of Feynman and the Princeton of the late thirties and early forties. John Tukey and Martin Gardner illuminated the history of flexagons. Robert R. Wilson discussed the isotron project and Feynman's initiation into the Manhattan Project, as well as much later history. The declassified documentary record of the isotron project, including a series of technical papers by Feynman, is in the Smyth papers at the American Philosophical Society.

93 A BLACK HOLE HAS NO HAIR: Wheeler and Ruffini 1971.
93 THERE IS NO LAW EXCEPT THE LAW: In Mehra 1973, 242.
93 I ALWAYS KEEP TWO LEGS GOING: John Archibald Wheeler, interview, Princeton, N.J.
93 IN ANY FIELD FIND THE STRANGEST THING: Boslough 1986, 109.
93 INDIVIDUAL EVENTS: Quoted in Dyson 1980, 54. As Dyson says, "It sounds like Beowulf, but it is authentic Wheeler."
94 SOMEWHERE AMONG THOSE POLITE FAÇADES: In Steuwer 1979, 214–15.
94 WHEN HE WAS A BOY: Bernstein 1985, 29; Wheeler 1979a, 221.
94 SLATER AND COMPTON PREFERRED: Slater 1975, 170–71.
94 WHEELER STILL REMEMBERED: Wheeler 1979a, 224.
94 WHEN WHEELER MET HIS SHIP: Ibid., 272.
95 IT WAS THIS LAST IMAGE: Bohr and Wheeler 1939.
95 THEY SPENT A LATE NIGHT TRYING: Bernstein 1985, 38.
96 WHEELER SAID THAT HE WAS TOO BUSY: H. H. Barschall, telephone interview.
96 YOU LOOK LIKE YOU'RE GOING TO BE: F-W, 209; Leonard Eisenbud, telephone interview.
96 THE NEXT TIME FEYNMAN SAW BARSCHALL: Barschall, interview.
96 WHEELER'S POINTED DISPLAY: F-W, 194 and 215–16.
97 LAZY AND GOOD-LOOKING: Mizener 1949, 34 and 38.
97 A QUAINT CEREMONIOUS VILLAGE: Einstein to Queen Elizabeth of Belgium, 20 November 1933, quoted in Pais 1982, 453.
97 THE OBLIGATORY BLACK GOWNS: SYJ, 49.
97 WHEN THE MATHEMATICIAN CARL LUDWIG SIEGEL RETURNED: Dyson 1988b, 3.
97 SURELY YOU'RE JOKING: F-W, 209; SYJ, 48–49.
97 IT BOTHERED HIM THAT THE RAINCOAT: Feynman to Lucille Feynman, 11 October 1939, PERS.
98 HE TRIED SCULLING: Ibid.
98 WHEN HE ENTERTAINED GUESTS: Feynman to Lucille Feynman, [?] October 1940, PERS.

98 HE EARNED FIFTEEN DOLLARS A WEEK: Feynman to Lucille Feynman, 3 March 1940, PERS.

98 THEY LISTENED WITH AWE: Edward Maisel, telephone interview; cf. F-W, 254.

98 AS WHEELER'S TEACHING ASSISTANT: Feynman to Lucille Feynman, 11 October 1939; Feynman notes on nuclear physics, H. H. Barschall papers, AIP.

99 IN CHOOSING A THEME: Schweber, forthcoming.

99 IT SEEMS THAT SOME ESSENTIALLY NEW: Dirac 1935, 297; NL, 434.

99 WILHELM RÖNTGEN, THE DISCOVERER OF X RAYS: Dresden 1987, 11.

100 EVEN NOW FEYNMAN DID NOT QUITE UNDERSTAND: F-W, 230.

100 HE PROPOSED—TO HIMSELF: NL, 434.

100 SHAKE THIS ONE: Ibid.

101 IT IS FELT TO BE MORE ACCEPTABLE: Bridgman 1952, 14–15.

102 THE TENSION IN THE MEMBRANE: Weinberg 1977a, 19.

102 WHEELER, TOO, HAD REASONS: Wheeler, interview.

102 HE ENJOYED TRYING TO GUESS: SYJ, 69–71.

103 ALTHOUGH HE TEASED THEM: F-L, for SYJ, 71.

104 "FLEXAGONS" LAUNCHED GARDNER'S CAREER: Gardner 1989; Albers and Alexanderson 1985.

104 SIRS: I WAS QUITE TAKEN: Quoted in Gardner 1989, 13–14.

104 FEYNMAN SPENT SLOW AFTERNOONS: SYJ, 77.

105 DON'T BOTHER ME: F-L; WDY, 56.

105 HUMAN SPERMATOZOA: Maisel, interview.

105 THEY DECIDED THAT THEIR BRAINS: WDY, 55–57.

105 WE WERE INTERESTED AND HAPPY: John Tukey, interview, Princeton, N.J.

105 HE READ SOME POEMS ALOUD: Maisel interview.

105 RHYTHM IS ONE OF THE PRINCIPAL TRANSLATORS: "Some Notes on My Own Poetry," in Sitwell 1987, 131.

105 WHILE A UNIVERSE GROWS IN MY HEAD: "Tattered Serenade," in Sitwell 1943, 19.

106 IT'S CLEAR TO EVERYBODY AT FIRST SIGHT: F-L.

106 WHEELER WAS ASKED FOR HIS OWN VERDICT: SYJ, 51; Wheeler 1989, 2–3.

106 THE PALMER PHYSICAL LABORATORY: Princeton University Catalogue: General Issue, 1941–42. PUL.

107 PRINCETON'S GAVE FEYNMAN A SHOCK: SYJ, 49–50.

107 THE HEAD OF THE CYCLOTRON BANISHED FEYNMAN: Wheeler 1989, 3.

107 IT DOES NOT TURN AT ALL: A sound explanation—with a description of a safer experiment than Feynman's—is in Mach 1960, 388–90. But physicists have never stopped arguing for either of the other answers, and there is an ongoing literature.

109 THERE IS NO SIGNBOARD: Eddington 1940, 68.

109 UNFORTUNATELY HE HAD MEANWHILE LEARNED: F-W, 233; NL, 435.

110 A BROADCASTING ANTENNA, RADIATING ENERGY: Cf. Feynman's later discussion of radiation resistance, Lectures, I-32-1.

110 HE ASKED WHEELER: F-W, 233–34; NL, 436.

111 TIME DELAY HAD NOT BEEN A FEATURE: Wheeler and Feynman 1949, 426.

111 THE WAVES WERE NOW RETARDED: Lectures, I-28-2.

111 VIEWED IN CLOSE-UP: Morris 1984, 137.

112 SHAKE A CHARGE HERE: F-W, 237.

112 OH, WHADDYAMEAN, HOW COULD THAT BE?: Feynman 1965*b*.

112 THE WORK REQUIRED INTENSE CALCULATION: He wrote his parents in November: ". . . last week things were going fast & neat as all heck, but now I'm hitting some mathematical difficulties which I will either surmount, walk around, or go a different way—all of which consumes all my time—but I like to do very much & and am very happy indeed. I have never thought so much so steadily about one problem . . . I'm just beginning to see how far it is to the end & how we might get there (altho aforementioned mathematical difficulties loom ahead)— SOME FUN!" Feynman to Lucille Feynman, November 1940, PERS.

112 FOR THOSE WHO WERE SQUEAMISH: Feynman 1941*a*, fig. 3 caption.

112 THEN THE EFFECT OF THE SOURCE: Feynman 1948*b*, 941.

113 HE DESCRIBED IT TO HIS GRADUATE STUDENT FRIENDS: F-W, 237–38.

113 FOR EXAMPLE, COULD ONE DESIGN A MECHANISM: Wheeler and Feynman 1949, 426–27; Hesse 1961, 279.

113 AS LONG AS THE THEORY RELIED ON PROBABILITIES: Feynman 1941*a*, 20.

113 HE CONTINUED TO CHERISH A NOTION: Wheeler, oral-history interview, 17 November 1985, 12, AIP.

113 EARLY IN 1941 HE TOLD FEYNMAN: Cf. Recommendation of Richard Phillips Feynman for Appointment as Porter Ogden Jacobus Fellow for 1941–1942, PUL.

113 AS THE DAY APPROACHED: F-W, 242–44; SYJ, 64–66.

115 PAULI DID OBJECT: Wheeler 1989, 26. Much later Feynman said of Pauli's objection: "It's too bad that I cannot remember what, because the theory is not right and the gentleman may well have hit the nail right on the bazeeto." F-W, 244. Pauli also presumably saw that the theory could not be quantized.

115 DON'T YOU AGREE, PROFESSOR EINSTEIN: F-W, 244.

115 HIS OWN EQUIVOCAL BALANCE SHEET: Feynman to Lucille Feynman, 3 March 1940, PERS.

115 LECTURE HIS FRIENDS: Simeon Hutner, telephone interview.

116 HOURS WHEN I HAVEN'T MARKED DOWN: Feynman to Lucille Feynman, November 1940, PERS.

116 BEFORE REVEALING IT TO ARLINE: Paul Olum, telephone interview.

116 SHE SENT HIM A BOX OF PENCILS: WDY, 43–44.

116 IF YOU DON'T LIKE THE THINGS I DO: Arline Greenbaum to Feynman, n.d., PERS.

117 THIS STYLE OF TREATMENT: Teller 1988, 97.

117 AN OLD FRATERNITY FRIEND PICKED HER UP: Robbins, interview.

117 HE CERTAINLY BELIEVES IN PHYSICAL SOCIETY: Ibid.

117 STILL, HE WORRIED: F-W, 252–53; Feynman 1941*a* is the manuscript on which he based the talk. Feynman and Wheeler 1941 is the published abstract.

117 THE ACCELERATION OF A POINT CHARGE: Feynman 1941*a*.

118 WHEELER NEEDED LITTLE ENCOURAGEMENT: Feynman (F-W, 243) thought the visit to Einstein "probably" came before his lecture; Wheeler remembers it coming after, and the acknowledgments in Feynman 1941*a* and Wheeler and Feynman 1945 suggest that Wheeler must be right.

118 EINSTEIN RECEIVED THIS PAIR: Wheeler 1989, 27.

118 FEYNMAN WAS STRUCK: F-W, 254.

118   AN OBSTINATE HERETIC: Quoted in Pais 1982, 462.

118   THE STRANGE LITTLE PAPER: *Physikalische Zeitschrift* 10(1909):323; Wheeler 1989, 27; Pais 1982, 484.

119   WE MUST DISTINGUISH BETWEEN TWO TYPES: Feynman 1941a, 13; Schweber 1986a, 459.

119   "PROF WHEELER," HE WROTE: Feynman 1941a, 13.

120   THE SUN WOULD NOT RADIATE: *Zeitschrift für Physik* 10(1922):317, quoted in Wheeler and Feynman 1945, 159–60.

120   LEWIS, TOO, WORRIED: Stuewer 1975, 485 and 499.

120   I AM GOING TO MAKE: Lewis, "The Nature of Light," *Proceedings of the National Academy of Sciences* 12(1926):22, quoted in Wheeler and Feynman 1945, 159 n.

121   THESE WERE DEAD ENDS: F-W, 260.

121   IT PROVED POSSIBLE TO COMPUTE PARTICLE INTERACTIONS: The first application of the least-action principle in this context came in work of which Wheeler and Feynman were not yet aware: a paper by A. D. Fokker in *Zeitschrift für Physik* 58(1929):386.

121   IN THE ABSORBER THEORY: NL, 438–39.

121   THE MORE FEYNMAN WORKED: Ibid., 440.

121   WE HAVE, INSTEAD: Ibid.

122   AN IMAGE, SO TO SPEAK: Minkowski, "Space and Time," in Weaver 1987, 2:156; Galison 1979.

122   FEYNMAN, I KNOW WHY: NL, 441.

122   IT WAS THE FIRST ANTIPARTICLE: Dirac, however, was reluctant to accept the idea of a new antiparticle; he first assumed that this positively charged particle must be the proton, despite the enormous discrepancy in mass.

123   EINSTEIN HAD WORRIED ABOUT THIS: Park 1988, 234.

123   A PHILOSOPHER, ADOLPH GRÜNBAUM: "The Anisotropy of Time," in Gold 1967, 149; Adolph Grünbaum, telephone interview.

123   MR. X: Feynman was enraged at the postconference suggestion that the proceedings be published; he surprised the other participants by declaring that there was no such subject as "the nature of time." Grünbaum said later: "Who was he worried about? If he was worried about people in the know then this device failed. I don't see how a man of his towering eminence could feel his reputation would be jeopardized." Grünbaum, interview.

123   GRÜNBAUM: I WANT TO SAY: Gold 1967, 178–79.

124   WHATEVER HIDDEN BRAIN MACHINERY: Ibid., 183.

124   ONE'S SENSE OF THE NOW: Morris 1984, 146.

124   ONE CAN SAY EASILY ENOUGH: Park 1988, 234.

125   IT'S A POOR MEMORY: Gold 1967, 235.

125   THIS PROCESS LEADS: Ibid., 4.

126   THREE ARROWS OF TIME: Ibid., 13–14.

126   IT'S A VERY INTERESTING THING: Ibid., 186.

126   HE HAD COME TO BELIEVE: F-W, 301.

127   HE READ UP ON TYPHOID: Ibid., 303; WDY, 34–35.

127   FEYNMAN HAD FELT FROM THE BEGINNING: F-W, 246.

127 SOMETIMES WHEELER TOLD FEYNMAN: Ibid., 268.

127 "OH?" PAULI SAID: Ibid., 245–46; cf. *SYJ*, 66.

127 WHEELER CANCELED THE LECTURE: F-W, 255 ("Q: The culmination of this grand paper was what? A: The culmination was, his grand paper has never come out").

128 DIRAC HAD PUBLISHED A PAPER: Dirac 1933.

128 THE NEXT DAY JEHLE AND FEYNMAN: NL, 443.

129 YOU AMERICANS!: F-W, 272; Schweber 1986*a*.

130 HERE IS A GREAT MAN: Robert R. Wilson, interview, Ithaca, N.Y.

130 NOTEBOOK OF THINGS: Feynman 1940; F-W, 287–88.

130 FEYNMAN WAS ASKED WHICH COLOR: F-W, 289–90.

130 FEYNMAN HAD BEEN FRUSTRATED: Ibid., 220–21.

131 AS I'M TALKING: WDY, 59.

132 IN FEYNMAN'S MIND A SEQUENCE: F-W, 273–74.

133 ALEXANDER FLEMING HAD NOTICED: Macfarlane 1984; Root-Bernstein 1989, 166–68.

134 THAT PATHOLOGICALLY LUXURIANT MORBID GROWTH: Mann 1927, 286–87.

134 FEYNMAN WAS BACK IN THE LIBRARY: The account of Feynman's relationship with Arline Greenbaum is based in part on two versions by Feynman: in F-W, 304; and in WDY, 35. Though more than twenty years apart, these are not independent versions; their wording is so consistent that Feynman must have reviewed his copy of the AIP interview before tape-recording the version that was then published, with further editing, in Feynman 1988.

135 GOODBYE LOVE LETTER: WDY, 38.

135 WHEN SHE CONFRONTED RICHARD: Cf. Arline Greenbaum to Feynman, 3 June 1941, PERS.

135 HE WAS SUPPORTING HIMSELF: Fellowship records, PUL.

135 WHEN HE TOLD A UNIVERSITY DEAN: F-W, 309.

136 A PHYSICISTS' WAR: Kevles 1987.

136 A NUMBER IN THE AIR: Wilson, interview.

136 THE HUNGARIAN CONSPIRACY: Rhodes 1987, 308.

136 I NEVER THOUGHT OF THAT: Ibid., 305.

136 WILSON AND SEVERAL OTHER PHYSICISTS: Wilson, interview.

137 THE BRITISH HAD INVENTED: Rigden 1987, 130.

137 IT'S SIMPLE—IT'S JUST A KIND OF WHISTLE: Edward U. Condon, quoted in Kevles 1987, 304.

137 OFFERED TO JOIN THE SIGNAL CORPS: Feynman 1981.

137 FROM THEIR WINDOWS THE BELL RESEARCHERS: *SYJ*, 83–84.

137 IT WAS A CHANCE TO SERVE: F-W, 294.

138 ONE-FOURTH OF THE NATION'S SEVEN-THOUSAND-ODD PHYSICISTS: Kevles 1987, 320. He estimates that the number included "three quarters of [the physics profession's] eminent leadership."

138 THE FIELD OF MECHANISMS, DEVICES: Compton, "Scientists Face the World of 1942," quoted in Schweber, forthcoming.

138 A PRIMITIVE SORT OF ANALOG COMPUTER: F-W, 294–95; *SYJ*, 85–87.

138 FEYNMAN FOUND HIMSELF DRAWN: Mitchell Feigenbaum, interview, New York.

139   HE CONSIDERED THE CASE OF TWO PARTICLES: Feynman 1941*b*.

139   THIS PREOCCUPATION WITH: Ibid.

139   THERE WAS A POSSIBILITY: Wilson, interview.

139   AN EXPATRIATE GERMAN CHEMIST: Peierls 1985, 169.

140   ONE MORNING HE HAD GONE INTO HIS KITCHEN: Rhodes 1987, 340.

140   STUDENTS WERE ASKED TO CHOOSE: Lavatelli, interview.

140   IF THERE WAS ANY BALONEY: Wilson, interview.

140   TO HIS DISMAY: Ibid.; F-W, 297.

140   SLIGHTLY DISILLUSIONED WITH WAR WORK: "I guess my patriotism had disintegrated or something." F-W, 297.

140   LONG AFTERWARD, AFTER ALL THE BOMB MAKERS: Ibid.

141   TO GET HELP WITH THE ELECTRONICS: Wilson, interview.

141   THE SENIOR THEORETICIAN CRUMPLED: Olum, interview.

142   WHAT'S HAPPENING HERE?: Ibid.

142   IT WAS LIKE A CARTOON: F-W, 298.

142   ERNEST LAWRENCE WAS CALLING A COMPETING DEVICE: Heilbron and Seidel 1989, 515–16.

143   WHEN EXPERIMENTERS TRIED HIGHER VOLTAGES: F- W, 320.

143   THE PHYSICISTS HAD TO INVENT: Ernest D. Klema, n.d., Response to Nuclear Physics Questionnaire. AIP.

143   MEANWHILE THE PROJECT'S WORST ENEMY: R. Wilson 1972, 474–75.

143   WHEN GENERAL LESLIE R. GROVES: Groueff 1967, 36–38.

144   FEYNMAN CARRIED THE ISOTRON'S FLYSPECK: F-W, 325–26.

144   THE FIRST SCIENTIFIC LECTURE HE HAD EVER HEARD: Ibid., 325.

144   WILSON WAS STUNNED: He wrote Smyth nearly a year later, from Los Alamos: "I am still not able to think objectively about the closing down of our project. It was certainly a hysterical move for the committee to shut the project down before the completion of the contract." Wilson to Smyth, 27 November 1943, LANL.

144   SMYTH AND WIGNER BOTH FELT PRIVATELY: Davis 1968, 136.

144   LAWRENCE'S CALUTRON SIMPLY USED: Lavatelli, quoted in Davis 1968, 135.

144   FEYNMAN HAD PRODUCED DETAILED CALCULATIONS: Feynman 1942*f*; Feynman 1943*a*; Smyth and Wilson 1942, 5.

145   MY WIFE DIED THREE YEARS AGO: Olum, interview.

146   IT WAS TIME TO FINISH HIS THESIS: Wheeler to Feynman, 26 March 1942, AIP.

146   LATER HE REMEMBERED: F-W, 281.

146   GREAT DIFFICULTIES HAVE ARISEN: Feynman 1942*a*.

146   MESON FIELD THEORIES HAVE BEEN SET UP: Feynman 1942*b*, 1 n.

146   DERIVED CONCEPT: Feynman 1942*a*.

146   WE CAN TAKE THE VIEWPOINT: Ibid.

147   IS IN FACT INDEPENDENT OF THAT THEORY: Feynman 1942*b*, 5.

147   WHEN HE WAS DONE: Wheeler and Wigner 1942.

147   FEYNMAN CONCLUDED WITH A BLUNT CATALOG: Feynman 1942*b*, 73–74.

147   IN THE MATHEMATICS WE MUST DESCRIBE: Ibid.

148   HONORARY ELECTRICIAN'S LICENSE: Feynman to George W. Beadle, 4 January 67, CIT. Turning down the first honorary degree he was offered, he told the president

of the University of Chicago that he remembered "the guys on the same platform receiving honorary degrees without work—and felt an 'honorary degree' was a debasement of the idea of a 'degree which confirms certain work has been accomplished.' . . . I swore then that if by chance I was ever offered one I would not accept it. Now at last (25 years later) you have given me a chance to carry out my vow."

148 THE RELATIONSHIP BETWEEN HUSBAND AND WIFE: Flick 1903, 289.

148 MANY A YOUNG CONSUMPTIVE MOTHER: Ibid., 288.

148 MARRIAGE IS APT TO BE: Underwood 1937, 342.

149 THEY WERE BOTH SO YOUNG: Solomon 1952, 122.

150 YOUR HEALTH IS IN DANGER: Lucille Feynman to Feynman, "Why I object to your marriage to Arline at this time," n.d., PERS.

150 HE TOLD HIS FATHER: Feynman to Melville Feynman, 15 June 1942, PERS.

150 BUT JUST A FEW DAYS LATER: Feynman to Lucille Feynman, "Why I want to get married," June 1942, PERS.

150 IN NO TIME FLAT: Arline Greenbaum to Feynman, June 1942, PERS.

151 SHE WALKED DOWN: Jules Greenbaum, telephone interview.

151 THEY MARRIED IN A CITY OFFICE: WDY, 42–43.

151 FEARFUL OF CONTAGION: "I knew not to kiss her . . . because the disease, I was afraid to catch it" (F-L); by contrast, the edited version, in SYJ, 43, says that Feynman, "bashful," kissed Arline on the cheek.

LOS ALAMOS

I did not seek the security clearance necessary to make direct use of the archives of the Los Alamos National Laboratory; however, the archives eventually provided a body of declassified material, including the notebook Feynman began keeping in his first days on the site, portions of his personnel record, and many technical documents—critical-mass calculations, analyses of computing issues, and notes and diagrams from Feynman's inspections of the Oak Ridge plant. Lillian Hoddeson and Gordon Baym shared their interview with Feynman about many of his classified notes. Also declassified is Feynman's manuscript for the account of the theoretical-physics division in what became the Smyth report, *Atomic Energy for Military Purposes*, and a related correspondence between Smyth, Oppenheimer, and Groves. Mary D. Lee had preserved a copy of Feynman's 9 August 1945 letter to his mother, describing the Trinity test. Feynman had saved Arline's personal papers, including their correspondence, her correspondence with her family, and other items. Much has been written about the Manhattan Project and the scientists who participated in it. Still, one or two things may remain to be said. Many individual memoirs are available. The best overall history is Richard Rhodes's *Making of the Atomic Bomb*. Hawkins et al. 1983 is extremely useful for its technical detail. If there was ever a time when eyewitness accounts could be obtained uncontaminated by hindsight and by many previous tellings, it is long past. I reinterviewed some participants and friends of Feynman anyway (Bethe, Weisskopf, Wilson, Olum, Welton, Rose Bethe, Philip Morrison, Robert Bacher, Robert Christy,

Robert Walker, Dorothy Walker). Nicholas Metropolis expanded on his published recollections of the laboratory's nascent computer science. Other sources on computation
include Alt 1972, Asprey 1990, Bashe et al. 1986, Goldstine 1972, Nash 1990, and
Williams 1985. Feynman retold his best stories in a talk (1975) at the University of
California, Santa Barbara. The tone of his letters in 1943–45 is very different, and I
have relied most heavily on these.

153   HE SWEATED: Feynman to Lucille Feynman, 9 August 1945, PERS.
153   THEN, SUDDENLY, MUSIC: Ibid.; Weisskopf, interview. But one of the oddities in
        the memories of that moment is how many different scientists heard different
        music. James W. Kunetka, for example, (1979) heard "The Star-Spangled Ban-
        ner."
154   MINUS THIRTY MINUTES: Feynman to Lucille Feynman, 9 August 1945.
154   AND THEN, WITHOUT A SOUND: Frisch 1979, 164.
154   IT BLASTED; IT POUNCED: Talk at Boston Institute for Religious and Social Studies,
        3 January 1946. In Rabi 1970, 138–39.
155   WHAT WAS THAT?: Peierls 1985, 202; Feynman 1975, 131. The correspondent
        was William L. Laurence. Eventually he came to terms with the sound he
        heard: "Then out of the great silence came a mighty thunder . . . the blast
        from thousands of blockbusters going off simultaneously . . . the big boom
        . . . earthquake . . . the first cry of a newborn world." Laurence 1959,
        117.
155   ENRICO FERMI, CLOSER TO THE BLAST: E.g., Kunetka 1979, 169.
155   ANOTHER PHYSICIST THOUGHT FEYNMAN: Jette 1977, 105.
155   NOW HE HAD BEEN DRIVEN SO LOW: Frisch 1979, 155.
156   A CHILL, WHICH WAS NOT THE MORNING COLD: Quoted in Rhodes 1987, 675.
156   IT'S A TERRIBLE THING THAT WE MADE: SYJ, 118.
156   WE JUMPED UP AND DOWN: Feynman to Lucille Feynman, 9 August 1945.
157   IT IS A WONDERFUL SIGHT: Ibid.
157   WE BECAME THEN: R. Wilson 1972, 475.
157   HAVE THEM DESCRIBE TO YOU: F-W, 328; Wilson, interview.
157   HE DID GATHER INFORMATION: F-W, 329.
157   WE ALL CAME TO MEET THIS BRASH CHAMPION: Morrison 1988, 42; also Morrison,
        oral-history interview, 7 February 1967, AIP, 34: "He was already heralded as
        this very clever fellow from Princeton who knew everything. And he did know
        everything, you know."
157   FEYNMAN SAW THAT THE PROBLEM: F-W, 330.
158   SCHWINGER, WHO WAS AMBIDEXTROUS: Bernard Feld, quoted in Schweber, forth-
        coming.
158   SOMEDAY WHEN THEY MAKE A MOVING PICTURE: F-W, 332; Olum, interview.
159   OPPENHEIMER'S FORMULA: Peierls, quoted in Heilbron and Seidel 1989, 256.
159   A PHYSICS OF BANK SHOTS: Rhodes 1987, 149.
159   WHY DON'T YOU HAVE FISH: Peierls 1985, 190.
159   HE CALLED LONG-DISTANCE: F-W, 337.
160   NOBODY COULD THINK STRAIGHT: Davis 1968, 163.
160   THE STATE OF SECRECY WAS SUCH: F-W, 332.

160 FEYNMAN'S CONTRARIETY WARRED: Feynman 1975, 108.

160 SHE HAD BEGGED RICHARD: Arline Feynman to Feynman, 26 March 1943, PERS.

160 ARLINE CRIED NIGHT AFTER NIGHT: Ibid. and Arline Feynman to Feynman, 19 March 1943, PERS.

161 YET ONE POSSIBILITY WAS PLAYING ITSELF OUT: F-H, 5.

161 AT FIRST THE ONLY TELEPHONE LINK: John H. Manley, "A New Laboratory Is Born," in Badash et al. 1980, 31.

161 WATER BOILER: Hawkins et al. 1983, 104–5; F-H, 4–6.

162 A TABLE BEHIND A HEAVY CONCRETE WALL: Groueff 1967, 210.

162 THE DRIVER'S LICENSE OF A NAMELESS ENGINEER: State of New Mexico Operator's License no. 185, 1944, PERS.

162 WELCOME TO LOS ALAMOS: Frisch 1979, 150.

163 TALKS ARE NOT NECESSARILY ON THINGS: Notebook, "A-83-002 7-7," LANL.

163 REFLECT NEUTRONS . . . KEEP BOMB IN: Ibid.

164 MOST OF WHAT WAS TO BE DONE: Feynman 1944.

164 THE GHOSTWRITER WAS FEYNMAN: Smyth to Oppenheimer, 1 February 1945, and Oppenheimer to Smyth, 14 April 1945, LANL.

164 FEYNMAN, GIVING SMYTH A TOUR: SYJ, 118; Groueff 1967, 326.

164 A REQUEST FOR OSMIUM: Groueff 1967, 326.

164 THE FIRST DOT OF PLUTONIUM: Hawkins et al. 1983, 72.

165 LISTED THE MAIN QUESTIONS: Feynman 1944. Feynman's references to tamper materials, along with some other sensitive technical details, were deleted from the report as published.

165 WHEN THEY HEARD THAT LAUGH: E.g., Joseph O. Hirschfelder, "Scientific-Technological Miracle at Los Alamos," in Badash et al. 1980, 81.

165 BETHE AND FEYNMAN—STRANGE PAIR: Frisch 1979, 154.

165 YOU'RE CRAZY: F-W, 339; Bethe, interview; Groueff 1967, 205.

166 IF FEYNMAN SAYS IT THREE TIMES: Schweber, forthcoming.

166 HE HAD WORKED ON: Groueff 1967, 207.

166 A WESTERN UNION KIDDIEGRAM: Rhodes 1987, 416.

166 BETHE HAD LEARNED HIS PHYSICS: Bernstein 1980, 29.

166 AT ROME: L. Fermi 1954, 217.

166 LIGHTNESS OF APPROACH: Bernstein 1980, 31.

168 BETHE LEFT THE INITIAL LECTURES: F-H, 40; Bethe, interview.

168 THE DANGEROUS PRACTICALITIES: Hawkins et al. 1983, 13.

168 FEYNMAN SPENT A LONG TIME THINKING: F-H, 12–13.

168 BRANCHING-PROCESSES THEORY: Ulam 1976, 153; Harris 1963; David Hawkins, "The Spirit of Play," in Cooper 1989.

169 HE ARRIVED AT A PRACTICAL METHOD: Bethe, interview.

169 BEGAN TO LOVE HANS BETHE: F-W, 409–10.

169 HE HAD INVITED ONE OF HIS MIT FRATERNITY FRIENDS: Feynman to Daniel Robbins, 24 June 1942, PERS.

169 HE WOULD BE PARTLY OUT OF THE RUSH: Feynman to Lucille Feynman, 24 June 1943, PERS.

169 WHEN HE WAS INVITED TO MEET A STRANGER: Welton 1983, 7.

170   DO YOU KNOW WHAT WE'RE DOING HERE?: Ibid.

170   IT STINKS: Davis 1968, 215.

170   AS WELTON LISTENED: Welton, interview.

170   HE WAS AMUSED AND IMPRESSED: Welton 1983, 8–9; Welton, interview.

171   WELTON BECAME THE FOURTH PHYSICIST: Along with Frederick Reines, Julius Ashkin, and Richard Ehrlich.

171   DEFINITELY UNGENTLE HUMOR: Welton 1983, 9.

171   ALL RIGHT, PENCILS: F-H, 42–43.

172   BY DEFINITION, AT CRITICAL MASS: Hawkins et al. 1983, 77.

172   FOR A SPHERICAL BOMB: Welton 1983, 11; Welton, interview.

173   BETHE HAD TOLD THEM: Bethe, interview; F-H, 23.

173   WHEN THE LOS ALAMOS METALLURGISTS: Hawkins et al. 1983, 139.

173   IT PUSHED THE THEORISTS PAST THE LIMITS: Welton 1983, 13.

173   FEYNMAN SOLVED THAT PROBLEM: Feynman and Welton 1947, a book-length report, draws together the chief findings of Feynman and his group on critical-mass calculations and neutron scattering. Feynman's own contribution to the version of the problem in which neutrons are assumed to have a single characteristic velocity—a practical simplification of methods developed by others—appears in Feynman 1946b.

173   THE EXPERIENCE OF ACTUAL COMPUTATION: F-H, 23–24.

173   AS HE DROVE THE MEN: Welton 1983, 14.

173   THAT SEEMED AN IMPOSSIBLE LEAP: Ashkin, Ehrlich, and Feynman 1944. Welton recalled wryly (1983, 14): "Only a short period of reflection was . . . required before Feynman announced that we were going to take the accumulated computational results from T-2, put them through the meat grinder, season them with some further insights (yet to be produced) and extrude this mixture as a handy interpolation-extrapolation formula."

174   UNFORTUNATELY CANNOT BE EXPECTED: Feynman 1946b, 3.

174   UNFORTUNATELY THE FIGURES CONTAINED: Ashkin, Ehrlich, and Feynman 1944, 4.

174   THESE METHODS ARE NOT EXACT: Feynman and Welton 1947, 6.

174   AN INTERESTING THEOREM WAS FOUND: Feynman 1946b, 3.

174   IN ALL CASES OF INTEREST: Feynman and Welton 1947, 6.

175   BETHE'S DEPUTY, WEISSKOPF: Weisskopf, oral-history interview, 31, AIP.

176   HE TOLD THEM HE COULD SPOT: F-H, 18.

176   WELL, FOUR HOURS AND TWENTY MINUTES AGO: Nicholas Metropolis, interview, Los Alamos, N.M.

176   YOU KNOW HOW IT IS WITH DAYLIGHT SAVING TIME: Morrison 1988, 42.

177   YOU WANT TO KNOW EXACTLY?: Feynman 1975, 109.

177   THAT'S 1.35: F-H, 41.

178   ALL RIGHT. IT'S PI TO THE FOURTH: Ibid., 39.

178   THEN PAUL OLUM SPOKE UP: Olum, interview; F-L for SYJ, 176.

178   SIMILARLY, WORKING WITH BETHE: Bethe, interview.

179   THEY WERE RARELY USED: Metropolis and Nelson 1982, 348–49.

180   LET'S LEARN ABOUT THESE DAMNED THINGS: Metropolis, interview.

180 THEY SPENT HOURS TAKING APART: Bethe, interview; Metropolis 1990, 237; Metropolis and Nelson 1982, 349.

180 ESCALATION OF THE COMPUTATION EFFORT: Metropolis and Nelson 1982, 350.

181 SO MUCH MORE POWERFUL WERE THEY: Weisskopf 1991, 134.

181 EVEN BEFORE THE IBM MACHINES ARRIVED: F-W, 362–63; Brode 1960; Feynman 1975, 125.

182 HE LEFT FEYNMAN WITH TWO ENDURING MEMORIES: Feynman 1975, 129.

182 FEYNMAN THOUGHT AT FIRST: F-H, 55–56: "We discovered a very annoying thing that we didn't understand. . . . When we set up the differential equation, we solved it numerically and the numbers seemed to come out irregularly. Then we would check and it would be the same thing. . . . The points would sort of wiggle around irregularly, and [von Neumann] explained that that was correct, that was all right, that was very interesting. . . . And there was nothing we could do about it. We just had to live with it, and we did. . . . We were terribly surprised by the fact that we would do the numbers over again, and it was the same crazy irregularities."

182 EACH TIME IT IS TURNED ON: T. Reid 1984, 14; Alt 1972, 693; Metropolis and Nelson 1982, 352.

182 SOME INTERESTING PROPERTIES OF NUMBERS: The lecture later became an element of his course at Cornell in mathematical methods and then, refined once again, became a remarkable set piece in his *Feynman Lectures on Physics*. Feynman to Lucille Feynman, 29 February 1944; *Lectures*, I-22.

182 ALL THE MIGHTY MINDS: Feynman to Lucille Feynman, 29 February 1944.

184 HE IS BY ALL ODDS THE MOST BRILLIANT: Oppenheimer to Birge, 4 November 1943, in Smith and Weiner 1980, 269.

184 HE IS A SECOND DIRAC: Oppenheimer to Birge, November 4 1943, in Smith and Weiner 1980, 269.

184 SHE LAUGHED, ASKING: Arline Feynman to Lucille and Melville Feynman, 28 June 1943; F-L for SYJ, 46.

185 ONCE, IN A FANCIFUL CONVERSATION: Moss 1987, 68.

185 A PIONEER PEOPLE STARTING A NEW TOWN: Brode 1960, 7.

185 THE MOST EXCLUSIVE CLUB IN THE WORLD: James Tuck, quoted in Davis 1968, 184.

185 WHAT EXACTLY IS SQUARE: Tuck, quoted in Brode 1960, 7.

185 ONE PARTY FEATURED AN ORIGINAL BALLET: Brode 1960, 6.

186 CODES, CIPHERS OR ANY FORM OF SECRET WRITING: Reprinted in Jette 1977, 130.

186 IT'S VERY DIFFICULT WRITING: Feynman 1975, 112.

187 RICHARD AND ARLINE TALKED ABOUT: Ibid.

187 THERE ARE CAPTAINS: Feynman to Lucille Feynman, 10 December 1943.

187 I EXPLAINED IT TO HIM: Feynman to Arline Feynman, 8 March 1945.

187 THE SECURITY STAFF TOLERATED THEM: Hirschfelder, in Badash et al. 1980, 79.

187 THE NEW DISPENSER STRUCK FEYNMAN: Olum, interview.

187 HE HAD GOT SO DRUNK: Feynman to Arline Feynman, 9 May 1945, PERS.

188 MORALLER AND MORALLER: Ibid.

188 TWO MEN ARRIVED: Feynman to Arline Feynman, 4 April 1945, PERS.

188 BECAUSE I LIKE PUZZLES SO MUCH: Ibid.

188   THAT ONE INSIGHT: *SYJ*, 124.

189   THE LOS ALAMOS PHYSICISTS: E.g., Frisch 1979, 154.

189   I OPENED THE SAFES: F-L for *SYJ*, 121–22.

189   THIS LAST INSIGHT ALONE: *SYJ*, 133.

189   BY FIDDLING WITH HIS OWN SAFE: Ibid., 124–25.

190   FEYNMAN REVELED IN THE CLOUDS: Feynman to Lucille Feynman, 10 December 1943, PERS.

190   SEE, I'M GETTING AN AESTHETIC SENSE: Ibid.

191   NOT CERTAIN WHETHER THIS TIME: Ian McEwan, *The Innocent* (New York: Doubleday, 1990), 85.

191   HE WOULD SIT IN A GROUP: F-W, 317.

191   HE FOUND A WAY: Feynman to Arline Feynman, 3 April 1945, PERS.

191   YOU ARE A STRONG AND BEAUTIFUL WOMAN: Ibid.

192   DON'T GET SCARED THO: Feynman to Arline Feynman, 24 April 1945.

192   YOU'RE NEVER THAT: Arline Feynman to Feynman, February 1945.

192   THE SCIENTISTS WERE IN AN UPROAR: R. Wilson 1974, 160.

193   I DO NOT KNOW—ALTHOUGH THERE ARE THOSE: Cohn 1943, 56–57; Arline Feynman, handwritten notes, PERS.

193   DARLING I'M BEGINNING: Arline Feynman to Feynman, 16 January 1945.

193   SHE REMINDED HIM OF THE FUTURE: Arline Feynman to Feynman, 17 January 1945.

194   WE HAVE TO FIGHT HARD: Ibid.

194   DRINK SOME MILK: Feynman to Arline Feynman, 2 May 1945, PERS.

194   HER WEIGHT HAD FALLEN: Arline Feynman, notebook of medical records and expenses, PERS.

194   YOU ARE A NICE GIRL: Feynman to Arline Feynman, 2 May 1945.

194   TIME PASSES FAST: Feynman to Arline Feynman, 19 April 1945, PERS.

194   HE HAD HEARD ABOUT A NEW DRUG: Feynman to Richard Gubner, draft, n.d., and Gubner to Feynman, 29 August 1944 and 14 November 1944, PERS.

194   JOAN WAS DAZZLED: Joan Feynman to Arline Feynman, 29 April 1945, PERS.

195   HE THOUGHT HE SAW SYMPTOMS: Feynman to Arline Feynman, 18 May 1945.

195   HAVE IT DONE BY A SPECIALIST: Henry Barenblatt to Arline Feynman, 19 April 1945 and 23 April 1945, PERS.

195   A DOCTOR AT LOS ALAMOS TOLD RICHARD: Feynman to Arline Feynman, 3 May 1945.

195   THE SAME DOCTOR: Ibid.

195   P.S. 59-TO-BE: Feynman to Arline Feynman, 15 May 1945.

195   HE DRIFTED THROUGH THE PAGES: Feynman to Arline Feynman, 17 May 1945.

195   KEEP HANGING ON: Feynman to Arline Feynman, 9 May 1945.

195   ENTIRE NATION CELEBRATES: Feynman to Arline Feynman, 10 May 1945.

195   AT THE MAYO CLINIC: Waksman 1964, 127–28.

196   THE DOCTOR WHO FIRST ISOLATED: Ibid., 115–18.

196   WORKERS HANDLING PLUTONIUM: Hawkins et al. 1983, 163–64.

196 One man, Harry Daghlian: L. Fermi 1980, 99; de Hoffman 1974, 166–67; Frisch 1979, 159–60.

197 Feynman himself proposed a safer experiment: Hawkins et al. 1983, 89.

197 At Teller's request: E. Teller to R. F. Bacher, 27 March 1944, LANL.

198 He became responsible for calculating: E.g., K. T. Bainbridge to Members of Committee on Fabrication and Assembly of Active Materials, 20 July 1944 and 5 September 1944, LANL.

198 It is expected that a considerable fraction: Bethe to Oppenheimer, 8 November 1944, and Bethe to Bacher, 3 January 1945, LANL; Robert F. Bacher, interview, Santa Barbara, Calif.

198 complete authority: Bethe to Oppenheimer, 26 January 1945, LANL.

198 Dear Sir, At the present time: J. L. Patterson to Major W. E. Kelley, 19 September 1944, and W. E. Kelley to Feynman, 21 September 1944, LANL.

198 As Segrè had discovered: Feynman 1975, 119–21.

198 Feynman began by retracing: F-H, 33.

199 He realized that the plant was headed: F-W, 353–54.

199 In answer to the Eastman superintendent's question: Feynman to Major W. E. Kelley, 27 September 1944, LANL.

199 During centrifuging some peculiar motion: Feynman to Colonel Arthur E. Peterson, 18 September 1945.

199 Is CT-1 empty when we drop: Notes, LANL.

199 He also invented a practical method: Feynman 1945.

199 A few people, long afterward: E.g.: "Unknowingly, he saved my life and the lives of everyone at Oak Ridge in those challenging years . . ." Irwin H. Goodwin to Ralph Leighton, 8 December 1988.

199 Feynman's first visit to Oak Ridge: F-L for SYJ, 104.

199 You should say: Los Alamos cannot accept: Feynman 1975, 122.

200 he had to grow up fast: Ibid.

200 Sometime that spring it struck him: F-H, 14.

200 Hitchhiking back one Sunday night: Feynman to Arline Feynman, 24 May 1945, PERS.

200 But they were kind of ugly: Ibid.

200 My Wife: I am always: Feynman to Arline Feynman, 6 June 1945, PERS.

201 One night he awoke: Feynman to Arline Feynman, 14 June 1945, PERS.

201 The group's productivity had risen: Bethe, interview.

201 He had invented a system: F-W, 371–74.

201 When he reached her room: Ibid., 343–46; F-L for WDY, 50–53.

202 The nurse recorded: Certificate of Death, PERS.

202 He came in and sat down: Robert and Dorothy Walker, interview, Tesuque, N.M.

202 When he comes in: Joan Feynman, interview.

202 An army car met him: Feynman to Lucille Feynman, 9 August 1945, PERS.

203 If a man had merely calculated: De Hoffman 1974, 171–72.

203 created not by the devilish inspiration: Smyth 1945, 223.

204   NO MONOPOLY: Notes, n.d., PERS.

204   MOST WAS KNOWN: Ibid.

204   IT WOULD SEEM TO ME THAT UNDER THESE CIRCUMSTANCES: Oppenheimer to Birge, 26 May 1944, in Smith and Weiner 1980, 276.

204   BIRGE FINALLY CAME THROUGH: Oppenheimer informed Birge of Feynman's choice in a blisteringly formal tone: "I am glad that you are going to take steps to increase the strength of the department. . . . Several months ago Dr. Feynman accepted a permanent appointment with the Physics Department at Cornell University. I do not know details of salary and rank, but they are presumably satisfactory to him. I shall of course do my best to call to your attention any men who are available . . ." (5 October 1944, in Smith and Weiner 1980, 284). The California offer did prompt Cornell, at Bethe's urging, to raise Feynman's salary before he arrived. His "potential" salary was $3,000; when Berkeley offered $3,900, Cornell agreed to $4,000. Bethe had written: "I know that it is unusual to raise a man's salary before he has even seen the University at which he is employed. The justification, I believe, is given by the unusual times and by the intimate knowledge that we here have acquired of Feynman's qualities." Bethe to R. C. Gibbs, 24 July 1945, and Gibbs to Feynman, 3 August 1945, CIT.

205   FEYNMAN BECAME THE FIRST OF THE GROUP LEADERS: Hawkins et al. 1983, 304.

205   IT WAS ON HIS LAST TRIP: WYD, 53.

CORNELL

Bethe provided access to his papers. Dyson shared copies of his remarkable letters home during these years (my portrait of him relies on these, on his various memoirs, on Brower 1978, and on Schweber, forthcoming). Schwinger collected the key scientific texts (1958) and gave his own rich perspective (1983). They and the other central figures in the postwar development of quantum electrodynamics all provided their oral recollections, as did Theodore Shultz, Michel Baranger, Evelyn Frank, Arthur Wightman, Abraham Pais, and others. Paul Hartman (1984) shared his entertaining history of the Cornell physics department and correspondence with Feynman about space flight. My discussion of scientific visualization is indebted to Arthur Miller 1984 and 1985, Bruce Gregory 1988, Schweber 1986a, Park 1988, essays by (and a conversation with) Gerald Holton, and Feynman's own introspection. My accounts of Feynman's relationships with women, in this chapter and the next, are based on correspondence in his personal papers and on my interviews with each of the women whose relationships are described in any detail; however, in the notes that follow, I usually omit individual citations of these letters and interviews for reasons of privacy.

207   AMONG THE DIVINITIES: Charles Clayton Morrison, "The Atomic Bomb and the Christian Faith," The Christian Century, 13 March 1946, 330.

207   WHAT OPPENHEIMER PREACHED: Oppenheimer 1945, 316.

208   IT'S A TERRIBLE THING: SYJ, 118.

208   AND RIGHTLY SO: Oppenheimer 1945, 317.

208 When you come right down to it: Ibid.

209 The events of the past few years: Truman, "Problems of Post-War America," 6 September 1945, in *Vital Speeches* 11(1945):23.

209 Before the war the government had paid: Kevles 1987, 341.

209 the quiet times when physics: R. Wilson 1958, 145.

210 The nature of the work: Oppenheimer 1945, 315–16.

210 In the first, he sat down: Hartman 1984, 202.

210 In the second, two months after Hiroshima: Bishop 1962, 560; Hartman 1984, 238.

211 He debarked with a single suitcase: F-W, 415.

211 The week before Feynman arrived: Bishop 1962, 556.

211 Huge raked piles of leaves: F-W, 417.

212 Look, buddy: Ibid., 419; cf. *SYJ*, 149–51.

212 Speech patterns struck him: "It was completely—like the nervousness of working during the war. And this university in the backwoods . . . was going at the typical university rate . . . he's talking so slowly and batting the breeze about the weather." F-W, 418.

212 Outside, three tennis courts: Hartman 1984, 204–5.

212 Morrison had been lured: Philip Morrison, interview, Cambridge, Mass.

212 Feynman depressed is just a little more cheerful: Quoted in Schweber 1986a, 468; Feynman said, "I got deeper and deeper into a kind of——I wouldn't say depression, because I wasn't depressed. I'm a lively and happy fellow. . . ." F-W, 425.

212 He spent time in the library: *SYJ*, 155.

212 His dance partners looked askance: F-W, 423; *SYJ*, 154.

212 Even before leaving Los Alamos: E.g. Olum, interview; Walker, interview. One physicist's wife said, "He exploded like a sexual firecracker."

213 now I want you to know: Lucille Feynman to Feynman, 17 June 1945, PERS.

213 Begging him to come home: Lucille Feynman to Feynman, 21 June 1945, PERS.

213 This is the Princeton Triangle: Lucille Feynman to Feynman, 8 August 1945, PERS.

213 I felt thrilled & frightened: Ibid.

213 By the way: Ibid.

214 Richard, what has happened: Lucille Feynman to Feynman, n.d., PERS.

215 He prided himself on speaking: Schwinger, interview.

215 a man possessed: Polkinghorne 1989, 14.

215 I abandoned my bachelor quarters: Schwinger 1983, 332.

215 their first encounter: E.g., Crease and Mann 1986, 129.

215 Are you a mouse or a man?: Norman Ramsey and Rabi, quoted in Schweber, forthcoming; Bernard T. Feld, talk at Julian Schwinger's 60th birthday celebration, February 1978, AIP.

216 Even before Schwinger got his college diploma: Schweber, forthcoming.

216 Schwinger made one tour: Schwinger, interview.

216 When he had long since: Feynman 1978.

216 THE HARVARD COMMITTEE: Schweber, forthcoming.

217 PHENOMENA COMPLEX—LAWS SIMPLE: "Methods of Math Phys 405," Notebook, PERS.

218 WHETHER HE WOULD SUCCEED: Robert Walker, interview.

218 IN AN ATOM BOMB: "Methods of Math Phys 405."

218 ANNOUNCER: LAST WEEK DR. FEYNMAN: "The Scientist Speaks," transcript, radio broadcast, WHCU, 26 April 1946, OPP.

218 THE RAYS EMITTED: Ibid.

218 AT LOS ALAMOS HE HAD INVENTED: Hawkins et al. 1983, 308.

218 I BELIEVE THAT INTERPLANETARY TRAVEL: Feynman to Paul Hartman, 5 December 1945, PERS.

219 FLYING UPSIDE DOWN: Ibid.

220 HE RETURNED HOME AND OCCASIONALLY SNEAKED OUT: Joan Feynman, interview.

220 ONE DAY FEYNMAN SAW HIM: F-L.

220 IT IS NOT SO EASY FOR A DOPE: Melville Feynman to Feynman, 10 September 1944, PERS.

220 THE DREAMS I HAVE OFTEN HAD: Ibid.

221 ON FEYNMAN'S FACE WAS A LOOK: Joan Feynman, interview.

221 CORNELL'S 1946 FALL-TERM ENROLLMENT: Bishop 1962, 555.

221 D'ARLINE, I ADORE YOU: Feynman to Arline Feynman, 17 October 1947, PERS.

223 FEYNMAN'S VERSION OF THE STORY: F-W, 620; SYJ, 137. The latter was dictated more than twenty years later but sometimes tracks the first version with uncanny, verbatim precision. The Selective Service files were destroyed, as the FBI discovered in assembling its dossier on Feynman. FOI.

226 FEYNMAN WAS INVITED: Princeton University 1946; F-W, 433–34; Wigner 1947; Feynman to Dirac, 23 July 1947, PERS.

226 DIRAC'S PAPER: Dirac 1946.

226 WE NEED AN INTUITIVE LEAP: Princeton University 1946, 15.

226 FEYNMAN LOOKED OUT THE WINDOW: F-W, 437.

226 HE HAD A QUESTION: Ibid., 272–73 and 437; Feynman 1948a, 378. Feynman cared about this detail of historical priority. He later emphasized it in his Nobel lecture: "I thought I was finding out what Dirac meant, but, as a matter of fact, had made the discovery that what Dirac thought was analogous, was, in fact, equal" (NL, 10). Schwinger, however, in a tribute delivered at a memorial service to Feynman, made a subtle point of dismissing the possibility that Dirac might not have understood the implications of his paper: "Now, we know, and Dirac surely knew, that to a constant factor the 'correspondence' . . . is an equality. . . . Why, then, did Dirac not make a more precise, if less general, statement? Because he was interested only in a general question." Schwinger 1989, 45.

226 OPPENHEIMER HAD INVITED HIM: Feynman to Oppenheimer, 5 November 1946, CIT.

226 THE CHAIRMAN OF THE UNIVERSITY OF PENNSYLVANIA'S PHYSICS DEPARTMENT: G. P. Harnwell to Bethe, 25 February 1947, and Bethe to Harnwell, 4 March 1947, BET.

227 OPPENHEIMER HAD NOW BEEN NAMED: SYJ, 155; Smyth to Feynman, 23 October 1946 and 22 April 1947, CIT.

227 HE EXPERIMENTED WITH VARIOUS TACTICS: F-W, 426.

227 FOR A MOMENT HE FELT LIGHTER: Ibid., 427–28.

227 DON'T WORRY SO MUCH: SYJ, 156; F-W, 428.

227 A CORNELL CAFETERIA PLATE: F-W, 430; SYJ, 157. Also, Benjamin Fong Chao, "Feynman's Dining Hall Dynamics," letter, Physics Today, February 1989, 15.

228 WELTON WAS NOW WORKING: Welton, interview.

228 I AM ENGAGED NOW: Feynman to Welton, 10 February 1947, CIT.

228 I AM FEYNMAN: Pais 1986, 23.

229 SPIN WAS A PROBLEM: Schweber 1986a, 469.

229 NO WONDER HIS EYE: "Within a week, altogether, this question of the rotation [of the plate] started me worrying about rotations, and then old questions about the spinning electron, and how to represent it in the path integrals and in the quantum mechanics, and I was in my work again. It just opened the gate." F-W, 430.

230 FEYNMAN DID NOT ATTEMPT TO PUBLISH: F-W, 444.

230 THE CHALLENGE WAS TO EXTEND: Ibid., 443; Schweber 1986a, 472.

232 THINKING I UNDERSTAND GEOMETRY: Feynman to Barbara Kyle, 20 October 1965, CIT.

232 THE LAST EIGHTEEN YEARS: K. K. Darrow diary, 14 April 1947, AIP.

232 THEORETICIANS WERE IN DISGRACE: Gell-Mann 1983a, 3.

232 THE THEORY OF ELEMENTARY PARTICLES: Weisskopf 1947.

233 SO TWO DOZEN SUIT-JACKETED PHYSICISTS: Schweber 1983, 313.

233 WHEN THEY GATHERED FOR BREAKFAST: Robert Marshak, telephone interview.

233 IT IS DOUBTFUL IF THERE HAS EVER BEEN: Stephen White, "Top Physicists Map Course at Shelter Island," New York Herald Tribune, 3 June 1947, 23.

233 FEYNMAN TRIED HIS METHODS OUT: Pais 1986, 452.

234 A CLEAR VOICE, GREAT RUSH OF WORDS: K. K. Darrow diary, 14 April 1947, AIP.

234 LAMB HAD GONE TO BED: Lamb 1980, 323.

234 TO SCHWINGER, LISTENING: Schwinger 1983, 337.

234 THE FACTS WERE INCREDIBLE: Quoted in Schweber, forthcoming.

234 AS THE MEETING ADJOURNED: Schwinger 1983, 332. Shortly afterward he was married; or, as he put it, "I abandoned my bachelor quarters and embarked on an accompanied, nostalgic trip around the country that would occupy the whole summer."

234 DEBACLE: Polkinghorne 1989, 12.

235 IT WAS HARDLY A COMMON NAME: Morrison, interview.

235 WHAT THEY DID THERE: Michel Baranger, interview, New York.

235 I EXPECT HER TO BE: Alice Dyson, quoted in Schweber, forthcoming.

235 I, SIR PHILLIP ROBERTS: Sir Philip Roberts's Erolunar Collision, in Dyson 1992, 3–4.

236 HE READ POPULAR BOOKS: Dyson 1979, 12.

236 THAT SAME YEAR, FRUSTRATED: Schweber, forthcoming.

236 SHE CONTINUED BY TELLING HIM: Dyson 1979, 15.

236 AT CAMBRIDGE HE HEARD: Brower 1978, 16.

236 DYSON'S WAR: Dyson 1979, 19–21.

237   AMONG THE BOOKS: Ibid., 4.

237   MY WISH FOR SOMETHING TO SERVE: D. H. Lawrence, *Study of Thomas Hardy*, quoted in Dyson 1988, 123.

237   THE NEWS OF HIROSHIMA: Brower 1978, 20.

237   YEARS LATER, WHEN DYSON: Ibid., 24.

238   BY HIS SOPHOMORE YEAR: Dyson 1987.

238   PROFESSOR LITTLEWOOD: Dyson 1944; Dyson, interview.

238   I AM LEAVING PHYSICS FOR MATHEMATICS: Kac 1985, xxiii; Dyson, interview.

238   HE PLAYED HIS FIRST GAME OF POKER: ". . . and found I was rather good at it," he wrote his parents, 11 June 1948.

238   HE EXPERIENCED THE AMERICAN FORM: Dyson to parents, 11 June 1948.

238   WE GO THROUGH SOME WILD COUNTRY: Dyson to parents, 19 November 1947.

239   HE HAS DEVELOPED A PRIVATE VERSION: Ibid.

239   HE TELEPHONED FEYNMAN: NL, 449.

239   IT WAS A BLUNT LOS ALAMOS–STYLE ESTIMATE: It diverged, but it only diverged logarithmically, heading ever higher, but ever more slowly, like the series $1 + \frac{1}{2} + \frac{1}{3} + \frac{1}{4} + \ldots$—after a million terms this has not even reached 15, but it never does stop rising. When the news reached Russia, the great Lev Landau said with obscure Slavic wisdom, "A chicken is not a bird, and a logarithm is not infinity." Weinberg 1977*a*, 30; Sakharov 1990, 84.

239   BUT THEY DID NOT COINCIDE: Bethe, interview.

240   KRAMERS PROPOSED A METHOD: Bethe had also talked with Schwinger and Weisskopf, both of whom had suggested forms of renormalization.

240   DYSON COULD SEE: Dyson, interview.

241   ONE-MAN PERCUSSION BAND: Dyson to parents, 19 November 1947.

241   DID YOU KNOW THERE ARE TWICE AS MANY NUMBERS: Henry Bethe to Gweneth, 17 February 1988, in *WDY*, 101.

241   FOR A WHILE, BECAUSE FEYNMAN: Dyson, interview.

241   HALF GENIUS AND HALF BUFFOON: Dyson to parents, 8 March 1948.

241   FEYNMAN IS A MAN WHOSE IDEAS: Dyson to parents, 15 March 1948.

242   THE THOUGHT THAT THE LAWS OF THE MACROCOSMOS: Quoted in Miller 1984, 129.

242   WE ARE THEREFORE OBLIGED TO BE MODEST: Bohr 1922, 338.

242   JUST IMAGINE THE ROTATING ELECTRON: Quoted in Miller 1984, 143.

242   I UNDERSTAND THAT WHEN AN ATOM: *WDY*, 18–19.

244   IT IS WRONG TO THINK THAT THE TASK: Quoted in Gregory 1988, 185.

244   FEYNMAN SAID TO DYSON: Dyson 1979, 62.

244   A CORNELL DORMITORY NEIGHBOR: Theodore Schultz, interview, Yorktown Heights, N.Y.

244   SPACE IS A SWARMING IN THE EYES: Pencil note, CIT. Vladimir Nabokov, *Pale Fire* (New York: Vintage, 1990), 40.

244   WHAT I AM REALLY TRYING TO DO: F-Sch.

245   WHEN I START DESCRIBING: *Lectures*, II-20-3.

245   AT ANY RATE DIAGRAMS HAD BEEN RARE: See Miller 1984.

246   WHEN HE FINALLY DID: Feynman 1948*a*.

247 HE STATED THE CENTRAL PRINCIPLE: Ibid., 367.

249 THE EDITORS NOW REJECTED THIS PAPER: Feynman and several other people recalled this, although the journal has no record of it. E.g., F-W, 485; Baranger, interview.

249 THERE IS A PLEASURE: Feynman 1948a, 367.

249 HE LATER FELT THAT HE WAS BETTER KNOWN: Kac 1985, 115–16.

250 THE ELECTRON DOES ANYTHING: Quoted by Dyson in "Comment on the Topic 'Beyond the Black Hole,' " in Woolf 1980, 376.

250 FEYNMAN FELT THAT HE HAD UNCOVERED: Feynman 1948a, 377–78.

250 THE TREACHEROUSLY INNOCENT EXERCISES: See QED.

251 SEMIEMPIRICAL SHENANIGANS: NL, 451; F-W, 459.

251 HE PROMISED BETHE AN ANSWER: NL, 449; F-W,459.

251 WHEN FEYNMAN HEARD LATE IN THE FALL: F-W, 462 f.

251 GOD IS GREAT: Rabi to Bethe, 2 December 1947, and Bethe to Rabi, 4 December 1947, BET. Quoted in Schweber, forthcoming.

252 THE SCHWINGER-WEISSKOPF-BETHE CAMP: Feynman to Corbens, 20 March 1948, CIT.

252 HIS LECTURE DREW A CROWD: "The Prodigy Who Grew," Newsweek, 23 February 1948, 45–46; William L. Laurence, "New Guide Offered on Atom Research," New York Times, 1 February 1948; Stephen White, "Physics Society Hears Theory of Electron Action," New York Herald Tribune, 1 February 1948, 51.

252 I DID IT TOO, DADDY: F-W, 463–64.

252 I'M SO SORRY: Rose McSherry to Feynman, 21 January, 1948, PERS.

253 MONTHS PASSED BEFORE FEYNMAN CALLED: Feynman formally acknowledged the error in a footnote to a paper published the next year; he juggled the footnotes so that the apology would fall in number 13: "That the result . . . was in error was repeatedly pointed out to the author, in private communication, by V. F. Weisskopf and J. B. French. . . . The author feels unhappily responsible for the very considerable delay in the publication of French's result occasioned by this error." Feynman 1949b, 777; Weisskopf, interview; Weisskopf 1991, 168.

253 A HOLE IF THERE WERE ONE: Dirac in Proceedings of the Royal Society of London A 133(1931):60.

253 SUPPOSE A BLACK THREAD: Feynman 1948f, 2–3.

254 A CORNELL STUDENT WHO HAD SERVED: Schweber 1986a, 488.

254 A BOMBARDIER WATCHING A SINGLE ROAD: Feynman 1948f, 4; cf. Feynman 1949a, 749.

254 HE KNEW FROM HIS OLD WORK: Feynman 1947, 1.

255 USUAL THEORY SAYS NO: Ibid., 4.

255 IT MAY PROVE USEFUL IN PHYSICS: Ibid.

256 FEYNMAN GLEEFULLY SAID: Marshak, interview. Instead these mesons were named muon and pion, after Greek letters.

256 AS THE POCONO MEETING OPENED: Wheeler 1948.

256 EACH SMALL VOLUME OF SPACE: Ibid.

257 SCHWINGER HATED THIS: Schwinger, interview.

257 BETHE NOTICED THAT THE FORMAL MATHEMATICS: F-W, 469; Bethe, interview.

257 FERMI, GLANCING ABOUT: Segrè 1970, 174.

257   THIS IS A MATHEMATICAL FORMULA: F-W, 470.

258   HE THOUGHT IT INTELLECTUALLY REPULSIVE: Schwinger, interview.

258   BOHR HAS RAISED THE QUESTION: Wheeler 1948.

258   BOHR CONTINUED FOR LONG MINUTES: F-W, 473; cf. Pais 1986, 459, but this conflates Teller's and Bohr's objections.

258   I HAD TOO MUCH STUFF: F-Sch.

258   ON HIS FIRST DAY BACK: Arthur Wightman, interview, Princeton, N.J.

259   YOU CAN IMAGINE THAT I WAS HIGHLY PLEASED: Quoted in Schweber, forthcoming.

259   ANONYMOUSLY, HE HOPED: Feynman 1948e, 9.

259   A MAJOR PORTION OF THE CONFERENCE: Ibid., 5.

259   WHEN SCHWINGER SAW FEYNMAN'S NAME: Schwinger 1983, 342.

260   TOMONAGA, A NATIVE OF TOKYO: Tomonaga 1966, 127–29.

260   AFTER SUPPER I TOOK UP MY PHYSICS: Quoted in Julian Schwinger, "Two Shakers of Physics," in Brown and Hoddeson 1983, 357–58.

260   HE MADE A HOME: Schweber, forthcoming.

261   JUST BECAUSE WE WERE ABLE: Oppenheimer to Members of the Pocono Conference, 5 April 1948, OPP.

261   OTHER PEOPLE PUBLISH TO SHOW HOW: Dyson 1965a, 428.

261   SCHWINGER OCCASIONALLY HEARD: Schwinger 1983, 341.

261   I GATHER I STAND ACCUSED: Ibid.

261   ARE WE TALKING ABOUT PARTICLES: Schwinger, interview.

262   YOUR SUDDEN DEPARTURE FROM ITHACA: Lloyd P. Smith to Feynman, 13 June 1947, CIT.

263   FEYNMAN THOUGHT DYSON: WDY, 65.

263   DYSON LIKED THE ROLE: Dyson to parents, 25 June 1948.

263   HE HAD DECIDED THAT MODERN AMERICA: Dyson to parents, 14 June 1948.

263   AN ATOMIC BLAST ON EAST 20th STREET: Morrison 1946; SYJ, 118.

264   DYSON SUDDENLY FELT THAT FEYNMAN: Dyson 1979, 59; Dyson to parents, 25 June 1948.

264   AS WE DROVE THROUGH CLEVELAND: Dyson 1979, 60–61.

264   SOMETIMES IT OCCURRED TO HIM: F-W, 532.

264   DYSON HAD NEVER SEEN RAIN: Dyson 1979, 59.

264   THE CAR RADIO REPORTED: Dyson to parents, 25 June 1948.

264   THIS HOTEL IS UNDER NEW MANAGEMENT: WDY, 65.

265   IN THAT LITTLE ROOM: Dyson 1979, 59.

265   THE ROOM WAS FAIRLY CLEAN: WDY, 66. Dyson, far from taking offense, merely commented that he had left out "the best part of the story." Dyson 1989, 38.

266   THAT STORMY NIGHT IN OUR LITTLE ROOM: Dyson 1979, 63.

266   ON WEDNESDAY OPPENHEIMER RETURNS: Dyson to parents, 10 October 1948, quoted in Schweber, forthcoming.

267   THERE WASN'T ENOUGH ROOM: Leonard Eyges to Dyson, quoted in Schweber, forthcoming.

267   A UNIFIED DEVELOPMENT OF THE SUBJECT: Dyson 1949a, 486.

267   THERE IS THE FATHER INCOMPREHENSIBLE: Dyson 1989, 35–36.

267   ANY MODERATELY INTELLIGIBLE ACCOUNT: Dyson to parents, 4 October 1948.

267 SO THE RESULT OF ALL THIS: Ibid.

268 THEIR EVALUATION GIVES RISE: Dyson 1949a, 491.

268 WRITE DOWN THE MATRIX ELEMENTS: Ibid., 495.

268 IT WAS "SCHWINGER'S THEORY": Oppenheimer 1948.

269 DEAR FREEMAN: I HOPE YOU DID NOT GO BRAGGING: Feynman to Dyson, 29 October 1948, CIT.

270 WELL, DOC, YOU'RE IN: Dyson to parents, quoted in Schweber, forthcoming.

270 FEYNMAN HAD NOT LEARNED: NL, 452.

271 HE BUTTONHOLED SLOTNICK: F-W, 489–92.; NL, 452; F-Sch; Schweber, forthcoming.

271 WHAT ABOUT SLOTNICK'S CALCULATION?: Later Case sent Feynman his manuscript, and Feynman found an algebraic error that undermined the proof. F-W, 495–96.

271 THERE WERE VISIONS AT LARGE: Schwinger 1983, 343.

272 THE REST MASS PARTICLES HAVE: Feynman 1948b, 943.

272 FEYNMAN AND I REALLY UNDERSTAND EACH OTHER: Dyson to parents, 1 November 1948.

272 FEYNMAN'S STUDENTS, HOWEVER, SOMETIMES NOTICED: Baranger, interview.

272 HE HAD STARTED HEARING ABOUT DYSON GRAPHS: Cf. F-W, 501.

272 WENTZEL HIMSELF: Crease and Mann 1986, 143.

273 FEYNMAN STRESSED HOW FREE: Feynman 1949b, 773.

275 FOR A WHILE IT WAS ALL SCHWINGER: F-W, 499.

275 IN THE SUMMER OF 1950: J. Ashkin, T. Auerbach, and R. Marshak, "Notes on a Possible Annihilation Process for Negative Protons," Physical Review 79(1950):266.

275 A TECHNIQUE DUE TO FEYNMAN: K. A. Brueckner, "The Production of Mesons by Photons," Physical Review 79(1950):641.

275 THE UNREASONABLE POWER OF THE DIAGRAMS: Segrè 1980, 274.

276 LIKE THE SILICON CHIP: Schwinger 1983, 343.

276 PEDAGOGY, NOT PHYSICS: Ibid., 347.

276 YES, ONE CAN ANALYZE EXPERIENCE: Ibid., 343.

277 ALTHOUGH 'ONE' IS NOT PERFECTLY: Bernstein 1987, 63.

277 THEY ALSO WORRIED ABOUT SCHWINGER'S ABILITY: Sheldon Glashow, interview, Cambridge, Mass.

277 MURRAY GELL-MANN LATER SPENT A SEMESTER: Murray Gell-Mann, interviews, Pasadena and Chicago.

278 THERE WAS A NEW NOTE: E.g., Virginia Prewett, "I Homesteaded in Brazil," Saturday Evening Post, 22 April 1950, 10, began, "It's going to be the first atomic-bomb shelter in the New World." Cf. F-W, 551.

278 AT LEAST A 40 PERCENT CHANCE OF WAR: Wheeler to Feynman, 29 March 1951, CIT.

278 WHEN A BRAZILIAN PHYSICIST: Lopes 1988; J. Leite Lopes, personal communication.

278 LATE THE NEXT WINTER HE IMPULSIVELY ASKED: Jayme Tiomno to Feynman, 6 March 1950, PERS. The Brazilians replied that a one-year appointment was the best they could offer at the time.

278   HE HAD ENDURED ONE TOO MANY DAYS: F-W, 546; Bacher, interview.
278   ALL THE INS AND OUTS: Feynman to Bacher, 6 April 1950, PERS.
278   I DO NOT LIKE TO SUGGEST: Ibid.
279   ONCE (AND IT WAS NOT YESTERDAY): Cvitanović 1983, 6.

CALTECH

Three local histories are Judith Goodstein's *Millikan's School*, Ann Scheid's *Pasadena: Crown of the Valley*, and Kevin Starr's *Inventing the Dream*; they were useful background, as was Robert Kargon's essay "Temple to Science." I've also relied on the recollections of many present and former Caltech professors, students, and administrators. Some information on Feynman's time in Brazil comes from the recollections of José Leite Lopes (1988 and personal communication), Cecile Dewitt-Morette, and others; from Feynman's 1951 correspondence with Fermi; from Brownell 1952; from Feynman's talk "The Problem of Teaching Physics in Latin America" (1963a), and from publications of the Centro Brasileiro de Pesquisas Físicas. Documentation of the government's secret scrutiny of Feynman and of his consultation with the State Department on the advisability of travel to the Soviet Union came through my Freedom of Information Act requests to the FBI, CIA, Department of the Army, and Department of Energy in 1988 and 1989. Some of the State Department correspondence is also in CIT. On superfluidity, Robert Schrieffer, Hans Bethe, Michael Fisher, and Russell Donnelly were especially helpful. Donnelly sent written reminiscences by several colleagues. Andronikashvili 1990 is a remarkable memoir from the Russian perspective. For the particle physics of the 1950s and 1960s: the Rochester conference proceedings; John Polkinghorne's witty memoir (1989) and Jeremy Bernstein's "informal history" (1989); Robert Marshak's account (1970); Brown, Dresden, and Hoddeson's symposium proceedings *Pions to Quarks: Particle Physics in the 1950s*; and interviews with the various scientists cited. Again, some material on personal relationships is based on letters and interviews that I cannot cite specifically for reasons of privacy. Feynman's thinking on gravitation can be seen in a fifteen-page letter to Victor Weisskopf written in January and February 1961 (WHE) and in his Faraday lecture (1961b), as well as his one published paper (1963b) and various lecture notes in CIT. The development of quarks and partons has been well chronicled from different points of view by Andrew Pickering (1984) and Michael Riordan (1987); Feynman kept his notes from this period in unusually good order (CIT); Riordan and Burton Richter provided useful on-site guidance at the Stanford Linear Accelerator Center; James Bjorken, George Zweig, Sidney Drell, Yung-Su Tsai, and, of course, Murray Gell-Mann were among those with especially helpful reminiscences. For the record of Feynman's illnesses I relied on notes and correspondence in his files and interviews with Drs. C. M. Haskell, William G. Bradley, and In Chang Kim. For the investigation into the *Challenger* accident: the hearing transcripts and documentation as published in the commission report; Feynman's personal notes and commission memorandums (CIT and PERS); Ralph Leighton's unedited transcript of Feynman's oral account (later published in *WDY*); interviews with commissioners, NASA officials and engineers, and others (only William P. Rogers refused to make himself available, despite my repeated requests for an interview). Carl Feynman shared the manuscript of the paper Feynman was working on until he entered the hospital for the last time.

281   The California Institute of Technology: J. Goodstein 1991, 180.

281   Pasadena is ten miles from Los Angeles: Morrow Mayo, quoted in Scheid 1986, 156.

281   every luncheon, every dinner: Letter to the Editor, Los Angeles Times, 6 March 31, quoted in J. Goodstein 1991, 100.

282   Could it be that nitrogen has two levels: F-W, 559.

282   Dear Fermi: Feynman to Enrico Fermi, 19 December 1951; Fermi to Feynman, 18 January 1952 and 28 April 1952, AIP. Some of Feynman's meson work that year emerges in Lopes and Feynman 1952.

282   Don't believe any calculation: Feynman to Fermi, 19 December 1951.

283   In recent years several new particles: Fermi and Yang 1949, 1739.

283   he could spend days at the beach: Lopes, personal communication.

283   I wish I could also refresh: Fermi to Feynman, 18 January 1952, AIP.

283   Feynman taught basic electromagnetism: Feynman 1963a.

284   Light impinging on a material: Ibid., 26.

284   But when he asked what would happen: SYJ, 192.

284   They could define "triboluminescence": Even in his sixties he continued to consider ways of intensifying this phenomenon in the substances he described as "WL (Wint-o-green Lifesavers) and S (sucrose)." Feynman to J. Thomas Dickenson, 13 May 1985, CIT.

284   Have you got science?: SYJ, 197.

284   What are the four types of telescope?: Feynman 1963a, 24.

284   He would sit idly at a café table: Joan Feynman, interview.

284   gives a feeling of stability: Feynman 1963a, 24.

285   Philip Morrison, who shared an office: Morrison, interview.

286   He joined a local school: SYJ, 185.

286   In the 1952 carneval: Lopes, personal communication.

287   He heard from hardly anyone: F-W, 564; Feynman to Oppenheimer, 27 May 1952, OPP.

287   He haunted the Miramar Hotel's outdoor patio bar: Bertram J. Collcutt to Feynman, 2 December 1985, CIT.

287   He took out Pan American stewardesses: SYJ, 183–84.

287   The old certainties of the past: Mead 1949, 4.

289   Tell me what it is like: Michels 1948, 16.

290   It seems to me that you go to lots of trouble: Feynman, note, n.d., PERS.

290   How is it possible: SYJ, 168.

290   You are worse than a whore: Ibid., 169–70.

292   Even before they married, they quarreled: Mary Louise Bell to Feynman, 30 May 1950 and 24 March 1952, PERS.

292   The pattern is that the girl: Bell to Feynman, 26 February 1952, PERS.

292 THEY HONEYMOONED IN MEXICO: *SYJ*, 286.

292 SHE DID NOT KNOW WHAT TO THINK: Mary Louise Bell, telephone interview.

292 SHE LIKED TO TELL PEOPLE: Bell, interview.

292 WHERE THERE'S SMOKE THERE'S FIRE: Gell-Mann, interview.

293 HAS WILFULLY, WRONGFULLY: Complaint for Divorce, 6 June 1956, Superior Court, Los Angeles County. "Final Adjustment of Property Settlement," handwritten agreement, 16 October 1956, PERS.

293 THE DRUMS MADE TERRIFIC NOISE: "Beat Goes Sour: Calculus and African Drums Bring Divorce," *Los Angeles Times*, 18 July 1956.

293 BEGGING FOR HIS OLD JOB BACK: Feynman to Bethe, 26 November 1954, BET.

293 SOON AFTERWARD, SOMEONE RUSHED UP: *SYJ*, 211–12.

293 MEANWHILE, ALTHOUGH BETHE HAD BEEN THRILLED: Bethe to Feynman, 3 December 1954, BET.

294 THE UNIVERSITY OF CHICAGO DECIDED: Goldberger, interview; *SYJ*, 213.

295 A HOST OF APPLIED SCIENCES: Cf. Forman 1987 and Kevles 1990.

295 WHEN SCIENCE IS ALLOWED TO EXIST: DuBridge, quoted in Forman 1987.

295 THESE WERE NOT SO MUCH CRUMBS: As the leading experimentalist Luis Alvarez told the physicist and historian Abraham Pais: "Right after the war we had a blank check from the military because we had been so successful. Had it been otherwise we would have been villains. As it was we never had to worry about money." Pais 1986, 19.

295 IN 1954 THE SECRETARY OF THE ARMY: Minutes of Executive Session, Army Scientific Advisory Panel, 17 November 1954, CIT; F-W, 599–601.

295 HOT DOG: Feynman to Lucille Feynman, n.d., PERS.

295 THE PUBLIC ANNOUNCEMENT CAME: "Einstein Award to Professor, 35," *New York Times*, 14 March 1954; F-W, 673.

296 THE AEC BEGAN FOUR WEEKS OF HEARINGS: Atomic Energy Commission 1954.

296 YOU SHOULD NEVER TURN A MAN'S GENEROSITY: A decade later, he was uncomfortable with his decision. "I knew what had happened to Oppenheimer, and that Strauss had something to do with it, and I didn't like it. . . . O.K.? And I thought—I'm going to fix him. I mean, I was not nice. I don't want to take it from him. The hell with it. And I thought: maybe I won't take the prize. All right? And I worried about it, because in a certain sense I felt that was unfair. The guy is offering the money—you know, he's trying to do something nice—and it isn't that he just did it because of this, because he's done it before. There were previous Einstein Awards, as far as I know, or something. . . . I was kind of confused." F-W, 673–74.

296 FEYNMAN'S OWN FILE AT THE FBI: 497 pages, partly expurgated, FOI.

296 PROFESSOR FEYNMAN IS ONE OF THE LEADING: Bethe to M. Evelyn Michaud, 7 April 1950, and Michaud to Bethe, 27 March 1950, BET.

297 ONE OF ITS PRINCIPAL ARCHITECTS: Sakharov 1990, 190–91.

298 I THOUGHT YOU WOULD BE INTERESTED: Feynman to Atomic Energy Commission, 14 January 1955, CIT. Also: "Is there danger that I would be kept there and not return?" Feynman to State Department, 14 January 1955, FOI.

298 PROPAGANDA GAINS: Walter J. Stoessel, Jr., "Invitation to United States Physicist to Attend Scientific Conference," confidential memorandum, Department of State, 21 January 1955, FOI; Stoessel, Jr., to Feynman, 15 March 1955, CIT.

298 CIRCUMSTANCES HAVE ARISEN: Feynman to A. N. Nesmeyarrov, 14 March 1955, CIT.

298 THIS IS A CLEAR CASE: "Scientist at Caltech Warned," Los Angeles Times, 8 April 1955.

299 WHEN FEYNMAN TALKED ABOUT FLUID FLOW: Lectures, II-40-1.

299 THEORISTS OF "DRY WATER": Lectures, II-40-3 and III-4-12.

300 TWO CITIES UNDER SIEGE: Feynman 1957a, 205.

300 NORWEGIAN I AND NORWEGIAN II: Donnelly 1991b.

300 THE MOST BASIC CLUE: Feynman 1955b, 18.

301 THE SPEAKERS HAD NO IDEA: Russell Donnelly, telephone interview.

301 HE HAD TRIED TO PICTURE: Feynman 1953c, 1302.

302 THE CHALLENGE WAS TO DRIVE: "The hardest part of the helium problem was done by physical reasoning alone, without being able to write anything. . . . it was very very interesting to be able to push through that doggoned thing without having stuff to write." F-W, 739.

301 HE LAY AWAKE IN BED: F-W, 693–95.

302 THE RINGS OF ATOMS WERE LIKE RINGS OF CHILDREN: Feynman 1958a, 21.

302 TYPICAL FEYNMAN: Donnelly, interview.

303 POSSIBLY I UNDERSTAND: Note, "Possibly I understand . . ." n.d., CIT.

303 THE YEAR BEFORE, SCHRIEFFER HAD LISTENED: Robert Schrieffer, telephone interview; Feynman 1957a.

303 WE HAVE NO EXCUSE: Feynman 1957a, 212.

304 BY THE FIRST OF THESE MEETINGS: Pais 1986, 461; Polkinghorne 1989, 20.

304 WITHIN A FEW YEARS PARTICLE TABULATIONS: Polkinghorne 1989, 21.

304 GENTLEMEN, WE HAVE BEEN INVADED: C. F. Powell, at a 1953 conference, quoted in Polkinghorne 1989, 48.

305 ONE EXPERIMENTALIST, MARCEL SCHEIN: Crease and Mann 1986, 178.

305 YOU HAVE A DIFFERENT THEORY: F-W, 603–5.

306 IF A CALTECH EXPERIMENTER: Barry Barish, interview, Pasadena.

307 HE THOUGHT PAIS WAS WRONG: Gell-Mann 1982, 399.

307 AT FOURTEEN HE HAD BEEN DECLARED: Columbiana 1944 (Columbia Grammar and Preparatory School), 28; Bernstein 1987, 20.

308 THE ONLY PERSON WHO WILL KNOW: Ralph Leighton, interview, Pasadena.

308 IT WAS THE CLOSEST TO SUCCESS: Gell-Mann, interview.

308 IT WAS HIS BROTHER: Gell-Mann 1989b, 3.

309 WHEN WEISSKOPF ADVISED HIM: Gell-Mann, interview.

309 FEYNMAN FELT A FLICKER OF ENVY: F-W, 670; SYJ, 223.

309 GELL-MANN, IN CHICAGO, FELT EVEN MORE: "Jealousy was another reason . . . I resented the publicity being given to the scheme of Pais, which I was convinced was wrong!" Gell-Mann 1982, 399.

310 THE EDITORS OF THE PHYSICAL REVIEW: Gell-Mann 1953; Gell-Mann 1982, 400.

310 WHY SHOULD A BROAD-MINDED THEORIST: Quoted in Polkinghorne 1989, 49. Similarly, the historian J. L. Heilbron: " 'Strangeness,' a word barely utterable in Romance languages and expressive of a surprise only briefly felt. . . . Does the new terminology express cynicism or disdain by particle theorists toward their own creations?" "An Historian's Interest in Particle Physics," in Brown et al. 1989, 53.

310 THE WINTER FERMI DIED: Gell-Mann, interview.

311 MOST OF HIS BODY WAS CREMATED: Thomas S. Harvey, telephone interview; William L. Laurence, "Key Clue Sought in Einstein Brain," New York Times, 20 April 1955; Steven Levy, "My Search for Einstein's Brain," New Jersey Monthly, August 1978, 43.

311 VARIOUS NINETEENTH-CENTURY RESEARCHERS: Gould 1981.

312 IS THERE A NEUROLOGICAL SUBSTRATE: Obler and Fein 1988, 6.

313 ENLIGHTENED, PENETRATING, AND CAPACIOUS MINDS: Duff 1767, 5.

313 RAMBLING AND VOLATILE POWER: Ibid., 9.

313 IMAGINATION IS THAT FACULTY: Ibid., 6–7.

314 IN POINT OF GENIUS: Gerard 1774, 13.

314 A QUESTION OF VERY DIFFICULT SOLUTION: Ibid., 18.

315 IT IS ONE OF THE HOPES: Quoted in Root-Bernstein 1989, 1.

315 A PHYSICIST STUDYING QUANTUM FIELD THEORY: Coleman, interview.

315 FROM GEOMETRY TO LOGARITHMS: Hood 1851, 10–11.

316 THE ASTROPHYSICIST WILLY FOWLER: Thorne, interview; Fowler, interview conducted by Charles Weiner, 30 May 1974, AIP: "I just thought Feynman's talking through his hat, what can he possibly mean, what can general relativity have to do with these objects?"

316 THAT FEYNMAN HAD SIGNED: John S. Rigden, interview, New York.

317 WHY DO I CALL HIM A MAGICIAN?: Quoted in Dyson 1979, 8–9.

317 MAGICAL MUMBO-JUMBO: Dyson 1979, 8.

318 BETWEEN THE PHYSIOLOGY OF THE MAN: Lombroso 1891, xiii.

319 LET EUROPEAN ROMANTICS CELEBRATE: Currie 1974.

319 I SPEAK WITHOUT EXAGGERATION: Quoted in Grattan 1933, 156.

319 MR. EDISON IS NOT A WIZARD: Quoted in LaFollette 1990, 97.

320 EDISON WAS NOT A WIZARD: Grattan 1933, 151.

320 HONEST CRAFTSMEN: Dyson 1979, 9.

321 HE WAS SEARCHING FOR GENERAL PRINCIPLES: Ibid., 62–63.

321 SO WHAT IS THIS MIND OF OURS?: WDY, 220.

322 KNOWING WHAT FERMI COULD DO: Zuckerman 1977.

323 I THINK IF HE HAD NOT BEEN SO QUICK: Coleman, interview.

324 THE WHOLE QUESTION OF IMAGINATION: Lectures, II-20-10.

325 NOT JUST SOME HAPPY THOUGHTS: Ibid.

325 OUR IMAGINATION IS STRETCHED: CPL, 127–28.

325 WE KNOW SO VERY MUCH: Feynman to Welton, 10 February 1947, CIT.

326 THERE ARE SO VERY FEW EQUATIONS: Ibid.

326 MAYBE THAT'S WHY YOUNG PEOPLE: Feynman 1965c.

326 Welton, too, was persuaded: Welton, interview: "I said, 'Dick, think in retrospect what would have happened if I had taught you the Q.E.D. that I knew—you would have known too much, and you wouldn't have been able to innovate as much,' and he said, 'You're right.'"

326 Would I had phrases: Attributed to Khakheperressenb, quoted in Lentricchia 1980, 318.

326 There are no large people: Quoted by Scott Spencer, "The Old Man and the Novel," New York Times Magazine, 22 September 1991, 47.

327 Giants have not ceded: Gould 1983, 224.

329 those countless footnotes: Merton 1961, 72.

329 I always find questions like that: Feynman to James T. Cushing, 21 October 1985, CIT.

330 Weisskopf declared at one meeting: Polkinghorne 1989, 61.

331 Feynman himself confessed: Millard Susman, personal communication, 29 May 1989.

331 Everything's really all right: Untitled videotape, n.d., recorded for the British Broadcasting Corporation; cf. Gardner 1969, 22–23.

331 Chemists can make them with either handedness: Feynman 1965e, 98–100.

332 Gell-Mann spent a long weekend: Gell-Mann, interview.

332 By the time the 1956 Rochester conference: Pais 1986, 524.

333 Be it recorded here that on the train: Ibid., 525.

333 An experimenter asked Feynman what odds: "I mention this story because I was prejudiced against thinking that parity wasn't conserved, but I knew it might not be. In other words, I couldn't bet one hundred to one, but just fifty to one." F-W, 721.

333 Pursuing the open-mind approach: Ballam et al. 1956, 27.

333 some strange space-time: Ibid., 28.

333 The chairman: Ibid.

334 I do not believe that the Lord: Quoted in Bernstein 1967, 59–60.

334 We are no longer trying to handle screws: Sheldon Penman, quoted in Gardner 1969, 244.

334 At the 1957 Rochester conference: Polkinghorne 1989, 65.

334 busy explaining that they personally: Ibid., 64–65.

335 he refused to referee papers: "To me there's an infinite amount of work involved. . . . I'm not built that way. I can't think his way. I can't follow and try to go through all these steps. If I want to worry about the problem, I read the paper to get the problem, and then maybe work it out some other way. . . . Now, to read and just check steps is——I can't do." F-W, 715.

335 Mr. Beard is very courageous: Feynman to Theodore Caris, 5 December 1961, CIT.

335 You've done it again and again,: F-W, 727–28; Joan Feynman.

336 As Lee pointed out: In Ascoli et al. 1957.

336 In reading Lee and Yang's preprint: F-W, 724.

336 He liked the idea enough: F-W, 725–26; SYJ, 228.

336   A TWO-COMPONENT EQUATION:

$$\left[\left(p_\mu - A_\mu\right)^2 + \underline{\sigma} \cdot \left(\underline{B} + i\underline{E}\right)\right]x = m^2 x$$

"and you can get ψ from x and vice versa":

$$\psi = \left(\frac{(\not{p} - \not{A}) + m}{2m}\right)\binom{x}{x}$$

$$x = (1 + i\gamma_5).$$

336   SUPPOSE THAT HISTORICALLY: Feynman 1957b, 43.

337   OF COURSE I CAN'T DO THAT: Ibid.

337   MARSHAK AND SUDARSHAN MET WITH GELL-MANN: An unhappy tangle of priority concerns followed. Marshak and Sudarshan were concerned to point out that Gell-Mann had learned of their work in progress in July; Gell-Mann was concerned to point out that he had been thinking about V–A "for all these years." Marshak and Sudarshan had missed the opportunity to speak at the Rochester meeting in April—when Feynman described his two-component Dirac equation—and forever after found themselves rehearsing their reasons for remaining silent. To their deep dismay, most physicists cited the Feynman–Gell-Mann paper, not the Marshak-Sudarshan paper (Sudarshan 1983, 486; Sudarshan and Marshak 1984, 15–20). They liked to quote a generous remark of Feynman's long afterward: "We have a conventional theory of weak interactions invented by Marshak and Sudarshan, published by Feynman and Gell-Mann, and completed by Cabibbo. . . ." Feynman 1974b.

337   I FLEW OUT OF THE CHAIR: F-W, 729–30.

337   GELL-MANN, HOWEVER, DECIDED: Gell-Mann, interview.

338   BEFORE THE TENSION BETWEEN THEM: Gell-Mann, Bacher, interviews.

338   COLLEAGUES STRAINED TO OVERHEAR: Matthew Sands, interview, Santa Cruz, Calif.

338   GELL-MANN SOMETIMES DISDAINED IT: Gell-Mann 1983b; Gell-Mann, interview: "He wrote his version using a two-component formalism, of which he was very proud. I disliked the approach: I found it clumsy and unnecessary. I added a lot of material to the paper, some good and some bad, but I didn't succeed in changing the emphasis on the two-component formalism. That was sort of unfortunate."

338   ONE OF THE AUTHORS HAS ALWAYS: Feynman and Gell-Mann 1958a, 194.

338   HAS A CERTAIN AMOUNT OF THEORETICAL: Ibid., 193.

338   THERE WAS A MOMENT WHEN I KNEW: Edson 1967, 64.

339   WE ARE WELL AWARE OF THE FRAGILITY: Feynman and Gell-Mann 1958b.

339   IMPRESSING LISTENERS WITH THE BODY LANGUAGE: Polkinghorne 1989, 72.

339   YOU SPEAK ENGLISH: Gweneth Feynman, interview.

340   FEYNMAN ARRIVED AT A PICNIC: Susman, personal communication.

340   A NEW ERA IN HISTORY: "Red Moon over U.S.," Time, 14 October 1957, 27.

340 ALL THE MASTERY THAT IT IMPLIES: "The Red Conquest," *Newsweek*, 14 October 1957, 38.

340 WELL, LET'S GET THIS STRAIGHT: Quoted in "The Feat That Shook the Earth," *Life*, 21 October 1957, 25.

340 OUR WAY OF LIFE IS DOOMED: Ibid., 23.

340 CURLY-HAIRED AND HANDSOME: "Bright Spectrum," *Time*, 18 November 1957, 24.

341 SCIENTIFIC AND TECHNICAL LEADERSHIP IS SLIPPING: "In Science," *Newsweek*, 20 January 1958, 65.

341 THEY WILL ADVANCE SO FAST: "Knowledge Is Power," *Time*, 18 November 1957, 21.

341 NO TIME FOR HYSTERIA: *Reader's Digest*, December 1957, 117.

341 A STATE DEPARTMENT OFFICIAL LET CALTECH KNOW: Feynman: ". . . someone from the State Department asked that Murray's name be on it also, in order to impress. This was very unfortunate altogether. I don't mind Murray's name on it, that's not the point, but this kind of crap. They call up—so many Russians are going to talk about this thing, they have to have more Americans talking about something scientific . . . this stuff about propaganda mixing up with the science, you know." F-W, 744.

341 IT REMINDED HIM OF THE FLOPHOUSES: WDY, 63–65.

341 SHE TOLD HIM SHE WAS MAKING HER WAY: Gweneth Feynman, interview.

342 I'VE DECIDED TO STAY HERE: Gweneth Howarth to Feynman, 13 October 1958, PERS.

343 HE CONSULTED A LAWYER: Sands, interview.

343 FEYNMAN CALCULATED FARES: Gweneth Howarth to Feynman, 1 November 1958, PERS.

343 YOU'LL WRITE & TELL ME: Gweneth Howarth to Feynman, 1 December 1958, PERS.

343 I'M IMPROVING, AM I NOT: Gweneth Howarth to Feynman, 2 January 1959, PERS.

344 YOU DO NEED SOMEONE: Gweneth Howarth to Feynman, 14 January 1959, PERS.

344 SHE IS AN INTELLIGENT GIRL: Feynman to American Consulate General, Zurich, 22 January 1959, PERS.

344 SHE HAD TO AVOID ENGELBERT: Gweneth Howarth to Feynman, 14 February 1959.

344 FROM WHAT MORAL HIGH GROUND: Gweneth Feynman, interview.

345 BUT FEYNMAN'S ATTORNEYS ADVISED HIM: Samuel C. Klein to Robert F. Diekman, 22 September 1959, and Robert F. Diekman to Feynman, 30 September 1959, PERS.

345 WELL, AT LAST: Feynman to Gweneth Howarth, 28 May 1959, PERS.

345 SHE SURREPTITIOUSLY INTRODUCED COLORED SHIRTS: Gweneth Feynman, interview.

346 AT FIRST HE KEPT HER PRESENCE SECRET: Gweneth Feynman, Gell-Mann, interviews.

346 SO THIS IS HOW WE'RE STARTING LIFE: Gweneth Feynman, interview.

346 MURRAY GELL-MANN, WHO HAD MARRIED: Gell-Mann 1989a, 50.

346 AN IMAGE LODGED IN GELL-MANN'S MEMORY: Ibid.

347   HELLO, MY SWEETHEART: Feynman to Gweneth Feynman, 11 October 1961, PERS.

347   AGE IS, OF COURSE, A FEVER CHILL: E.g. Kragh 1989, 347*n*.

348   TO CONVEY A SENSE OF HOW "DELICATELY": *QED*, 7.

348   WE HAVE BEEN COMPUTING TERMS: Feynman 1961*a*, 17.

348   NOTE THE CUNNING OF REASON AT WORK: Schweber, forthcoming.

348   WE VERY MUCH NEED A GUIDING PRINCIPLE: Weinberg 1977*a*, 33.

348   "DIPPY" AND "A SHELL GAME": *QED*, 128.

349   THE ELECTRON DISTORTS THE LATTICE: Feynman 1955*a*; Feynman et al. 1962.

349   HIS CALTECH SALARY: Salary records, Lee DuBridge papers, CIT.

349   HE STARTED TELLING PEOPLE: Susman, personal communication.

349   FEYNMAN TOLD HIMSELF THAT HE WOULD GO: F-W, 751.

349   FEYNMAN BEGAN IN THE SUMMER: Notebook, "Biochemical Techniques," CIT; F-W, 751.

350   UNDERSTANDING WHEN A THING IS REALLY KNOWN: F-W, 753.

350   HE FOCUSED ON A PARTICULAR MUTATION: Benzer 1962; Crick 1962; Crick 1966.

350   FEYNMAN COMPARED FINDING: Susman, personal communication.

350   FRIENDS OF HIS IN THE LABORATORY: Robert Sinsheimer to Feynman, n.d., "Dear Feyntron . . . ," CIT; "Mutual Suppression of *r*II Mutants of Bacteriophage T4D," draft by Robert Sinsheimer, CIT. F-W, 752: "I knew they were very interesting and unusual, but I didn't write it up." He did contribute to a group paper in *Genetics*, however: Edgar, Feynman, et al. 1961.

351   THE SPECIALISTS HAD AN ADVANTAGE: Crick et al. 1961; Crick 1962.

351   THE STORY OF THE GENETIC CODE: Crick 1966, 55–56.

351   QUANTUM-MECHANICAL SMEARING OF SPACE-TIME: Gell-Mann 1989*a*, 53.

352   THE GRAVITATIONAL FORCE IS WEAK: Alexander J. Glass, letter to *Physics Today*, May 1988, 136.

352   I HAVE NOT SEEN ANY PLANS: Feynman to Weisskopf, 4 January to 11 February 1961, WHE.

352   MAYBE GRAVITY IS A WAY: Ibid.

352   A CONFERENCE ON GRAVITATION: Feynman 1963*b*.

353   SINCE 1916 WE HAVE HAD A SLOW: Quoted in Schucking 1990, 486.

353   THEY TEASED EACH OTHER: Feynman to Gweneth Feynman, n.d., in WDY, 90.

353   WHAT HAVE YOU EVER DONE: Schucking 1990, 483.

353   THE "WORK" IS ALWAYS: Feynman to Gweneth Feynman, n.d., in WDY, 91–92.

354   THE REAL FOUNDATION OF QUANTUM MECHANICS: Gell-Mann 1989*a*, 54.

354   THERE IS A DEVICE ON THE MARKET: Feynman 1960*a*, 22–24.

355   HE ENVISIONED MACHINES THAT WOULD MAKE: The idea of ever-tinier servo-controlled robotic hands had been anticipated by the science fiction writer Robert Heinlein, who called them Waldoes. Cf. Regis 1990, 142.

355   HOW TO BUILD AN AUTOMOBILE: *Popular Science Monthly*, November 1960, 114.

356   NOT UNTIL 1985 DID FEYNMAN HAVE TO PAY: Thomas H. Newman to Feynman, 30 January 1986, CIT.

356   BY JUNE, WHEN HE HAD NOT HEARD: William McLellan, telephone interview; "McLellan Micromotor," note, CIT.

356 UH-OH: McLellan, interview.

356 HE HAD NEGLECTED TO MAKE ANY ARRANGEMENTS: Feynman to McLellan, 15 November 1960.

357 FUNDAMENTAL PHYSICS (IN THE SENSE OF: Feynman 1960a, 22.

357 WHAT WE ARE TALKING ABOUT IS REAL: Feynman to Ashok Arora, 4 January 1967, CIT.

357 PITH BALLS AND INCLINED PLANES: F-W, 760.

358 1. HISTORICAL DEVELOPMENT: Beyer and Williams 1957. Cf. Lindsay 1940, Bonner and Phillips 1957, and Mendenhall et al. 1950.

358 A GENERATION-OLD TEXT BY ITS OWN LUMINARY: Millikan et al.

358 SO, WHAT IS OUR OVER-ALL PICTURE: Lectures, I-1-2.

359 IF WATER—WHICH IS NOTHING: Lectures, I-1-9.

359 GUIDE TO THE PERPLEXED: F-W, 762.

359 A TEAM OF CALTECH PHYSICS PROFESSORS: Sands, interview; D. Goodstein 1989, 74.

360 PHYSICS BEFORE 1920: Lectures, I-2-3.

360 NOT THE PROBLEM OF FINDING NEW: Lectures, I-3-9.

360 IT IS THE ANALYSIS OF CIRCULATING: Ibid.

360 WHAT WE REALLY CANNOT DO: Lectures, I-3-10.

360 WELL, THE HOUR IS UP: F-W, 765.

360 HE TIMED HIS DIAGRAMS: Sands, interview.

361 IN THE GRADUAL INCREASE IN THE COMPLEXITY: Lectures, I-4-2.

362 THEY DEPEND UPON HOW: Lectures, I-17-2.

362 EVEN PHYSICISTS FELT THEY WERE LEARNING: Stabler 1967, 48; Lectures, I-20-7.

362 IT IS A WONDERFUL THING: Lectures, I-20-7.

363 THE RATCHET AND PAWL WORKS: Lectures, I-46-9.

363 AS THE MONTHS WENT ON: F-W, 766.

363 I'VE SPOKEN TO SOME: D. Goodstein 1989, 74.

363 IT IS ODD: CPL, 13.

364 GIVE A HUMAN APPROACH: Tord Pramberg to Feynman, 15 November 1966, and Feynman to Tord Pramberg, 4 January 1966, CIT.

364 WHEN YOU HAVE LEARNED: Feynman to Ashok Arora, 4 January 1967, CIT.

364 WITH THIS QUESTION PHILOSOPHY BEGAN: Heidegger 1959, 20.

365 ALL SATELLITES TRAVEL: CPL, 19.

365 THAT IS THE SAME: Ibid., 33.

365 EXACTLY THE SAME LAW: Ibid., 34.

365 MEANWHILE, WHY DOES AN OBJECT: Ibid., 19.

366 SCIENCE REPUDIATES PHILOSOPHY: Quoted in Ziman 1978, 1.

366 NONE OF THE ENTITIES THAT APPEAR: Park 1988, xx.

366 LIKE TOURISTS MOVING IN: CPL, 173.

366 QUESTIONS ABOUT A THEORY: Slater, "Electrodynamics of Ponderable Bodies," Journal of the Franklin Institute 225 (1938):277. Quoted by Schweber, forthcoming.

366 AN UNDERSTANDING OF THE LAW: CPL, 169.

367 AFTERWARD, MURRAY GELL-MANN "COUNTERED": SYJ, 290.

367   THEY COUNTED A CERTAIN NUMBER: *CPL*, 169.

367   "YES," SAYS THE ASTRONOMER: Ibid., 170.

368   TO DYSON'S ASTONISHMENT: Dyson to his parents, October 1948.

368   OH, NO, IT'S NOT SERIOUS: Dyson, interview; Dyson 1990.

368   HIS MOTIVATION WAS TO DISCOVER: Dyson 1990, 210.

368   DIFFERENT IDEAS FOR GUESSING: *CPL*, 168.

369   TO GET SOMETHING THAT WOULD PRODUCE: Ibid., 169.

369   WHAT CAN YOU EXPLAIN: Stephen Wolfram, telephone interview.

369   IF YOU GET HOLD OF TWO MAGNETS: Untitled videotape, n.d., recorded for the British Broadcasting Corporation.

371   I THINK THAT FOR SCIENTIFIC: Quine 1987, 109.

371   THE POST-SCHOLASTIC ERA: Ziman 1992.

371   THE SCIENTIST HAS A LOT OF EXPERIENCE: Feynman 1955c, 14.

372   GREAT VALUE OF A SATISFACTORY PHILOSOPHY: Notes, "The Uncertainty of Science," PERS.

372   THE KIND OF A PERSONAL GOD: Dan L. Thrapp, "Science, Religion Conflict Traced," *Los Angeles Times*, 30 June 1956. Cf. Feynman 1956a.

372   IT DOESN'T SEEM TO ME: Interview for "Viewpoint," with Bill Stout, transcript, CIT. Feynman complained to the station: "It was said at one time that my views might antagonize people. . . . I consider your refusal to utilize the program recorded with me as a direct censorship of the expression of my views." Feynman to Bill Whitley, 14 May 1959, CIT.

372   THE GROUND OF ALL THAT IS: Polkinghorne 1990.

373   POETS SAY SCIENCE TAKES AWAY: *Lectures*, I-3-6 n.

373   I HAVE ARGUED FLYING SAUCERS: Feynman 1963c, 62.

373   IF IT'S NOT A MIRACLE: Ibid., 64.

374   ORANGE BALLS OF LIGHT: Ibid., 61.

374   I HAD THE MOST REMARKABLE EXPERIENCE: Ibid., 66.

374   I WAS UPSTAIRS TYPEWRITING: Ibid.

375   A DESK-THUMPING, FOOT-STAMPING SHOUT: Fine 1991, 271.

375   THE GREAT LESSON OF TWENTIETH-CENTURY: Ibid., 274.

376   THE NOBEL COMMITTEE HAS AWARDED: "Nobel Prize for Einstein," *New York Times*, 10 November 1922, 4.

376   AS THE NE PLUS ULTRA OF HONORS: Zuckerman 1977, 11.

377   EACH FALL, AS THE ANNOUNCEMENT NEARED: "I always thought——I mean, I thought that there was a possibility that I might get a Nobel prize, because I thought somebody might think the work in helium, or maybe the beta decay, or even the electrodynamics might be something for the Nobel prize. . . . Each year when the Nobel prize talking comes around, of course you half think, maybe it's possible." F-W, 800–801.

378   THE WESTERN UNION "TELEFAX": Erik Rundberg to Feynman, 21 October 1965, PERS.

378   THE FIRST CALL HAD COME: F-W, 801; "Dr. Richard Feynman Nobel Laureate!" *California Tech*, 22 October 1965, 1.

378   WILL YOU PLEASE TELL US: F-W, 804.

378 WHAT APPLICATIONS DOES THIS PAPER: "Dr. Richard Feynman Nobel Laureate!"

378 LISTEN, BUDDY, IF I COULD TELL YOU: F-W, 804.

378 JULIAN SCHWINGER CALLED: Schwinger, interview.

378 I THOUGHT YOU WOULD BE HAPPY: Feynman to Lucille Feynman, n.d., PERS.

379 [FEYNMAN:] CONGRATULATIONS: "Dr. Richard Feynman Nobel Laureate!"

379 THERE WERE CABLES FROM SHIPBOARD: F-W, 806.

379 HE PRACTICED JUMPING BACKWARD: Ibid., 808–9.

380 FEYNMAN REALIZED THAT HE HAD NEVER READ: Ibid., 812.

380 HE BELIEVED THAT HISTORIANS: Feynman 1965a.

380 WE HAVE A HABIT IN WRITING: Ibid.

380 AS I WAS STUPID: Ibid.

381 THE CHANCE IS HIGH: Feynman 1965c.

381 I DISCOVERED A GREAT DIFFICULTY: Ibid.

382 THE ODDS THAT YOUR THEORY: Feynman 1965a.

382 DR. CRICK THANKS YOU: Quoted in Zuckerman 1977, 224.

383 MR. FEYNMAN WILL PAY THE SUM: Giuseppe Cocconi to Victor F. Weisskopf, 2 February 1976, CIT.

383 HE BEGAN BY SCRIBBLING A NOTE: Feynman to B. L. Kropp, 9 November 1960, CIT.

383 MY DESIRE TO RESIGN: Feynman to Detlev W. Bronk, 10 August 1961. CIT.

384 THANK YOU FOR YOUR WILLINGNESS: Detlev W. Bronk to Feynman, 26 October 1961, CIT.

384 SUPPOSE THAT WE TRULY: Philip Handler to Feynman, 25 June 1969, CIT.

384 I HAVE YOUR SOMEWHAT CRYPTIC NOTE: Philip Handler to Feynman, 31 July 1969, CIT.

384 HE TURNED DOWN HONORARY DEGREES: George W. Beadle to Feynman, 4 January 1967, and William J. McGill to Feynman, 16 February 1976, CIT.

384 INTRODUCE A DRAFT OF FRESH AIR: Martin Mann to Feynman, 13 September 1962, and reply, CIT.

384 HE REFUSED TO SIGN PETITIONS: E.g., Feynman to Margaret Gardiner, 15 May 1967, CIT.

385 THE COMMENT YOU SENT BACK WITH OUR QUESTIONNAIRE: R. Hobart Ellis, Jr., to Feynman, 25 August 1966, and reply, CIT.

385 FEYNMAN HID BEHIND HER DOOR: Helen Tuck, interview, Pasadena.

385 A DISCRETIONARY KITTY: Goldberger, interview.

386 IT MUST HAVE BEEN VERY DIFFICULT: Holton, interview.

386 HANS BETHE TURNED SIXTY: R. E. Marshak to Feynman, 11 May 1965, and reply, CIT.

386 DON'T LET ANYBODY CRITICIZE: Feynman to James D. Watson, 10 February 1967, CIT.

387 IT IS OF COURSE A YANG-MILLS THEORY: Gell-Mann 1983a, 3.

387 BY THE WAY, SOME PEOPLE: Ibid.

388 THE POINT WAS HARDLY LOST: As Gell-Mann said at a memorial service for Feynman in 1989: "Everybody knows that Richard didn't think one should be able to tell the difference between one bird and another. . . . He tried to show

in yet another way that he could stand out from the herd—like not being a birdwatcher." Talk at Feynman memorial, San Francisco, 18 January 1989.

388   SITS CALMLY BEHIND HIS DESK: Riordan 1987, 192.

389   MURRAY'S MASK WAS A MAN: Coleman, interview.

390   ZWEIG, FAR MORE VULNERABLE: Zweig 1981.

390   THEIR PALPITANT PIPING, CHIRRUP: Quoted in Crease and Mann 1986, 185.

391   THE CONCRETE QUARK MODEL: Zweig, interview.

391   IT IS FUN TO SPECULATE ABOUT THE WAY QUARKS: Gell-Mann 1964.

391   I ALWAYS CONSIDERED THAT TO BE A CODED MESSAGE: Polkinghorne 1989, 110.

391   FOR GELL-MANN THIS BECAME: "People have deliberately misunderstood this for twenty-seven years." Gell-Mann, interview.

391   I'VE ALWAYS TAKEN AN ATTITUDE: F-W, II-26.

391   AT FIRST HIS SYLLABUS CONTAINED: Zweig, interview; F-W,II-15.

392   A SINGLE BUBBLE CHAMBER: Traweek 1988, 52–53.

392   LIKE TRYING TO FIGURE OUT A POCKET WATCH: Quoted in Riordan 1987, 151–52.

392   THE PHYSICISTS WHO WOULD GATHER: Riordan 1987, 149.

393   HE ISOLATED A REMARKABLE REGULARITY: Bjorken 1989, 57; Bjorken, telephone interview.

393   OXFORD ENGLISH DICTIONARY: "Each of the hypothetical point-like constituents of the nucleon that were invoked by R. P. Feynman to explain the way the nucleon inelastically scatters electrons of very high energy." A Supplement to the Oxford English Dictionary, 279.

394   QUANTUM ELECTRODYNAMICS HAD ITS PARTONS: Feynman 1969b, 241.

394   HE CHOSE NOT TO DECIDE: Feynman to Michael Riordan, 26 February 1986, CIT.

394   WHEN FEYNMAN DIAGRAMS ARRIVED: Bjorken 1989, 56.

394   FEYNMAN TOOK ON A PROJECT IN 1970: Feynman et al. 1971.

395   CONVERTED INTO A QUARKERIAN: F-W, II-47.

395   A QUARK PICTURE MAY ULTIMATELY PERVADE: Feynman et al. 1971, 2727.

395   HE DISLIKED THE FANFARE: "These things were quarks and antiquarks (and sometimes gluons), but he didn't want to call them by their names. At first, he wasn't sure that that's what they were, but as time went on it became clearer, and it annoyed me that he still didn't acknowledge that he was talking about quarks. Eventually, some authors began to speak of 'quark partons,' but as if they were somehow different from ordinary current quarks.

"The so-called parton model was an approximate description of quarks and gluons that could apply in the appropriate high-energy limits if the interaction of the particles became weak at short distances (as turned out to be the case in quantum chromodynamics). Dick painted a naïve picture, which was taken not just as an approximation to an unknown theory, but as a kind of revealed truth.

"Physicists all over the world learned the 'parton' story, memorized it, and immediately began to use it to interpret experiments. In other words Dick has oversimplified the picture so that it could be used by everybody." Gell-Mann, personal communication.

395   WE HAVE BUILT A VERY TALL HOUSE OF CARDS: Feynman 1972*c*.

395   I'M A LITTLE BIT FRUSTRATED: F-W, II-86.

396   QUIETLY NOMINATED GELL-MANN AND ZWEIG: They never knew it. B. Wagel to Feynman, 26 January 1977, CIT. Gell-Mann, Zweig, interviews.

396   JEE-JEE-JEE-JU-JU. JEE-JEE-JEE-JU-JU: F-L.

396   IT TOOK YEARS FOR FEYNMAN'S CHILDREN: Michelle Feynman, Carl Feynman, Gweneth Feynman, interviews.

397   RICHARD, I'M COLD: Leighton, interview.

397   I COULD HAVE KILLED HIM: Feynman to Sheila Sorenson, 21 October 1974, CIT.

397   TRUMPET PLAYING—SOCIAL WORKER—ZYGOPHALATELIST: Feynman to Carl Feynman, 18 February 1980, PERS.

397   AFTER MUCH EFFORT AT UNDERSTANDING: Ibid.

398   WHAT MAKES IT MOVE: Feynman 1966*a*.

398   TO TELL A FIRST-GRADER: Ibid., 14.

398   YOU SAY, "WITHOUT USING": Ibid., 15.

398   SHOE LEATHER WEARS OUT: Ibid., 16.

398   FEYNMAN TAUGHT THIRTY-FOUR: D. Goodstein 1989, 73.

399   I COULDN'T REDUCE IT: Ibid. 75.

399   IT IS AN EXAMPLE OF THE USE OF WORDS: Feynman 1964*a*, 16.

400   I DOUBT THAT ANY CHILD: Ibid., 3.

401   MICHELLE LEARNED THAT HE HAD A THOUSAND: Michelle Feynman, interview.

401   OH YES, WE DO: Gweneth Feynman, interview.

401   TRAVELING IN THE SWISS ALPS: Gweneth Feynman, interview.

402   FEYNMAN'S TUMOR: C. M. Haskell, interview, Los Angeles.

402   FIVE-YEAR SURVIVAL RATES: Sheldon C. Binder, Bertram Katz, and Barry Sheridan, "Retroperitoneal Liposarcoma," *Annals of Surgery*, March 1978, 260.

402   YOU ARE OLD, FATHER FEYNMAN: "Father Feynman," n.d., CIT.

402   WITH A POSTDOCTORAL STUDENT: Feynman et al. 1977; Field and Feynman 1977; Field and Feynman 1978.

403   FEYNMAN DID NOT REALIZE THAT FIELD: Richard Field, telephone interview.

403   I DON'T GET ANY PHYSICS: Victor F. Weisskopf to Feynman, 23 March 1979, CIT.

403   QCD FIELD THEORY WITH SIX FLAVORS: "Qualitative Behavior," typescript for Feynman 1981, CIT.

404   VASCULAR INCIDENT: In Chang Kim, interview, Pasadena.

404   FEYNMAN NEEDED SEVENTY-EIGHT PINTS: Haskell, interview. Gweneth Feynman, interview.

404   IT'S IMPOSSIBLE TO TALK: Cvitanović, interview.

404   MY WIFE: Douglas R. Hofstadter, telephone interview.

405   I HAVE NOT ACCOMPLISHED ANYTHING: Feynman to Robert B. Leighton, 10 June 1974, CIT.

405   WHAT THE HELL IS FEYNMAN INVITED FOR: Feynman to Sidney Coleman, 13 August 1976, CIT.

405   ANOTHER PIECE OF EVIDENCE: Coleman to Feynman, 26 July 1976, CIT.

405   AGGRESSIVE DOPINESS: Carl Feynman, interview.

406   HE LISTENED PATIENTLY AS BABA RAM DAS: *SYJ*, 303–5.

406   PEOPLE IN HIGHER ECHELONS: He titled the talk, "Los Alamos from Below."
      Feynman 1975, 105.

406   STILL, HE WOULD EMERGE: SYJ, 306.

406   SPORADICALLY, HE WORKED: E.g., Jon N. Leonard to Feynman, 3 November
      1987, and Peter H. Hambling to Feynman, 4 August 1987, CIT.

407   ARE WE PHYSICS GIANTS: Feynman to Philip Morrison, 23 May 1972, CIT.

407   MYSTICISM, EXPANDED CONSCIOUSNESS: SYJ, 309.

407   IT HAS TO DO WITH THE QUESTION: Videotape, courtesy of Ralph Leighton.

407   PEACE OF MIND AND ENJOYMENT: Quoted in Leighton 1991, 83–84.

408   IT SEEMED TO GWENETH: Gweneth Feynman, interview; William G. Bradley,
      interview.

408   BUT YOU CAN'T SEE: Feynman to William G. Bradley, 13 July 1984, CIT.

409   OKAY, START YOUR WATCH: Weiner, interview.

409   A RECORD OF THE DAY-TO-DAY WORK: F-W, II-4.

410   TODAY I WENT OVER TO THE HUNTINGTON: F-L.

410   AND THE NEXT MORNING, ALL RIGHT: Ibid.

411   "LISTEN," I SAID TO THE DISPATCHER: SYJ, 236.

411   A NICE BROOKLYN RING: Edwin Barber to Feynman, 2 March 1984, CIT.

411   GELL-MANN'S RAGE COULD BE HEARD: E.g., Tuck, interview.

411   OF COURSE IT WASN'T TRUE: SYJ, 229. He also changed "Murray Gell-Mann and
      I wrote a paper on the theory" to "Murray Gell-Mann compared and combined
      our ideas and wrote a paper on the theory" (232). Gell-Mann still called it "that
      joke book." He knew that Feynman had not deliberately tried to take undeserved
      credit, but he was hurt nonetheless. "He was not at all a thief of ideas—even
      very generous in some ways," Gell-Mann said. "It's just that he was not always
      capable of regarding other people as really existing."

411   A NIFTY BLONDE: SYJ, 241 and 168.

412   OUT WITH HIS GIRL FRIEND: Lectures, I-3-7.

412   DEAR ROTHSTEIN: DON'T BUG ME: "Protest," mimeograph sheet, CIT.

412   HE HAD SPENT MANY PLEASANT HOURS: Jenijoy La Belle, interview, Pasadena;
      "Feynman Commends La Belle," letter to California Tech, 5 March 1976; La
      Belle 1989.

413   AND, LIKE FALLING IN LOVE: NL, 435.

413   SO WHAT HAPPENED TO THE OLD THEORY: NL, 456.

413   THERE IS IN THE WORLD OF PHYSICS: Feynman 1972e, 1.

414   GENERALLY MR. FEYNMAN IS NOT JOKING: Morrison 1985, 43.

414   NOT AN AUTOBIOGRAPHY: Feynman to Robert Crease, 18 September 1985, CIT.
      And Feynman to Klaus Stadler, 15 October 1985, CIT: "This shows a complete
      misunderstanding of the nature of my book. . . . It is not in any way a scientific
      book, nor a serious one. It is not even an autobiography. It is only a series of
      short disconnected anecdotes, meant for the general reader which, we hope, the
      reader will find amusing."

414   WHAT I REALLY WAS: Feynman to Crease.

414   A HALF-HOUR AFTER THE LAUNCH: Richard Witkin, "Canaveral Hopes for Success
      Fade," New York Times, 6 March 1958, 1.

414   THEY USED A ROOM-SIZE: Hibbs, interview.

416 AN OUTSIDE GROUP OF EXPERTS: "Reagan names panel on shuttle explosion," Walter V. Robinson, *Washington Post*, 4 February 1986, 1.

417 ARMSTRONG SAID ON THE DAY: "President Names 12-Member Panel in Shuttle Inquiry," Gerald Boyd, *New York Times*, 4 February 1986, 1.

417 WE ARE NOT GOING TO CONDUCT: Ibid. In the commission's first closed session, on February 10, he emphasized: "This is not an adversarial procedure. This commission is not in any way adversarial . . ." *Report*, IV, 244.

417 YOU'RE RUINING MY LIFE: William R. Graham, telephone interview.

417 FEYNMAN WAS NOW SUFFERING: Haskell, interview.

417 FEYNMAN HIMSELF REFUSED TO CONSIDER: Haskell, interview.

417 HE IMMEDIATELY ARRANGED A BRIEFING: Hibbs, interview; Charles Lifer, interview; Winston Gin, interview; WDY, 119–21.

419 ROGERS OPENED THE FIRST: *Report*, IV, 1.

419 IN RESPONSE, MOORE DENIED: Ibid., 21.

419 A CONCERN BY THIOKOL: Ibid., 97.

419 NEWSPAPER REPORTS THE NEXT DAY: Esp. David Sanger, "NASA Seems Surprised By Aggressive Queries," *New York Times*, 7 February 1986, A19.

419 THIS IS WHAT WE WOULD HAVE CALLED: *Report*, IV, 220.

419 EVERYTHING THAT I KNOW: Ibid., 221.

420 WHEN WE ASK QUESTIONS: Ibid., 222.

420 YOU SAID WE DON'T EXPECT IT: Ibid., 224.

420 CO-PILOT TO PILOT: Donald J. Kutyna, interview, Peterson Air Force Base, Colo.; WDY, 126.

420 I HAVE A PICTURE OF THAT SEAL: *Report*, IV, 224.

421 THE LACK OF A GOOD SECONDARY SEAL: "August 19, 1985 Headquarters Briefing," *Report*, I, 139; WDY, 135.

421 LOSS OF VEHICLE, MISSION, AND CREW: "NASA Had Warning of a Disaster Risk Posed by Booster," Philip Boffey, *New York Times*, 9 February 1986, 1.

422 YOU KNOW, THOSE THINGS LEAK: WDY, 139–40; Kutyna, interview. Feynman misremembered this as a telephone conversation.

422 I THINK IT GOES WITHOUT SAYING: *Report*, IV, 244.

422 LAWRENCE MULLOY, PROJECT MANAGER: Ibid., 291.

422 HOW ARE THESE MATERIALS, THIS PUTTY AND THE RUBBER: Ibid., 347.

423 IF THIS MATERIAL WEREN'T RESILIENT: Ibid., 345.

423 HE HAD MADE AN OFFICIAL REQUEST: WDY, 146.

423 FEYNMAN IS BECOMING A REAL PAIN: David Sanger, personal communication.

424 YOU DIDN'T, I ASSUME: *Report*, IV, 380–82.

424 MULLOY, UNDER FURTHER QUESTIONING: "NASA Acknowledges Cold Affects Boosters Seals," Philip Boffey, *New York Times*, 12 February 1986, 1.

424 THE PUBLIC SAW WITH THEIR OWN EYES: Dyson 1992, 284.

425 TO EXAGGERATE: TO EXAGGERATE HOW ECONOMICAL: WDY, 214.

425 ONE OF THE MOST PRODUCTIVE: *Report*, I, 1.

426 IT WAS A GREAT BIG WORLD: WDY, 158.

426 KUTYNA TOLD HIM HE WAS THE ONLY: Kutyna, interview; WDY, 156.

426 IN BETWEEN, HE MADE REPEATED VISITS: F-L.

426   I AM DETERMINED TO DO THE JOB: Feynman to Gweneth Feynman, 12 February
       1986, quoted in WDY, 157.

426   THE COMMISSION STRONGLY RECOMMENDS: WDY, 200–201.

427   HISTORY OF O-RING PROBLEMS HAD BEEN REPORTED: E.g. Report, I, Appendix H;
       Graham, interview.

427   OVERALL HE ESTIMATED: Feynman 1986, F-2.

427   A KIND OF RUSSIAN ROULETTE: Report, I, 148.

427   IT HAS TO BE UNDERSTOOD: Ibid., IV, 817.

428   A TEAM OF STATISTICIANS: Dalal et al. 1989; Bruce Hoadley, telephone interview.

428   FEYNMAN DISCOVERED THAT SOME ENGINEERS: WDY, 182–83.

428   FOR A SUCCESSFUL TECHNOLOGY: Feynman 1986, F-5.

EPILOGUE

429   RATHER THAN EMBARRASS THEM: Lectures, I-16-1.

430   DID THE UNCERTAINTY PRINCIPLE IMPOSE: Lectures, I-6-10.

430   THE UNCERTAINTY PRINCIPLE SIGNALED: Hawking 1987, 55.

430   IT IS USUALLY THOUGHT THAT THIS INDETERMINACY: Lectures, I-38-9.

430   IF WATER FALLS OVER A DAM: Ibid.

431   FIFTY YEARS OF PARTICLE PHYSICS: Cahn and Goldhaber 1989, ix.

432   A CONTRIVED INTELLECTUAL STRUCTURE: Schwartz 1992, 173.

432   PEOPLE SAY TO ME, "ARE YOU LOOKING: F-Sy.

432   WE MAY NOW BE NEAR THE END: Hawking 1987, 156.

432   I'VE HAD A LIFETIME OF THAT: Interview conducted by P. C. W. Davies, transcript, CIT.

433   YOUR CAREER SPANS THE PERIOD: Interview conducted by Robert Crease, 22 Febru-
       ary 1985; transcript, courtesy of Crease. Robert Crease to Feynman, 18 July 1985,
       CIT.

434   I SEE YOU'VE MET DICK: Robert Crease to Feynman, 18 July 1985, CIT.

435   FORGET ALL THAT "LOCAL MINIMA" STUFF: Hillis 1989, 82.

435   AND HE BEGAN TO PRODUCE MAVERICK RESEARCH: Feynman 1982; Feynman 1984.

435   THE PHYSICAL REVIEW OF THE BLIND MEN: Lectures, II-20-11.

435   WE ARE ALL REDUCTIONISTS TODAY: Weinberg 1987a, 66; Weinberg, personal
       communication.

436   THE INFINITE VARIETY AND NOVELTY: Lectures, II-41-12.

436   HE MAY ALSO BELIEVE IN THE EXISTENCE: Einstein and Infeld 1938, 31.

436   ONE OF THE GREAT PHILOSOPHERS: Mermin 1985, 47; Feynman 1982, 471.

437   I HAVE DECIDED IT IS NOT A VERY GOOD IDEA: Feynman to Lee Dye, 23 September
       1987, CIT.

437   IT IS REALLY LIKE THE SHAPE: Ulam 1976, xi.

437   I'M GOING TO DIE: Michelle Feynman, interview.

437   HE WAS WATCHED AND GUARDED: Joan Feynman, Gweneth Feynman, and Frances
       Lewine, interviews.

438   TAUGHT PEOPLE MOST OF THE GOOD STUFF: Hillis 1989, 83.

438   YOU SEE, ONE THING IS, I CAN LIVE: F-Sy.

438   I'D HATE TO DIE TWICE: Gweneth Feynman, interview.

A FEYNMAN
BIBLIOGRAPHY

· ÷ ·

Because almost all Feynman's work originated with the spoken word, and because its publication took so many shapes, formal and informal, no final bibliography will ever be compiled. Neither Feynman nor the Caltech libraries maintained more than a partial listing. Some lectures were published repeatedly, in journals and collections, in versions that vary slightly or not at all. Others exist only in the form of Feynman's notes before the fact, a student's handwritten notes after the fact, a university preprint, a typed transcript, an edited or unedited conference proceeding, a file on a computer disk, or a video- or audiotape. Some manuscripts are virtually intact and publishable; others are no more than notes on a placemat; and in between is an unbroken continuum.

The following is a guide to work of Feynman's that can be construed as published in any form; major unpublished work; and other important manuscripts and papers cited in this book.

1933–34. "The Calculus: Scribble-In Book." Notebook. AIP.

1935. "The Calculus of Finite Differences." *The f(x)*. Far Rockaway High School Mathematics Club. January, 1. CIT.

Feynman and Welton, T. A. 1936–37. Notebook. AIP.

1939*a*. "Forces and Stresses in Molecules." Thesis submitted in partial fulfillment of the requirements for the degree of bachelor of science in physics. AIP.

1939*b*. "Forces in Molecules." *Physical Review* 56:340.

Vallarta, M. S., and Feynman. 1939. "The Scattering of Cosmic Rays by the Stars of a Galaxy." *Physical Review* 55:506.

1940. "Notebook of Things I Don't Know About." Notebook. CIT.

1941*a*. "The Interaction Theory of Radiation." Typescript. AIP.

1941*b*. "Particles Interacting thru an Intermediate Oscillator." Draft pages toward Ph.D. thesis. PERS.

Feynman and Wheeler, John Archibald. 1941. "Reaction of the Absorber as the Mechanism of Radiative Damping. Abstract." *Physical Review* 59: 682.

1942a. Ph.D. thesis manuscript. CIT.

1942b. "The Principle of Least Action in Quantum Mechanics." Ph.D. thesis, Princeton University.

1942c. *Effects of Space Charge; Use of Sine Waves.* Isotron Report no. 2, 5 January. SMY.

1942d. *Kinematics of the Separator.* Isotron Report no. 7, 14 April. SMY.

1942e. *The Design of the Buncher and Analyzer.* Isotron Report no. 17, 26 August. SMY.

1942f. *A Note on the Cascade Operation of Isotrons.* Isotron Report no. 20, 8 September. SMY.

Wheeler, John Archibald, and Feynman. 1942. "Action at a Distance in Classical Physics: Reaction of the Absorber as the Mechanism of Radiative Damping." Typescript. AIP.

1943a. *The Operation of Isotrons in Cascade.* Isotron Report no. 29, 27 January. SMY.

1943b. *Factors Which Influence the Separation.* Isotron Report no. 35, 22 February. SMY.

1944. "Theoretical Department." Unsigned draft typescript for Smyth 1945. LANL.

Ashkin, J.; Ehrlich, R.; and Feynman. 1944. "First Report on the Hydride." Typescript, 31 January. LANL.

1945. "A New Approximate Method for Rapid Calculation of Critical Amounts of X." Typescript. LANL.

Wheeler, John Archibald, and Feynman. 1945. "Interaction with the Absorber as the Mechanism of Radiation." *Reviews of Modern Physics* 17:157.

1946a. *Amplifier Response.* Los Alamos Reports, LA-593. LANL.

1946b. *A Theorem and Its Application to Finite Tampers.* Los Alamos Reports, LA-608, Series B. LANL.

Feynman and Bethe, Hans A. 1946. Abstract for New York Meeting of the American Physical Society, 19–21 September. Typescript. CIT.

1947. "Theory of Positrons." Notes. CIT.

Feynman and Welton, T. A. 1947. *The Calculation of Critical Masses Including the Effects of the Distribution of Neutron Energies.* Los Alamos Reports, Series B, LA-524. LANL.

1948a. "Space-Time Approach to Non-Relativistic Quantum Mechanics." *Reviews of Modern Physics* 20: 367.

1948b. "A Relativistic Cut-Off for Classical Electrodynamics," *Physical Review* 74: 939.

1948c. "Relativistic Cut-Off for Quantum Electrodynamics." *Physical Review* 74: 1430.

1948d. "Pocono Conference." *Physics Today*, June, 8.

1948e. "Pocono Conference." Typescript. LOC.

1948f. Paper T5: "Theory of Positrons." Talk prepared for American Physical Society meeting in January 1949. CIT.

1949a. "The Theory of Positrons." *Physical Review* 76:749.

1949b. "Space-Time Approach to Quantum Electrodynamics." *Physical Review* 76:769.

Feynman; Metropolis, Nicholas; and Teller, Edward. 1949. "Equations of State of

Elements Based on the Generalized Fermi-Thomas Theory." *Physical Review* 75:1561.

Wheeler, John Archibald, and Feynman. 1949. "Classical Electrodynamics in Terms of Direct Interparticle Action." *Reviews of Modern Physics* 21:425.

1950. "Mathematical Formulation of the Quantum Theory of Electromagnetic Interaction." *Physical Review* 80:440.

1951a. "An Operator Calculus Having Applications in Quantum Electrodynamics." *Physical Review* 84:108.

1951b. "The Concept of Probability in Quantum Mechanics." Second Berkeley Symposium on Mathematical Statistics and Probability, University of California, Berkeley, 1950:533.

Brown, Laurie M., and Feynman. 1952. "Radiative Corrections to Compton Scattering." *Physical Review* 85:231.

Lopes, J. Leite, and Feynman. 1952. "On the Pseudoscalar Meson Theory of the Deuteron." Symposium on New Research Techniques in Physics, 15–29 July.

1953a. "The Lambda Transition in Liquid Helium." *Physical Review* 90:1116.

1953b. "Atomic Theory of Lambda Transition in Helium." *Physical Review* 91:1291.

1953c. "Atomic Theory of Liquid Helium near Absolute Zero." *Physical Review* 91:1301.

1953d. "Atomic Theory of Liquid Helium." Talk at the Theoretical Physics Conference in Tokyo, September 1953. In *Notas de Físicas* 12.

1954a. "Atomic Theory of the Two-Fluid Model of Liquid Helium." *Physical Review* 94:262.

1954b. "The Present Situation in Fundamental Theoretical Physics." *Academia Brasileira de Ciencias* 26:51.

Feynman; Baranger, Michel; and Bethe, Hans A. 1954. "Relativistic Correction to the Lamb Shift." *Physical Review* 92:482.

Feynman and Speisman, G. 1954. "Proton-Neutron Mass Difference." *Physical Review* 94:50.

1955a. "Slow Electrons in a Polar Crystal." *Physical Review* 97:660.

1955b. "Application of Quantum Mechanics to Liquid Helium." In *Progress in Low Temperature Physics*. Edited by C. J. Gorter. Amsterdam: North Holland.

1955c. "The Value of Science." Transcript of address at the autumn 1955 meeting of the National Academy of Sciences. In *Engineering and Science*, June, 3.

Feynman and Cohen, Michael. 1955. "The Character of the Roton State in Liquid Helium." *Progress in Theoretical Physics* 14:261.

1956a. "The Relation of Science and Religion." *Engineering and Science*, June, 20.

1956b. "Dr. Feynman Replies to Mr. Sohler's 'New Hypothesis.'" *Engineering and Science*, October, 52.

Feynman and Cohen, Michael. 1956. "Energy Spectrum of the Excitations in Liquid Helium." *Physical Review* 102:1189.

Feynman; de Hoffmann, Frederic; and Serber, Robert. 1956. "Dispersion of the Neutron Emission in U-235 Fission." *Journal of Nuclear Energy* 3:64.

1957a. "Superfluidity and Superconductivity." *Reviews of Modern Physics* 29:205.

1957b. "Alternative to the Two-Component Neutrino Theory." Remarks at the Seventh

Annual Rochester Conference on High-Energy Physics, 15–19 April. In Ascoli et al. 1957, IX-42.

1957c. "The Role of Science in the World Today." *Proceedings of the Institute of World Affairs* 33:17.

Feynman; Vernon, F. L.; and Hellwarth, Robert W. 1957. "Geometric Representation of the Schrodinger Equation for Solving Maser Problems." *Journal of Applied Physics* 28:49.

Cohen, Michael, and Feynman. 1957. "Theory of Inelastic Scattering of Cold Neutrons from Liquid Helium." *Physical Review* 107:13.

1958a. "Excitations in Liquid Helium." *Physica* 24:18.

1958b. "A Model of Strong and Weak Couplings." Typescript. CIT.

1958c. "Forbidding of $\pi - \beta$ Decay." Talk at Annual International Conference on High Energy Physics at CERN, Geneva, 30 June–5 July. In Ferretti 1958.

Feynman and Gell-Mann, Murray. 1958a. "Theory of the Fermi Interaction." *Physical Review* 109:193.

Feynman and Gell-Mann, Murray. 1958b. "Theoretical Ideas Used in Analyzing Strange Particles." Manuscript for Geneva Conference on the Peaceful Uses of Atomic Energy. CIT.

Feynman and Gell-Mann, Murray. 1958c. "Problems of the Strange Particles." *Proceedings of the Second Geneva Conference on the Peaceful Uses of Atomic Energy.*

1960a. "There's Plenty of Room at the Bottom: An Invitation to Enter a New Field of Physics." Talk at the annual meeting of the American Physical Society, 29 December 1959. In *Engineering and Science*, February, 22.

1960b. "The Status of the Conserved Vector Current Hypothesis." In Sudarshan et al. 1960, 501.

1961a. "The Present Status of Quantum Electrodynamics." Talk for 1961 Solvay Conference. Typescript. CIT. In *Extrait des Rapports et Discussions, Solvay.* Institut International de Physique, October.

1961b. "Theory of Gravitation." Faraday Lecture, 13 April. Transcript. PERS.

1961c. *Quantum Electrodynamics.* New York: W. A. Benjamin.

1961d. *Theory of Fundamental Processes.* New York: W. A. Benjamin.

Edgar, R. S.; Feynman; Klein, S.; Lielausis, I.; and Steinberg, C. M. 1961. "Mapping Experiments with r Mutants of Bacteriophage T4D." *Genetics* 47:179.

Feynman; Hellwarth, R. W.; Iddings, C. K.; and Platzman, P. M. 1962. "Mobility of Slow Electrons in a Polar Crystal." *Physical Review* 127:1004.

1963a. "The Problem of Teaching Physics in Latin America." Transcript of keynote speech given at the First Inter-American Conference on Physics Education in Rio de Janeiro. In *Engineering and Science*, November, 21.

1963b. "The Quantum Theory of Gravitation." *Acta Physica Polonica* 24:697.

1963c. "This Unscientific Age." John Danz Lectures. Transcript. CIT.

Feynman; Leighton, Robert B.; and Sands, Matthew. 1963. *The Feynman Lectures on Physics.* Reading, Mass.: Addison-Wesley.

Feynman and Vernon, F. L. 1963. "The Theory of a General Quantum System Interacting with a Linear Dissipative System." *Annals of Physics* 24:118.

1964a. "Comments on the New Arithmetic Textbooks." Typescript. PERS.

1964b. "Theory and Applications of Mercerau's Superconducting Circuits." Draft typescript. CIT.

Feynman; Gell-Mann, Murray; and Zweig, George. 1964. "Group $U(6) \times U(6)$ Generated by Current Components." *Physical Review Letters* 13:678.

1965a. "The Development of the Space-Time View of Quantum Electrodynamics." Nobel Prize in Physics Award Address, Stockholm, 11 December. In *Les Prix Nobel en 1965* (Stockholm: Nobel Foundation, 1966); in *Physics Today*, August 1966, 31; in *Science* (1966) 153:699; and in Weaver 1987, 2:433.

1965b. "The Development of the Space-Time View of Quantum Electrodynamics." Transcript, physics colloquium at California Institute of Technology, 2 December. CIT.

1965c. "The Development of the Space-Time View of Quantum Electrodynamics." Talk at CERN, Geneva, 17 December. Tape courtesy of Helen Tuck.

1965d. Address to Far Rockaway High School. Transcript. CIT.

1965e. *The Character of Physical Law.* Cambridge, Mass.: MIT Press.

1965f. "New Textbooks for the 'New' Mathematics." *Engineering and Science*, March, 9.

1965g. "Consequences of $SU(3)$ Symmetry in Weak Interactions." In *Symmetries in Elementary Particle Physics*, III. New York: Ettore Majorana Academic Press.

Feynman and Hibbs, Albert R. 1965. *Quantum Mechanics and Path Integrals.* New York: McGraw-Hill.

1966a. "What Is Science?" Address to National Science Teachers Association, 1–5 April. Corrected transcript. PERS.

1966b. "What Is and What Should Be the Role of Scientific Culture in Modern Society?" *Supplemento al Nuovo Cimento* 4:292.

1969a. "What Is Science?" *The Physics Teacher*, September, 313.

1969b. "The Behavior of Hadron Collisions at Extreme Energies." Talk at Third International Conference on High Energy Collisions, State University of New York, 5–6 September. In Yang et al. 1969, 237.

1969c. "Very High-Energy Collisions of Hadrons." *Physical Review Letters* 23:1415.

1970. "Partons." Talk at Symposium on the Past Decade in Particle Theory, University of Texas at Austin, 14–17 April. In Sudarshan and Ne'eman 1973, 773.

Thornber, K. K., and Feynman. 1970. "Velocity Acquired by an Electron in a Finite Electric Field in a Polar Crystal." *Physical Review* B10:4099.

Feynman; Kislinger, M.; and Ravndal, F. 1971. "Current Matrix Elements from a Relativistic Quark Model." *Physical Review* D3:2706.

1972a. "Closed Loops and Tree Diagrams." In Klauder 1972, 355.

1972b. "Problems in Quantizing the Gravitational Field, and the Massless Yang-Mills Field." In Klauder 1972, 377.

1972c. *Photon-Hadron Interactions.* New York: W. A. Benjamin.

1972d. *Statistical Mechanics: A Set of Lectures.* New York: W. A. Benjamin.

1972e. "The Proton Under the Electron Microscope." Oersted Medal Lecture. Manuscript. PERS.

1972f. "What Neutrinos Can Tell Us About Particles." In *Proceedings of Neutrino '72 Europhysics Conference.* Budapest: OMKD Technoinform.

1974. "Structure of the Proton." Talk at Dansk Ingeniørforening, Copenhagen. *Science* 183:601.

1974*b*. "Conference Summary." Talk at International Conference on Neutrino Physics and Astrophysics, Philadelphia, 28 April. Typescript. CIT.

1975. "Reminiscences of Wartime Los Alamos." Talk at University of California at Santa Barbara. Audio tapes. AIP. Edited version in *Engineering and Science*, January, 11. Also in Badash et al. 1980, 105. Excerpted in *SYJ*, 90. [Page references to Badash et al.]

1976. "Gauge Theories." Lecture at Les Houches, Session 29.

1977. "Correlations in Hadron Collisions at High Transverse Momentum." Talk at Orbis Scientiae 1977, University of Miami, Coral Gables, Florida.

Feynman, Field, Richard D.; and Fox, Geoffrey C. 1977. "Correlations among Particles and Jets Produced with Large Transverse Momenta." *Nuclear Physics* B128:1.

Field, Richard D., and Feynman. 1977. "Quark Elastic Scattering as a Source of High Transverse Momentum." *Physical Review* D15:2590.

1978. Talk at Julian Schwinger's 60th birthday celebration. AIP.

Field, Richard D., and Feynman. 1978. "A Parametrization of the Properties of Quark Jets." *Nuclear Physics* B136:1.

1981. "The Qualitative Behavior of Yang-Mills Theory in 2 + 1 Dimensions." *Nuclear Physics* B188:479.

1982. "Simulating Physics with Computers." *International Journal of Theoretical Physics* 21:467.

1984. "Quantum Mechanical Computers." Plenary talk at IQEC-CLEO Meeting, Anaheim, 19 June. Typescript.

1985*a*. *Surely You're Joking, Mr. Feynman! Adventures of a Curious Character.* New York: Norton.

1985*b*. *QED: The Strange Theory of Light and Matter.* Princeton: Princeton University Press.

1986. "Personal Observations of Reliability of Shuttle." In *Report of the Presidential Commission on the Space Shuttle Challenger Accident,* II-F.

1987*a*. "The Reason for Antiparticles." In Feynman and Weinberg 1987, 1.

1987*b*. "Negative Probability." In Hiley and Peat 1987, 235.

1987*c*. "Linear D Dimensional Vector Space." Manuscript. PERS.

Feynman and Weinberg, Steven. 1987. *Elementary Particles and the Laws of Physics: The 1986 Dirac Memorial Lectures.* Cambridge: Cambridge University Press.

1988. *What Do You Care What Other People Think? Further Adventures of a Curious Character.* New York: Norton.

BIBLIOGRAPHY

· ÷ ·

Albers, Donald J., and Alexanderson, G. L., eds. 1985. *Mathematical People: Profiles and Interviews.* Boston: Birkhäuser.

Albert, Robert S., ed. 1983. *Genius and Eminence: The Social Psychology of Creativity and Exceptional Achievement.* New York: Pergamon Press.

Alt, Franz L. 1972. "Archeology of Computers: Reminiscences, 1945–1947." *Communications of the Association for Computing Machinery,* July, 693.

Anderson, Philip W. 1972. "More Is Different." *Science* 177:393. In Weaver 1987, 3:586.

Andronikashvili, Elevter L. 1990. *Reflections on Liquid Helium.* Translated by Robert Berman. New York: American Institute of Physics.

Ascoli, G.; Feldman, G.; Koester, Jr., L. J.; Newton, R.; Riesenfeld, W.; Ross, M.; and Sachs, R. G., eds. 1957. *High Energy Nuclear Physics.* Proceedings of the Seventh Annual Rochester Conference, 15–19 April. New York: Interscience.

Aspray, William. 1990. *John von Neumann and the Origins of Modern Computing.* Cambridge, Mass.: MIT Press.

Atomic Energy Commission. 1954. "In the Matter of J. Robert Oppenheimer." Transcript of Hearings before Personnel Security Board.

Badash, Lawrence; Hirschfelder, Joseph O.; and Broida, Herbert P., eds. 1980. *Reminiscences of Los Alamos, 1943–1945.* Dordrecht: Reidel.

Ballam, J.; Fitch, V. L.; Fulton, T.; Huang, K.; Rau, R. R.; and Treiman, S. B., eds. 1956. *High Energy Nuclear Physics.* Proceedings of the Sixth Annual Rochester Conference, 3–7 April. New York: Interscience.

Bashe, Charles J.; Johnson, Lyle R.; Palmer, John H.; and Pugh, Emerson W. 1986. *IBM's Early Computers.* Cambridge, Mass.: MIT Press.

Battersby, Christine. 1989. *Gender and Genius: Towards a Feminist Aesthetics.* London: Women's Press.

Benzer, Seymour. 1962. "The Fine Structure of the Gene." *Scientific American*, January, 70.

Berenda, Carlton W. 1947. "The Determination of Past by Future Events: A Discussion of the Wheeler-Feynman Absorption-Radiation Theory." *Philosophy of Science* 14:13.

Berkeley, George. 1952. *The Principles of Human Knowledge*. Chicago: University of Chicago Press.

Berland, Theodore. 1962. *The Scientific Life*. New York: Coward-McCann.

Bernal, J. D. 1939. *The Social Function of Science*. New York: Macmillan.

Bernstein, Jeremy. 1967. *A Comprehensible World: On Modern Science and Its Origins*. New York: Random House.

———. 1980. *Hans Bethe: Prophet of Energy*. New York: Basic Books.

———. 1985. "Retarded Learner: Physicist John Wheeler." *Princeton Alumni Weekly*, 9 October, 28.

———. 1987. *The Life It Brings: One Physicist's Beginnings*. New York: Ticknor and Fields.

———. 1989. *The Tenth Dimension: An Informal History of High Energy Physics*. New York: McGraw-Hill.

Bethe, Hans A. 1979. "The Happy Thirties." In Stuewer 1979, 11.

———. 1988. "Richard Phillips Feynman (1918–1988)." *Nature* 332:588.

Bethe, Hans A.; Bacher, Robert F.; and Livingston, M. Stanley. 1986. *Basic Bethe: Seminal Articles on Nuclear Physics, 1936–1937*. Los Angeles: Tomash/American Institute of Physics.

Bethe, Hans A., and Christy, Robert F. 1944. "Memorandum on the Immediate After Effects of the Gadget," March 30. LANL.

Beyer, Robert T., and Williams, Jr., A. O. 1957. *College Physics*. Englewood Cliffs, N.J.: Prentice-Hall.

Bishop, Morris. 1962. *A History of Cornell*. Ithaca, N.Y.: Cornell University Press.

Bjorken, James D. 1989. "Feynman and Partons." *Physics Today*, February, 56.

Bloch, Felix. 1976. "Reminiscences of Heisenberg and the Early Days of Quantum Mechanics." *Physics Today*, December, 23.

Blumberg, Stanley A., and Owens, Gwinn. 1976. *The Life and Times of Edward Teller*. New York: Putnam.

Boden, Margaret A. 1990. *The Creative Mind: Myths and Mechanisms*. New York: Basic Books.

Bohm, David, and Peat, F. David. 1987. *Science, Order, and Creativity*. New York: Bantam.

Bohr, Niels. 1922. Nobel Prize in Physics Award Address, 11 December. In Weaver 1987, 2:315.

———. 1928. "New Problems in Quantum Theory: The Quantum Postulate and the Recent Development of Atomic Theory." *Nature* 121:580.

———. 1935. "Can Quantum-Mechanical Description of Physical Reality Be Considered Complete?" *Physical Review* 48:696.

Bohr, Niels, and Wheeler, John Archibald. 1939. "The Mechanism of Nuclear Fission." *Physical Review* 56:426.

Boltzman, Ludwig. 1974. *Theoretical Physics and Philosophical Problems*. Edited by Brian McGuinness. Boston: Reidel.

Bondi, Hermann. 1967. *Assumption and Myth in Physical Theory*. Cambridge: Cambridge University Press.

Bonner, Francis T., and Phillips, Melba. 1957. *Principles of Physical Science*. Reading, Mass.: Addison-Wesley.

Born, Max. 1971. *The Born-Einstein Letters: Correspondence between Albert Einstein and Max and Hedwig Born from 1916 to 1955*. Translated by Irene Born. New York: Walker.

Bosanquet, Bernard. 1923. *Three Chapters on the Nature of Mind*. London: Macmillan.

Boscovitch, Roger G. 1922. *A Theory of Natural Philosophy*. Translated by J. M. Child. Chicago: Open Court.

Boslough, John. 1986. "Inside the Mind of John Wheeler." *Reader's Digest*, September, 106.

Bowerman, Walter G. 1947. *Studies in Genius*. New York: Philosophical Library.

Boyd, Richard; Gasper, Philip; and Trout, J. D., eds. 1991. *The Philosophy of Science*. Cambridge, Mass.: MIT Press.

Boyer, Paul. 1985. *By the Bomb's Early Light: American Thought and Culture at the Dawn of the Atomic Age*. New York: Pantheon.

Bridgman, Percy. 1950. *Reflections of a Physicist*. New York: Philosophical Library.

———. 1952. *The Nature of Some of Our Physical Concepts*. New York: Philosophical Library.

———. 1961. *The Logic of Modern Physics*. New York: Macmillan.

Briggs, John. 1988. *Fire in the Crucible: The Alchemy of Creative Genius*. New York: St. Martin's Press.

Brillouin, Leon. 1964. *Scientific Uncertainty and Information*. New York: Academic Press.

Brode, Bernice. 1960. "Tales of Los Alamos." *LASL Community News*, 11 August.

Broglie, Louis de. 1951. "The Concept of Time in Modern Physics and Bergson's Pure Duration." In *Bergson and the Evolution of Physics*. Knoxville: University of Tennessee Press.

Bromberg, Joan. 1976. "The Concept of Particle Creation before and after Quantum Mechanics." *Historical Studies in the Physical Sciences* 7:161.

Brower, Kenneth. 1978. *The Starship and the Canoe*. New York: Holt, Rinehart, and Winston.

Brown, Laurie M.; Dresden, Max; and Hoddeson, Lillian, eds. 1989. *Pions to Quarks: Particle Physics in the 1950s*. Cambridge: Cambridge University Press.

Brown, Laurie M., and Hoddeson, Lillian, eds. 1983. *The Birth of Particle Physics*. Cambridge: Cambridge University Press.

Brown, Lawrason. 1934. *Rules for Recovery from Pulmonary Tuberculosis*. Philadelphia: Lea and Feberger.

Brownell, G. L. 1952. "Physics in South America." *Physics Today*, July, 5.

Bunge, Mario. 1979. *Causality and Modern Science*. New York: Dover.

Cahn, Robert N., and Goldhaber, Gerson. 1989. *The Experimental Foundations of Particle Physics*. Cambridge: Cambridge University Press.

Casimir, Hendrik B. G. 1983. *Haphazard Reality: Half a Century of Science*. New York: Harper and Row.

Chamber of Commerce of the Rockaways. 1934. *Annual Year Book of the Rockaways*.

Chandrasekhar, Subrahmanyan. 1987. *Truth and Beauty: Aesthetics and Motivations in Science*. Chicago: University of Chicago Press.

Chase, W. Parker. 1932. *New York: The Wonder City*. Facsimile edition. New York: New York Bound, 1983.

Chown, Marcus. 1989. "The Heart and Soul of Richard Feynman." *New Scientist*, 25 February, 65.

Churchland, Paul M., and Hooker, Clifford A. *Images of Science*. Chicago: University of Chicago Press.

Clark, Ronald W. 1971. *Einstein: The Life and Times*. New York: World.

Close, F. E. 1979. *An Introduction to Quarks and Partons*. London: Academic Press.

Cohen, I. Bernard, ed. 1981. *The Conservation of Energy and the Principle of Least Action*. New York: Arno Press.

Cohen, I. Bernard. 1985. *Revolution in Science*. Cambridge, Mass.: Belknap Press.

Cohen, Michael. 1991. "It Never Passed *Him* By." Typescript.

Cohn, David L. 1943. *Love in America*. New York: Simon and Schuster.

Colodny, Robert G. 1965. *Beyond the Edge of Certainty: Essays in Contemporary Science and Philosophy*. Englewood Cliffs, N.J.: Prentice-Hall.

Compton, Arthur. 1956. *Atomic Quest: A Personal Narrative*. Chicago: University of Chicago Press.

Cooper, Necia Grant. 1989. *From Cardinals to Chaos: Reflections of the Life and Legacy of Stanislaw Ulam*. Cambridge: Cambridge University Press.

Corey, C. L. 1988. *Diary of a Safeman*. Facsimile edition. Streamwood, Ill.: National Publishing.

Crease, Robert P., and Mann, Charles C. 1986. *The Second Creation*. New York: Macmillan.

Crick, Francis H. C. 1962. "The Genetic Code." *Scientific American*, October, 66.
————. 1966. "The Genetic Code: III." *Scientific American*, October, 55.

Crick, Francis H. C.; Barnett, Leslie; Brenner, Sydney; and Watts-Tobin, R. J. 1961. "General Nature of the Genetic Code for Proteins." *Nature* 192:1227.

Currie, Robert. 1974. *Genius: An Ideology in Literature*. New York: Schocken.

Curtin, Deane W. 1980. *The Aesthetic Dimension of Science*. New York: Philosophical Library.

Curtin, Deane W., ed. 1982. *The Aesthetic Dimension of Science: 1980 Nobel Conference*. New York: Philosophical Library.

Cvitanović, Predrag. 1983. *Field Theory*. Copenhagen: Nordita Classics Illustrated.

Dalal, Siddhartha R.; Fowlkes, Edward B.; and Hoadley, Bruce. 1989. "Risk Analysis of the Space Shuttle: Pre-*Challenger* Prediction of Failure." *Journal of the American Statistical Association* 84:945.

Davies, John D. 1973. "The Curious History of Physics at Princeton." *Princeton Alumni Weekly*, 2 October, 8.

Davies, P. C. W. 1974. *The Physics of Time Asymmetry*. Berkeley: University of California Press.

Davis, Nuel Pharr. 1968. *Lawrence and Oppenheimer*. New York: Simon and Schuster.

De Hoffmann, Frederic. 1974. "A Novel Apprenticeship." In J. Wilson 1975, 162.

De Sitter, Willem. 1932. *Kosmos*. Cambridge, Mass.: Harvard University Press.

Dedmon, Emmett. 1953. *Fabulous Chicago*. New York: Random House.

Dembart, Lee. 1983. "Nobel Prize: Another Side of the Medal." *Los Angeles Times*, 4 February, 20.

Descartes, René. 1955. *The Philosophical Works of Descartes*. Translated by E. S. Haldane and G. R. T. Ross. New York: Dover.

D'Espagnat, Bernard. 1976. *Conceptual Foundations of Quantum Mechanics*. Reading, Mass.: Addison-Wesley.

———. 1979. "The Quantum Theory and Reality." *Scientific American*, November, 158.

Dirac, P. A. M. 1928. "The Quantum Theory of the Electron." *Proceedings of the Royal Society of London* A117:610.

———. 1933. "The Lagrangian in Quantum Mechanics." *Physikalische Zeitschrift der Sowjetunion* 2:64. In Schwinger 1958.

———. 1935. *The Principles of Quantum Mechanics*. Second edition. Oxford: Clarendon Press.

———. 1946. "Elementary Particles and Their Interactions." Typescript. PUL.

———. 1971. *The Development of Quantum Theory*. J. Robert Oppenheimer Memorial Prize Acceptance Speech. New York: Gordon and Breach.

———. 1975. *Directions in Physics*. New York: Wiley and Sons.

Dobkowski, Michael N. 1979. *The Tarnished Dream: Basis of American Anti-Semitism*. Westport, Conn.: Greenwood Press.

Dodd, J. E. 1984. *The Ideas of Particle Physics*. Cambridge: Cambridge University Press.

Donnelly, Russell. 1991a. *Quantized Vortices in Helium II*. Cambridge: Cambridge University Press.

———. 1991b. "The Discovery of Superfluidity." Manuscript.

Dresden, Max. 1987. *H. A. Kramers: Between Tradition and Revolution*. New York: Springer-Verlag.

Duff, William. 1767. *An Essay on Original Genius*. A facsimile reproduction edited with an introduction by John L. Mohoney. Gainesville, Pa.: Scholars' Facsimiles and Reprints, 1964.

Duga, René. 1955. *A History of Mechanics*. Translated by J. R. Maddox. Neuchatel: Griffon.

Dye, Lee. 1988. "Nobel Physicist R. P. Feynman of Caltech Dies." *Los Angeles Times*, 16 February, 1.

Dyson, Freeman. 1944. "Some Guesses in the Theory of Partitions." *Eureka* 8:10.

———. 1949a. "The Radiation Theories of Tomonaga, Schwinger, and Feynman." *Physical Review* 75:486. In Schwinger 1958.

———. 1949b. "The S-Matrix in Quantum Electrodynamics." *Physical Review* 75:1736. In Schwinger 1958.

———. 1952. "Divergence of Perturbation Theory in Quantum Electrodynamics." *Physical Review* 85:631.

———. 1965a. "Tomonaga, Schwinger, and Feynman Awarded Nobel Prize for Physics." *Science* 150:588. In Weaver 1987, 1:427.

————. 1965*b*. "Old and New Fashions in Field Theory." *Physics Today*, June, 23.

————. 1979. *Disturbing the Universe*. New York: Basic Books.

————. 1980. "Manchester and Athens." In Curtin 1980, 41.

————. 1984. *Weapons and Hope*. New York: Harper and Row.

————. 1987. "A Walk through Ramanujan's Garden." Lecture at the Ramanujan Centenary Conference, University of Illinois, 2 June.

————. 1988*a*. *Infinite in All Directions*. New York: Harper and Row.

————. 1988*b*. "The Lemon and the Cream." Talk prepared for Gemant Award ceremonies, 25 October. Institute for Advanced Study.

————. 1989. "Feynman at Cornell." *Physics Today*, February, 32.

————. 1990. "Feynman's Proof of the Maxwell Equations." *American Journal of Physics* 58:209.

————. 1992. *From Eros to Gaia*. New York: Pantheon.

Earman, John. 1989. *World Enough and Space-Time*. Cambridge, Mass.: MIT Press.

Eddington, A. S. 1940. *The Nature of the Physical World*. New York: Macmillan.

Edson, Lee. 1967. "Two Men in Search of the Quark." *New York Times Magazine*, 8 October, 54.

Einstein, Albert. 1909. "Development of Our Conception of the Nature and Constitution of Radiation." *Physikalishe Zeitschrift* 22:1909. In Weaver 1987, 2:295.

Einstein, Albert, and Infeld, Leopold. 1938. *The Evolution of Physics: From Early Concepts to Relativity and Quanta*. New York: Dover.

Erwin, G. S. 1946. *A Guide for the Tuberculous Patient*. New York: Grune and Stratton.

Far Rockaway High School. 1932. *History of the Rockaways*. Monograph by the students of Far Rockaway High School. Brooklyn Historical Society.

Feinberg, Gerald. 1977. *What Is the World Made Of? The Achievements of Twentieth Century Physics*. Garden City, N.Y.: Anchor Press.

Fermi, Enrico. 1932. "Quantum Theory of Radiation." *Reviews of Modern Physics* 4:87.

Fermi, Enrico, and Yang. C. N. 1949. "Are Mesons Elementary Particles?" *Physical Review* 76:1739.

Fermi, Laura. 1954. *Atoms in the Family: My Life with Enrico Fermi*. Chicago: University of Chicago Press.

————. 1971. *Illustrious Immigrants: The Intellectual Migration from Europe 1930–41*. Chicago: University of Chicago Press.

————. 1980. "The Fermis' Path to Los Alamos." In Badash et al. 1980.

Ferretti, B., ed. 1958. *Annual International Conference on High Energy Physics at CERN*. Geneva, 30 June–5 July. Geneva: CERN.

Ferris, Timothy. 1988. *Coming of Age in the Milky Way*. New York: Morrow.

Fine, Arthur. 1986. *The Shaky Game: Einstein, Realism, and the Quantum Theory*. Chicago: University of Chicago Press.

————. 1991. "The Natural Ontological Attitude." In Boyd et al. 1991, 271.

Flick, Lawrence F. 1903. *Consumption a Curable and Preventable Disease*. Philadelphia: McKay.

Foley, H. M., and Kusch, P. 1948. "On the Intrinsic Moment of the Electron." *Physical Review* 73:412.

Forman, Paul. 1987. "Behind Quantum Electronics: National Security as a Basis for

Physical Research in the United States, 1940–1960." *Historical Studies in the Physical Sciences* 18:149.

Fox, David. 1952. "The Tiniest Time Traveler." *Astounding Science Fiction* (magazine).

Francis, Patricia. 1989. "Science as a Way of Seeing: The Case of Richard Feynman." Manuscript, University of Maryland.

Franklin, Allan. 1979. "The Discovery and Nondiscovery of Parity Nonconservation." *Studies in History and Philosophy of Science* 10:201.

————. 1990. *Experiment, Right or Wrong.* Cambridge: Cambridge University Press.

Frisch, Otto B. 1979. *What Little I Remember.* Cambridge: Cambridge University Press.

Galdston, Iago. 1940. *Progress in Medicine: A Critical Review of the Last Hundred Years.* New York: Knopf.

Galison, Peter Louis. 1979. "Minkowski's Space-Time: From Visual Thinking to the Absolute World." *Historical Studies in the Physical Sciences* 10:85.

————. 1987. *How Experiments End.* Chicago: University of Chicago Press.

Galton, Francis. 1869. *Hereditary Genius: An Inquiry into Its Laws and Consequences.* New York: Horizon Press.

Gamow, George. 1966. *Thirty Years That Shook Physics: The Story of Quantum Theory.* Garden City, N.Y.: Doubleday.

Gardner, Martin. 1969. *The Ambidextrous Universe.* New York: Mentor.

————. 1989. *Hexaflexagons and Other Mathematical Diversions.* Chicago: University of Chicago Press.

Gay, Peter. 1988. *Freud: A Life for Our Time.* New York: Norton.

Gell-Mann, Murray. 1953. "Isotopic Spin and New Unstable Particles." *Physical Review* 92:833.

————. 1964. "A Schematic Model of Baryons and Mesons." *Physics Letters* 8:214.

————. 1982. "Strangeness." *Journal de Physique* 43:395.

————. 1983a. "From Renormalizability to Calculability?" In Jackiw et al. 1983, 3.

————. 1983b. "Particle Theory from S-Matrix to Quarks." Talk presented at the First International Congress on the History of Scientific Ideas at Sant Feliu de Guixols, Catalunya, Spain.

————. 1989a. "Dick Feynman—The Guy Down the Hall." *Physics Today*, February, 50.

————. 1989b. Remarks at a Conference Celebrating the Birthday of Murray Gell-Mann, 27–28 January.

Gell-Mann, Murray, and Ne'eman, Yuval. 1964. *The Eightfold Way.* New York: Benjamin.

Gemant, Andrew. 1961. *The Nature of the Genius.* Springfield, Ill.: Charles C. Thomas.

Gerard, Alexander. 1774. *An Essay on Genius.* London: Strahan.

Gieryn, Thomas F., and Figert, Anne E. 1990. "Ingredients for a Theory of Science in Society: O-Rings, Ice Water, C-Clamp, Richard Feynman and the New York Times." In *Theories of Science and Society.* Edited by Susan E. Cozzens and Thomas F. Gieryn. Bloomington, Ind.: Indiana University Press.

Gilbert, G. Nigel, and Mulkay, Michael. 1984. *Opening Pandora's Box: A Sociological Analysis of Scientists' Discourse.* Cambridge: Cambridge University Press.

Glashow, Sheldon. 1980. "Towards a Unified Field Theory: Threads in a Tapestry." *Science*, 19 December, 1319.

———. 1988. *Interactions: A Journey through the Mind of a Particle Physicist and the Matter of This World*. With Ben Bova. New York: Warner Books.

Gold, Thomas, ed. 1967. *The Nature of Time*. Ithaca, N.Y.: Cornell University Press.

Goldstine, Herman H. 1972. *The Computer from Pascal to Von Neumann*. Princeton: Princeton University Press.

Golovin, N. E. 1963. "The Creative Person in Science." In Taylor and Frank 1963, 7.

Goodstein, David. 1989. "Richard P. Feynman, Teacher." *Physics Today*, February, 70.

Goodstein, Judith R. 1991. *Millikan's School: A History of the California Institute of Technology*. New York: Norton.

Gould, Stephen Jay. 1981. *The Mismeasure of Man*. New York: Norton.

———. 1983. "Losing the Edge." In *The Flamingo's Smile*. New York: Norton.

Grattan, C. Hartley. 1933. "Thomas Alva Edison: An American Symbol." *Scribner's Magazine*, September, 151.

Greenberg, Daniel S. 1967. *The Politics of Pure Science*. New York: New American Library.

Greenberger, Daniel M., and Overhauser, Albert W. 1980. "The Role of Gravity in Quantum Theory." *Scientific American*, May, 66.

Gregory, Bruce. 1988. *Inventing Reality*. New York: Wiley and Sons.

Groueff, Stephane. 1967. *Manhattan Project: The Untold Story of the Making of the Atomic Bomb*. Boston: Little, Brown.

Groves, Leslie. 1975. *Now It Can Be Told*. New York: Da Capo.

Grünbaum, Adolph. 1963. *Philosophical Problems of Space and Time*. New York: Knopf.

Hanson, Norwood Russell. 1963. *The Concept of the Positron: A Philosophical Analysis*. Cambridge: Cambridge University Press.

Harris, Theodore E. 1963. *The Theory of Branching Processes*. Englewood Cliffs, N.J.: Prentice-Hall.

Hartman, Paul. 1984. "The Cornell Physics Department: Recollections and a History of Sorts." Typescript. Cornell University.

Hawking, Stephen W. 1987. *A Brief History of Time: From the Big Bang to Black Holes*. New York: Bantam.

Hawkins, David; Truslow, Edith C.; and Smith, Ralph Carlisle. 1983. *Project Y: The Los Alamos Story*. Los Angeles: Tomash.

Heidegger, Martin. 1959. *An Introduction to Metaphysics*. Translated by Ralph Manheim. New Haven, Conn.: Yale University Press.

Heilbron, J. L., and Seidel, Robert W. 1989. *Lawrence and His Laboratory: A History of the Lawrence Berkeley Laboratory*. Berkeley: University of California Press.

Heisenberg, Werner, ed. 1946. *Cosmic Radiation: Fifteen Lectures*. Translated by T. H. Johnson. New York: Dover.

———. 1971. *Physics and Beyond*. New York: Harper and Row.

Hempel, Carl G. 1965. *Aspects of Scientific Explanation*. New York: The Free Press.

Herbert, Nick. 1985. *Quantum Reality*. Garden City, N.Y.: Anchor Press.

Hesse, Mary B. 1961. *Forces and Fields: The Concept of Action at a Distance in the History of Physics*. London: Nelson.

Hewlett, Richard G., and Anderson, Jr., Oscar E. 1962. *The New World, 1939/1946*:

A History of the United States Atomic Energy Commission. Volume 1. University Park, Pa.: Pennsylvania State University Press.

Hiley, B. J., and Peat, F. David. 1987. *Quantum Implications: Essays in Honour of David Bohm.* London: Routledge and Kegan Paul.

Hillis, W. Daniel. 1989. "Richard Feynman and the Connection Machine." *Physics Today*, February, 78.

Hofstadter, Douglas R. 1979. *Gödel, Escher, Bach.* New York: Basic Books.

———. 1985. *Metamagical Themas.* New York: Basic Books.

———. 1991. "Thinking about Thought." *Nature* 349:378.

Holton, Gerald, ed. 1965. *Science and Culture.* Boston: Houghton Mifflin.

———. 1972. *The Twentieth-Century Sciences: Studies in the Biography of Ideas.* New York: Norton.

———. 1978. *The Scientific Imagination: Case Studies.* Cambridge: Cambridge University Press.

———. 1988. *Thematic Origins of Scientific Thought: Kepler to Einstein.* Revised edition. Cambridge, Mass.: Harvard University Press.

Hood, Edwin Paxton. 1851. *Genius and Industry: The Achievements of Mind Among the Cottages.* London: Partridge and Oakey.

Hudson, Liam, ed. 1970. *The Ecology of Human Intelligence.* London: Penguin.

Huxley, Thomas H. 1897. *Discourses: Biological and Geological Essays.* New York: Appleton.

Hyman, Anthony. 1982. *Charles Babbage: Pioneer of the Computer.* Oxford: Oxford University Press.

Infeld, Leopold. 1950. *Albert Einstein: His Work and Influence on Our World.* New York: Scribner.

Jaki, Stanley L. *The Relevance of Physics.* Chicago: University of Chicago Press.

Jackiw, Roman; Khuri, Nicola N.; Weinberg, Steven; and Witten, Edward, eds. 1983. *Shelter Island II: Proceedings of the 1983 Shelter Island Conference on Quantum Field Theory and the Fundamental Problems of Physics.* Cambridge, Mass.: MIT Press.

James, William. 1917. *Selected Papers on Philosophy.* London: Dent.

Jammer, Max. 1966. *The Conceptual Development of Quantum Mechanics.* New York: McGraw-Hill.

———. 1974. *The Philosophy of Quantum Mechanics: The Interpretations of Quantum Mechanics in Historical Perspective.* New York: Wiley and Sons.

Jeans, Sir James. 1943. *Physics and Philosophy.* Cambridge: Cambridge University Press.

Jette, Eleanor. 1977. *Inside Box 1663.* Los Alamos, N.M.: Los Alamos Historical Society.

Johnson, Charles W., and Jackson, Charles O. 1981. *City Behind a Fence: Oak Ridge, Tennessee, 1942–1946.* Knoxville: University of Tennessee.

Jourdain, Philip E. B. 1913. *The Principle of Least Action.* Chicago: Open Court.

Judson, Horace Freeland. 1979. *The Eighth Day of Creation: The Makers of the Revolution in Biology.* New York: Simon and Schuster.

Jungk, Robert. 1956. *Brighter Than a Thousand Suns.* New York: Harcourt.

Jungnickel, Christa; and McCormmach, Russell. 1986. *The Now Mighty Theoretical Physics.* Chicago: University of Chicago Press.

Kac, Mark. 1985. *Enigmas of Chance*. New York: Harper and Row.

Kamen, M. D. 1985. *Radiant Science, Dark Politics*. Berkeley: University of California Press.

Kargon, Robert H. 1977. "Temple to Science: Cooperative Research and the Birth of the California Institute of Technology." *Historical Studies in the Physical Sciences* 8:3.

Kazin, Alfred. 1951. *A Walker in the City*. New York: Harcourt, Brace.

Kevles, Daniel. 1987. *The Physicists: The History of a Scientific Community in Modern America*. Cambridge, Mass.: Harvard University Press.

———. 1990. "Cold War and Hot Physics: Science, Security, and the American State, 1945–56." *Historical Studies in the Physical Sciences* 20:239.

Klauder, John R., ed. 1972. *Magic Without Magic: John Archibald Wheeler*. San Francisco: W. H. Freeman.

Kragh, Helge. 1989. *Dirac: A Scientific Biography*. Cambridge: Cambridge University Press.

Kroll, Norman M., and Lamb, Willis E. 1949. "On the Self-Energy of a Bound Electron." *Physical Review* 75:388.

Kuhn, T. S. 1962. *The Structure of Scientific Revolutions*. Chicago: University of Chicago Press.

———. 1977. *The Essential Tension*. Chicago: University of Chicago Press.

———. 1978. *Black Body Theory and the Quantum Discontinuity, 1894–1912*. Oxford: Clarendon Press.

Kunetka, James W. 1979. *City of Fire*. Albuquerque: University of New Mexico Press.

Kursunoglu, Behram N., and Wigner, Eugene P. 1987. *Reminiscences about a Great Physicist: Paul Adrien Maurice Dirac*. Cambridge: Cambridge University Press.

La Belle, Jenijoy. 1989. "The Piper and the Physicist." *Engineering and Science*, Fall, 25.

LaFollette, Marcel C. 1990. *Making Science Our Own: Public Images of Science 1910–1955*. Chicago: University of Chicago Press.

Lamb, Willis. 1980. "The Fine Structure of Hydrogen." In Brown and Hoddeson 1983, 311.

Landsberg, P. T., ed. 1982. *The Enigma of Time*. Bristol: Adam Hilger.

Laurence, William L. 1959. *Men and Atoms*. New York: Simon and Schuster.

Leighton, Ralph. 1991. *Tuva or Bust! Richard Feynman's Last Journey*. New York: Norton.

Lentricchia, Frank. 1980. *After the New Criticism*. Chicago: University of Chicago.

Leplin, J., ed. 1984. *Scientific Realism*. Berkeley: University of California Press.

Lewis, Gilbert N. 1930. "The Symmetry of Time in Physics." In Landsberg 1982, 37.

Lifshitz, Eugene M. 1958. "Superfluidity." *Scientific American*, June, 30.

Lindsay, Robert Bruce. 1940. *General Physics for Students of Science*. New York: Wiley and Sons.

Lipset, Seymour Martin, and Ladd, Jr., Everett Carll. 1971. "Jewish Academics in the United States." *American Jewish Yearbook*, 89.

Lombroso, Cesare. 1891. *The Man of Genius*. London: Walter Scott.

Lopes, J. Leite. 1988. "Richard Feynman in Brazil: Recollections." Manuscript.

Lopes, J. Leite, and Feynman, Richard. 1952. "On the Pseudoscalar Meson Theory of the Deuteron." *Symposium on New Research Techniques in Physics*, 251.

Macfarlane, Gwyn. 1984. *Alexander Fleming: The Man and the Myth*. London: Hogarth Press.

Mach, Ernst. 1960. *The Science of Mechanics: A Critical and Historical Account of Its Development*. Translated by Thomas J. McCormack. Lasalle, Ill.: Open Court.

Maddox, John. 1988. "The Death of Richard Feynman." *Nature* 331:653.

Mann, Thomas. 1927. *The Magic Mountain*. Translated by H. T. Lowe-Porter. New York: Modern Library.

Marshak, Robert E. 1970. "The Rochester Conferences." *Bulletin of the Atomic Scientists*, June.

Masters, Dexter, and Way, Katharine, eds. 1946. *One World or None: A Report to the Public on the Full Meaning of the Atomic Bomb*. New York: McGraw-Hill.

Maugham, W. Somerset. 1947. "Sanatorium." In *Creatures of Circumstance*. London: Heinemann.

Mead, Margaret. 1949. *Male and Female: A Study of the Sexes in a Changing World*. New York: Morrow.

Medawar, Peter Brian. 1969. *Induction and Intuition in Scientific Thought*. Philadelphia: American Philosophical Society.

Mehra, Jagdish, ed. 1973. *The Physicist's Conception of Nature*. Dordrecht: Reidel.

Mehra, Jagdish. 1988. "My Last Encounter with Richard Feynman." Talk at Department of Physics, Cornell University, 24 February.

Melsen, Andrew G. van. 1952. *From Atomos to Atom: The History of the Concept "Atom."* Translated by Henry J. Koren. Pittsburgh, Duquesne University Press.

Mendenhall, C. E.; Eve, A. S.; Keys, D. A.; and Sutton, R. M. 1950. *College Physics*. Boston: Heath.

Menge, Edward J. v. K. 1932. *Jobs for the College Graduate in Science*. New York: Bruce.

Mermin, N. David. 1985. "Is the Moon There When Nobody Looks? Reality and the Quantum Theory." *Physics Today*, April, 38.

Merton, Robert K. 1961. "The Role of Genius in Scientific Advance." *New Scientist* 259: 306. In Hudson 1970, 70.

———. 1973. *The Sociology of Science: Theoretical and Empirical Investigations*. Chicago: University of Chicago Press.

Metropolis, Nicholas. 1990. "The Los Alamos Experience, 1943–1954." In Nash 1990.

Metropolis, Nicholas, and Nelson, E. C. 1982. "Early Computing at Los Alamos." *Annals of the History of Computing* 4:348.

Michels, Walter C. 1948. "Women in Physics." *Physics Today*, December, 16.

Miller, Arthur I. 1984. *Imagery in Scientific Thought: Creating Twentieth-Century Physics*. Boston: Birkhäuser.

———. 1985. "Werner Heisenberg and the Beginning of Nuclear Physics." *Physics Today*, November, 60.

Millikan, Robert Andrews. 1947. *Electrons (+ and −) Protons, Photons, Neutrons, Mesotrons, and Cosmic Rays*. Chicago: University of Chicago Press.

Millikan, Robert Andrews; Roller, Duane; and Watson, Earnest Charles. 1937. *Mechanics, Molecular Physics, Heat, and Sound*. Boston: Ginn.

Mizener, Arthur. 1949. *The Far Side of Paradise*. Boston: Houghton Mifflin.

Morris, Richard. 1984. *Time's Arrows: Scientific Attitudes toward Time*. New York: Simon and Schuster.

Morrison, Philip. 1946. "If the Bomb Gets Out of Hand." In Masters and Way 1946.

———. 1985. Review of *Surely You're Joking, Mr. Feynman!* In *Scientific American*, May, 41.

[Morrison, Philip.] 1988. "Richard P. Feynman 1918–1988." *Scientific American*, June, 38.

Morse, Philip. 1977. *In at the Beginnings: A Physicist's Life*. Cambridge, Mass.: MIT Press.

Moss, Norman. 1987. *Klaus Fuchs*. London: Grafton.

Murray, Francis J. 1961. *Mathematical Machines*. New York: Columbia University Press.

Nash, Stephen G., ed. 1990. *A History of Scientific Computing*. New York: ACM.

New Yorker. 1988. "Richard Feynman." 14 March, 30.

Nisbet, Robert. 1980. *History of the Idea of Progress*. New York: Basic Books.

Noyes, H. P.; Hafner, E. M.; Yekutieli, G.; and Raz, B. J., eds. *High Energy Nuclear Physics*. Proceedings of the Fifth Annual Rochester Conference, 31 January–2 February. New York: Interscience.

Nye, Mary Jo, ed. 1984. *The Question of the Atom*. Los Angeles: Tomash.

Obler, Loraine K., and Fein, Deborah, eds. 1988. *The Exceptional Brain: Neuropsychology of Talent and Special Abilities*. New York: Guilford Press.

Ochse, R. 1990. *Before the Gates of Excellence: The Determinants of Creative Genius*. Cambridge: Cambridge University Press.

Oppenheimer, J. Robert. 1945. Speech to the Association of Los Alamos Scientists, 2 November. In Smith and Weiner 1980, 315.

———. 1948. "Electron Theory." In Schwinger 1958.

Osgood, Charles G., ed. 1947. *The Modern Princeton*. Princeton: Princeton University Press.

Osgood, Charles G. 1951. *Lights in Nassau Hall*. Princeton: Princeton University Press.

Pais, Abraham. 1982. *"Subtle Is the Lord": The Science and the Life of Albert Einstein*. Oxford: Oxford University Press.

———. 1986. *Inward Bound*. Oxford: Oxford University Press.

———. 1991. *Niels Bohr's Times, in Physics, Philosophy, and Polity*. Oxford: Oxford University Press.

Park, David. 1988. *The How and the Why*. Princeton: Princeton University Press.

Peierls, Rudolf. 1985. *Bird of Passage*. Princeton: Princeton University Press.

Perutz, Max. 1989. *Is Science Necessary?* Oxford: Oxford University Press.

Pickering, Andrew. 1984. *Constructing Quarks: A Sociological History of Particle Physics*. Edinburgh: Edinburgh University Press.

Polkinghorne, John C. 1980. *Models of High Energy Processes*. Cambridge: Cambridge University Press.

———. 1989. *Rochester Roundabout: The Story of High Energy Physics*. New York: W. H. Freeman.

————. 1990. "Chaos and Cosmos: A Theological Approach." Talk at Nobel Symposium, St. Peter, Minn.

Pollard, Ernest C. 1982. *Radiation: One Story of the MIT Radiation Laboratory.* Durham, N.C.: Woodburn Press.

Popper, Karl. 1958. *The Logic of Scientific Discovery.* London: Hutchinson.

Presidential Commission on the Space Shuttle Challenger Accident. 1986. *Report of the Presidential Commission on the Space Shuttle Challenger Accident.* Washington, D.C.

Princeton University. 1946. *The Future of Nuclear Science.* Princeton University Bicentennial Conferences: Series I, Conference I. Princeton: Princeton University Press.

Putnam, Hilary. 1965. "A Philosopher Looks at Quantum Mechanics." In Colodny 1965, 75.

Quine, W. V. 1969. *Ontological Relativity.* New York: Columbia University Press.

————. 1987. *Quiddities: An Intermittently Philosophical Dictionary.* Cambridge, Mass.: Belknap Press.

Rabi, Isidor Isaac. 1970. *Science: The Center of Culture.* New York: World.

Regis, Ed. 1987. *Who Got Einstein's Office?: Eccentricity and Genius at the Institute for Advanced Study.* Reading, Mass.: Addison Wesley.

————. 1990. *Great Mambo Chicken and the Transhuman Condition.* Reading, Mass.: Addison-Wesley.

Reichenbach, Hans. 1956. *The Direction of Time.* Berkeley: University of California Press.

Reid, Hiram Alvin. 1895. *History of Pasadena.* Pasadena: Pasadena History Company.

Reid, R. W. 1969. *Tongues of Conscience: Weapons Research and the Scientists' Dilemma.* New York: Walker and Company.

Reid, T. R. 1984. *The Chip: How Two Americans Invented the Microchip and Launched a Revolution.* New York: Simon and Schuster.

Reingold, Nathan, and Reingold, Ida H., eds. 1981. *Science in America: A Documentary History 1900–1939.* Chicago: University of Chicago Press.

Rhodes, Richard. 1987. *The Making of the Atomic Bomb.* New York: Simon and Schuster.

Rigden, John S. 1987. *Rabi: Scientist and Citizen.* New York: Basic Books.

Riordan, Michael. 1987. *The Hunting of the Quark.* New York: Simon and Schuster.

Root-Bernstein, Robert Scott. 1989. *Discovering: Inventing and Solving Problems at the Frontiers of Scientific Knowledge.* Cambridge, Mass.: Harvard University Press.

Sakharov, Andrei. 1990. *Memoirs.* Translated by Richard Lourie. New York: Knopf.

Salam, Abdus, and Strathdee, J. 1972. "The Path-Integral Quantization of Gravity." In *Aspects of Quantum Theory.* Edited by Abdus Salam and E. P. Wigner. Cambridge: Cambridge University Press.

Sánchez, George I. 1961. *Arithmetic in Maya.*

Scheid, Ann. 1986. *Pasadena: Crown of the Valley.* Northridge, Calif.: Windsor.

Schlossberg, David. 1988. *Tuberculosis.* Second edition. New York: Springer-Verlag.

Schrödinger, Erwin. 1967. *What Is Life?* Cambridge: Cambridge University Press.

Schucking, Engelbert L. 1990. "Views from a Distant Past." In *General Relativity and Gravitation 1989*. Edited by Neil Ashby. Cambridge: Cambridge University Press.

Schwartz, Joseph. 1992. *The Creative Moment: How Science Made Itself Alien to Modern Culture*. New York: HarperCollins.

Schweber, Silvan S. 1983. "A Short History of Shelter Island I." In Jackiw et al. 1983, 301.

———. 1986a. "Feynman and the Visualization of Space-Time Processes." *Reviews of Modern Physics*, 58:449.

———. 1986b. "The Empiricist Temper Regnant: Theoretical Physics in the United States 1920–1950." *Historical Studies in the Physical and Biological Sciences* 17:1.

———. 1986c. "Shelter Island, Pocono, and Oldstone: The Emergence of American Quantum Electrodynamics after World War II." *Osiris* 2:265.

———. 1989. "The Young Slater and the Development of Quantum Chemistry." *Historical Studies in the Physical and Biological Sciences* 20:339

———. Forthcoming. *QED: 1946–1950: An American Success Story*. Manuscript.

Schwinger, Julian. 1934. "On the Interaction of Several Electrons." Typescript. Courtesy of Schwinger.

———, ed. 1958. *Selected Papers on Quantum Electrodynamics*. New York: Dover.

———. 1973. "A Report on Quantum Electrodynamics." In Mehra 1973, 413.

———. 1983. "Renormalization Theory of Quantum Electrodynamics: An Individual View." In Brown and Hoddeson 1983.

———. 1989. "A Path to Quantum Electrodynamics." *Physics Today*, February, 42.

Segrè, Emilio. 1970. *Enrico Fermi: Physicist*. Chicago: University of Chicago.

———. 1980. *From X-Rays to Quarks: Modern Physicists and Their Discoveries*. Berkeley, Calif.: W. H. Freeman.

Sharpe, William. 1755. *A Dissertation Upon Genius*. Reprint, with introduction by William Bruce Johnson. Delmar, N.Y.: Scholars' Facsimiles and Reprints, 1973.

Sheppard, R. Z. 1985. "The Wonderful Wizard of Quark." *Time*, 7 January, 91.

Shryock, Richard Harrison. 1947. *The Development of Modern Medicine*. New York: Knopf.

Shuttle Criticality Review Hazard Analysis Audit Committee. 1988. *Post- Challenger Evaluation of Space Shuttle Risk Assessment and Management*. Washington, D.C.: National Academy of Sciences Press.

Silberman, Charles E. 1985. *A Certain People: American Jews and Their Lives Today*. New York: Summit.

Simonton, Dean Keith. 1984. *Genius, Creativity, and Leadership: Historiometric Inquiries*. Cambridge, Mass.: Harvard University Press.

———. 1989. *Scientific Genius*. Cambridge: Cambridge University Press.

Sitwell, Edith. 1943. *Street Songs*. London: Macmillan.

———. 1987. *Façade*. London: Duckworth.

Sklar, Lawrence. 1974. *Space, Time, and Spacetime*. Berkeley: University of California Press.

Slater, John C. 1955. *Modern Physics*. New York: McGraw-Hill.

———. 1963. *Quantum Theory of Molecules and Solids*. Vol. 1. New York: McGraw-Hill.

————. 1975. *Solid-State and Molecular Theory: A Scientific Biography.* New York: Wiley and Sons.

Slater, John C., and Frank, Nathaniel H. 1933. *Introduction to Theoretical Physics.* New York: McGraw-Hill.

Smith, Alice Kimball. 1965. *A Peril and a Hope: The Scientists' Movement in America 1945–47.* Chicago: University of Chicago Press.

Smith, Alice Kimball, and Weiner, Charles. 1980. *Robert Oppenheimer: Letters and Reflections.* Cambridge, Mass.: Harvard University Press.

Smith, Cyril Stanley. 1981. *A Search for Structure.* Cambridge, Mass.: MIT Press.

Smith, F. B. 1988. *The Retreat of Tuberculosis 1850–1950.* London: Croom Helm.

Smyth, H. D. 1945. *Atomic Energy for Military Purposes.* Princeton: Princeton University Press.

Smyth, H. D., and Wilson, Robert R. 1942. "The Isotron Method." Isotron Report no. 18, 20 August. SMY.

Snow, C. P. 1981. *The Physicists.* Boston: Little, Brown.

Solomon, Saul. 1952. *Tuberculosis.* New York: Coward-McCann.

Sopka, Katherine Russell. 1980. *Quantum Physics in America: 1920–1935.* New York: Arno Press.

Stabler, Howard P. 1967. "Teaching from Feynman." *Physics Today*, March, 47.

Starr, Kevin. 1985. *Inventing the Dream: California through the Progressive Era.* Oxford: Oxford University Press.

Steinberg, Stephen. 1971. "How Jewish Quotas Began." *Commentary*, September.

Stigler, George. 1985. "Sex and the Single Physicist." *Wall Street Journal*, 3 May, 21.

Storch, Sylvia. 1966. "A Nobel-Prize Winner Comes Home." *Highpoints* (Board of Education of the City of New York), June, 5.

Stückelberg, E. C. G. 1941. "Remarque à propos de la Création de Paires de Particules en Théorie de Relativité." *Helvetica Physica Acta* 14:588.

Stuewer, Roger H. 1975. "G. N. Lewis on Detailed Balancing, the Symmetry of Time, and the Nature of Light." *Historical Studies in the Physical Sciences* 6:469.

Stuewer, Roger H., ed. 1979. *Nuclear Physics in Retrospect: Proceedings of a Symposium on the 1930s.* Minneapolis: University of Minnesota Press.

Sudarshan, E. C. G. 1983. "Midcentury Adventures in Particle Physics." In Brown et al. 1989, 40.

Sudarshan, E. C. G., and Marshak, Robert E. 1984. "Origin of the V–A Theory." Talk at International Conference on Fifty Years of Weak Interactions, Racine, Wis., 29 May–1 June.

Sudarshan, E. C. G., and Ne'eman, Yuval, eds. 1973. *The Past Decade in Particle Theory.* London: Gordon and Breach.

Sudarshan, E. C. G.; Tinlot, J. H.; and Melissinos, A. C., eds. 1960. *Proceedings of the 1960 Annual International Conference on High Energy Physics at Rochester.* 25 August–1 September. New York: Interscience.

Taylor, Calvin, and Barron, Frank, eds. 1963. *Scientific Creativity: Its Recognition and Development.* New York: Wiley and Sons.

Taylor, J. C. 1976. *Gauge Theories of Weak Interactions.* Cambridge: Cambridge University Press.

Teich, Malvin C. 1986. "An Incessant Search for New Approaches." *Physics Today*, September, 61.

Telegdi, Valentine L. 1972. "Crucial Experiments on Discrete Symmetries." In Mehra 1973, 457.

Teller, Michael E. 1988. *The Tuberculosis Movement: A Public Health Campaign in the Progressive Era*. New York: Greenwood Press.

Tobey, Ronald C. 1971. *The American Ideology of National Science 1919–1930*. Pittsburgh: University of Pittsburgh Press.

Tomonaga, Shin'ichirō. 1966. "Development of Quantum Electrodynamics: Personal Recollections." In *Nobel Lectures: Physics 1963–1970*. Amsterdam: Elsevier.

Torretti, Roberto. 1990. *Creative Understanding: Philosophical Reflections on Physics*. Chicago: University of Chicago Press.

Toulmin, Stephen. 1953. *The Philosophy of Science*. New York: Harper and Row.

Traweek, Sharon. 1988. *Beamtimes and Lifetimes: The World of High Energy Physicists*. Cambridge, Mass.: Harvard University Press.

Tricker, R. A. R. 1966. *The Contributions of Faraday and Maxwell to Electrical Science*. Oxford: Pergamon Press.

Trigg, George L. 1975. *Landmark Experiments in Twentieth Century Physics*. New York: Crane, Russak.

Ulam, Stanislaw M. 1976. *Adventures of a Mathematician*. New York: Scribner.

Underwood, E. Ashworth. 1937. *Manual of Tuberculosis for Nurses and Public Health Workers*. Second edition. Edinburgh: E. & S. Livingstone.

Von Neumann, John. 1955. *Mathematical Foundations of Quantum Mechanics*. Princeton: Princeton University Press.

Waksman, Selman A. 1964. *The Conquest of Tuberculosis*. Berkeley: University of California Press.

Watson, James D. 1968. *The Double Helix*. New York: Atheneum.

Weart, Spencer R. 1988. *Nuclear Fear: A History of Images*. Cambridge, Mass.: Harvard University Press.

Weaver, Jefferson Hane, ed. 1987. *The World of Physics*. 3 volumes. New York: Simon and Schuster.

Weinberg, Steven. 1977a. "The Search for Unity: Notes for a History of Quantum Field Theory." *Daedalus* 106:17.

———. 1977b. *The First Three Minutes: A Modern View of the Origin of the Universe*. New York: Basic Books.

———. 1981. "Einstein and Spacetime: Then and Now." *Proceedings of the American Philosophical Society* 125:20.

———. 1987. "Towards the Final Laws of Physics." In Feynman and Weinberg 1987, 61.

Weisskopf, Victor F. 1947. "Foundations of Quantum Mechanics: Outline of Topics for Discussion." Typescript. OPP.

———. 1980. "Growing Up with Field Theory: The Development of Quantum Electrodynamics." In Brown and Hoddeson 1983, 56.

———. 1991. *The Joy of Insight: Passions of a Physicist*. New York: Basic Books.

Welton, T. A. 1983. "Memories." Manuscript. CIT.

Weyl, Hermann. 1922. *Space—Time—Matter*. Translated by Henry L. Brose. New York: Dover.

———. 1949. *Philosophy of Mathematics and Natural Science*. Princeton: Princeton University Press.

———. 1952. *Symmetry*. Princeton: Princeton University Press.

Wheeler, John Archibald. 1948. "Conference on Physics: Pocono Manor, Pennsylvania, 30 March–1 April, 1948." Mimeographed notes.

———. 1979a. "Some Men and Moments in the History of Nuclear Physics." In Stuewer 1979, 217.

———. 1979b. "Beyond the Black Hole." In Woolf 1980, 341.

———. 1985 "Not Consciousness, but the Distinction between the Probe and the Probed, as Central to the Elemental Quantum Act of Observation." Lecture at the annual meeting of the American Association for the Advancement of Science, 8 January.

———. 1989. "The Young Feynman." *Physics Today*, February, 24.

Wheeler, John Archibald, and Ruffini, Remo. 1971. "Introducing the Black Hole." *Physics Today*, 24.

Wheeler, John Archibald, and Wigner, Eugene P. 1942. Report of the Readers of Richard P. Feynman's Thesis on "The Principle of Least Action in Quantum Mechanics." Typescript, PUL.

Wheeler, John Archibald, and Zurek, Wojciech Hibert. 1983. *Quantum Theory and Measurement*. Princeton: Princeton University Press.

White, D. Hywel; Sullivan, Daniel; and Barboni, Edward J. 1979. "The Interdependence of Theory and Experiment in Revolutionary Science: The Case of Parity Violation." *Social Studies of Science* 9:303.

Whitrow, G. J. 1980. *The Natural Philosophy of Time*. Oxford: Clarendon Press.

Wiener, Norbert. 1956. *I Am a Mathematician: The Later Life of a Prodigy*. Garden City, N.Y.: Doubleday.

Wigner, Eugene P., ed. 1947. *Physical Science and Human Values*. Princeton Bicentennial Conference on the Future of Nuclear Science. Princeton: Princeton University Press.

Williams, L. Pearce. 1966. *The Origins of Field Theory*. New York: Random House.

Williams, Michael R. 1985. *A History of Computing Technology*. Englewood Cliffs, N.J.: Prentice-Hall.

Williams, Robert Chadwell. 1987. *Klaus Fuchs, Atom Spy*. Cambridge, Mass.: Harvard University Press.

Wilson, Jane, ed. 1975. *All in Our Time: The Reminiscences of Twelve Nuclear Pioneers*. Chicago: Educational Foundation for Nuclear Sciences.

Wilson, Robert R. 1942. "Isotope Separator: General Description." Isotron Report no. 1. SMY.

———. 1958. Review of *Brighter Than a Thousand Suns*. In *Scientific American*, December, 145.

———. 1972. "My Fight Against Team Research." In Holton 1972, 468.

———. 1974. "A Recruit for Los Alamos." In J. Wilson 1975, 142.

Woolf, Harry, ed. 1980. *Some Strangeness in the Proportion: A Centennial Symposium to Celebrate the Achievements of Albert Einstein*. Reading, Mass.: Addison-Wesley.

Wright, Kenneth W.; Monroe, James; and Beck, Frederick. 1990. "A History of the Ray Brook State Tuberculosis Hospital." *New York State Journal of Medicine* 90, 406.

Yang, Chen Ning. 1957. "The Law of Parity Conservation and Other Symmetry Laws in Physics." Nobel lecture, 11 December 1957. In *Nobel Lectures: Physics*. Amsterdam: Elsevier, 1964.

———. 1962. *Elementary Particles: A Short History of Some Discoveries in Atomic Physics*. Princeton: Princeton University Press.

———. 1983. "Particle Physics in the Early 1950s." In Brown et al. 1989, 40.

Yang, Chen Ning.; Cole, J. A.; Good, M.; Hwa, R.; and Lee-Franzini, J., eds. 1969. *High Energy Collisions: Third International Conference*. London: Gordon and Breach.

Yukawa, Hideki. 1973. *Creativity and Intuition: A Physicist Looks East and West*. Translated by John Bester. Tokyo: Kodansha.

Zeman, Jiří, ed. 1971. *Time in Science and Philosophy*. Amsterdam: Elsevier.

Ziman, John. 1978. *Reliable Knowledge: An Exploration of the Grounds for Belief in Science*. Cambridge: Cambridge University Press.

———. 1992. "Unknotting Epistemology." *Nature* 355:408.

Zuckerman, Harriet. 1977. *Scientific Elite: Nobel Laureates in the United States*. New York: The Free Press.

Zurek, Wojciech Hubert; van der Merwe, Alwyn; and Miller, Warner Allen, eds. 1988. *Between Quantum and Cosmos: Studies and Essays in Honor of John Archibald Wheeler*. Princeton: Princeton University Press.

Zweig, George. 1981. "Origins of the Quark Model." In *Baryon '80: Proceedings of the Fourth International Conference on Baryon Resonances*. Toronto: University of Toronto Press.

INDEX

· ÷ ·

ILLUSTRATION CREDITS

· ÷ ·

TEXT ILLUSTRATIONS

P. 35—courtesy of the American Institute of Physics; p. 57—Robin Brickman; p. 103—Gardner 1989; p. 107—Robin Brickman; p. 113—Wheeler and Feynman 1945; p. 230—courtesy of the Archives, California Institute of Technology; p. 246—Victor Weisskopf and E. Wigner, "Berechnung der natürlichen Linienbreite auf Grund der Diracschen Lichttheorie," *Zeitschrift für Physik* 63 (1930); p. 248—Robin Brickman; p. 254—Robin Brickman; p. 268—Dyson 1949a; p. 273—Stückelberg 1941; p. 274—Feynman 1949b; p. 275—Feynman 1949b; p. 276—Cvitanović 1983; p. 418—Feynman 1985a.

PHOTOGRAPHS

Insert between pp. 118 and 119: p. 1 of insert—courtesy Joan Feynman; p. 2, top—courtesy Joan Feynman; p. 2, bottom—courtesy Michelle Feynman; p. 3—courtesy Michelle Feynman; p. 4—Los Alamos Scientific Library, courtesy AIP Niels Bohr Library; p. 5—Los Alamos Scientific Library, courtesy AIP Niels Bohr Library; p. 6, top—S. A. Goudsmit, courtesy AIP Niels Bohr Library; p. 6, bottom—AIP Niels Bohr Library; p. 7, top—AIP Niels Bohr Library, Marshak Collection; p. 7, bottom—AIP Niels Bohr Library, Physics Today Collection; p. 8, top—Richard Hartt Photography, courtesy of Caltech; p. 8, bottom—AIP Niels Bohr Library

Insert between pp. 310 and 311: p. 1 of insert, top—AIP Niels Bohr Library, Physics Today Collection; p. 1, bottom—Joe Munroe, courtesy of Caltech; p. 2—courtesy of the Archives, California Institute of Technology; p. 3, top—UPI/Bettmann; p. 3, bottom—courtesy Gweneth Feynman; p. 4, top and bottom—courtesy Michelle Feynman; p. 5, top—courtesy Michelle Feynman; p. 5, bottom—Consulate General of Japan, N.Y., courtesy AIP Niels Bohr Library; p. 6, top—courtesy Michelle Feynman; p. 6, bottom—Robert Walker, courtesy Michelle Feynman; p. 7, top and bottom—courtesy of Caltech; p. 8—UPI/Bettmann.